Preface

The first edition of *The Science of Photobiology* was published in 1977, and was the first textbook to cover all of the major areas of photobiology. The science of photobiology is currently divided into 14 subspecialty areas by the American Society for Photobiology. In this edition, however, the topics of phototechnology and spectroscopy have been combined in a new chapter entitled "Photophysics." The other subspecialty areas remain the same, i.e., Photochemistry, Photosensitization, UV Radiation Effects, Environmental Photobiology, Photomedicine, Circadian Rhythms, Extraretinal Photoreception, Vision, Photomorphogenesis, Photomovement, Photosynthesis, and Bioluminescence.

This book has been written as a textbook to introduce the science of photobiology to advanced undergraduate and graduate students. The chapters are written to provide a broad overview of each topic. They are designed to contain the amount of information that might be presented in a one- to two-hour general lecture. The references are not meant to be exhaustive, but key references are included to give students an entry into the literature. Frequently a more recent reference that reviews the literature will be cited rather than the first paper by the author making the original discovery. The chapters are not meant to be a repository of facts for research workers in the field, but rather are concerned with demonstrating the importance of each specialty area of photobiology, and documenting its relevance to current and/or future problems of man.

Although written as a basic text for introductory courses in photobiology and as a vehicle for encouraging students to enter the field, this book will also be of interest to scientists outside of the area of photobiology and to interested laypersons, since it is now becoming more apparent, even to the general public, that light, both natural and artificial, has important consequences to man other than just as an aid to vision. Photobiology has come of age as a major scientific discipline.

The reader should be aware of the major sources of literature and information relevant to the science of photobiology.

For review articles, there are two major sources: *Photophysiology,* Volumes 1–8, (A. C. Giese, ed.), Academic Press, New York, covering the years 1964–1973; and *Photochemical and Photobiological Reviews,* Volumes 1–7 (K. C. Smith, ed.), Plenum Press, New York, covering the years 1976–1983.

For research papers, the major source is the international journal *Photochemistry and Photobiology* (Pergamon Press, London). Short yearly reviews on many areas of photobiology appear in the June and December issues of *Photochemistry and Photobiology.* This journal was inaugurated in 1962, and is now the official organ of the American Society for Photobiology.

The American Society for Photobiology was founded in 1972 to (1) promote original research in photobiology, (2) facilitate the integration of different disciplines in the study of photobiology, (3) promote the dissemination of knowledge of photobiology, and (4) provide information on the photobiological aspects of national and international problems. Membership in the society is open to persons who share the stated purpose of the society and who have educational, research, or practical experience in photobiology or an allied scientific field.

The name of the society was chosen to encompass both North and South America, but members from all parts of the world are welcome. The journal is included in the membership dues. The American Society for Photobiology holds an annual scientific meeting (usually in June) and publishes frequent newsletters of interest to photobiologists. Further information may be obtained by writing to the Executive Secretary, American Society for Photobiology, 8000 Westpark Drive, Suite 400, McLean, VA 22102. A history of the American Society for Photobiology has been published.[1]

The European Society for Photobiology was founded in 1986, and holds meetings every other year. The next meeting is scheduled for 1989. The journal for the European Society is the *Journal of Photochemistry and Photobiology* (Elsevier, Lausanne). For information about this society, write to Dr. R. M. Tyrrell, Treasurer ESP, Swiss Institute for Cancer Research, 1066 Epalinges/Lausanne, Switzerland.

The Association Internationale de Photobiologie sponsors an international congress on photobiology every four years. The most recent congress was held in Israel in 1988. The proceedings of these congresses are published and serve as an excellent source of review articles. At these congresses, the Finsen Medal is awarded for "outstanding fundamental contributions to photobiology." Two histories of the Association Internationale de Photobiologie have been published.[2,3]

Kendric C. Smith

REFERENCES

1. K. C. Smith, History of the American Society for Photobiology (ASP) (The first 10 years, and before), *Photochem. Photobiol.* **35,** 597–614 (1982).
2. D. Vince-Prue and D. O. Hall, International cooperation in photobiology, Comite International de Photobiologie. A History of CIP, *Photochem. Photobiol.* **22,** 77–82 (1975).
3. L. O. Bjorn, International cooperation in photobiology, Association Internationale de Photobiologie, part 2, 1975–1984, *Photochem. Photobiol.* **41,** 497–499 (1985).

Contents

1

Photophysics

1.1. INTRODUCTION

Photophysics is concerned with the physical interactions of light and matter. The first part of this chapter reviews the general nature of light, light sources, and the devices used to

Leonard I. Grossweiner • Physics Department, Illinois Institute of Technology, Chicago, Illinois 60616.

TABLE 1-1. SI Units[a]

	Base units	
Length	meter	m
Time	second	s
Mass	kilogram	kg
Temperature	kelvin	K
Electric current	ampere	A
Amount of substance	mole	mol
Luminous intensity	candela	cd
Plane angle[b]	radian	rad
Solid angle[b]	steradian	sr
	Some SI-derived units	
Area	square meter	m^2
Volume	cubic meter	m^3
Density	kilogram per cubic meter	kg/m^3
Frequency	hertz	Hz
Speed	meters per second	m/s
Force	newton	N
Pressure	pascal	Pa
Work/energy	joule	J
Power	watt	W
Quantity of electricity	coulomb	C
Potential difference	volt	V
Electric field strength	volt per meter	V/m
Electric resistance	ohm	Ω
Capacitance	farad	F
Magnetic flux	weber	Wb
Inductance	henry	H
Magnetic flux density	tesla	T
Magnetic field strength	ampere per meter	A/m
Specific heat capacity	joule per kilogram kelvin	J/(kg-K)
Thermal conductivity	watt per meter kelvin	W/(m-K)

[a]Radiometric units are given in Table 1-5.
[b]Supplementary unit.

modify and measure light properties. This is followed by a discussion of the interactions of light with matter at the atomic and molecular levels. Spectroscopic measurements are then considered, emphasizing the aspects relevant to photobiology. The international system of units is used (abbreviated SI from the French "Le System International d'Unites") (Table 1-1).

1.2. THE NATURE OF LIGHT

By strict definition, "light" is the form of radiant energy perceived by the human eye. However, the spectrum of sunlight at the surface of the earth includes invisible rays, adjacent to the visible spectrum, that are absorbed by and affect biological systems (Fig. 1-1). Light is a form of electromagnetic radiation. The nominal wavelengths of *visible light* are 400–760 nm. This is a very small region of the overall electromagnetic spectrum,

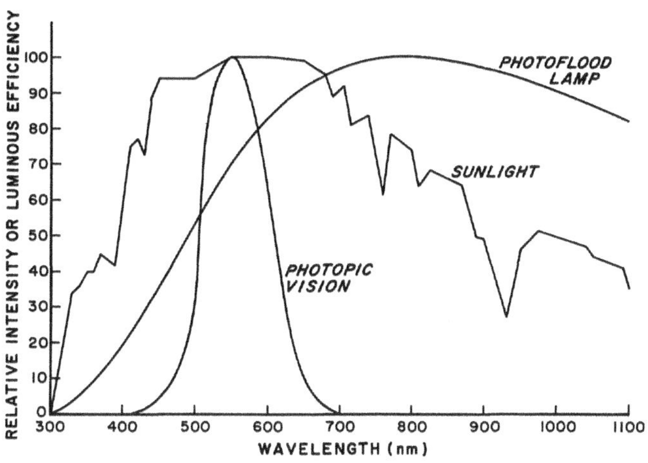

Fig. 1-1. Spectrum of sunlight from the zenith at the surface of the earth. The dips in the red and infrared regions are due to absorption by water vapor [300–365 nm from R. D. Cadle and E. R. Allen, *Science* **167,** 243–249 (1970); 650–1100 nm from R. B. Withrow and A. P. Withrow, Chap. 3 in *Radiation Biology,* Vol. 3 (A. Hollaender, ed.), McGraw-Hill, New York (1956)]. Spectrum of a 250-W photoflood lamp (color temperature, 3475°K) [From R. B. Withrow and A. P. Withrow, Chap. 3 in *Radiation Biology,* Vol. III (A. Hollaender, ed.), McGraw-Hill, New York (1956)]. Spectrum of average human photopic visual sensitivity. (Adapted from Ref. 1.)

which ranges from short-wavelength γ rays to long-wavelength radio waves (Table 1-2). The *ultraviolet* (UV) radiations important in photobiology lie in the wavelength band from approximately 160 nm, below which oxygen is strongly absorbing, to the threshold of violet light at 400 nm. *Infrared* (IR) radiation begins at the long-wavelength limit of red light at 760 nm. Wavelengths longer than 1 μm are essentially "heat waves," whose biological effects are initiated by temperature elevation.

Electromagnetic radiation is propagated in the form of waves. A perfectly monochro-

TABLE 1-2. Electromagnetic Spectrum

Spectral band[a]	Wavelength range	Photon energy (eV)
Microwave	100–0.1 cm	10^{-6}–10^{-3}
IR-C	1000–3 μm	0.001–0.4
IR-B	3–1.4 μm	0.4–0.9
IR-A	1.4–0.76 μm	0.9–1.6
Visible	760–400 nm	1.6–3.1
UV-A	400–320 nm	3.1–3.9
UV-B	320–280 nm	3.9–4.4
UV-C	280–100 nm	4.4–12.4
Vacuum UV	100–10 nm	12.4–124
X rays	10–0.01 nm	10^2–10^5
γ rays	0.1–0.0001 nm	10^5–10^7

[a]IR = infrared radiation; UV = ultraviolet radiation.

matic light wave is characterized by a single wavelength (λ) and a single frequency (ν). The wave propagates at a wave velocity (V) that depends on the refractive index of the medium (n) according to the fundamental relationship:

$$V = \lambda\nu = c/n \tag{1-1}$$

where c is the speed of light in vacuum (3.00×10^8 m/s). Monochromatic light is an idealized concept because a real electromagnetic wave must have a finite "bandwidth." Light may be considered as essentially monochromatic if the bandwidth is sufficiently small that it does not affect the interactions of the light with its environment. Polychromatic light contains a distribution of wavelengths, e.g., the spectrum of sunlight in Fig. 1-1. An electromagnetic wave carries with it coupled electric and magnetic fields, which are responsible for its interactions with matter. According to the classical electromagnetic theory of Maxwell, the electric and magnetic fields lie in the plane transverse to the direction of wave motion. (In optics, this direction is the "ray" and the perpendicular plane is the "wavefront.") The electric and magnetic fields of electromagnetic radiations are mutually perpendicular, and their amplitudes oscillate with the frequency of the light. With the advent of quantum optics in the late 19th and early 20th centuries, it was deduced that certain properties of electromagnetic radiation may be described more accurately by considering the radiation to be composed of small energy packets or *quanta*. The energy content of a light quantum, which is also known as a *photon*, is related to the frequency by the equation:

$$E = h\nu = hc/\lambda \tag{1-2}$$

where h is Planck's constant (6.63×10^{-34} J-s). Equation (1-2) specifies the energy content (J) of a photon at a given frequency (s^{-1}) or wavelength (m). The quantum energy associated with different wavelengths of electromagnetic radiations are given in Table 1-2 in units of eV, where 1 eV = 1.60×10^{-19} J. When electromagnetic radiation interacts with matter, the microscopic transfer of energy is limited by the photon energy, e.g., X rays can eject strongly bound inner-shell electrons from atoms, while visible light interacts only with the less strongly bound valence electrons. The macroscopic properties of electromagnetic radiation may be described by wavelike and quantum concepts. The wave description is based on the radiant power or "flux" carried by the waves, and the corresponding quantum description is based on the flow of discrete photons. In general, the quantum energy distribution determines the types of microscopic processes that may be initiated by a given electromagnetic radiation, and the radiant power limits the rate at which a process actually does take place under given conditions.

1.3. PHOTOTECHNOLOGY

A photobiological system consists of a coupled light source and a receiver. Artificial light sources are generally employed for research because sunlight is too variable for good reproducibility, although the scaling of biological responses obtained with lamps to those expected for sunlight can be a difficult problem. Whatever the size of the receiver, the ultimate absorbers are molecular entities referred to as *chromophores,* which may range in size from individual molecules (e.g., chlorophyll) to subunits of a macromolecule (e.g., aromatic amino acid residues in proteins). This section is primarily concerned with the methods of light generation, control, and characterization. Much of the research in photobiology utilizes precision optical and electronic equipment, assembled so as to accomplish a specific task. In order to understand and carry out experiments in photobiology, it is necessary to acquire a basic knowledge of optical instruments and measurements.

The First Law of Photochemistry states that photons must be absorbed before photochemistry can occur. Therefore, the wavelength distribution is a key determinant of biological responses to light. An important objective in photobiology involves measuring and understanding the responses of biological systems to light of different wavelengths. (The ubiquitous term "system" is used to designate the object of interest, which may range from a single molecule to the entire earth.) The key role of wavelength has led to the designation of certain wavelength bands by the Committee on Photobiology of the CIE (Commission Internationale de l'Eclairage). Ultraviolet radiation of biological importance has been divided into three subbands (Table 1-2). UV-C is the band from 100 to 200 nm. All cellular constituents absorb UV-C. However, sunlight at the earth's surface contains virtually no UV-C because of strong filtering by the stratospheric ozone layer. The UV-B band ranges from 280 to 320 nm. The nucleic acids and aromatic amino acids are the major cellular absorbers of UV-B. UV-A ("black light") ranges from 320 to 400 nm. Biological effects induced by UV-A may involve either direct energy absorption, e.g., in the long-wavelength absorption tail of proteins and DNA, or photosensitization by endogenous or exogenous substances (Chapters 3 and 6). Visible light is the band from 400 to 760 nm, and is responsible for most natural phenomena in photobiology, including photosynthesis (Chapter 12) and vision (Chapter 9). Notable exceptions are the effects of UV radiation on human skin, leading to sunburn and vitamin D production (Chapter 6). The designated bands of IR radiation are IR-A (760–1400 nm), IR-B (1.4–3 μm), and IR-C (3–1000 μm). IR-A is sometimes called "near-IR" radiation. Except for the short-wavelength edge of IR-A, the photon energy of IR radiation is too low for the initiation of photochemical reactions, and the dominant effect of IR absorption is heating. Water is the major biological absorber of IR radiation above 900 nm, e.g., absorption by water vapor accounts for the sharp decreases in the near-IR spectrum of sunlight (Fig. 1-1).

1.3.1. Light Sources[1–4]

The important properties of a light source are the spectral power output (watts of radiant energy emitted in a given wavelength band), the dimensions of the emitting surfaces, lifetime, stability, and cost. The spectral power output determines whether a specific lamp is useful for the intended purposes. The source dimensions should be matched to the light delivery system and the receiver. Light from a small source, e.g., a laser, can be delivered to a small surface area with good efficiency. An extended source, e.g., a fluorescent tube, is useful for illumination of large areas. The stability of a lamp should be appropriate for the duration of the experiments. Cost evaluation should include the replacement of lamps as well as the initial cost of the entire system. In general, photochemical sources are "bright," because they are intended to accelerate processes that are normally slow or inactive at ambient conditions. It is usually desired to irradiate in

TABLE 1-3. Operating Characteristics of Typical Photochemical Lamps

Lamp type	Range: UV	Range: Vis	Range: Near-IR	Output power	Initial cost	Lamp lifetime
Tungsten-halogen		+	+	Medium	Low	Short
Deuterium	+			Medium	High	Long
Mercury arc	+	+		High	High	Long
Xenon arc	+	+	+	High	High	Long
Fluorescent	+	+	+	Low	Low	Long
Lasers	+	+	+	High	High	Varies

a specific spectral region, which may vary from selected monochromatic wavelengths to broad-band radiation. The radiation must be coupled efficiently to the receiver. The general properties of light sources used as photochemical lamps are summarized in Table 1-3.

1.3.1.1. Incandescent Lamps[1]

Incandescent lamps utilize light emission from an electrically heated filament. This radiation approximates the *blackbody* spectrum of a hot object. According to the Stefan–Boltzman law, the total radiant power output of an ideal blackbody increases with the fourth power of the absolute temperature, e.g., an ideal blackbody emits 92 W/cm^2 at 2000K and 466 W/cm^2 at 3000K. According to Planck's radiation law, the spectral distribution of blackbody radiation shifts toward the blue with increasing temperature. Therefore, a hotter object emits more light at all wavelengths, with an increase in the relative fraction of UV radiation. The blackbody spectra in Fig. 1-2 illustrate both laws.

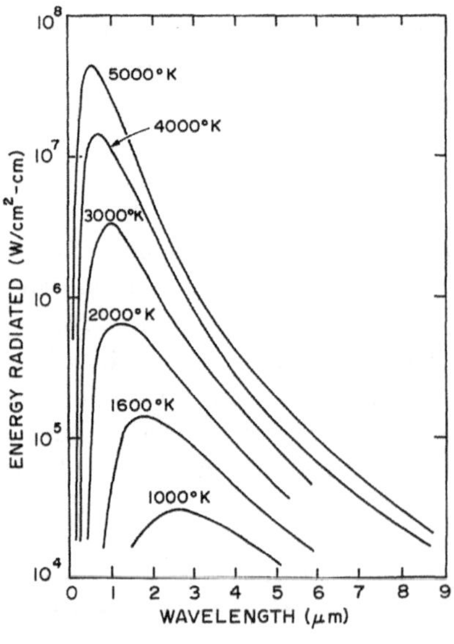

Fig. 1-2. Spectral irradiance of blackbody radiation. The ordinate is the spectral power emitted (W/cm^2) per unit wavelength (cm). The total radiation emitted per unit surface area increases with the fourth power of the absolute temperature (Stefan–Boltzmann law). The product of the absolute temperature and the wavelength of maximum emission is constant (Wien's displacement law). (Adapted from Ref. 1.)

''Infrared thermometers'' are highly sensitive, noninvasive devices for measuring surface temperature, based on the intensity of the emitted blackbody radiation. The concept of ''color temperature'' is used in the lighting industry to relate the spectral output of a lamp (usually an incandescent lamp) to a blackbody radiator. It is defined as the temperature of an ideal blackbody that emits approximately the same visible spectral distribution as the lamp. The color temperature of an incandescent lamp is always lower than the physical temperature, and it may be shifted with color filters. Tungsten filament lamps can operate under nitrogen gas at 3360°K color temperature for approximately 10 h, e.g., a photoflood lamp. The output of a photoflood lamp is ''redder'' than sunlight, with approximately 0.5% UV and 16% visible radiation (Fig. 1-1). The useful lifetime can be extended to approximately 100 h by employing a quartz envelope, and by introducing a small quantity of halogen vapor, which reacts with the black tungsten film deposited on the envelope and redeposits it back on the filament. The spectral output of a 100-W quartz-tungsten-halogen lamp is shown in Fig. 1-3. This type of lamp is a useful photochemical source for the visible region, with the advantages of broad-band emission, and low initial and replacement costs.

1.3.1.2. Arc Lamps

Arc lamps are probably the most frequently employed type of photochemical source. A typical arc lamp has a quartz envelope containing a filling gas, and two main electrodes that carry a high current. In some arc lamps the cathode is heated with a separate filament (''hot cathode'' type), and in others the cathode operates at ambient temperature (''cold cathode'' type). The radiation emitted by an arc is characteristic of the filling gas and consists of strong lines at specific wavelengths superimposed on a continuous radiation ''background'' (Fig. 1-3). The mechanism of arc action is complex and has been the subject of much investigation.[1] Arc lamps require a starting device because the voltage required to initiate an arc is considerably higher than the normal operating voltage. The arc lamps used as photochemical sources are usually filled with mercury vapor, an inert gas such as argon, or both. The low-pressure mercury arc is the standard source for nearly

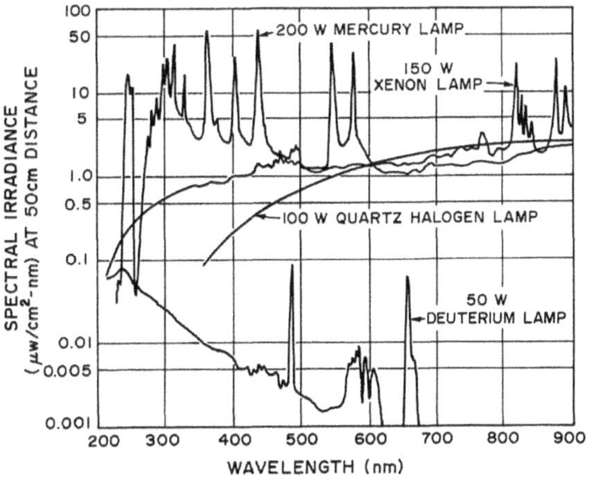

Fig. 1-3. Spectral irradiance of several lamps used as photochemical sources. The ordinate is the power in a 1-nm bandwidth incident on a 1-cm^2 receiver located at 50 cm from the lamp. The spectra are for a 200-W high–pressure mercury arc lamp (Osram HBO 200W/2), a 150-W high-pressure xenon arc lamp (Hanovia 901-C1), a 100-W quartz-halogen-tungsten lamp (Oriel 6333), and a 50-W deuterium lamp (Oriel 6316). (Adapted from 1985 Oriel Catalog, Vol. 25, *Light Sources, Monochromators, Detection Systems.*)

monochromatic 254-nm radiation. It operates at low power, and 90% of the output is at 254 nm, with the remainder distributed among the other primary mercury emission lines at 313, 365, 405, 436, 546, and 577 nm. Germicidal lamps (e.g., General Electric Co.) are hot cathode, low-pressure mercury arcs (Fig. 1-4). The most intense, general-purpose arc lamps available for research are cold cathode, high-pressure xenon and xenon-mercury arcs. They are made with closely spaced electrodes in a quartz envelope, and operate at 50–80 times atmospheric pressure to permit a high current density. The emission lines of a high-pressure arc are quite broad compared to low-pressure arcs, and there is a strong continuous background, providing useful intensities at all wavelengths transmitted by quartz (Fig. 1-3). High-pressure arcs are relatively expensive, requiring a regulated, high-current power supply, an air-cooled or water-cooled lamp housing, and an ignition device. They are long-lived (typically 1000 h) with moderately high replacement cost. Arc lamps with quartz envelopes generate significant amounts of toxic ozone, and must be vented or operated in a well-ventilated area.

1.3.1.3. Fluorescent Lamps[1,2]

Fluorescent lamps are low-pressure mercury arcs, in which the 254-nm emission is absorbed by a *phosphor* coating on the inside of the glass envelope, and is re-emitted as a longer wavelength *fluorescence* (Section 1.4). The output consists of one or several overlapping broad bands with superimposed "spikes" at the mercury emission lines. The spectral range is determined by the chemical composition of the phosphor coating and the glass envelope (Fig. 1-5). About 20% of the electrical energy is converted to useful visible or near-UV radiations. Fluorescent lamps are inexpensive, low-power sources, useful for long-term irradiation of large areas. Many special types of fluorescent lamps are available. The phosphor in a fluorescent "sunlamp" emits from 260 to 400 nm with the peak at 320 nm. The envelope is made of a glass that limits the output to wavelengths greater than 320 nm. "Black-light" fluorescent lamps emit from 325 to 440 nm, with the peak near 360 nm (Fig. 1-6). "Daylight" fluorescent lamps give approximately the same visual response as sunlight, without the UV and IR radiations. Fluorescent lamps from

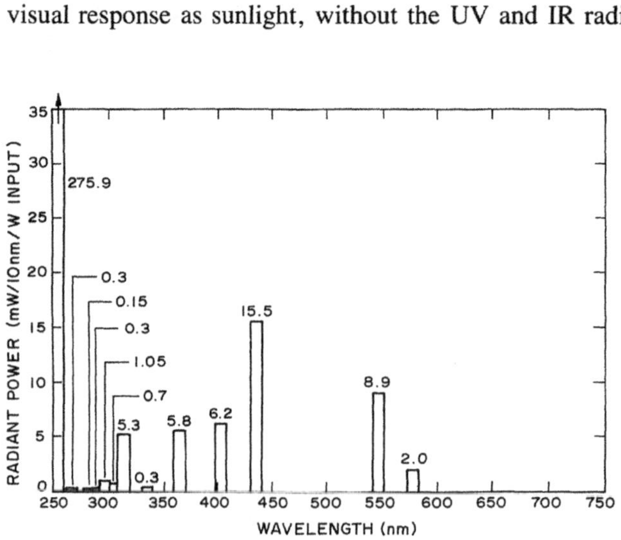

Fig. 1-4. Spectrum of a "germicidal" low-pressure mercury arc (General Electric Co. G30T8) after 100 h of operation. The numbers adjacent to each narrow line indicate the radiant power output, in units of mW per 10 nm bandwidth per W of input power. Approximately 85% of the total radiant power is emitted in the 254-nm "resonance line." (Courtesy of General Electric Co.)

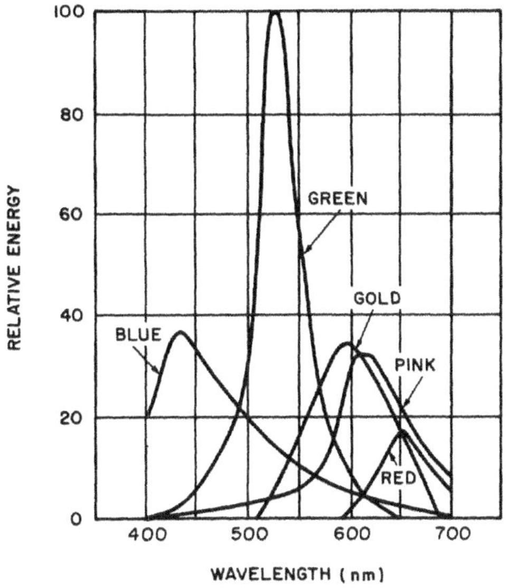

Fig. 1-5. Spectral output of various colored fluorescent lamps of equal wattage. The mercury spectral lines are omitted. (Courtesy of Westinghouse Electric Corp.)

different suppliers with similar designations are usually not identical, e.g., black light. Furthermore, a specific lamp desigination does not guarantee that the phosphor or envelope will not be changed by a manufacturer. Readers who use fluorescent lamps for research are advised to measure the emission spectrum or obtain the current specifications from the manufacturer.

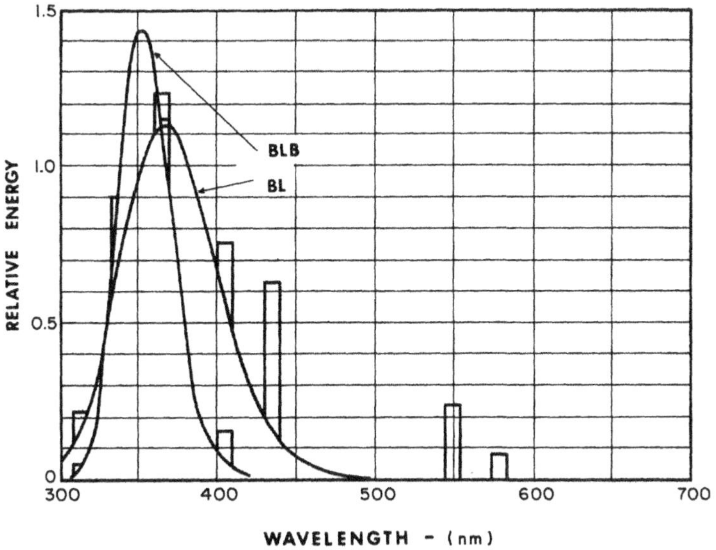

Fig. 1-6. Spectral output of 40-W T12 black-light (BL) and black-light-blue (BLB) fluorescent lamps after 100 h of operation. (Courtesy of General Electric Co.)

1.3.1.4. *Lasers*[5,6]

A laser is a special type of light source that emits intense, monochromatic radiation in a highly collimated beam. The first practical laser was constructed in 1960 by Maiman, utilizing a ruby rod (chromium ions in a matrix of crystalline aluminum oxide), which was followed rapidly by a remarkable diversity of new laser designs. A laser is described by its "active medium," i.e., the chemical substances and their physical state in which laser action takes place. The more common types of lasers include solid state (e.g., ruby, Nd-yttrium-aluminum garnet [Nd-YAG]), gas (e.g., He-Ne), molecular (e.g., CO_2), gas ion (e.g., Ar, Kr, Xe), metal vapor (e.g., He-Cd, He-Cu, He-Au), excimer (e.g., ArF, KrF, XeCl), chemical (e.g., HF, HCl), dye (e.g., fluorescein, rhodamine 6G), and diode (e.g., GaAs, GaInAs). Laser action orginates in the fundamental principle of atomic physics which states that the internal energies of atoms are quantized in discrete energy levels (Section 1.4.1). When a substance is in thermal equilibrium with its surroundings at ordinary temperatures, almost all atoms of the substance are in their lowest energy state. Atoms can be elevated to higher energy states by heating, carrying out a chemical reaction, establishing an electrical discharge, or irradiating with light at wavelengths absorbed by the substance. In laser terminology, this process is known as "pumping." (The reader should be warned that laser science is replete with jargon, some of which must be introduced if this section is to be of practical use.) It is difficult to pump more atoms to higher energy states than remain in lower energy states, e.g., heat alone cannot accomplish this *population inversion*. However, population inversion can be achieved in special circumstances by proper selection of the active medium and the operating conditions, which is the basis of laser action. When population inversion occurs, the absorption of one photon at a wavelength corresponding to the energy difference between an upper state and a lower state can initiate the emission of more than one photon. This process is known as *stimulated emission,* leading to the acronym LASER, for "light amplification by stimulated emission of radiation." The photons emitted by stimulated emission have the same wavelength, polarization (the direction of the electric field, Section 1.3.2.4), and move in the same direction as the initiating photons. Laser light is coherent, i.e., the light waves move together in space and time. The properties of very narrow bandwidth, collimation, and coherence make it possible to transmit laser light over long distances with very little spreading, and to focus a laser beam to a very small, extremely intense spot. Although the physical appearance of various types of lasers may be quite different, all lasers employ an optical cavity to contain the active medium and promote amplification of those light waves propagating along the cavity axis. The helium-neon laser, depicted in Fig. 1-7, was the first practical gas laser, constructed in 1961 by Javan, Bennett, and Harriott. The optical cavity is a glass or metal cylinder, containing a mixture of helium and neon at low pressure, with mirrors at each end. One mirror is highly reflecting at the emission wavelength, and the other has a low transmittance to permit extraction of laser light after many "round trips" in the cavity. The helium-neon laser is pumped by a high-voltage discharge, and may be designed to emit a single line at 632.8 nm. Table 1-4 summarizes the operating characteristics of some lasers used in photobiology.

The two most important properties of a laser are the emission wavelengths and the time span over which laser action occurs. Most lasers emit strong lines at one or a few wavelengths. A laser can be tuned to emit single wavelengths by locating a wavelength

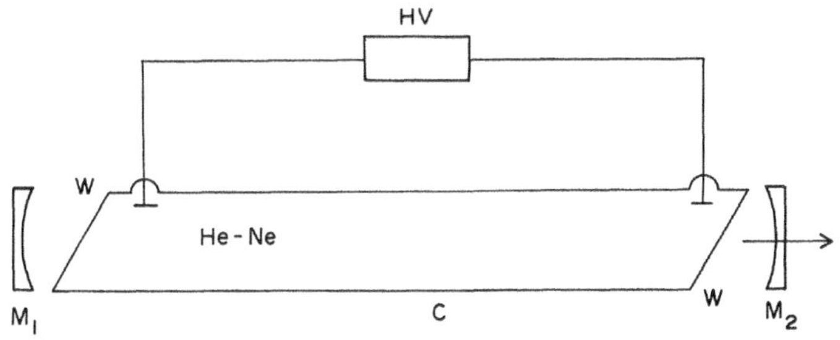

Fig. 1-7. Schematic diagram of a helium-neon laser. The cylindrical cavity (C) is filled with a mixture of helium and neon at low pressure. A gaseous discharge established by the high-voltage power supply (HV) excites helium atoms to metastable (long-lived) states, which transfer their energy to neon atoms by collisions. This selective population of neon-excited states leads to population inversion. Laser action is initiated by radiative transitions of excited neon atoms to lower energy excited states. The photons emitted parallel to the longitudinal axis are trapped in the optical cavity formed by a totally reflecting mirror (M_1) and a partially reflecting mirror (M_2). The longitudinal light waves corresponding to the neon emission wavelengths are amplified by stimulated emission. The strongest visible line is at 632.8 nm. The glass windows (W) of the cavity are tilted at the Brewster angle of approximately 57°, which tunes the laser to a single transverse mode parallel to the plane of incidence, and polarizes the emitted laser radiation.

control element in the optical cavity, e.g., a prism. The dye laser is a notable exception, in which the emission wavelength can be tuned over the fluorescence band of a dye, typically 100 nm (Section 1.5.2). A wider range of wavelengths is available with interchangeable dyes. The wavelength (or frequency) generated in normal laser operation is known as the "fundamental." Wavelengths shorter than the fundamental can be obtained by "frequency doubling." This is accomplished with useful efficiency (typically 5 to 30%) in high-power lasers by passing the laser light through a crystal with "nonlinear"

TABLE 1-4. Properties of Lasers Useful in Photobiology

Type	Principal wavelengths (nm)	Typical operation: CW	Typical operation: Pulsed	Average power[a]
Helium-neon	633, 1153	+		Very low
Krypton ion	458, 568, 647	+		Medium
Argon ion	458, 488, 515	+		High
Nitrogen	337		+	Low
Copper vapor	510, 578		+	High
Gold vapor	628		+	High
Excimer (rare gas halide)	193, 248, 308, 351		+	High
Ruby	694[b]		+	Low
Nd:YAG	1064[c]		+	High
Dye	260–900	+	+	High
Carbon dioxide	10,600	+		High
Semiconductor	Red, near-IR		+	Very low

[a]Power levels vary considerably with laser construction and applications. The ranges are based on <0.1 W (very low), 0.1–1 W (low), 1–10 W (medium), 10–100 W (high).
[b]347 nm with frequency doubling.
[c]532 nm with frequency doubling, 355 nm with frequency tripling, 266 nm with frequency quadrupling.

dielectric properties, e.g., potassium dihydrogen phosphate (KDP). Halving the wavelength is referred to as "doubling," dividing the wavelength by three is "tripling," etc. The helium-neon laser operates in the "continuous wave" (CW) mode, i.e., light is emitted continuously during pumping. (The output of a CW laser has a substructure of very short pulses, induced by standing waves generated in the optical cavity, which do not affect its usual applications as a continuous light source.) The power output of a CW laser is not remarkable, e.g., a 10-W laser corresponds to intermediate power (Table 1-4). However, good collimation makes it possible to focus a laser beam to a very small area, providing high *irradiance* (Table 1-5), e.g., the irradiance of a 1-cm-diameter, 10-W laser beam focused to 1 mm is 1.3 kW/cm^2. By comparison, the total visible light irradiance in the collimated beam of a 150-W high-pressure xenon arc is approximately 0.2 mW/cm^2. Since laser light is highly monochromatic, the *spectral irradiance* (W/m^2-nm) exceeds that of strong arc lamps by many orders of magnitude. Pulsed operation is the other general type of laser action. It is initiated by pulsed pumping, e.g., by pumping a Nd-YAG laser with a gas-filled flash lamp. If pulsed pumping is carried out repetitively, the output consists of a sequence of relatively short pulses separated by longer time intervals. Some lasers can operate only in CW mode, others only in pulsed mode, and some in either mode, depending on the construction and type of pumping (Table 1-4).

There has been a continuing effort to develop light sources with the shortest possible

TABLE 1-5. Radiometric and Photometric Units

Radiometric quantity[a]	Concept	SI unit	Photometric equivalent[b]
Energy (Q)	Quantity of light	J	lm-s[c]
Flux (F)	Power	W	lm
Energy density (W)	Energy content per unit volume	J/m^3	lm/m^3
Irradiance (E)	Flux per unit area incident on a small-plane surface	W/m^2	Illuminance[d] lm/m^2
Excitance (M)	Flux per unit area leaving a small-plane surface	W/m^2	lm/m^2
Intensity (I)	Flux emitted into a unit solid angle	W/sr	Candlepower[e] cd (lm/sr)
Radiance (L)	Flux in a given direction per unit solid angle and per unit area normal to the direction of propagation	W/m^2-sr	Luminance[f] cd/m^2
Exposure (Q)	Energy per unit area incident on one side of a small-plane area	J/m^2	lx-s

[a]These terms should be prefixed by "radiant" to avoid confusion with the photometric units, e.g., radiant energy.

[b]These terms should be prefixed by "luminous" to avoid confusion with the radiometric units, e.g., luminous flux.

[c]The *lumen* (lm) is the basic unit of luminous flux. It is the luminous flux emitted into a unit solid angle by a uniform point source whose luminous intensity is 1 *candela* (cd). The radiant flux in the lumen is weighted according to its efficiency in producing a visual response. One lumen is equivalent to approximately 682 W at 555 nm and to fewer watts at other wavelengths, where the phototopic response is lower.

[d]Illuminance is the basic unit of illumination. An illuminance of 1 $lumen/m^2$ is also called 1 *lux* (lx). In the British system, 1 $lumen/ft^2$ is 1 *foot-candle*.

[e]The *candela* is the SI unit of luminous intensity. It is based on the luminous intensity of a blackbody at the melting temperature of platinum. One lumen of luminous power per steradian corresponds to 1 cd. The luminous of a point source intensity in candela equals the candlepower.

[f]Luminance is luminous intensity per unit projected area. One candela/m^2 square meter is also known as 1 "nit," and 1 cd/cm^2 is 1 "stilb."

pulse durations in order to study fast processes in photochemistry and photobiology, e.g., for *flash photolysis* (Section 1.5.1) and fluorescence lifetime measurements (Section 1.5.2.2). Shortly after the invention of lasers, it was discovered that the pulse duration of ruby and Nd-YAG lasers is reduced from the order of 10^{-3} to 10^{-8} s by blocking the optical path until the pumping light flash is nearly complete. This technique is known as "Q switching" and can be accomplished with a fast mechanical or electro-optical shutter. "Mode-locking" was a subsequent development, in which the output consists of a pulse "train" of approximately 10 ns overall duration, with a substructure of pulses ranging from less than 1 ps to several ns duration. Single mode-locked pulses of about 10 ps duration have been obtained by amplifying one pulse of the train. (A laser "amplifier" is another laser of the same type, pumped by the emission from the laser "oscillator.") The *peak power* is the energy in a single pulse divided by the average pulse duration, and the *average power* is the energy in a single pulse divided by the time between the pulses. The peak power is important for interactions of the laser light with matter that occur in time durations comparable to that of the laser pulse. The average power is applicable for relatively slow processes that do not respond to the individual pulses. Some applications of lasers in medicine will be described in Section 1.3.5.

1.3.2. Optical Components

Equation (1-1) indicates that the velocity of light in a medium equals c/n, where "empty space" has a refractive index of unity. When propagating monochromatic light encounters a change of refractive index, e.g., on going from air to polished glass, the incident beam is partially *reflected* at the interface (specular reflection) and partially *refracted* (bent) on entering the new medium. (Diffuse reflection at a rough surface is considered in Section 1.3.5.) Polychromatic light introduces the additional complication that the different wavelengths are propagated at different speeds in a uniform material. This effect is referred to as *dispersion,* and is responsible for the ability of a prism to separate "white" light into its component wavelengths. *Diffraction* phenomena occur when light passes through apertures with dimensions comparable to the wavelength, e.g., diffraction limits the ability of a light microscope to image objects smaller than about 0.2 μm. (However, an electron microscope has a much smaller effective wavelength and can image objects of atomic dimensions.) *Interference* is a related phenomenon, in which light waves combine in a regular pattern of intensity variations or "fringes." The propagation of light through a material with directional properties can affect its *polarization,* i.e., the orientation of the electric field in the plane of the wave front. Polarizing devices are used to polarize ordinary light and alter the properties of polarized light. With this brief introduction into the language of classical optics, we proceed to consider the optical elements most widely used in photobiology. Additional information about "physical" optics and the optical properties of materials is available in standard texts.[7–9]

1.3.2.1. Lenses and Mirrors

Lenses are used to relay light between optical elements. The most important properties of a lens or lens system are the focal length, aperture, and wavelength transmission. The focal length is the distance from the lens at which collimated rays are brought to

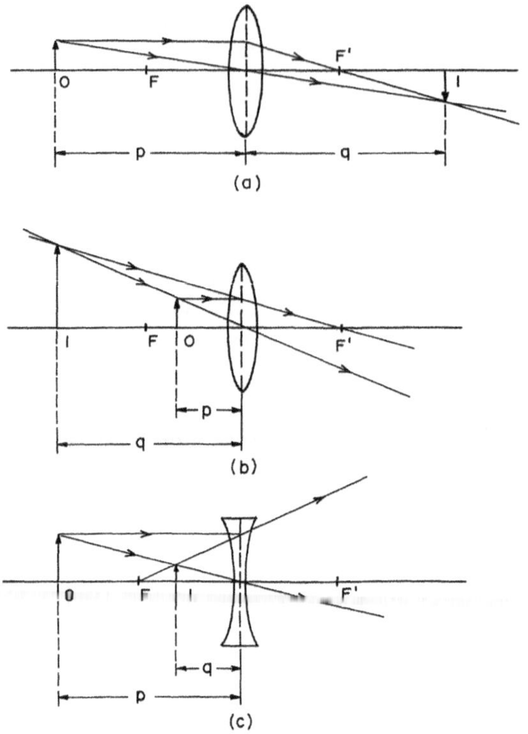

Fig. 1-8. Ray diagrams for spherical thin lenses. In ray tracing, a ray parallel to the lens axis is refracted through a focal point (F and F′); a ray diverging from a focal point is refracted as parallel light; a ray passing through the lens center is not deflected. The convergence of rays from a point on the object (O) produces an equivalent point on the real image. (a) An object located outside the front focal plane of a convex lens (F) gives a real, inverted image. (b) An object located inside the front focal plane of a convex lens gives a virtual, upright image. (An image is virtual if the rays appear to diverge from image points not actually reached by the light.) (c) An object located outside the front focal plane of a concave lens gives an upright, virtual image. The distance of the image plane from the lens (q) is related to the distance of the object plane to the lens (p) and the focal length of the lens (F) by: $1/p + 1/q = 1/F$, and the linear magnification is q/p. The distances q and p are taken as positive for real objects and images and negative for virtual objects and images. The linear magnification is negative for an inverted image and positive for an upright image.

convergence. The fundamentals of image formation by thin lenses are described in Fig. 1-8. A ''converging'' lens can be used to collect light from a small source (a ''condenser''), to match a collimated light beam to a receiver (''relay'' lens), and to focus a light beam. Diverging lenses are used as intermediate elements in compound lenses, e.g., a telephoto lens, and to correct the human eye for near-sightedness. The light-gathering ability of a lens is limited by its *aperture*. The relative aperture or ''*f* number'' is the ratio of the focal length to the lens diameter. Light-gathering effectiveness is inversely proportional to the square of the *f* number. Thus, an *f*/1.0 lens collects four times more light than *f*/2.0. The *resolving power* of a lens is the ability to separate two closely spaced objects. For a magnifying lens (e.g., a microscope objective), the resolving power equals $\lambda/2n_0 \sin \alpha$, where λ is the wavelength, n_0 is the refractive index of the medium from which the light enters the lens, and α is the half-angle of the acceptance cone. (The acceptance cone of an optical fiber is shown in Fig. 1-9.) The wavelength transmissivity of the lens material should be given special attention because ordinary optical glass transmits only from about 370 nm to 2.5 μm, and cannot be used in the UV and far-IR regions. The short-wavelength limit is lowered to 330 nm with Pyrex glass, and below 220 nm with fused silica or crystal quartz. Lenses for use in the IR region can be ground from ionic crystals, e.g., NaCl (0.25–16 μm). The IR ''heat waves'' generated by light sources are usually undesirable, and can be eliminated with heat absorbing glass and a water filter. Although a multielement lens system has superior imaging properties compared to a simple lens, reflections at the air–glass interfaces lead to significant losses of light, which are the order of 25–30% overall. Front-surfaced mirrors provide an alter-

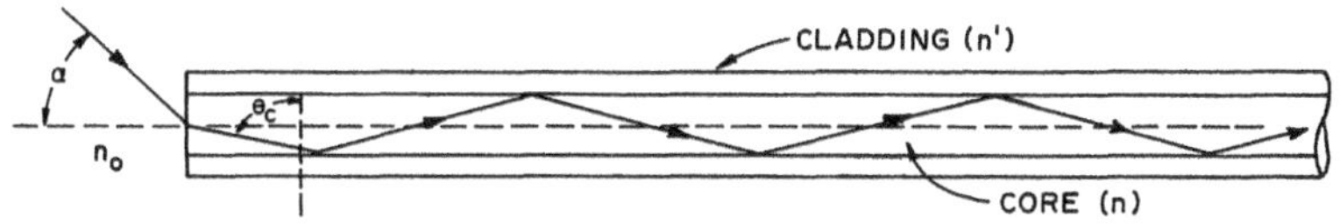

Fig. 1-9. Propagation of light by a cylindrical optical fiber. A ray striking the input face within the maximum acceptance angle α is totally reflected at the interfaces between the fiber (refractive index n) and the cladding (refractive index n') because the angle of incidence exceeds the critical angle Θ_c, where $\sin \Theta_c = n'/n$. This process is repeated until the ray exits at the opposite face. The numerical aperture (N.A.) is defined as $n_o \sin \alpha$, where n_o is the refractive index of the external medium (unity for air). It is related to the fiber properties by: N.A. $= \sqrt{(n^2 - n'^2)}/n_o$. A typical optical fiber with a fused silica core and plastic cladding in air has a N.A. of 0.22, corresponding to a lens "*f* number" of approximately 2.

native method for collimating and focusing light with lower losses than obtained with glass lenses. The reflectivity of an evaporated aluminum mirror exceeds 90% from 240 nm to the far-IR. Unlike lenses, mirrors are not subject to *chromatic aberration* (dependence of focal length on wavelength) because specular reflection does not depend on the refractive index of the surface. Focusing mirrors are fabricated with spherical, elliptical, parabolic, and aspheric surfaces, depending on the application.

1.3.2.2. Optical Fibers[10]

Optical fibers are efficient devices for transporting light to awkward-to-reach locations and over long distances. An optical fiber consists of a glass or fused-silica core, covered by a lower refractive index cladding. Light rays incident at one end are transmitted down the fiber by *total internal reflection* at the interfaces between the core and cladding (Fig. 1-9). The light-gathering power of a flat-ended fiber depends on the numerical aperture and compares quite favorably with a fast (i.e., low *f* number) lens. Glass fibers transmit from 340 to 1300 nm. Fused-silica fibers transmit down to 220 nm but are about four times more costly. Fiber bundles or "light guides" can transmit higher optical powers than single fibers and have greater mechanical strength. Useful optical devices can be fabricated by utilizing fibers to transmit several optical signals. A bifurcated cable or "Y guide" can act as an optoelectronic sensor of surface reflection or motion. The incident light is transmitted down one arm of the Y and out of the common leg, and the reflected light is collected by the common leg and measured at the other arm of the Y. This principle is used in the "oximeter," a device that measures the oxygen content of blood *in vivo*, by detecting the diffusely scattered light from hemoglobin at several wavelengths.[11] An ordinary light guide cannot transmit an image because the spatial arrangement of the fibers at the input is not retained at the exit. Image transmission has been accomplished with "coherent bundles," where the relative positions of the individual fibers at the two ends are in a simple relationship. Medical endoscopy is a major biological application of coherent fiber bundles.

1.3.2.3. Filters and Monochromators[7–9]

Filters and monochromators are used to restrict radiation to selected wavelengths. Glass filters are prepared from polished optical glass, by incorporating additives that limit

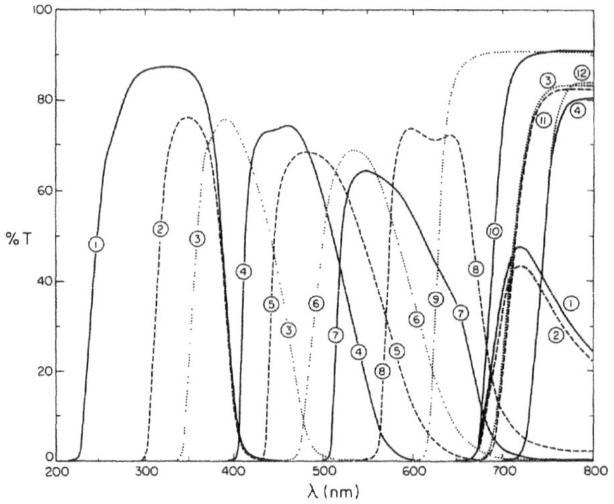

Fig. 1-10. Percent transmission of some high-transmission combinations of glass filters. (1) Corning C.S. No. 7-54; (2) Corning C.S. No. 7-54 + C.S. No. 0-54; (3) Wratten 39 + Corning C.S. No. 0-52; (4) Corning C.S. No. 5-56 + C.S. No. 3-74; (5) Schott BG-14 + Corning C.S. No. 4-97 + C.S. No. 3-72; (6) Corning C.S. No. 4-97 + C.S. No. 3-71; (7) Schott BG-38 + Kodak infrared cut-off, No. 301 + Corning C.S. No. 3-69; (8) Kodak 301 + Corning C.S. No. 3-66; (9) Schott RG630; (10) Schott RG695; (11) Kodak 39 + Schott RG695; (12) Corning C.S. No. 5-56 + Schott RG695. [Adapted from W. D. Herkstroeter, A series of sharp-cut and bandpass glass filters for the range 252-800 nanometers. *Mol. Photochem.* **4**, 551–557 (1972).] The Corning glass numbers in the original reference have been replaced by the currently used "color specification" (C.S.) numbers.

the transmission to certain bands (bandpass filters) or above certain wavelengths (short-wavelength cutoff filters). Hundreds of glass filter types are available from the major suppliers. Filter combinations can be employed to obtain high transmission in selected wavelength bands (Fig. 1-10). Glass filters are relatively inexpensive and are available in large sizes. *Interference filters* are the other major type of filter. The simplest type of interference filter consists of two parallel, partially reflecting mirrors (usually on a glass base) separated by a layer of dielectric material. This arrangement preferentially transmits light of wavelength: $\lambda = 2dn$, where n is the refractive index of the dielectric layer and d is its thickness. (The incident ray and a ray that reflects twice before emerging add constructively, and other wavelengths are attenuated by destructive interference.) Interference filters are more expensive than glass filters, but they can be made to much more exact specifications. However, a given design is exact only for perpendicular collimated rays, and off-axis rays are transmitted with a wavelength shift, which may be employed to tune the filter over a small bandwidth.

A *monochromator* is a precision optical device that separates broad-band light into its composite wavelengths. The standard dispersing elements in monochromators are glass or quartz prisms and diffraction gratings. A prism disperses polychromatic light because the velocity of light in a transparent material depends on the wavelength. A longer wavelenth ray moves more rapidly than a shorter wavelength ray, and therefore it takes a shorter path through the prism, and exits at a smaller angle to the prism face. The wavelengths are selected by rotating the prism (or a mirror) to pass a narrow bandwidth through an exit slit. A diffraction grating disperses polychromatic light by multiple interference of the light waves reflected from closely spaced grooves in a metal coating (usually aluminum) on a glass base. Figure 1-11 shows the arrangement in a small diffraction grating monochromator designed for high optical throughput (light gathering power). The *dispersion* of a monochromator indicates its ability to spread the wavelengths of light. The *angular*

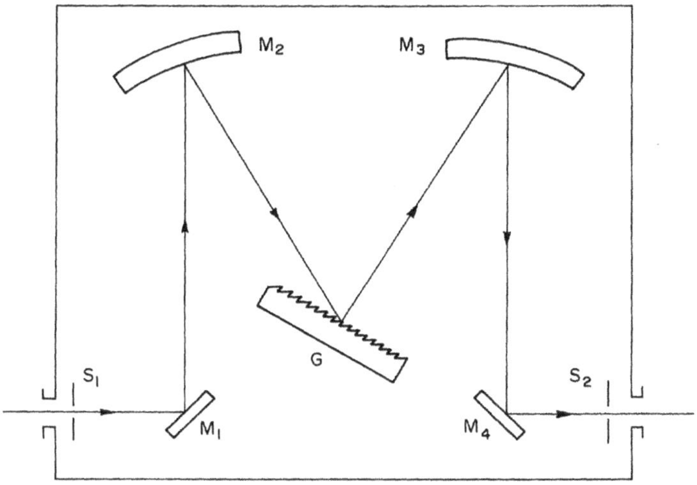

Fig. 1-11. Optical paths in a small monochromator with symmetrical, in-line optics. The incident light is focused on the entrance slit (S_1), deflected by a 45° mirror (M_1), collected by a collimating mirror (M_2), and directed to the diffraction grating (G). The different wavelengths are reflected at different angles by the grating, which are imaged by mirror (M_3) and deflected to the exit slit (S_2) by a 45° mirror (M_4). The exiting bandwidth equals the product of the inverse linear dispersion of the monochromator (nm/mm) and the exit slit width (mm). The grating is rotated to select different wavelengths.

dispersion (β) is defined as the angular difference (in radians) between two rays leaving the prism or diffraction grating that differ in wavelength by 1 nm. The *linear dispersion* (m) is the physical width at the plane of the exit slit occupied by a 1-nm bandwidth. They are related by $m = \beta F$, where F is the focal length of the lens or mirror (M_3 in Fig. 1-11) that collects the light from the dispersing element and focuses it on the exit slit. The "inverse linear dispersion" is usually specified in units of nm/mm at the exit slit. Prism monochromators were used in most laboratory instruments until the 1960s because high-quality diffraction gratings must be ruled with a grating engine and are quite expensive. Prism monochromators have the disadvantage that the dispersion is nonuniform and must be calculated at each wavelength from data provided by the manufacturer. Grating monochromators are designed to have linear dispersion, i.e., the bandwidth for a given exit slit width is the same at all wavelengths. The development of high-quality plastic copies of ruled gratings (replica gratings) has led to the dominance of gratings in optical instruments. Holographic gratings are a recent advance, in which interference fringes from two laser beams are recorded on a photosensitive surface, which is then selectively etched. Holographic gratings are more expensive than replica gratings, but they have the advantages of higher efficiency (the ratio of the diffracted energy to the incident energy at a given wavelength), less "stray light" (the background of unwanted wavelengths in the exit beam), and complete absence of "ghosts" (spurious spectral lines). In addition to linear dispersion, grating monochromators have higher light-gathering power than prism instruments of comparable cost, and the convenience of interchangeable gratings to cover a wide wavelength range.

The key determinants in monochromator selection are the wavelength range, the dispersion, the throughput, and the scattered light. Wavelength selection is the most frequent use of monochromators in optical instruments, e.g., in absorption and luminescence spectrophotometers (Section 1.5), and irradiation systems. The desirable characteristics for such applications are high throughput, small size, and moderate cost, which are obtained at the expense of high dispersion and high resolving power. Diffraction-grating theory shows that multiple "orders" of spectra are diffracted into the same directions, such that the first-order wavelength λ coincides with the second-order wavelength $\lambda/2$, the third-order wavelength $\lambda/3$, etc. (The complication of multiple orders does not occur with prism monochromators.) A high fraction of the output energy can be localized in a given order (usually the first order) by ruling the grooves at an angle with respect to the grating surface or "blazing." A diffraction grating is most efficient at the blaze wavelength (typically 50–70% for small gratings), which decreases by a factor of 0.7 at 2/3 and 3/2 times the blaze wavelength. The overlapping spectral orders at the same angle can be eliminated with short-wavelength cutoff "blocking" filters. Most monochromator applications in photobiology can be handled within the wavelength range of 200–1000 nm, which requires at least two gratings—one blazed in the UV (e.g., 250 nm) and one blazed in the visible region (e.g., 500 nm). The dispersion and the cost of a diffraction grating increase with the number of lines/mm. A replica or holographic grating with 1200 lines/mm provides a reverse linear dispersion in the order of 5 nm/mm in a typical small, high-aperture instrument, which corresponds to useful bandwidths from 0.05 to 20 nm with variable slits.

The efficient use of monochromators is described in standard texts.[12] The blaze of the grating should be optimized for the working wavelength range. Ideally, the monochromator and light source should have the same *f* number for the most efficient coupling. Energy is lost and stray light increases if the grating is overfilled (i.e., the *f* number of the source is lower than that of the monochromator), and the resolving power is lower if the grating is underfilled. Similarly, the exit beam should underfill the detector area to avoid loss of light. A relay lens or mirror can be used with very small detectors, to focus the diverging exit cone from the monochromator on the detector. Appropriate blocking filters should be used if the detector is sensitive to higher order spectra. For symmetrical optical systems, the entrance and exit slit widths should be the same. For a broad-band source, the throughput increases with the *square* of the entrance slit width, although a larger slit width also increases the bandwidth. The full entrance slit height should be illuminated for maximum throughput. The signal-to-noise ratio is limited by stray light. The stray light in a typical small, high-aperture-grating monochromator is 0.1%. The use of a "double monochromator" (two monochromators in tandem) is the most effective way of reducing stray light, and may be required for spectral resolution of broad-band, low-intensity signals.

1.3.2.4. Polarizers[7–9]

Spectroscopic measurements with polarized light provide information about the directional properties of molecules, e.g., optical activity, dichroism, and fluorescence polarization (Section 1.5). The electric field of electromagnetic radiation is located in a plane perpendicular to the ray and oscillates with the frequency of the light. The magnitude and direction of the electric field are conveniently represented by a vector, depicted as an "arrow" of length proportional to the strength of the electric field and pointing in the direction of the field. (Since the electric field is moving through space and oscillating in time, a two-headed vector is used, whose length indicates the *amplitude* of the oscillation.) The magnetic vector is perpendicular to the electric vector in the same plane, and oscillates with the same phase. (The phase of an oscillation of frequency f is $2\pi ft$, where t is the time measured from an arbitrary zero. Two oscillations are "in phase" if they reach their minimum and maximum amplitudes at the same time.) In "ordinary" light, the electric vector has a random orientation and the light wave has no inherent directionality perpendicular to the ray. (Alternatively, the electric vector can be resolved into two perpendicular vibrations, of the same amplitudes and without phase coherence.) This picture of ordinary light (O) is shown in Fig. 1-12a. In *linearly polarized* light the direction of the electric vector is fixed. One way of producing linearly polarized light is to pass ordinary light through a material with a different absorption for the two polarization components of ordinary light. This property, referred to as *dichroism,* is exhibited by

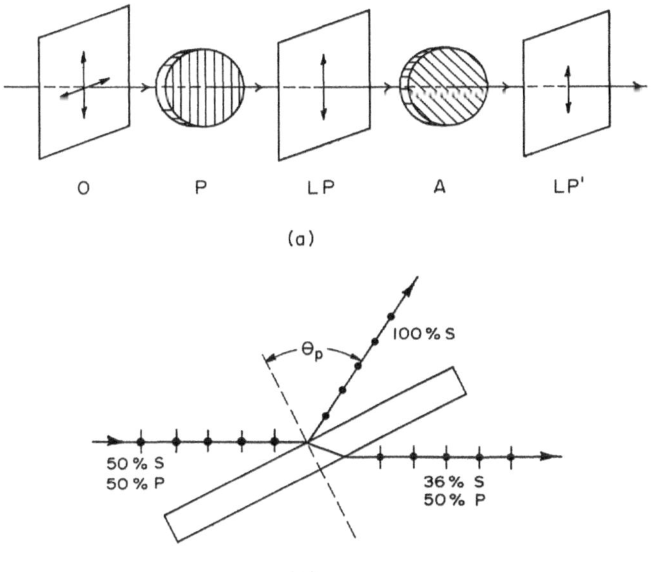

Fig. 1-12. (a) Polarization of light by a thin-film polarizer. The electric vectors in ordinary light (O) have a random orientation, as depicted by their equal projections in two perpendicular directions. The polarizer (P) is a dichroic material, with its maximum transmission for the polarization depicted by the parallel lines. The transmitted light is linearly polarized (LP). If the linearly polarized light is passed through another polarizer (the "analyzer"), the transmission is maximum if the two polarizers are aligned and minimum if they are "crossed." In the case shown, the analyzer is set at an intermediate angle, and the resultant linearly polarized light is attenuated (LP′). (b) Polarization by reflection and refraction. The electric vectors of the incident light are resolved into equal parallel (P) and perpendicular (S) components, relative to the plane containing the incoming ray and the normal direction to the glass plate (dashed line), which is referred to as the "plane of incidence." For incidence at Brewster's angle (θ_P), the reflected ray is 100% polarized in the S direction, and the transmitted light is partially linearly polarized. Brewster's angle is given by $\tan \theta_P = n_o/n$, where n_o is the refractive index of the external medium and n is the refractive index of the glass. It is approximately 57° for optical glass in air.

some materials, e.g., tourmaline, and by plastic films containing aligned organic molecules, e.g., a Polaroid film. The polarization of ordinary light by a dichroic type of polarizer is shown in Fig. 1-12a. The linearly polarized light is detected by passing it through a second polarizer, designated by "analyzer." The intensity of the light transmitted by the analyzer varies from zero, when the two polarizers are "crossed," to a maximum when they are aligned. Linearly polarized light is also produced by passing ordinary light through an optically anisotropic crystal with the property of *birefringence* or "double refraction," e.g., crystal quartz or calcite. "Nicol" and "Glan Taylor" polarizers are calcite crystals cut in such a way that beams of each polarization are separated. The passage of linearly polarized light through a thin plate cut from a birefringent crystal leads to interference effects that alter the polarization properties. A "half-wave" plate rotates the direction of linearly polarized light by an angle that depends on the relative directions of the electric vector and the crystal axes. Passing linearly polarized light through a "quarter-wave" plate at the correct angle generates *circularly polarized* light, in which the electric vector rotates rapidly in its plane. The polarization of ordinary light after reflection by a glass plate was discovered by Malus in 1808. At a specific angle of incidence, "Brewster's angle," the reflected light is linearly polarized and the transmitted component is partially polarized (Fig. 1-12b). Brewster's angle is approximately 57° for optical glass in air. The windows on a laser cavity are usually inclined at Brewster's angle, to selectively amplify a single transverse mode and obtain a smooth beam cross-section (Fig. 1-7).

1.3.3. Photodetectors

The two basic types of photodetectors are thermal detectors and quantum detectors. *Thermal detectors* respond to radiant power and are independent of the wavelength. Thermopiles and pyroelectric detectors are of this type. An illuminated thermopile generates a low voltage, induced by the absorption of radiant energy in a series array of dissimilar metal junctions (thermocouples). Thermopiles are very accurate in all spectral regions, but have low sensitivity and a slow time response. A "pyroelectric" detector is fabricated from a crystal in which a surface voltage is generated while its temperature is changing. (Pyroelectricity is a property of certain "ferroelectric" crystals, e.g., lithium niobate, with an inherent dielectric polarization or charge separation). Pyroelectric detectors have a faster time response than thermopiles, with equally wide spectral response and comparable sensitivity. However, they respond only to pulsed or chopped radiation, necessitating an AC measurement system. Since thermal detectors respond to average optical power, the readings can be converted to photons conveniently only for monochromatic radiation.

Quantum detectors respond to the average photon absorption rate, usually with a different sensitivity at each wavelength. (A dye fluorescence "quantum counter" is an exception; Section 1.5.2.2.) The standard types of quantum detectors are vacuum photodiodes, photomultipliers, and solid state photodetectors. A vacuum photodiode consists of an evacuated glass bulb, containing a photocathode that emits electrons when exposed to light and an anode that collects the emitted photoelectrons. In a photomultiplier, a set of auxiliary "dynodes" are located between the photocathode and anode. Dynodes have the property of "secondary electron emission," i.e., several electrons are emitted by a dynode for each electron absorbed. The effect is cumulative, and the output current can be

TABLE 1-6. Properties of Photodetectors

Type	Spectral range (μm)	Time response (μs)	Typical gain[a]	D^* ($\sqrt{\text{Hz}}$-cm/W)[b]
Vacuum photodiode	0.2–0.7	< 0.01	0.04 A/W	10^{11}
Photomultiplier	0.2–0.7	< 0.01	2500 A/W	10^{14}
	0.4–1.1	< 0.01	40 A/W	
Thermopile	0.2–50	10^5	10 V/W	10^8
Pyroelectric	0.2–100	1	900 V/W	10^8
Silicon photodiode	0.3–1.1	0.01	0.5 A/W	10^{12}

[a]The "gain" of a photodetector is defined as the output electrical signal in amperes (A) or volts (V) per unit incident radiant flux in watts (W).
[b]*D* is a standard figure of merit for solid state photodetectors. It is based on the ratio of the output signal to the random noise signal per unit bandwidth, scaled to a uniform detector area. The values of D^* are estimates at the spectral sensitivity maxima. The higher the value of D^*, the more sensitive the detector.

orders of magnitude higher than in a vacuum photodiode. Vacuum photodiodes and photomultipliers have excellent stability and fast time response. However, the spectral sensitivity is nonuniform and depends on the type of photocathode. Solid state photodetectors are a less expensive alternative to photomultipliers, useful in the visible and near-IR regions. A "photovoltaic" cell operates at zero bias (no applied voltage) and generates a photovoltage when exposed to light. A photoconductive cell is connected to an external voltage and generates a photocurrent when exposed to light. The useful sensitivity of a photodetector is limited by the random noise power generated in the dark. The "noise equivalent power" (NEP) is defined as the radiant flux required to give an output signal (in volts) equal to the detector noise. NEP has the unusual units of W/$\sqrt{\text{Hz}}$, because electrical noise power is proportional to the bandwidth and noise voltage is proportional to the square root of the noise power. The reciprocal of NEP is the "detectivity" (*D*). A standard figure of merit for comparing solid state photodectors is referred to as D^*; it equals *D* multiplied by the square root of the detector area (Table 1-6).

1.3.4. Radiometry and Actinometry[3,13,14]

Radiometry refers to the measurement of radiant power emitted by a source or incident on a receiver. The total radiometric power can be measured with a thermal detector. More frequently, it is desired to measure the radiant power in a given wavelength band. This may be done with inexpensive devices, consisting of band-pass filters, a solid state photodetector, and a readout meter. Accurate measurements of the spectral power distribution require a spectroradiometer, utilizing a monochromator instead of filters. The instrument must be calibrated at each wavelength, e.g., with a standard lamp of known spectral irradiance. The applications of radiometry require a working knowledge of *radiometric* units, a subject that has been complicated by a diversity of definitions and concepts. Radiometric units are based on the following concepts: radiant flux emitted by a small area of a source in a given direction (*radiance*), radiant flux emitted by a point source in a given direction (*intensity*), radiant flux received by a small area on a surface (*irradiance*), and radiant flux leaving a small area of an emitting or reflecting surface (*exitance*). Each radiometric unit has a *photometric* equivalent, which is scaled to the phototopic (black and white) response of the human eye. The SI radiometric units and their photometric equivalents are summarized in Table 1-5. Photometric units are used extensively in the lighting and photographic industries. However, they are not applicable in the UV and IR regions, and should be avoided whenever possible.

Actinometry is based on the incidence and absorption of photons, usually in connection with photochemistry or photobiology. The *fluence rate* is defined as the number of photons per second incident in a perpendicular direction on a unit area of the receiver. The *fluence* is the corresponding total number of photons incident on a unit area. In photobiology, fluence rate and fluence are frequently referred to as "dose rate" and "dose," respectively. However, this usage is ambiguous because "dose" is also employed for the number of photons *absorbed* per unit volume of the medium. In order to avoid confusion, "dose" and "dose rate" should always be defined when these terms are used. There is no simple method for measuring absolute fluence rates with polychromatic light because processes initiated by photon absorption usually depend on the wavelength of the radiation. The *quantum efficiency* is the microscopic efficiency of a photoprocess, defined as the probability that the absorption of one photon at a given wavelength will lead to single observed event, e.g., photolysis of one molecule, emission of one photon, etc. (Quantum efficiency is often called quantum "yield.") A useful actinometer should have a constant quantum efficiency over the wavelength range of application, or at least known accurate values of the dependence of the quantum efficiency on wavelength. Since only absorbed radiation can initiate a photoprocess, the fraction of incident light absorbed by the actinometer must be known at each wavelength. This complication is avoided if the actinometer is completely absorbing over the desired wavelength range. An actinometer should have the properties of "linearity" (e.g., doubling the fluence doubles the response) and "reciprocity" (e.g., doubling the fluence rate and halving the time gives the same response). Finally, the measurements must be reproducible, and preferably convenient. In view of these requirements, it is not surprising that relatively few practical actinometers have been devised. One of the most widely used chemical actinometers is based on the photochemical decomposition of aqueous potassium ferrioxalate.[12,15] This system has a constant quantum yield of 1.2 ± 0.1 for ferrous ion production from 210 to 440 nm, and can be "pushed" to about 550 nm with a correction for the decreasing quantum yield above 440 nm. The measurements are made in darkness or under weak red light, and must be corrected for the fractional absorption by the actinometer solution and reflection losses at the windows. The calculation of absorbed dose rates from incident fluence rates in an actual photochemical system may be quite difficult, especially for polychromatic radiation.

1.3.5. Light–Tissue Interactions

The interactions of light and tissue are involved in many aspects of photobiology, e.g., plant growth and optical dosimetry in photomedicine. When collimated light impinges on a highly turbid material, the independent concepts of reflection, refraction, and diffraction are no longer distinguishable, and the interactions of the light with the material are described as *scattering*. The "cloudy" appearance of biological tissue is caused by multiple scattering at diverse inhomogeneities, including macromolecules, cell organelles, microscopic "pools" of water, etc. Scattering leads to the spreading of a collimated light beam, and loss of its initial directionality. The fraction of incident radiation that exits at the back face of a scattering layer is the *diffuse transmittance,* and the fraction that exits at the front face is the *remittance* or "diffuse reflectance." In addition to scattering centers, biological tissues contain chromophores with significant absorptions in certain wavelength regions. The attenuation of light within tissue depends on both scattering and absorption. The *optical penetration depth* (δ) provides a measure of how far light can penetrate into tissue. For a nonscattering medium, δ equals the depth at which the transmissivity is $1/e$ or 0.37. The initial irradiance is reduced by approximately 90% at 2δ. The situation is more complicated for a scattering medium because the radiant flux inside the material is no longer collimated and the irradiance does not provide an adequate measure of the power incident on a molecule. The concept of *space irradiance* (χ) has been introduced for noncollimated light. It is defined as the radiant flux incident from all directions on a small sphere inside the material divided by the cross-sectional area of the sphere. For completely isotropic radiation, $\chi = 4E$. In highly scattering materials, the flow of photons parallels particle diffusion in a fluid medium, where χ is the analog of particle concentration. The application of the "diffusion approximation" to a thick, uniform tissue layer, exposed to a collimated beam of monochromatic light, leads to a relatively simple expression for the space irradiance at depth x from the illuminated face:

$$\chi(x) = 2E_0(1 + R)\ \mathrm{Exp} - (x/\delta) \qquad (1\text{-}3)$$

where E_0 is the incident irradiance, R is the remittance, and δ is the optical penetration depth.(16) The radiant flux entering the tissue layer is augmented by back-scattered radiation, and may be up to fourfold higher near the front surface compared to nonscattering media. The value of δ depends on the nature of the tissue and the wavelength. Important absorbers in tissues include proteins, hemoglobin, melanin, and water.(17) In general, δ decreases with the vascularity (blood content) of the tissue and is lower for blue light than red light, e.g., for adult brain tissue, δ is 0.45 mm at 488 nm and 1.5 mm at 710 nm.(16) An earlier approach to tissue optics is based on the theory of Kubelka and Munk, which assumes that the flux in a layer of turbid material can be divided into a forward beam and a backward beam, each of which undergoes back-scattering and absorption.(17,18) The transmitted light is taken as the forward beam that exits at the rear face, and the remitted light is taken as the backward beam that exits at the front face. The Kubelka–Munk theory was developed for layers of finite thickness and has been used extensively in the pigment and paper industries. Butler pointed out that the optical path for light absorption may be lengthened significantly in highly scattering tissues because the photons are scattered repeatedly in different directions until a chance encounter occurs with a chromophore. He exploited this phenomenon to obtain remarkably good absorption spectra of very weak chromophores in seeds and tissue slices, using a special spectrophotometer in which the photodetector measured the light transmitted over a wide angular distribution.(19,20) Another method for measuring absorption and reflection spectra of turbid materials is based on an "integrating sphere." The sample and photodetector are located inside a hollow sphere, coated with a white, highly reflecting powder. When the sample is illuminated with diffuse radiation, most of the transmitted or remitted light exiting the sample at different angles eventually reaches the photodetector.

The use of lasers in medicine is one of fastest growing applications of photobiology.

Lasers are used in surgery primarily as a heat source. The beam is usually focused to an image less than 2 mm in diameter, which provides a high power density and good control. Although the detailed analysis of laser tissue heating is quite complicated, considerations based on the wavelength dependence of δ provide insight into the basic principles. For CW laser operation, the heating effect originates in tissues with the higher values of δ. Oxyhemoglobin in blood is the dominant tissue absorber for the blue-green emission of the argon laser, which passes readily through water and colorless tissue. The argon laser is well suited for inducing coagulation of hypervascularized tissue, which requires heating to approximately 60°C. There are major applications in ophthalmology because the radiation can pass through the cornea, the lens, and vitreous media, and induce heating in the highly pigmented retina, e.g., for treatment of diseases of the retina in which abnormal blood vessels are formed. The 1064-nm emission of the Nd-YAG laser lies in a "window" between strongly absorbing blood at shorter wavelengths and strongly absorbing water at longer wavelengths. Consequently, Nd-YAG laser radiation is highly penetrating and may be employed for heating relatively large tissue regions. A major clinical application is tissue coagulation with good hemostatic properties (stopping of blood flow), e.g., for endoscopic surgery of the gastrointestinal tract. The 10.6-μm emission of the carbon dioxide laser is strongly absorbed by water. The high density of energy absorption leads to vaporization of tissue water. The heat is carried away by the steam and cannot damage the adjacent tissue. The carbon dioxide laser acts as a "laser scalpel," and it has been used for a wide variety of surgical procedures, especially for tissue of high water content. The temperature reached in laser heating of tissue depends on a balance between the rate at which energy is introduced by the laser and the rate at which heat is carried away by blood flow. However, light absorption and heat transfer are not the only factors when laser energy is delivered in short, high-power pulses, e.g., with a Q-switched Nd-YAG laser. In this case, the high electrical fields associated with the electromagnetic radiation induce a sequence of very fast electromechanical events, leading to strongly localized disruption of the tissue. A recent clinical application involves ophthalmic treatments in the anterior regions of the eye with good protection against retinal damage. Some types of lasers induce photochemical reactions in tissues with potential clinical applications. Excimer lasers emit high-intensity pulses in the UV region that are strongly absorbed by tissue proteins (Table 1-4). This radiation produces a very sharp, well-defined cut with minimal thermal damage in the neighboring tissue, which is referred to as "photoablation." Laser–tissue interactions may be initiated by "photosensitization," in which the energy absorbed by a low concentration of a colored exogeneous agent is responsible for the effects (Chapter 3). "Photodynamic therapy" of malignant tumors is a recent development, in which the photosensitizer is a mixture of porphyrins referred to as "hematoporphyrin derivative."[(21)] This drug has the property of localizing preferentially in tumor tissue after intravenous administration. A pink fluorescent light emission induced by illumination of the tissue surface with UV-A makes it possible to identify a tumor and visualize its dimensions. Exposure of the tumor to strong red light, e.g., from a dye laser, initiates the sequence of events leading to eradication of the tumor in favorable cases. Clinical trials on photodynamic therapy of superficial and internal tumors have been in progress since the late 1970s, and an evaluation of the efficacy of this treatment for certain types of cancer should be available within the next few years (Chapter 6).

1.4. MOLECULAR STRUCTURE AND EXCITED STATES

1.4.1. Interactions of Atoms with Light

The exchange of energy between light and matter is involved in all phenomena of photobiology. Electromagnetic radiation carries electric and magnetic fields, which can interact with the electrons and nuclei of atoms and molecules. The strongest interaction is between the electric field of the radiation and the electric charge of the electrons. An electric field exerts a force on a free electron, causing a change in its kinetic energy, e.g., the electrons released by light at the photocathode of a vacuum photodiode are accelerated to the anode by the applied electric field. The situation is more complicated for the interactions of light with atoms, where the electrons move in stable orbits corresponding to the discrete energy levels of the atom. The transfer of one photon from electromagnetic radiation to an atom raises one electron to a new stable orbit, subject to certain "selection rules," involving the properties of the initial and final states of the atom and the polarization of the light. Similarly, an electronically excited atom loses energy by emitting one photon, either by "spontaneous" emission in which light is not required, or by "stimulated" emission, in which the atom interacts with radiation. The most important consequence of the interactions between atoms and radiation is that *radiative transitions are efficient only when the photon energy of the radiation is close to the energy difference of the atom in the initial and final states.* Therefore, an absorption or emission spectrum provides direct information about the atomic energy levels because each strong line corresponds to the energy difference between two allowed energy states. The photon energies in the UV, visible, and IR bands correspond to radiative transitions of the valence electrons, typically 0.1–10 eV (Table 1-2). The inner-shell electrons are more strongly bound to the nuclei, and transitions leading to the occupation or depletion of these levels correspond to the lines in X-ray spectra.

In addition to its electric charge, an electron has an intrinsic magnetic field, identified with electron "spin." This term originated in early quantum theory, when it was postulated that an electron is similar to a small rotating sphere with an electric charge on the surface, and the magnetic field is generated by the equivalent electric current. (In classical physics, a charged particle possessing angular momentum, e.g., in circular or elliptical motion, is equivalent to a current loop, and therefore it generates a magnetic field.) The present view is that spin is an intrinsic angular momentum of the electron, in addition to any angular momentum it acquires from its orbital motion, which is the source of its magnetic field. Quantum theory shows that the electron spin of a free electron can be oriented in only two directions relative to an external magnetic field, which are referred to as "spin up" ($\uparrow$) and "spin down" ($\downarrow$). Quantization of electron spin direction also applies to electrons in atoms. Each electronic state in an atom is identified with a set of four "quantum numbers," three of which specify its orbital motion and the fourth specifies its spin orientation. The spins of electrons in a complete shell are "paired" ($\uparrow\downarrow$), which maximizes the number of electrons that can occupy an electron shell. (The Pauli Exclusion Principle states that no two electrons in an atom can have the same set of four quantum numbers.) The valence shell in free atoms is unfilled, except for the rare gases, and therefore a free atom can have one or more unpaired electrons. The spin state of an atom is designated by the quantity $(2S + 1)$, where S is the total spin quantum number. [The laws of quantum physics show that an atom with total spin S can orient itself in $(2S + 1)$ directions relative to a magnetic field. This "spin multiplicity" can be determined experimentally from atomic spectra.] For a single unpaired electron, $S = ½$ and $(2S + 1) = 2$. This spin state is referred to as "doublet," which indicates that a single electron can orient parallel or antiparallel to an applied magnetic field. A hydrogen atom is always in a doublet state because it has only one electron. If all spins in an atom are paired, $S = 0$ and the multiplicity is 1; this state is referred to "singlet." The rare gas "ground" (lowest energy) states are always singlet. If two spins are unpaired, $S = 1$ and therefore $(2S + 1) = 3$; this is a "triplet state." A helium atom, with two electrons, can exist in singlet states and triplet states. The ground state of the helium atom is a singlet, and the strong absorption lines show only transitions to higher energy singlet states. This is an example of a selection rule, limiting allowed optical transitions to states of the same spin multiplicity. The spin selection rule does not apply to states populated by collisions of an atom with electrons, e.g., the helium arc spectrum has emission lines corresponding to transitions between singlet states and between triplet states. (Neon atoms in a helium-neon laser are pumped by collisions with relatively long-lived (metastable) excited helium atoms.) The spin selection rule may be "relaxed" in heavy atoms, where interactions between the spin and orbital motions of an electron, induced by the electric field of the nucleus ("spin-orbit" interaction), leads to a "mixing" of singlet character into triplet states and vice versa. The 254-nm emission line of the low-pressure mercury arc (Fig. 1-4) is a transition from an excited triplet state to the singlet ground state. This line has high intensity because the same transition occurs in absorption, leading to a "resonance" between emission and absorption. The repetitive emission and reabsorption of 254-nm radiation, prior to escape of the photons through the lamp envelope, directs a high fraction of the energy into this line. The strong yellow emission lines of the sodium vapor arc lamp (589.0 and 589.6 nm) are also resonance lines.

1.4.2. Molecular Orbitals[22,23]

The total energy of a molecule in its ground state includes contributions from the motions of the electrons and the atomic nuclei. According to "semiclassical" physics, the nuclei have translational, vibrational, and rotational motions in space, and the electrons move in discrete orbits about one or more nuclei. This general picture is retained in wave mechanics, except the discrete electron orbits are replaced by "smeared out" electron density distributions. Each set of electron and nuclear motions corresponds to a possible energy state of the entire molecule. The transitions between energy states induced by the absorption or emission of light may be identified by optical spectroscopy. Molecular spectra in the UV and visible regions involve transitions between different electronic energy levels. The sharp absorption lines in Fig. 1-13a correspond to discrete electronic transitions. Electronic spectra usually have a *vibronic* substructure, resulting from the vibrational motions associated with each electronic level, and leading to broadening of the electronic transitions into bands (Fig. 1-13b). In an "allowed" (high-probability) absorption process, a molecule absorbs one photon and is raised to a higher energy state. Similarly, in an allowed emission process, the molecule falls to a lower energy state and emits one photon. As with atoms, these transitions are efficient only when the quantum

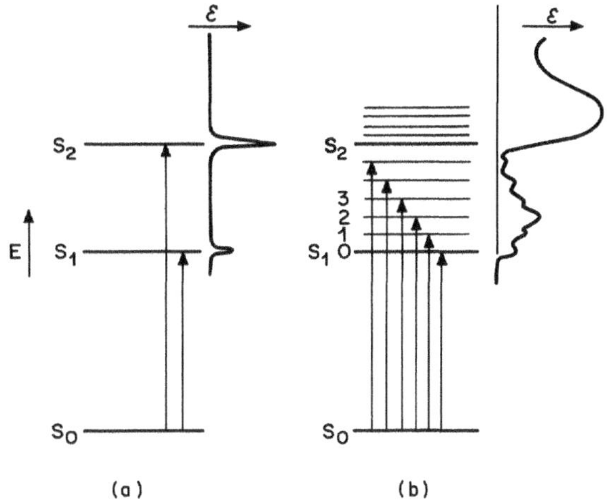

Fig. 1-13. (a) Atomic-like electronic energy levels. The optical transitions are very narrow because only light of wavelengths corresponding to the electronic energy differences between the sharp levels can be absorbed. The lower absorption line results from the transition $S_o \rightarrow S_1$, and the higher absorption line results from the transition $S_o \rightarrow S_2$. The strength of each transition is indicated by the extinction coefficient (ϵ), and the line width is determined by the resolution of the spectrometer. (b) Molecular-like energy levels. The lighter horizontal lines are the vibrational energy states associated with the excited electronic states S_1 and S_2. The vibrational levels of the ground state S_o are not shown. The interactions of the electronic and vibrational energy levels broadens the electronic transitions into bands. The structure in the absorption spectrum $S_o \rightarrow S_1$ results from the vibronic transitions 0-0′, 0-1′, 0-2′, etc. Only the envelope of the vibronic transitions is resolved in the $S_o \rightarrow S_2$ spectrum.

energy of the photon is close to the difference between the energy of the molecule in the initial and final states. Therefore, the character of absorption and emission spectra provides information about the structure and properties of molecules.

Quantum mechanics leads to a powerful description of the electron energies and motions in molecules. The inner-shell electrons of atoms larger than hydrogen are tightly bound to their own nuclei, and they can be considered as localized in *atomic orbitals* (AO), which do not participate in optical transitions, except at vacuum UV wavelengths. The valence electrons are less strongly bound, and can be excited by absorbing radiation from UV-C through the visible and near-IR spectral regions. The orbits of the valence electrons usually extend over two or more atoms. They can be described by *molecular orbitals* (MO), formed by combining the AO. The combination of atomic orbitals to form molecular orbitals is described in standard texts on molecular physics.(22,23) There are several basic types of MO in common organic molecules, as illustrated for ethylene and formaldehyde in Fig. 1-14. A bonding sigma orbital (σ) is strongly localized around the line between the two atoms. A covalent single bond consists of one σ orbital occupied by two electrons. Each σ orbital has a counterpart (σ^*), in which the electron density extends away from the two carbon atoms along the molecular axis. This MO is called "antibonding" because a molecule is less stable when the σ^* orbital is occupied, compared to the corresponding σ orbital. The bonding pi MO (π) is localized in two parallel lobes, above and below the axis, with a node (no electron density) on the axis. The antibonding pi MO (π^*) has an additional node halfway between the atoms and perpendicular to the axis. A covalent double bond contains four electrons, two electrons in one σ MO and two electrons in one π MO. Atoms such as oxygen and nitrogen have "lone pair" π electrons that may not participate in chemical bonding, e.g., the lone pair electrons on the oxygen atom in formaldehyde (Fig. 1-14b) and on the nitrogen atom in pyridine. The correspond-

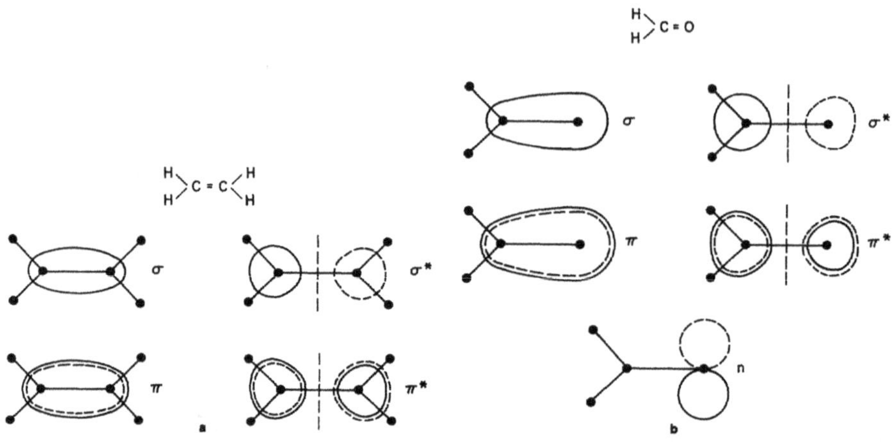

Fig. 1-14. (a) Schematic representation of some molecular orbitals of ethylene. Ethylene is a planar molecule; all six atoms are located in the plane of the page. Each carbon atom has four valence electrons. The σ molecular orbital (MO) connecting the two carbon atoms has cylindrical symmetry with to the C–C axis. The σ* MO is the counterpart of the σ MO, with a node halfway between the two carbon atoms. It is referred to as "antibonding", because the molecule is less stable if this MO is occupied by an electron, compared to the σ MO. A carbon–carbon π MO has two lobes; one lobe is located above the plane (solid line) and the other lobe is located below the plane (dashed line). There is a node in the molecular plane. An equivalent π MO is oriented at 90° to the one shown (perpendicular to the page). In the C–C double bond, one electron from each carbon atom, of opposite spins, occupies the σ MO, and one electron from each carbon atom, of opposite spins, occupies a π MO. The other two valence electrons on each carbon atom form σ orbitals with the hydrogen atoms, accounting for four single bonds (not shown). (b) Formaldehyde has carbon–oxygen orbitals similar to the carbon–carbon MOs of ethylene, labeled as σ, σ*, π, and π*. Two of the four valence electrons on the carbon atom and the two valence electrons on the oxygen atom form the carbon–oxygen double bond. The other two valence electrons on the carbon atom form single bonds with the two hydrogen atoms (not shown). Oxygen has two paired π electrons in the outer shell that are not used for bonding in formaldehyde, referred to as "lone-pair" electrons. They are localized in a nonbonding MO (n), oriented perpendicular to the C–O axis.

ing MOs are referred to as nonbonding (n). However, the nitrogen lone pair electrons in aniline and pyrrole are conjugated with the π orbitals in the rings, and they are not nonbonding.

The "configuration" of a given molecule is described by specifying the number of electrons and their spin directions in each occupied MO. In almost all stable molecules, each MO is occupied by two electrons with paired spins. Since the inner shells are fully occupied by electrons with paired spins, $S = 0$ for the molecule, corresponding to a singlet state. A few stable molecules exist with two unpaired electrons, i.e., $S = 1$ and the molecule is a triplet state. Ordinary molecular oxygen is one of these unusual stable molecules with unpaired electron spins. An allowed optical absorption elevates one electron in an occupied MO to a vacant MO, without a change of electron spin direction. One way of describing this process is to specify the MOs involved, e.g., $\pi \rightarrow \pi^*$ indicates that an electron is a π MO is excited to a π* MO. Alternatively, the effect of the transition on the spectroscopic state of the entire molecule may be specified, e.g., $S_0 \rightarrow S_1$ indicates that a molecule in its singlet ground state makes a transition to the first excited singlet

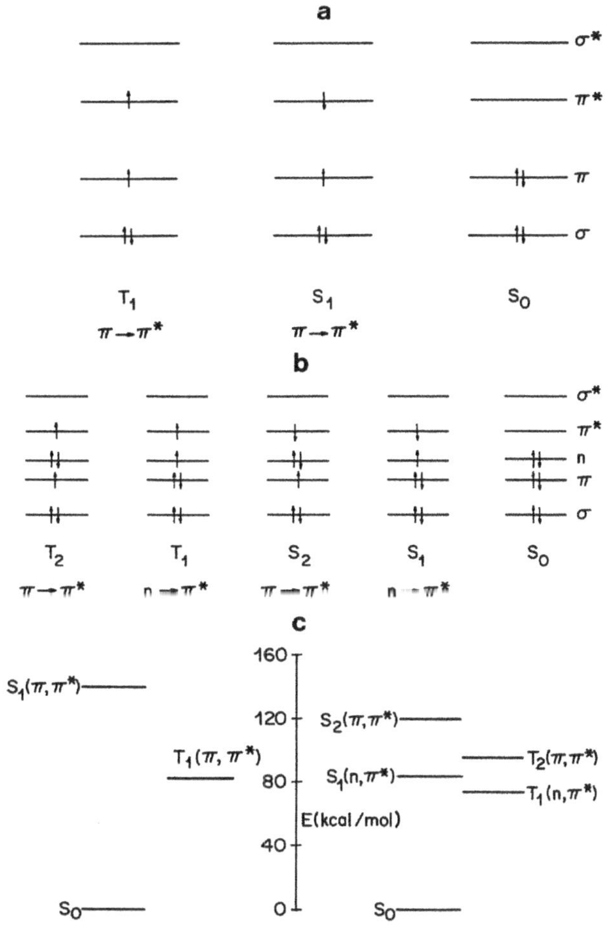

Fig. 1-15. (a) The occupation of molecular orbitals in the lower electronic states of ethylene. Each arrow represents the spin direction of the electron. In the ground state (S_o), two electrons with paired spins occupy the lowest σ and π orbitals. In the first excited singlet state (S_1), one electron in the π orbital is elevated to the empty π^* orbital, without a change of spin direction. In the lowest triplet state, one electron in the π orbital is elevated to the empty π^* orbital, with a reversal of spin direction. The electrons in the four carbon–hydrogen bonds and the inner shell atomic core electrons are not depicted. (b) The occupation of molecular orbitals in the lower electronic states of formaldehyde. The two lone-pair electrons of the oxygen atom occupy the n orbital in the ground state (S_o). In the lowest excited singlet state (S_1), one electron in the n orbital is elevated to the empty π^* orbital, without a change of spin direction. In the second excited singlet state (S_2), one electron in the π orbital is elevated to the empty π^* orbital, without a change of spin direction. A reversal of spin direction takes place in the corresponding excitations to the lowest triplet state (T_1) and second triplet state (T_2). (c) State energy diagrams for ethylene (left) and formaldehyde (right). The energy scale shows the energy of the electronically excited molecules relative to the corresponding ground states. The one-electron transitions are indicated in parentheses, e.g., $S_1(\pi,\pi^*)$ designates the excited singlet state of ethylene formed when one electron in the highest filled π orbital is elevated to the lowest empty π^* orbital, without a change of spin direction. The singlet and triplet states are displaced horizontally to clarify the representation. The energy scale of kcal/mol is used frequently in molecular photophysics; 1 kcal = 4184 J.

state. The occupation of individual MOs are shown in Figs. 1-15a and b for ethylene and formaldehyde, respectively. Figure 1-15c is a "state energy" diagram, showing the energy of the excited states relative to the ground state of each molecule.

The concept of *transition moment* is used in molecular photophysics to describe the interactions between light and molecules. The transition moment can be viewed as the shift of electric charge in a molecule when it makes a transition between two electronic states. This charge separation is induced by the electric field of the light, and it oscillates with the light frequency. In classical physics, an entity consisting of two separated charges of equal magnitude and opposite sign is known as a *dipole*. The *dipole moment* is

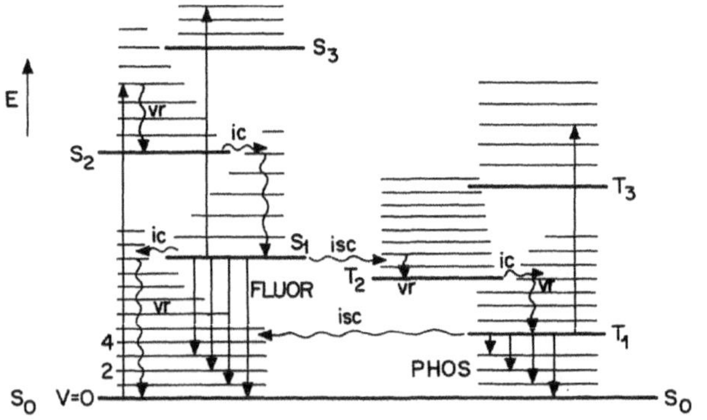

Fig. 1-16. Jablonski diagram for an organic molecule, showing the transitions between the singlet (S_o, S_1, S_2, . . .) and triplet (T_1, T_2, T_3, . . .) electronic levels (heavy horizontal lines). The energy levels of only one vibrational mode are portrayed for each electronic state (light horizontal lines). Radiative transitions are shown as straight arrows. Radiationless relaxation processes are shown as wavy arrows: vr, vibrational relaxation; ic, internal conversion; isc, intersystem crossing. The radiative decay processes depicted are fluorescence (FLUOR) and phosphorescence (PHOS). The diagram exemplifies three spin-allowed absorption processes: the transition from the thermally equilibrated, ground state, $S_o(\nu = 0)$, to a vibrationally excited level of the second excited singlet state, $S_2(\nu = 4')$; the transition from the thermally equilibrated, first excited singlet state, $S_1(\nu = 0)$, to a vibrationally excited level of the third excited singlet state, $S_3(\nu = 3')$; the transition from the thermally equilibrated, lowest triplet state, $T_1(\nu = 0)$, to a vibrationally excited level of the second excited triplet state, $T_3(\nu = 3')$. Fluorescence is a spin-allowed, radiative transition from the thermally equilibrated, first excited singlet state, $S_1(\nu = 0)$, to different vibronic levels of the ground state, $S_1(\nu = 0, 1', 2', 3', \ldots)$. Phosphorescence is a spin-forbidden, radiative transition from the thermally equilibrated, lowest triplet state, $T_1(\nu = 0)$, to different vibronic levels of the ground state, $S_1(\nu = 0', 1', 2', 3', \ldots)$.

defined as the product of the charge magnitude and the separation distance. This terminology has been carried over into radiation physics, and the transition moment connecting two molecular energy states is also called the "transition dipole." In an allowed optical transition, the magnitude of the transition moment is comparable to the displacement of one electronic charge (1.6×10^{-19} C) by a distance the order of atomic dimensions (10^{-10} m). The strength of an optical transition depends on the square of the transition dipole connecting the two states, and the radiative lifetime of an excited state varies inversely with the square of the transition dipole connecting it to the lower state. *Large transition dipoles are associated with strong absorption, strong emission, and short-lived excited states, and vice versa.* The concept of the transition dipole as a measure of light interactions with the electrons in a molecule will be used to help explain the general properties of excited states.

The electronic transitions in typical organic molecules can be represented by a "Jablonski diagram" (Fig. 1-16), which is especially useful for relating the photophysical properties to optical spectra. The heavy horizontal lines are the lowest vibrational levels of the singlet states (S_0, S_1, S_2, . . .) and the triplet states (T_1, T_2, . . .). The vibronic levels associated with one vibrational mode of each electronic level are indicated by the sets of lighter horizontal lines. The straight vertical arrows indicate allowed radiative transitions, and the wavy vertical arrows indicate the very fast loss of excess vibrational energy by *vibrational relaxation* (vr). The horizontal wavy lines indicate two important types of energy-conserving, internal processes: In *internal conversion* (ic), the system shifts from the lowest vibronic level of one electronic state to an upper vibronic level of another electronic state. In *intersystem crossing* (isc), the system shifts from the lowest vibronic level of the S_1 state to an upper vibronic level of the T_1 state. The isc process is promoted by the spin-orbit interaction, which gives some triplet character to singlet states and vice versa. After ic or isc, the system relaxes (via vr) to the lowest vibronic state of the electronic level within 10^{-11} s. Normal light absorption involves the transition $S_0 \rightarrow S_n$, where $n = 1, 2, \ldots$. The radiative lifetime of an S_1 state is typically 10^{-9} s, and the light emitted in the radiative process $S_1 \rightarrow S_0$ is *fluorescence*. The diagram shows a competing, nonradiative transition to S_0, which is relatively inefficient in many aromatic and heterocyclic molecules. Intersystem crossing is another competing process, leading to population of the lowest triplet state (T_1). The T_1 state is long-lived because of the small transition dipole connecting it with the ground state S_0. Light emission by a triplet state is *phosphorescence*. Thermal excitation can induce the transition $T_1 \rightarrow S_1$ if the levels are close in energy, typically less than 0.1 eV. This process leads to a luminescence with the lifetime of phosphorescence and the emission spectrum of fluorescence; it is referred to as "delayed fluorescence." Triplet states are important photochemical intermediates because they are energy-rich, long-lived species that can participate in many types of chemical reactions with unexcited molecules.

The processes shown in Fig. 1-16 are applicable to (π,π^*) and (n,π^*) transitions. However, there are significant differences between the two types of transitions, because (n,π^*) transition dipoles are much smaller than (π,π^*). In general, $n \rightarrow \pi^*$ absorptions are more than 10-fold weaker than $\pi \rightarrow \pi^*$ absorptions in the same molecule and show more vibrational structure. Solvents of higher polarity shift $n \rightarrow \pi^*$ absorptions to shorter wavelengths and $\pi \rightarrow \pi^*$ absorptions to longer wavelengths. (The interactions of an electric charge with a dielectric medium tends to stablize the charge by lowering the energy. The extended electron distribution in a $\pi \rightarrow \pi^*$ excited state interacts more strongly with a polar medium than the ground state, leading to a smaller energy difference and a "red shift" of the absorption band. On the other hand, the $n \rightarrow \pi^*$ ground state interacts with a polar medium more strongly than the excited state, leading to a higher transition energy and a "blue shift.") The fluorescence of (n,π^*) excited singlet states is less efficient and longer lived than (π,π^*) singlet states. An important difference in the photochemistry of (n,π^*) and (π,π^*) excited states is the high efficiency of hydrogen abstraction by (n,π^*) states. Consequently, photoreduction may be much more important for carbonyl compounds than similar aromatic hydrocarbons (Section 2.1.2).

1.4.3. Quantum Yield, Lifetime, and Relaxation of Excited States[23]

The absorption of light at ordinary intensities leads to the population of thermally equilibrated S_1 states at a rate proportional to the incident fluence rate. The relaxation (energy loss) processes obey *first-order kinetics,* i.e., the loss rate at any time is proportional to the number of surviving S_1 molecules. (The radioactive decay of atoms follows the same rate law.) A first-order process is characterized by its *rate constant* (k) or, equivalently, the *mean lifetime* (τ), where $\tau = 1/k$. The first-order rate constants for transitions out of S_1 are k_f, k_{ic}, and k_{isc}; the corresponding transitions are indicated in Fig. 1-16. The fraction of fluorescence events equals k_f/k, where $k = (k_f + k_{ic} + k_{isc})$. This fraction is the quantum efficiency (or quantum yield) of fluorescence (Φ_f). The experimental fluorescence lifetime (τ) equals $1/k$ because the fluorescence rate monitors the overall decay of S_1 excited states. The theoretical radiative lifetime of the S_1 state (τ_f) is defined as $1/k_f$. The ratio of the experimental fluorescence lifetime to the radiative lifetime is the fluorescence efficiency:

$$\Phi_f = \tau/\tau_f \tag{1-4}$$

Therefore, when an excited state decays by both radiative and nonradiative processes, the lifetime of the excited state is reduced and the fluorescence efficiency is less than unity. A similar kinetics analysis can be carried out for metastable triplet states.[23]

1.4.4. Fluorescence Quenching and Electronic Energy Transfer

Many substances "quench" (inhibit) the fluorescence of organic molecules. Fluorescence quenching in liquid media can take place by either of two general mechanisms. In "dynamic quenching" the molecules of the fluorescent species (*fluorophore*) interact with the molecules of the quencher via intermolecular collisions. This process provides an additional pathway for the relaxation of S_1 states and leads to a reduction in the experimental fluorescence decay lifetime. Alternatively, the presence of a quencher molecule close to a fluorophore may induce the immediate deactivation of S_1 states. This process is "static quenching." Some quenchers exhibit dynamic quenching at low concentrations and static quenching at high concentrations. Dynamic quenching follows the well-known Stern–Volmer equation:

$$\Phi_f^0/\Phi_f = 1 + \tau k_Q C_Q \tag{1-5}$$

where Φ_f^0 and Φ_f are the fluorescence efficiencies in the absence and presence of the quencher, respectively, τ is the fluorescence lifetime, k_Q is the bimolecular rate constant for reactions of the quencher with the fluorophore, and C_Q is the quencher concentration. When this model is applicable, a plot of Φ_f^0/Φ_f vs. C_Q gives a straight line with its slope equal to τk_Q. The rate constant κ_Q may be measured, or estimated from molecular theory.

Electronic energy transfer is a special type of quenching in which the electronic excitation energy of a *donor* molecule is transferred to an *acceptor* molecule via a radiationless process. It is characterized by a shift of the fluorescence band from that of the donor to that of the acceptor, and is called "sensitized fluorescence." Several mechanisms of energy transfer have been identified in liquid and rigid media. One of the most important is *resonance excitation transfer,* which is also referred to as "dipole–dipole" or "long-range" energy transfer. The theory of resonant transfer, as developed by Förster, is based on two coupled radiationless transitions, leading to deactivation of the donor and simultaneous excitation of the acceptor (Fig. 1-17). The strength of the interaction depends on the product of the emission dipole of the donor and the absorption dipole of the acceptor. The detailed calculation of the energy transfer rate shows that it depends on the overlap between the fluorescence band of the donor and the absorption band of the acceptor, and varies with the inverse sixth power of their separation.[3,23,24] The range of Förster-type energy transfer attains 1–10 nm in favorable cases, which is far greater than the collision diameters of small molecules. Energy transfer by this mechanism has been

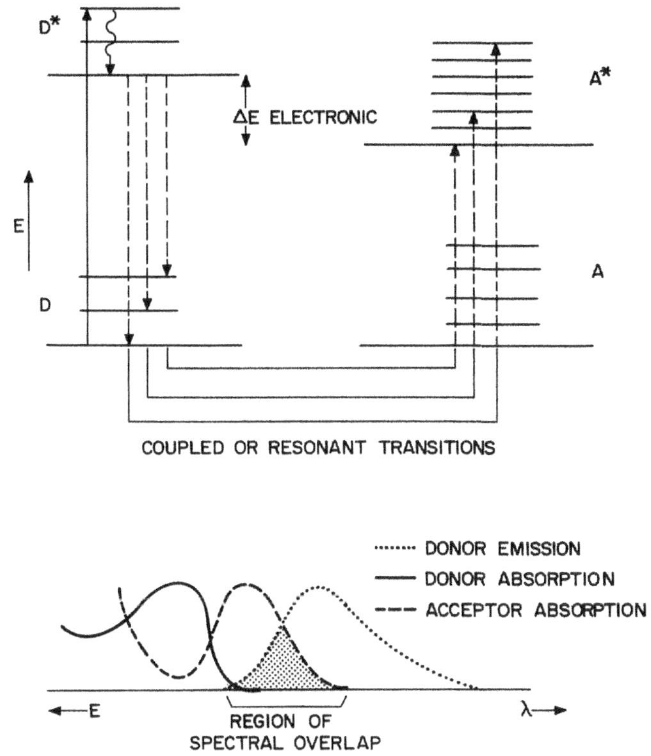

Fig. 1-17. Electronic energy transfer via coupled transitions for a donor–acceptor pair. The downward transitions of the thermally equilibrated, excited donor (D*) to different vibronic levels of the donor ground state (D) match the upward transitions of the acceptor ground state (A) to different vibronic levels of the acceptor excited state (A*). The electronic energy difference ΔE is the excess energy of D* relative to A*, in their thermally equilibrated, excited singlet states. The overlap of many broadened vibrational levels in a condensed medium ensures resonance, if the fluorescence band of the donor overlaps the absorption band of the acceptor (lower diagram).

observed in fluid and rigid media, and between chromophores in a macromolecule. An important example of the latter takes place in proteins, where the direction of excitation energy transfer is phenylalanine → tyrosine → tryptophan. In human serum albumin, with one tryptophan and 18 tyrosine residues, approximately 50% of the potential tyrosine fluorescence excited at 275 nm is transferred to the tryptophan residue. Resonance energy transfer between triplet states is less efficient than between singlet states. The long lifetime of triplet states favors an alternative short-range energy transfer process, known as "triplet–triplet" energy transfer. The *triplet energy* is the key parameter, defined as the energy difference between the T_1 and S_0 states (Fig. 1-16). Triplet–triplet energy transfer is always "downhill", i.e., the triplet energy of the donor must be higher than that of the acceptor. The formation of singlet oxygen in photosensitized chemical reactions takes place by triplet–triplet energy transfer (Section 3.2.3).

1.5. SPECTROSCOPIC MEASUREMENTS

A major objective of molecular spectroscopy is to deduce fundamental information about the structure and properties of molecules. Spectroscopy is used also as a tool in many laboratory procedures, including chemical analysis, biochemical assays, enzyme kinetics, etc. A basic understanding of the underlying photophysics is required for all but the most routine spectroscopic measurements in order to utilize the instrumentation properly and interpret the results.

1.5.1. Absorption Spectroscopy

The measurement of spectra with photoelectric detectors is referred to as spectrophotometry. In absorption spectrophotometry, the fraction of incident light transmitted by a sample is measured over a selected range of wavelengths. Modern instruments measure the *transmittance* (T) and the *optical density* (O.D.), which are related by

$$\text{O.D.} = \log_{10} (1/T) \tag{1-6}$$

An O.D. range of 0–4 corresponds to a change of T from 1 to 0.0001. Measurements of O.D. must be corrected for the spectral intensity distribution of the source, and the dependence of the photodetector sensitivity on wavelength. This correction is made by subtracting the O.D. of a "reference" from that of the "sample" at each wavelength. Any process that alters the light signal reaching the detector contributes to O.D., e.g., absorption, light scattering, and luminescence. The *absorbance* (A) is used to designate the contribution of absorption alone to optical density. The ideal absorbance of a dilute solute in a transparent solvent follows the Lambert–Bouguer–Beer law:

$$A = \epsilon c d \tag{1-7}$$

where ϵ (liters/mole-cm) is the *molar extinction coefficient* of the solute at the wavelength of measurement, c (moles/liter) is the molarity of the solute, and d (cm) is the optical pathlength. The concentration of a solute can be calculated by measuring A, when ϵ is known at the measurement wavelength. The linear dependence of A on c in Eq. (1-7) may not hold if there are concentration-dependent effects, e.g., aggregation or acid–base dissociation of the solute. The applicability of "Beer's law" is usually tested by measuring the dependence of O.D. on solute concentration. A linear dependence confirms that Beer's law is valid for the given conditions. This "standard curve" can then be used to relate O.D. to solute concentration for analytical purposes. A typical two-beam, recording spectrophotometer is shown in Fig. 1-18. In this arrangement, the light transmitted by the sample and the reference are measured alternately by means of a rotating sector, the O.D. difference is calculated electronically, and displayed at each wavelength. The more recent instruments utilize microprocessors to store and analyze the data. The spectra can be displayed in many formats, including plots of the first and higher derivatives of the absorption bands vs. wavelength, and the sum and difference of several spectra. Flash photolysis is a major application of absorption spectroscopy. In this technique, the system of interest is irradiated with a short, intense light flash, and the short-lived intermediates

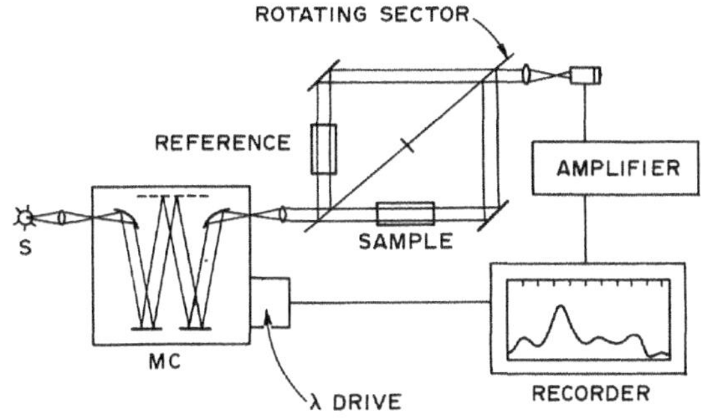

Fig. 1-18. A schematic representation of a dual-beam recording spectrophotometer. Light from the source (S) is focused on the entrance slit of the monochromator (MC). The exiting monochromatic radiation is collimated by a lens, and passed alternately through the sample cuvette and the reference cuvette by a partially silvered, rotating sector. The amplifier processes the photomultiplier signals, and outputs either the transmission or the optical density to the y axis of the recorder. The wavelength (λ) drive on the monochromator generates an electrical signal that drives the x axis of the recorder. More than one source may be used to cover a wide wavelength range, e.g., a deuterium lamp for the UV region and a tungsten-halogen lamp for the visible-IR regions.

are identified and followed in real time by making fast optical absorption measurements.[25] A pulsed laser flash photolysis apparatus is shown schematically in Fig. 1-19. This type of apparatus has been used to identify short-lived free radicals and triplet states and to investigate their fast chemical reactions.

1.5.2. Luminescence Spectroscopy[12,24,26]

Luminescence spectra lead to information about optical transitions from higher to lower energy states, which augments the information about transitions from lower to

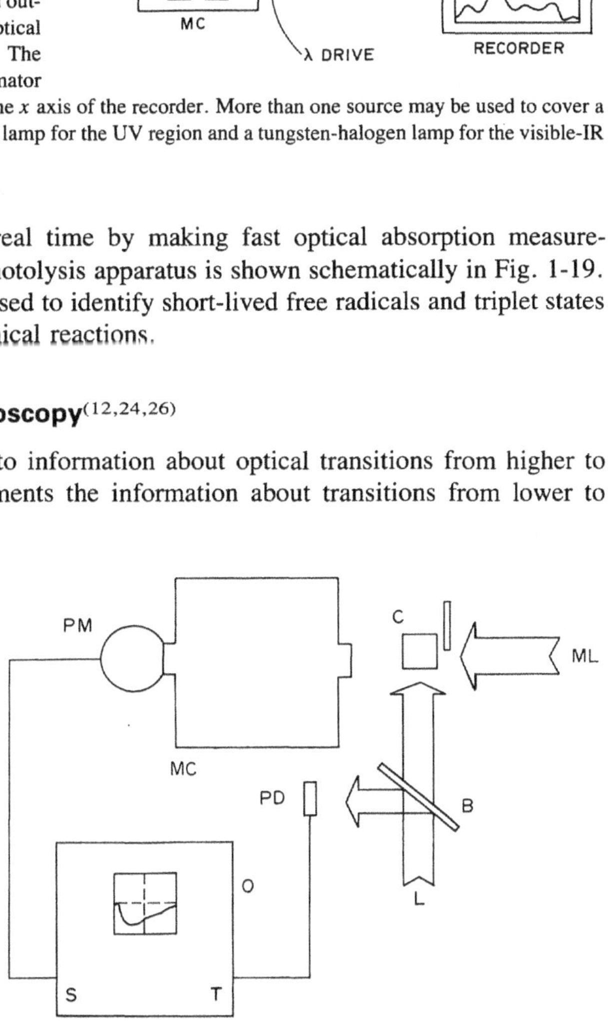

Fig. 1-19. Schematic diagram of a laser flash photolysis apparatus. The sample in the cuvette (C) is irradiated with a photolytic flash from a laser (L). The photochemical intermediates are detected by passing a monitoring light beam (ML) through the cuvette, and to a monochromator (MC) with a photomultiplier (PM) at the exit slit. The monitoring source may be a continuous lamp, e.g., an arc lamp, or a flash lamp of relatively long duration on the time scale of the intermediates. The photomultiplier output (S) is displayed on an oscilloscope (O), which is triggered (T) by a signal from an auxiliary photodiode (PD), activated by the laser pulse with a beam splitter (B). The oscilloscope trace records the changes in transmission induced by the laser pulse at the wavelength setting of the monochromator. Absorption spectra of the transient intermediates can be determined by repeating the measurements at different wavelengths. The time resolution in flash photolysis ranges from microseconds with gas-filled photolytic flash lamps, to nanoseconds with Q-switched lasers, and picoseconds with mode-locked lasers. The effective spectral range depends on the choice of monitoring lamp, monochromator, and photodector, typically 200–900 nm. In an alternative arrangement, the photolysis beam and the monitoring beam are passed through the sample in the same direction.

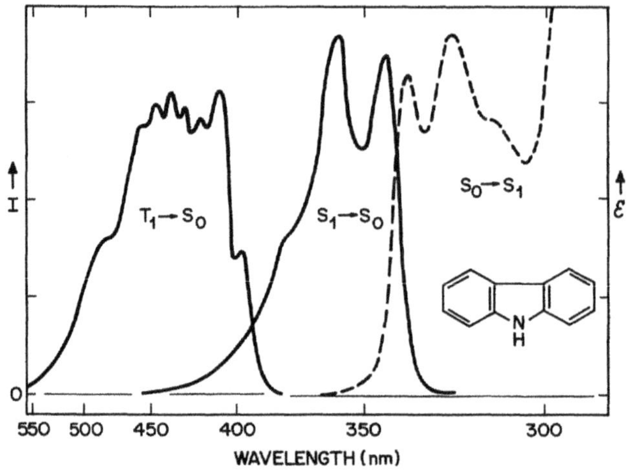

Fig. 1-20. Absorption and emission spectra of carbazole in ethanol. The intensities of the various spectra are not to scale. The spectra are portrayed on a linear energy scale (increasing from left to right), in which case the wavelength scale is nonuniform. This display best demonstrates the mirror image relationship between the vibronic structure of the absorption band ($S_o \rightarrow S_1$) and that of the fluorescence band ($S_1 \rightarrow S_o$). The phosphorescence band ($T_1 \rightarrow S_o$) is located at higher wavelengths (lower energy) than the fluorescence band, because nonradiative relaxation processes are involved in the intersystem crossing from the fluorescent state to the triplet state. The weak $S_o \rightarrow T_1$ absorption (not shown) is expected to commence near 400 nm.

higher energy states provided by absorption spectra. The relationship between absorption and emission spectra is exemplified in Fig. 1-20, showing the first absorption band ($S_0 \rightarrow S_1$) and fluorescence band ($S_1 \rightarrow S_0$) of carbazole in ethanol. The fluorescence band is always centered at longer wavelengths than the absorption band, which is referred to as the *Stokes' shift*. The Stokes' shift occurs because some energy is lost via nonradiative relaxation processes prior to the population of the thermally equilibrated S_1 state from which fluorescence takes place (Fig. 1-16). Excitation into the higher excited states S_2, S_3, . . . with shorter wavelength light almost always leads to the $S_1 \rightarrow S_0$ fluorescence because the higher excited states are depopulated by fast nonradiative relaxation processes. (The $S_2 \rightarrow S_0$ fluorescence can be detected with low quantum efficiency in many molecules.) Luminescence measurements are not routine. They require careful attention to the experimental conditions, and may involve sophisticated instrumentation and considerable skill in data analysis. However, the wealth of information provided by luminescence justifies this effort, and every researcher in photobiology should be familiar with the basic aspects of the subject and its potential applications.

1.5.2.1. Measurement of Luminescence Spectra

Steady-state fluorescence spectra can be measured with a luminescence spectrophotometer. In a typical arrangement (Fig. 1-21), light from the source is passed through an excitation monochromator onto the sample; the fluorescence is collected at 90° to the excitation beam, analyzed by the emission monochromator, and measured with a photodetector. Two types of fluorescence spectra may be obtained. In a *fluorescence emission spectrum,* the excitation wavelength is fixed and the emission monochromator is scanned over the fluorescence band. This mode is frequently employed for the identification of fluorescent substances, energy transfer studies, and concentration measurements. In a *fluorescence excitation spectrum,* the emission wavelength is fixed and the excitation

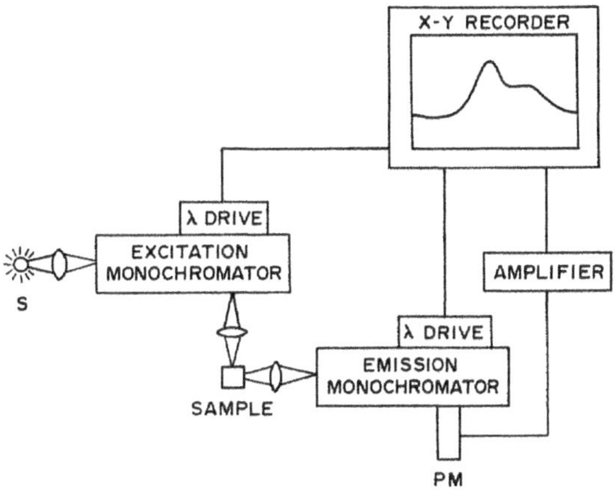

Fig. 1-21. Schematic representation of a luminescence spectrophotometer. Light from the source (S) is focused on the entrance slit of the excitation monochromator, and the exiting monochromatic radiation is collimated by a lens and relayed to the sample cuvette. The fluorescence is collected at 90° to the excitation beam and focused on the entrance slit of the emission monochromator. A photomultipler (PM) located behind the exit slit provides an electrical signal proportional to the emission intensity, that drives the y axis of the x–y recorder. For fluorescence emission spectra, the wavelength (λ) drive on the excitation monochromator is fixed, and the λ drive on the emission monochromator is scanned over the fluorescence band, and drives the x axis of the recorder; for fluorescence excitation spectra, the λ drive on the emission monochromator is fixed, and the λ drive on the excitation monochromator is scanned over the absorption spectrum and drives the x axis of the recorder. The fluorescence spectra as recorded are "uncorrected," i.e., the excitation bands are not a true representation of the absorption bands responsible for the fluorescence, and the emission bands are not a true representation of the fluorescence. Procedures for measuring corrected spectra are described in the literature. [26,27]

monochromator is scanned over the absorption band of the sample. The excitation spectrum selects the absorption of fluorophores from the overall absorption spectrum.

Fluorescence measurements are subject to instrumental errors that must be eliminated or corrected. The "inner filtering" of fluorescence by the absorption of the sample is minimized by keeping the absorbance low in the region of the spectrum where fluorescence and absorption overlap. Another source of error is distortion of excitation spectra when the extent of light absorption is not proportional to the absorbance of the sample. According to Eq. (1-6), $T = 10^{-\text{O.D.}}$, and the fractional absorption $(1 - T)$ is an exponential function of O.D. The error is minimized if O.D. < 0.02 at all excitation wavelengths, in which case the absorption by the sample will be almost proportional to its O.D. Fluorescence excitation spectra must be corrected for variations in the spectral intensity distribution of the light that reaches the sample. This can be done by measuring the output of the lamp and excitation monochromator with a calibrated power meter as a function of wavelength and dividing the measured excitation spectrum by this calibration curve to obtain the corrected spectrum. Some instruments make this correction internally with a "quantum counter," which is a device that measures the relative photon flux incident on the sample at each excitation wavelength. A useful quantum counter is a concentrated dye solution, e.g., rhodamine B in ethylene glycol, that absorbs all incident light from 220 to 600 nm and emits a fluorescence band centered at 630 nm that is proportional to the incident fluence rate, independent of the excitation wavelength.[12] Emission spectra must be corrected for the spectral response of the photomultiplier, and the efficiency and bandwidth of the emission monochromator. A simple but reliable procedure for correcting fluorescence emission spectra involves calibrating the instrument with standard fluorophores for which corrected spectra are available, e.g., quinine bisulfite in 0.1 N sulfuric acid and anthracene in ethanol.[12] The fluorescence of a standard is measured at exactly the same conditions as the sample (including the same solvent), and the ratio of the measured emission spectrum to the corrected spectrum provides correction factors at wavelengths where both substances emit. A similar method can be used to determine the relative fluorescence efficiency, by comparing the emission spectrum of a given sample to that of a standard fluorophore with a correction for the relative absorbance by the two fluorophores at the excitation wavelength. Since the two emission bands are different, it is necessary to integrate over the corrected emission spectra of the sample and standard in order to compare the relative fluorescence intensities. Some expensive spectrophotometers make an automatic correction of emission spectra with microprocessors or by interfacing to a microcomputer.

Although phosphorescence is weaker than fluorescence, it persists longer and can be measured by chopping the incident radiation and synchronizing the detector to read only during the dark periods between the light pulses. This was first accomplished with a mechanical device by Becquerel in 1871. Two circular disks with holes cut around the circumference were mounted on the same axle, the holes in the first disk being offset from those in the second. The specimen was placed between the disks, and when the axle was rotated, the specimen was illuminated intermittently by a beam of exciting light passing through the holes in the first disk. The phosphorescence was viewed by looking at the sample through the holes in the second disk, and the lifetime could be estimated by varying the relative position of the two sets of holes. Modern instruments use high-speed choppers or repetitively pulsed flash lamps to screen the photodetector from fluorescence. Phosphorescence may be detectable in an ordinary spectrofluorimeter in rigid media at low temperature, which suppress delayed fluorescence, and quenching of triplet states by their diffusional reactions with unexcited fluorophore molecules and impurities. Much of the available information about triplet energy levels has been obtained by phosphorescence measurements.

1.5.2.2. Fluorescence Lifetime

Fluorescence lifetimes provide information about the excited state properties of fluorophores and their interactions with the environment. Several different techniques are used for measurements of fluorescence lifetimes. In a single-pulse method, the sample is excited with a strong light flash of nanosecond or shorter duration, and the fluorescence is recorded in real time on a fast oscillocope or a transient recorder. The method is applicable for relatively strong emissions, that can be recorded with adequate signal-to-noise in a single pulse. When the fluorescence decay of a given sample is highly reproducible, the signal-to-noise ratio can be improved by using repetitive pulses and an integrating measurement system (stroboscopic method) because the random noise tends to average out. One approach utilizes "pulse sampling," which measures the signal in short-time "windows" at various delay times after the excitation pulse. A "sampling" oscilloscope can be employed, or the decay signals can be measured with a "boxcar" integrator and printed out on a chart recorder. Time-correlated single-photon counting is an alternative technique in which the time delay between the excitation pulse and the arrival of the first photon emitted by the fluorophore at the photodetector is measured for a large number of very weak pulses. A histogram showing the number of counts in narrow time delay channels reproduces the fluorescence decay because the probability of detecting a single photon at a certain time after excitation is directly proportional to the intensity of the sample emission at the same time. The light source may be a high repetition rate (typically 2–50 kHz) gas-filled lamp of nanosecond duration or a repetitively pulsed laser. Single-photon counting requires sophisticated and expensive electronics equipment. The decay profiles measured with any pulse method must be corrected for the instrument response, which includes the finite width of the excitation pulses and the bandwidth of the detector electronics. This process of "deconvolution" is usually carried out with a computer. Fluorescence decay profiles are usually a single exponential function of time or, less frequently, the sum of two exponential functions. The computer analysis leads to the decay lifetime of each component and its relative contributions to the observed decay profile. A different approach to the measurement of fluorescence decay is based on excitation with a very high-frequency, sinusoidally modulated lamp (phase modulation method). (A lamp is "modulated" by superimposing a repetitive, time-varying pattern on the intensity, e.g., the light emitted by an arc lamp operated from a poorly filtered power supply is partially modulated at twice the power line frequency.) The basis of this method is that the time delay between excitation and light emission leads to a phase shift corresponding to the decay lifetime, and the extent of sinusoidal modulation is reduced by mixing with the decay profile. These parameters can be measured with a phase-sensitive detector (a "lock-in" amplifier), in which the phase and amplitude of a signal generated by the modulated source are compared with the fluorescence signal. Lock-in amplifiers can recover weak signals several orders of magnitude below the input noise level. The phase modulation method can measure subnanosecond fluorescence lifetimes at modulation frequencies in the order of 30 MHz. However, the method is based on the assumption that the decay profile is either exponential or the sum of two exponentials, which is not always the case. *Time-resolved fluorescence spectroscopy* is a field of rapidly growing importance in photobiology. A time-resolved spectrum shows the changes of the fluorescence spectrum over a time span comparable to several decay lifetimes, providing information about the interactions of the excited state, e.g., proton dissociation, the relaxation of the medium immediately after the excited fluorophore reverts to the ground state, and the decay of molecular complexes formed in the excited state (exciplexes).

1.5.2.3. Polarization of Fluorescence

Fluorescence spectra are inherently polarized because the absorption and emission dipoles have specific directions relative to the geometry of the molecule. When a group of *rigid* molecules in random orientations are exposed to linearly polarized light, the probability of light absorption by a given molecule increases with the alignment of the absorption dipoles with the electric vector of the light. Therefore, the polarization of the light "photoselects" an absorbing population of molecules. Since the emission dipole has a fixed orientation relative to an absorption dipole, the fluorescence will be partially polarized in a direction that depends on the molecular structure and the specific absorption transition. Polarization is measured by placing a polarizer between the excitation monochromator and the sample and an analyzer in front of the emission monochromator. The *polarization* is defined as

$$P = (I_{\parallel} - I_{\perp})/(I_{\parallel} + I_{\perp}) \tag{1-8}$$

where $I_{\parallel}$ and $I_{\perp}$ are the fluorescence signals with the analyzer parallel and perpendicular to the analyzer, respectively. The maximum value of P for a random distribution of molecular directions depends on the angle between the absorption and transition dipoles in the molecule, e.g., $P = +(1/2)$ when the absorption and emission moments are parallel, and $P = -(1/3)$ when they are perpendicular.(24,26) The polarization of fluorescence excitation spectra (in rigid media) provides information about the orientation of the absorption dipoles of different excited states relative to the emission dipole. However, fluorescence polarization cannot be observed if the molecule is rotating so rapidly that all directionality is lost prior to emission. This effect is *fluorescence depolarization.* The rotation rate of a molecule depends on its size and shape, the temperature, and the viscosity of the medium. The rate of rotation is characterized by a "rotational relaxation time," which can be calculated from hydrodynamic theory. The dependence of P on temperature and viscosity leads to values of the rotational relaxation time and information about the molecular dimensions. Measurements on small molecules (e.g., 10^{-10}-m radius) require a viscous solvent, such as glycerol. Larger molecules (e.g., $> 10^{-9}$-m radius) show fluorescence depolarization in fluid water and other low-viscosity solvents. Dynamic information about the interactions of fluorophores with the medium can be obtained by measuring the polarization of fluorescence decay. In the "fluorescence probe" technique, a small fluorophore is attached to a larger entity, such as a macromolecule or a biomembrane. The probe fluorescence provides information about the "microenvironment" in the vicinity of the probe, e.g., depolarization measurements lead to values of the motional relaxation times for the probe and the larger entity. Certain probes have the additional convenient property that their fluorescence efficiency depends markedly on the polarity of the microenvironment, e.g., the fluorescence of the dye ethidium bromide is weak in water and very intense when it intercalates into a double-helical region of a nucleic acid.

1.5.3. Optical Activity(27)

Nearly all molecules synthesized by living organisms are optically active, i.e., they alter the properties of polarized light. Optical activity results from asymmetry of small molecules, especially asymmetric carbon atoms, and from the conformation of macromolecules, e.g., helical structures. The rotation of the plane of linearly polarized light exemplifies optical activity. The dependence of optical rotation on wavelength is referred to as *optical rotatory dispersion* (ORD) and is usually measured in a spectral region where the material is nonabsorbing. The method is equivalent to Fig. 1-12a, where the first polarizer is the optically active substance. Another consequence of optical activity is a difference in the absorption of right-handed and left-handed circularly polarized light, where the "handedness" corresponds to the rotation direction of the electric vector. (In circularly polarized light, the electric vector rotates with the frequency of the electromagnetic wave.) This phenomenon is called *circular dichroism* (CD). A CD spectrum is based on the differential absorption of a sample for right-handed and left-handed circularly polarized light, e.g., as obtained by passing linearly polarized light through a quarter-wave plate. ORD and CD are different manifestations of the same physical property, and an ORD spectrum can be transformed into a CD spectrum by a mathematical operation, and vice versa. CD measurements on proteins from 190 to 230 nm have led to estimates of the relative extent of α-helical, β-sheet, and random coil conformations in their secondary structures. CD measurements on polynucleotides from 200 to 300 nm provide information about the extent of single-stranded, double-stranded, and random coil regions.

1.5.4. Photoacoustic Spectroscopy[28,29]

Photoacoustic spectroscopy (PAS) is based on the generation of sound waves by the absorption of modulated or chopped light. Although this phenomenon was discovered by Alexander Graham Bell in 1880, there has been a great resurgence of interest in recent years, with the development of highly sensitive sound detectors and intense light sources, especially lasers. Heat generation in a medium initiates a pressure wave that travels with the velocity of sound. The heat sources are the *radiationless* relaxation processes initiated by light absorption. Since PAS responds only to absorbed light, it permits measurements in highly scattering or opaque materials, including living tissues. In a typical PAS spectrometer, the sample is illuminated with modulated or chopped monochromatic light and the detector is a microphone or a transducer attached to the walls of the sample holder. The acoustic signal is measured with a phase-sensitive detector tuned to the chopping frequency. A plot of the signal strength vs. wavelength gives a photoacoustic spectrum. The phase shift between the incident light and the acoustic signal depends on the transit time required for the acoustic pulse to arrive at the detector, which provides information about the physical location of the absorbing chromophores relative to the position of the detector. Photoacoustic spectroscopy has been used to separate the absorption spectra of stratified chromophores by exploiting the effect of depth on the phase shift. Figure 1-22 shows the PA spectrum of an oat seedling. Similar measurements have been made on animal tissues, including intact human skin. A major application of PAS involves photochemical energy conversion in biological systems. Since photochemical reactions compete with the decay of metastable states, a decrease of heat production can be related to the energy stored in chemical products. Photoacoustic spectroscopy has been used to study photosynthetic energy storage in chloroplasts and intact leaves. Very weak absorptions have been measured with PAS, including the visible absorption spectra of water, absorbed substances on various substrates, and aerosols.

Other techniques, related to PAS, are based on alternative detection methods. "Pho-

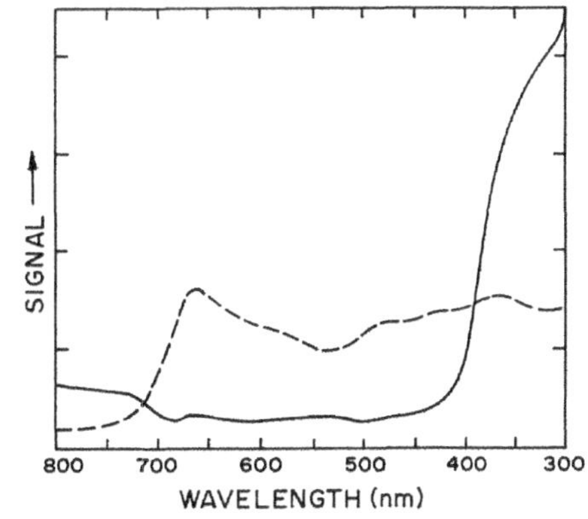

Fig. 1-22. Photoacoustic spectra of an oat seedling[28]. The sample was illuminated with modulated light (12 Hz) at different wavelengths (10-nm bandwidth), and the rate of energy absorption was measured with a detector of acoustic waves. The phase shift between the photoacoustic signal and the exciting light depends on the distance the sound wave must travel from the chromophore to the detector. The solid line was obtained by adjusting the phase shift of the lock-in amplifier to null the absorption of chloroplasts from 500 to 650 nm; the strong absorption below 400 nm corresponds to the cuticular coat. The dashed line was obtained with the phase shift adjusted to emphasize the chlorophyll absorption in the chloroplasts.

tothermal radiometry'' is the measurement of the IR radiation generated by optical absorption, with the advantage that there is no physical contact between the sample and the IR detector. ''Photorefractive'' methods are based on the changes in the refractive index of the medium induced by local heating. In a variation of this method, the defocusing of an auxiliary CW laser beam by the refractive index gradient in the medium is monitored with a photodetector located behind a pinhole; it is referred to as ''thermal lensing.'' A new generation of PAS instrumentation is under development, utilizing the analytical techniques currently employed in Fourier transform IR spectrometers. The sample is illuminated with polychromatic light and the resultant intensity of the photoacoustic signal is measured as a function of the path differences in a Michelson interferometer.[(8)] The ''spectrum'' of intensity vs. path difference is converted to the desired spectrum of intensity vs. frequency by mathematical operation, the Fourier transformation.

1.5.5. Action Spectra

The concept of *action spectra* dates to the 19th century, when biologists observed that the rate of plant growth depends on the spectral region of the ambient light. The quantitative aspects of action spectroscopy have been developed during the past 30 years, leading to a better understanding of the acceptable conditions and practices required to draw reliable conclusions from the results. An action spectrum is obtained by measuring the light dose required to evoke the same biological response at different wavelengths. Assuming that a higher light dose implies a weaker absorption by the responsible chromophores and vice versa, a plot of the reciprocal of the light dose against wavelength should parallel the absorption spectrum of the chromophores. Jagger[(4)] elucidated the conditions whereby action spectra may be expected to parallel absorption spectra. These include

1. The action mechanism must be the same at all wavelengths.
2. The quantum yield must be the same at all wavelengths.
3. The absorption spectrum of the chromophore must be the same *in vitro* and *in vivo*.
4. Absorption of light by inactive chromophores and light scattering must be negligible.
5. Only a small fraction of the incident light must be absorbed by the sample in the wavelength range of interest.
6. Reciprocity of the exposure time and fluence rate must hold.

Action spectra are obtained by comparing the number of photons required for the same biological effect at different wavelengths. It is convenient to borrow the concept of *cross-section* from atomic physics as a measure of the response of a single microscopic object to particulate radiation. The absorption cross section (σ_{abs}) of a molecule is proportional to the probability of light absorption at a given wavelength. The absorption cross-section and molar extinction coefficient are two different ways of specifying the same molecular property. At a given wavelength they are related by

$$\sigma_{abs} = 3.82 \times 10^{-21}\ \epsilon\ \text{cm}^2 \qquad (1\text{-}9)$$

The absorption cross-section of a strongly absorbing molecule is comparable to its physical cross-section. According to Eq. (1-9), the extinction coefficient of a molecule of approximately 0.1-nm radius is the order of 10^5 liters/mol-cm, which is typical of a strongly absorbing pigment molecule. The product of the absorption cross-section and the photon fluence is the probability that one photon will be absorbed by a molecule exposed to that fluence. For a thin slab of thickness *d* and unit cross-sectional area, the number of absorbed photons *N* after exposure to fluence *D* at a given wavelength is

$$N = D\sigma_{abs}C_{abs}d \quad (1\text{-}10)$$

where C_{abs} is the average concentration of chromophores in the material. If the observed response is the same at two wavelengths, the incident fluence levels required for this response are related by

$$D_1/D_2 = \sigma_2\Phi_2/\sigma_1\Phi_1 \quad (1\text{-}11)$$

where Φ is the quantum yield of the biological effect. The left side of Eq. (1-11) is the relative efficiency of the photoeffect at two wavelengths. A plot of the relative efficiency vs. wavelength is the action spectrum. This plot parallels the absorption spectrum of the chromophores involved in the biological effect, if $\Phi_1 = \Phi_2$, and the system meets the other criteria specified above.

The inactivation of cells by UV radiation represents an important application of action spectra. The major UV chromophores in cells are nucleic acids and proteins, with absorption maxima near 265 and 280 nm, respectively (Chapter 2). Because of spectral overlap, it is virtually impossible to irradiate a cell with light absorbed only by nucleic acids or proteins. However, when the action spectrum for cell inactivation, or another

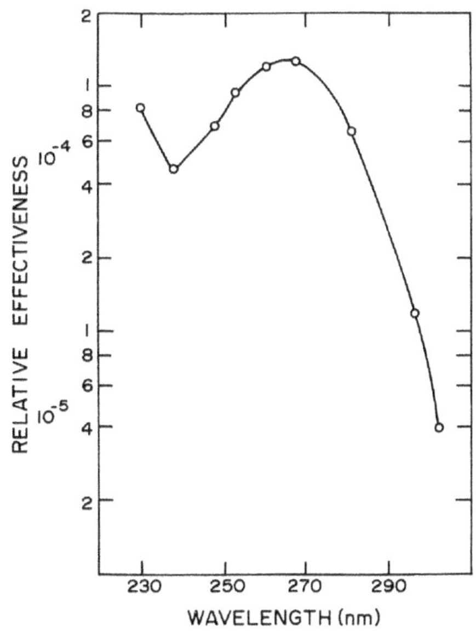

Fig. 1-23. Action spectrum for bacteria killing. The points are the reciprocal of the incident energy required for 50% killing of *E. coli* at each wavelength. The 260-nm band corresponds to absorption by nucleic acids. [Adapted from F. L. Gates, A study of the bacteriacidal action of ultraviolet light. III. The absorption of ultraviolet light by bacteria, *J. Gen. Physiol.* **14,** 31–41 (1930).]

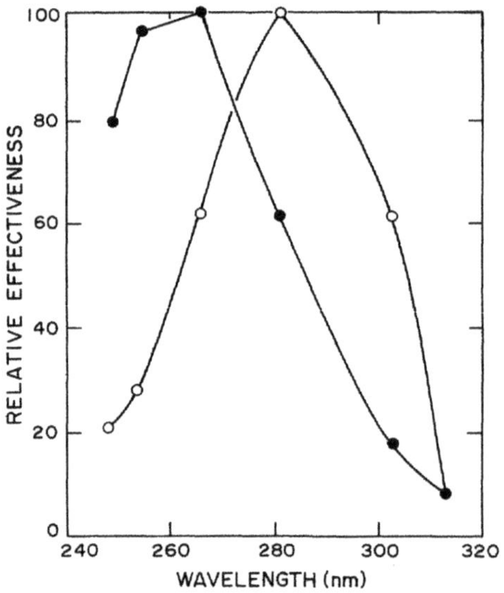

Fig. 1-24. Action spectra for retardation of cleavage in the sea urchin for irradiated sperm (●) and irradiated eggs (○). The 280-nm band corresponds to protein absorption and the 260-nm band corresponds to nucleic acid absorption. [Adapted from A. C. Giese, Radiation and cell division, *Q. Rev. Biol.* **22,** 253–282 (1947)].

photoprocess, matches that of either absorber, it can be concluded that damage to the corresponding chromophore was a significant factor in the photoprocess under observation. An action spectrum for killing of bacteria, reported more than 50 years ago, is reproduced in Fig. 1-23. The peak near 260 nm indicates that the nucleic acids are a major lethal target. The results in Fig. 1-24, also based on early work, show that nucleic acid damage is a major factor in the retardation of cleavage when sea urchin sperm are UV-irradiated, but protein damage is the dominant factor when sea urchin eggs are UV-irradiated. The action spectra for UV-irradiated mammalian cells have been reviewed.(30)

1.6. REFERENCES

1. L. R. Koller, *Ultraviolet Radiation,* 2nd ed., Wiley, New York (1965).
2. J. A. Parrish, R. R. Anderson, F. Urbach, and D. Pitts, *UV-A,* Plenum Press, New York (1980).
3. J. G. Calvert and J. N. Pitts, *Photochemistry,* Wiley, New York (1966).
4. J. Jagger, *Introduction to Research in Ultraviolet Photobiology,* Prentice-Hall, Englewood Cliffs, N.J. (1967).
5. O. Svelto, *Principles of Lasers,* 2nd ed., Plenum Press, New York (1982).
6. J. T. Verdegen, *Laser Electronics,* Prentice-Hall, Englewood Cliffs, N.J., (1981).
7. R. S. Longhurst, *Geometrical and Physical Optics,* 2nd ed., Wiley (1967).
8. R. W. Ditchburn, *Light,* 2nd ed., Interscience (1963).
9. F. A. Jenkins and H. E. White, *Fundamentals of Optics,* 4th ed., McGraw-Hill, New York (1976).
10. W. B. Allan, *Fiber Optics,* Plenum Press, London (1973).
11. L. Reynolds, C. C. Johnson, and A. Ishimaru, Diffuse reflectance from a finite blood medium: Applications to the modeling of fiber optic catheters, *Appl. Opt.* **15,** 2059–2067 (1976).
12. C. A. Parker, *Photoluminescence of Solutions,* Elsevier, Amsterdam (1968).
13. C. S. Rupert, Dosimetric concepts in photobiology, *Photochem. Photobiol.* **20,** 203–212 (1974).
14. D. Sliney and M. Wolbarsht, *Safety with Lasers and Other Optical Sources,* Plenum Press, New York (1980).

15. C. G. Hatchard and C. A. Parker, A new sensitive chemical actinometer. II. Potassium ferrioxalate as a standard chemical actinometer, *Proc. Roy. Soc.* (London) **A235,** 518–536 (1956).
16. L. O. Svaasand and R. Ellingsen, Optical penetration in human intercranial tumors, *Photochem. Photobiol.* **41,** 73–76 (1985).
17. R. R. Anderson and J. A. Parrish, in: *The Science of Photomedicine* (J. D. Regan and J. A. Parrish, eds.), pp. 147–194, Plenum Press, New York (1982).
18. G. Kortum, *Reflectance Spectroscopy,* Springer-Verlag, Berlin (1969).
19. W. L. Butler and K. H. Norris, The spectrophotometry of dense light scattering material, *Arch. Biochem. Biophys.* **87,** 31–40 (1960).
20. W. L. Butler, Absorption of light by turbid materials, *J. Opt. Soc. Am.* **52,** 291–299 (1962).
21. T. J. Dougherty, in *Methods in Porphyrin Photosensitization* (D. Kessel, ed.), pp. 313–328, Plenum Press, New York (1985).
22. G. W. King, *Spectroscopy and Molecular Structure,* Holt, Rinehart and Winston, New York (1964).
23. N. J. Turro, *Modern Molecular Photochemistry,* Benjamin/Cummings, Menlo Park, Cal. (1978).
24. A. J. Pesce, C.-G. Rosen, and T. L. Pasby, *Fluorescence Spectroscopy,* Marcel Dekker, New York (1971).
25. R. V. Bensasson, E. J. Land, and T. G. Truscott, *Flash Photolysis and Pulse Radiolysis,* Pergamon Press, Oxford (1983).
26. J. R. Lakowicz, *Principles of Fluorescence Spectroscopy,* Plenum Press, New York (1983).
27. C. R. Cantor and P. R. Schimmel, *Biophysical Chemistry,* Part II, W. H. Freeman, San Francisco (1980).
28. T. A. Moore, in: *Photochemical and Photobiological Reviews,* Vol. 7 (K. C. Smith, ed.), pp. 187–221, Plenum Press, New York (1983).
29. A. C. Tam, in: *Ultrasensitive Laser Spectroscopy* (D. S. Kliger, ed.), pp. 1–108, Academic Press, New York (1983).
30. T. P. Coohill, Action spectra for mammalian cells in vitro. In *Topics in Photomedicine* (K. C. Smith, ed.), Plenum Press, New York (1984).

2

Photochemistry

2.1. FUNDAMENTALS OF PHOTOCHEMISTRY

The central principle in photochemistry, first stated by Grotthus and Draper in 1818, is that *only absorbed light can produce a chemical change.* The initial step in a pho-

Leonard I. Grossweiner • Physics Department, Illinois Institute of Technology, Chicago, Illinois 60616. **Kendric C. Smith** • Department of Radiation Oncology, Stanford University School of Medicine, Stanford, California 94305.

tochemical reaction at ordinary light intensities is the absorption of a single photon by a molecule, and the production of an excited state in which one electron of the absorbing molecule is raised to a higher energy level (Chapter 1). The subsequent events depend on the molecular structure, the wavelength of the light, and the specific reaction conditions. A certain fraction of the excited molecules may release their excess energy by a sequence of "relaxation" processes, leading back to the original unexcited molecules. Alternatively, a photochemical process may take place at some stage of the relaxation chain, either spontaneously or via interactions with other molecules. The *quantum yield* of a given photochemical product is the probability that the absorption of one photon by the system leads to one molecule of that product. The first chemically reactive species formed by light absorption are referred to as "primary products," typically metastable excited molecular states and free radicals. In most cases light is not involved in the subsequent reactions, and they are referred to as "dark" or "thermal" reactions. However, some complex photobiological processes may utilize light in more than one stage, e.g., there are two light-activated steps in photosynthesis by green plants (Chapter 12). A photobiological process takes place in stages, starting with the very fast processes at the physical level, followed by dark chemical reactions, and terminating with relatively slow events at the biological level (Fig. 2-1). The photophysical processes of biomolecules are described in Chapter 1. Their photochemical reactions may be described by the well-developed concepts of organic reaction mechanisms.[1] An alternative approach, to be followed here, emphasizes the photochemistry of proteins and nucleic acids, which are the two most important classes of functional biomolecules in cells. These "colorless" mac-

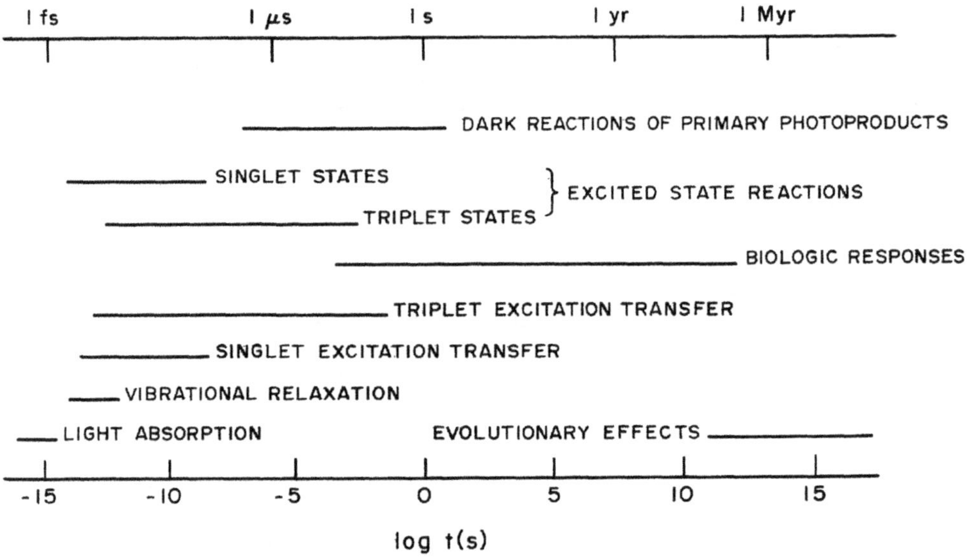

Fig. 2-1. Time scales in photophysics, photochemistry, and photobiology. The lower scale is the logarithm of the time (s) at which the events indicated by the horizontal lines take place. The upper scale locates the time intervals in ordinary units. [Adapted from N. J. Turro and A. A. Lamola, Chap. 3, in *The Science of Photobiology,* 1st ed. (K. C. Smith, ed.), Plenum Press, New York (1977).]

romolecules absorb in the UV region of the spectrum. Although many different types of proteins and nucleic acids are present in a given cell, characteristic photochemical reactions have been identified that are relevant to biological damage induced by the exposure of organisms to UV radiation. In addition, the effects induced by exposing cells to UV radiation under controlled conditions provide a sensitive probe of biological responses to molecular alterations, e.g., DNA repair. Other important classes of biomolecules absorb both UV and visible radiations, including photosensitizing drugs and the chromophores of natural processes. The photochemistry of "colored" molecules is considered in Chapter 3 and subsequent chapters on specific photoprocesses in biology.

2.1.1. Overall Reactions in Photochemistry

An overall photochemical reaction specifies only the initial reactants and stable products. Some important overall reactions with examples in photobiology are as follows:

1. Linear addition to an unsaturated molecule:

(2-1)

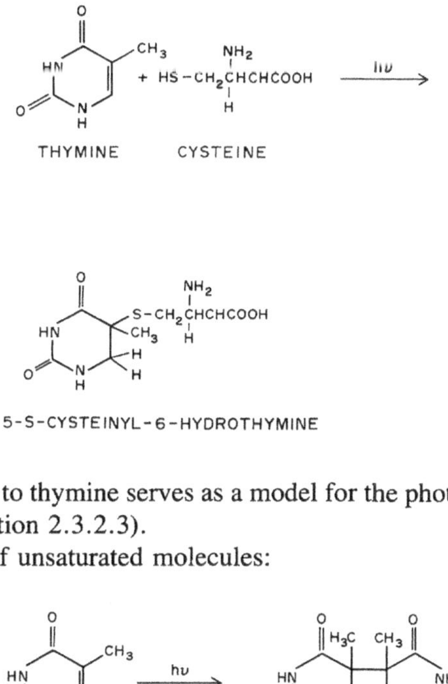

The addition of cysteine to thymine serves as a model for the photochemical cross-linking of DNA to protein (Section 2.3.2.3).

2. Cycloaddition of unsaturated molecules:

(2-2)

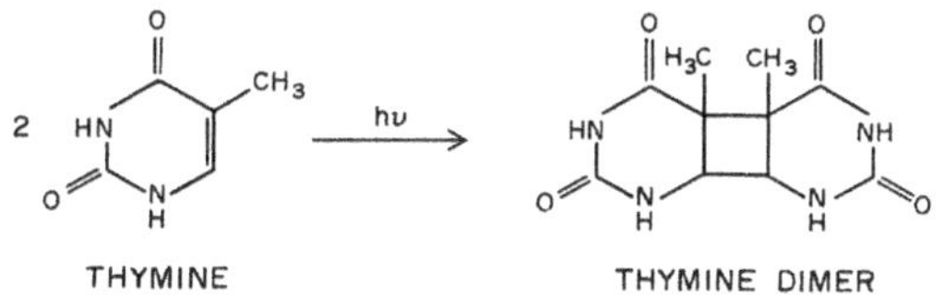

In cycloaddition reactions, two molecules or molecular groups on a macromolecule react to form a ring product. Thymine dimers are an important class of molecular lesions generated by the action of UV radiation on DNA (Section 2.3.1.3).

3. Photofragmentation:

(2-3)

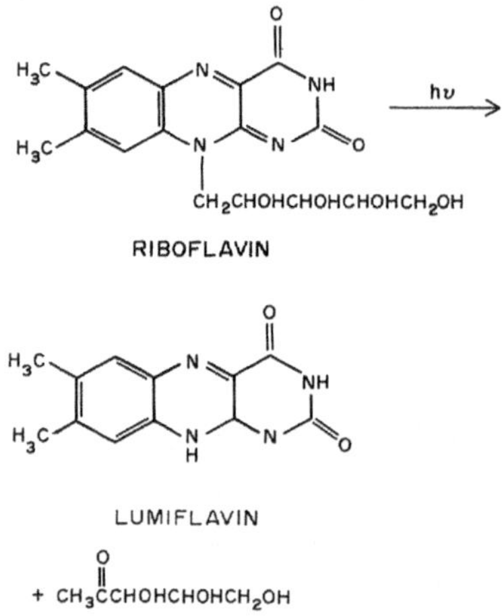

In photofragmentation reactions, the original molecule is converted to smaller products by photolytic cleavage at one or a few chemical bonds. These reactions take place in essentially all biomolecules, if exposed to UV radiation of sufficiently short wavelengths.

4. Photooxidation:

(2-4)

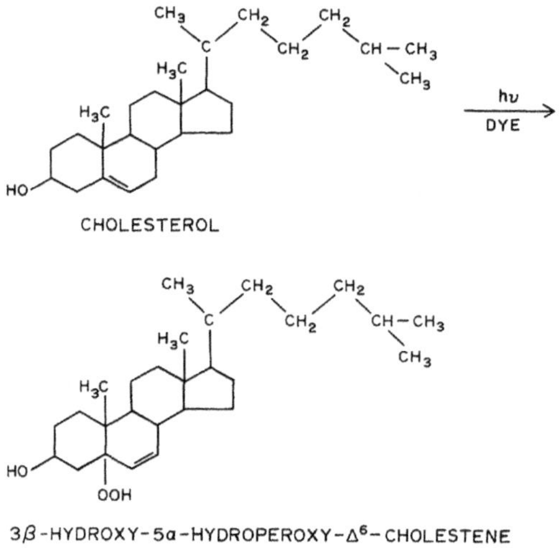

The formation of this "5α" photooxidation product of cholesterol is a selective test for the involvement of singlet molecular oxygen (Chapter 3).

5. Photohydration:

(2-5)

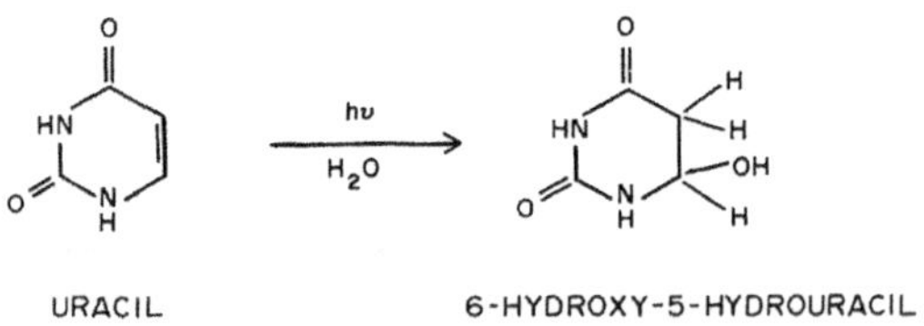

Photohydration products are another class of lesions induced by the UV irradiation of DNA (Section 2.3.1.3).

6. *Cis-trans* isomerization:

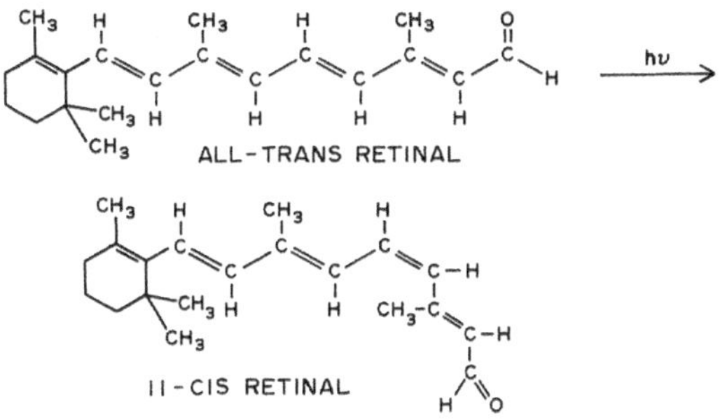

(2-6)

This process leads to a 180° rotation around a double bond. Photoisomerization of 11-*cis*-retinal is the first photochemical step in retinal vision (Chapter 9).

7. Photorearrangement:

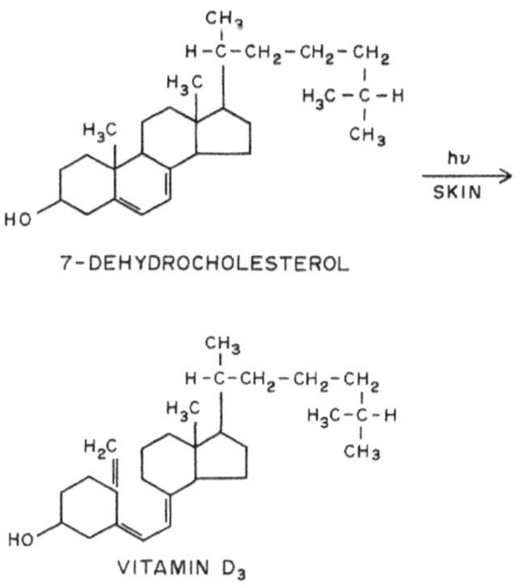

(2-7)

In a photorearrangement reaction, the original and product molecules are made up of the same atoms. The production of vitamin D_3 in mammalian skin is initiated by the UV irradiation of 7-dehydrocholesterol (Chapter 6).

2.1.2. Primary Reactions in Photochemistry

The overall photochemical reactions indicated in Section 2.1.1 lead to stable products, except when the products themselves undergo subsequent photochemical reactions, e.g., thymine dimers are photodissociated by UV-C radiation (Section 2.3.1.3). The primary processes usually lead to unstable intermediates that undergo additional reactions under dark conditions. The excitation pathways of a typical organic biomolecule are shown in a Jablonski diagram (Fig. 1-16). In addition to the physical relaxation processes indicated on the diagram, competing photochemical reactions may take place in the excited states. In a *unimolecular* reaction, an excited state reacts without the intervention of another molecule. The solvent may be involved in apparent unimolecular reactions in condensed media, e.g., in the dissociation of an electronically excited acidic molecule, the proton is transferred from the excited molecule to a molecule of the solvent. Most elementary reactions in fluid media are *bimolecular,* and require proximity between the two reacting species. The short lifetime of excited singlet states, typically 1–20 ns, limits their bimolecular reactions to nearby reaction partners. The lifetime of triplet states may exceed 1 ms, and they may diffuse significant distances prior to reacting. Excited state reactions usually lead to short-lived free radical products, whose subsequent reactions are "diffusion-limited," i.e., the slowest step is the time required for diffusional encounters of the reaction partners. Diffusion-limited reactions are very fast in water and other low-viscosity solvents, and may be quite slow in high-viscosity media such as organic glasses. In contrast, many chemical reactions between stable molecules require "thermal activation" in order to provide the energy necessary to overcome potential energy barriers between the reactants and the products. A small temperature rise can increase the rate of a thermally activated reaction many fold, but it has a relatively small effect on a diffusion-limited reaction, i.e., only to the extent that it affects the viscosity of the medium.

Two general methods have been used to study primary photochemical processes. Fast reaction techniques make it possible to study short-lived species in real time, e.g., in flash photolysis, detectable concentrations of the intermediates are generated by an intense light flash, and their subsequent reactions are followed in real time by making fast optical absorption measurements (Fig. 1-19). The picosecond time resolution available with mode-locked lasers can probe virtually all photophysical and photochemical processes. Alternatively, dry or frozen systems may be used to slow down the initial reactions of photochemical intermediates, facilitating the use of conventional analysis methods, e.g., optical spectroscopy and magnetic resonance. Some important types of primary photophysical and photochemical processes are:

1. Singlet–singlet excitation energy transfer:

$$^{1}D^{*} + {}^{1}A \rightarrow {}^{1}D + {}^{1}A^{*} \quad (2\text{-}8)$$

where * labels an electronically excited state, and the superscript indicates the spin multiplicity (Section 1.4.1). Reactions of this type can occur by short-range (exchange) and long-range (Förster) mechanisms. Exchange transfer is efficient only when the two molecules are in contact, and when the excitation energy of the donor exceeds that of the acceptor, i.e., the energetics are "downhill." The rate of long-range transfer depends on the spectral overlap between the fluorescence band of the donor (D) and the absorption band of the acceptor (A), the distance between the two species, and their relative orientiations. The effective range can be 5–10 nm for favorable conditions (Section 1.4.4).

2. Triplet–triplet energy transfer:

$$^{3}D^{*} + {}^{1}A \rightarrow {}^{1}D + {}^{3}A^{*} \quad (2\text{-}9)$$

In this process, an excited triplet state of the donor generates an excited triplet state of the acceptor. Triplet–triplet transfer takes place by an exchange mechanism, requiring close contact and downhill energetics, i.e., the triplet energy of the donor must be larger than the triplet energy of the acceptor. The long lifetime of triplet states compared to excited singlet states can make triplet–triplet transfer important in photochemical systems. Two triplet states can react also, leaving one molecule in an excited singlet state and the other in the ground state. This process is referred to as "triplet–triplet annihilation," and generally requires two photons per transfer in order to generate the two triplet states. However, the ground state of molecular oxygen is a triplet, and the excited triplet states of many photosensitizers can transfer energy to molecular oxygen, thereby generating electronically excited molecular oxygen or "singlet oxygen" (Chapter 3).

3. Excited state electron transfer:

$$D^* + A \rightarrow D^+ + A^- \quad \text{(2-10a)}$$

$$A^* + D \rightarrow A^- + D^+ \quad \text{(2-10b)}$$

Many excited singlet and triplet states of photosensitizers can act as electron donors (reducing agents) or electron acceptors (oxidizing agents). The configuration diagram in Fig. 1-15a shows that the S_1 and T_1 states have unpaired electrons in two energy levels. It may be energetically favorable to either add or release an electron, leading to anionic or cationic free radicals, respectively. The electron transfer act may be followed by a fast proton transfer in polar solvents, in order to form the most stable acidity state of the radical ion. The triplet states of some sensitizers can act as both energy and electron transfer agents, e.g., the triplet state of the common dye, eosin Y, transfers energy to molecular oxygen via Eq. (2-9), oxidizes aromatic compounds (e.g., tryptophan) via Eq. (2-10a), and reduces strong oxidants such as ferricyanide ion via Eq. (2-10b).[2]

4. Photolytic production of free radicals:

$$R_1\text{–}R_2 + h\nu \rightarrow R_1^{\cdot} + R_2^{\cdot} \quad \text{(2-11)}$$

Photocleavage may take place via different reaction pathways. One possible mechanism involves a nonradiative transition from an unrelaxed (vibrationally excited) excited singlet state to a "dissociative" state of the molecule, followed by bond rupture.

5. Photoionization:

$$A + h\nu \rightarrow A^+ + e^- \quad \text{(2-12)}$$

Most molecules are photoionized in the gas phase by sufficiently energetic UV radiation, typically in excess of 7 eV. Photoionization of inorganic anions (e.g., I^-, Br^-, NO_3^-), small aromatic compounds (e.g., phenol, tyrosine, tryptophan), and some dyes occurs in solution at photon energies lower than 4 eV. The ejected electron is temporarily stabilized in the medium as a "solvated electron" (Section 2.2.1). Photoionization of aromatics and dyes also takes place at low temperature in frozen aqueous and hydrocarbon media.

6. Hydrogen atom abstraction by carbonyl compounds:

$$R_1-\underset{\underset{O}{\|}}{C}-R_2 + h\nu \longrightarrow R_1-\underset{\underset{O^*}{\|}}{C}-R_2 \qquad (2\text{-}13a)$$

$$R_1-\underset{\underset{O^*}{\|}}{C}-R_2 + R_3H \longrightarrow R_1-\underset{\underset{OH}{|}}{C^{\cdot}}-R_2 + R_3^{\cdot} \qquad (2\text{-}13b)$$

This type of reaction involves organic ketones and an oxidizable constituent (R_3H). Carbonyl derivatives have photochemical properties derived from (n,π^*) excited states (Section 1.4.2), in addition to the usual (π,π^*) excited states of unsaturated hydrocarbons. The lowest triplet state has (n,π^*) character in some carbonyl derivatives (e.g., acetone, acetophenone, benzophenone) and (π,π^*) character in others (e.g., 2-acetylnaphthalene, 4-phenylbenzophenone, fluorenone). Hydrogen abstraction is an important reaction of (n,π^*) triplet states, e.g., irradiation of benzophenone in isopropanol from 320 to 360 nm leads to the disappearance of benzophenone with a quantum yield close to 2. In the first step, the (n,π^*) excited state of benzophenone abstracts an H atom from the alcohol, thereby reducing the ketone to a ketyl radical and oxidizing the alcohol to a free radical. Each alcohol radical then reduces another molecule of the ketone in a dark reaction. This reaction exemplifies a process whereby the absorption of one photon leads to the disappearance of more than one reactant molecule. In more complicated processes, known as "chain reactions," an intermediate capable of further reacting with an initial reactant is continuously restored by a sequence of intermediate steps. Photopolymerization reactions are an important class of chain reactions.

2.1.3. Photochemical Kinetics

The *order* of a chemical reaction specifies how the overall rate depends on the concentrations of the reactants. The rate of a first-order reaction increases linearly with a concentration, the rate of a second-order reaction increases with the square of a concentration, etc. Reaction order is a macroscopic concept, and it is not always equivalent to the molecularity of the elementary processes that combine to give the overall reaction. For example, $A \rightarrow B$ is first order overall and first order with respect to species A. However, the elementary reaction may not be unimolecular if solvent molecules are involved. The reaction $A + B \rightarrow C$ is second order overall and first order with respect to A or B. It may also be bimolecular if the reaction actually takes place by collisions between species A and B. The *rate constant* of a reaction specifies the rate at unit concentrations of the reactants, and given conditions of the solvent, temperature, and any other parameters that affect the reactivity. Concentrations in dilute solution are usually expressed in *molar* units, in which case first-order rate constants have units of s^{-1}, and second-order rate constant have units of liters mol^{-1} s^{-1}. The basic unit of light quantity in photochemical dosimetry is *fluence,* defined as the number of photons of specified wavelengths in a collimated beam incident in a perpendicular direction on a unit area of the reaction vessel. The relationship between incident fluence and absorbed dose depends on the conditions of the experiment. If a well-stirred photochemical system absorbs all incident light, then the absorbed dose equals the total number of incident quanta divided by the volume of the system. The other extreme is an "optically dilute" system, whereby only a small fraction of the incident photons are absorbed. In this case, the absorbed dose is proportional to the product of the concentration of the absorbing molecules and the optical pathlength. (This result follows from Beer's law; Section 1.5.1). The calculation of absorbed dose is more complicated in intermediate cases where neither approximation applies.

The conventional analysis of photochemical kinetics will be illustrated with a typical reaction, in which constant optical excitation of a reactant (A) generates an excited state (A*), which may decompose to give a permanent product (A′) in competition with the relaxation of A* back to A:

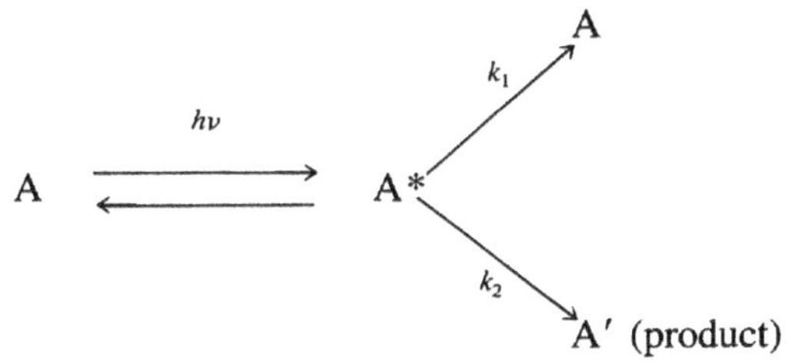

(2-14)

The kinetics analysis may be simplified by assuming that the processes by which the short-lived intermediate A* is formed and relaxes back to A are much faster than the reactions leading to permanent products, in this case species A′. This approximation is usually valid when unstable intermediates are generated at ordinary light intensities, e.g., for metastable triplet states and reactive free radicals. Reaction kinetics are expressed mathematically by equating the concentration changes of each species to the difference between the rate at which that species is formed and the rate at which it disappears. For species A* this relation is

$$d[\mathrm{A}^*]/dt = \Phi_0(dD/dt)/N - (k_1 + k_2)[\mathrm{A}^*] \simeq 0 \qquad (2\text{-}15)$$

where $(dD/dt)/N$ is the rate of light absorption by species A in units of quanta per liter per second divided by the Avogadro number (N), Φ_0 is the quantum yield for the photochemical production of A*, k_1 (s^{-1}) is the first-order rate constant for the decay of A* to A, and k_2 (s^{-1}) is the first-order rate constant for the formation of the final product A′ from A*. The left side of Eq. (2-15) is approximately zero, because the concentration of short-lived species A* is essentially constant. The "steady-state" concentration of A* from Eq. (2-15) is

$$[\mathrm{A}^*] = \Phi_0(dD/dt)/(k_1 + k_2) \qquad (2\text{-}16)$$

where [A*] is much smaller than [A] and [A′] in the present approximation. The photolysis rate of the initial reactant A at any instant in time is given by

$$-d[\mathrm{A}]/dt = k_2[\mathrm{A}^*] = k_2\Phi_0(dD/dt)/(k_1 + k_2) \qquad (2\text{-}17)$$

The initial quantum yield for the disappearance of A, Φ_{-A}, is defined as $d[A]/dD$ at $t = 0$, i.e.,

$$\Phi_{-A} = k_2\Phi_0/(k_1 + k_2) \tag{2-18}$$

Equation (2-17) shows that the fraction of absorption events leading to the photolysis of A equals the ratio of the rate constant for product formation to the sum of the rate constants for product formation and relaxation of A* back to A. The relationship between absorbed dose (D) and incident fluence (F) depends on the reaction conditions. If all of the incident light is absorbed by species A, integration of Eq. (2-17) leads to a linear dependence of [A] on the incident fluence:

$$[A] = [A]_0 - a\Phi_{-A}F \tag{2-19}$$

where a is a constant equal to $1000/Nd$, d is the pathlength of the irradiation vessel, $[A]_0$ is the initial concentration of A, and F is the incident fluence (quanta cm^{-2} s^{-1}). (Centimeter units are frequently employed in photochemical kinetics because extinction coefficients are expressed in units of liters mol^{-1} cm^{-1}.) The other limiting approximation applies to "optically dilute" solutions of A in an inert medium, such that the *absorbance* of A is much smaller than unity (Section 1.5.1). In this case, integration of Eq. (2-17) shows that the disappearance of A follows an exponential function of the incident fluence:

$$[A]/[A]_0 = \exp - b\Phi_{-A}F \tag{2-20}$$

where b is a constant equal to $(1000 \log_e 10)\epsilon_A/N$, and ϵ_A is the molar extinction coefficient of A at the irradiation wavelength. [The quantity b is the absorption cross-section of species A, as defined by Eq. (1-9).]

The experimental implimentation of this kinetics analysis requires much effort and attention to details. The incident fluence must be determined with appropriate radiometry, and assays must be calibrated for the loss of reactant A and the formation of product A′. The kinetics scheme of Eq. (2-14) can be tested by the fit of the yield data (i.e., A vs. F or A′ vs. F) to either Eq. (2-19) for completely absorbing solutions or Eq. (2-20) for "optically dilute" solutions. The fitting procedure also leads to the value of Φ_{-A}, which should be constant for different values of $[A]_0$. The ratio of k_1 to k_2 can be calculated from Eq. (2-19), if Φ_0 is known from other measurements, e.g., if A* is a triplet state, then Φ_0 is the triplet yield of A. In this case, k_1 is the reciprocal of the triplet state lifetime, and the value of k_2 can be determined. In a thorough study, the measurements would be carried out over a range of temperature, solvent properties, and possibly at several wavelengths. Relatively few photochemical reactions have been completely analyzed. For most applications in photobiology, a minimally satisfactory level of understanding requires the identification of the more important photochemical intermediates and permanent products, measurements of the key quantum yields and rate constants, and specification of reasonable reaction mechanisms.

2.2. PHOTOCHEMISTRY OF PROTEINS AND AMINO ACIDS[2]

Proteins are the most abundant molecules in cells, except for water, and the most diverse of all biomolecules in size, form, and function. Functional proteins may be highly sensitive to light because their biological properties are controlled by relatively small regions of the entire macromolecule, e.g., the activity center of an enzyme. The absorption of light by a protein is localized at those chromophoric groups with significant extinction coefficients at the incident wavelengths. However, the subsequent events may not be confined to the initial sites because excitation energy and electrons can migrate within a large molecule. Protein photochemistry has been investigated with many techniques, including assays of permanent residue destruction, flash photolysis, electron spin resonance detection of photochemical intermediates, and circular dichroism studies on conformation changes. A reasonable starting point for the study of protein photochemistry is the photochemistry of the chromophoric amino acids that are responsible for most of the light absorption. For wavelengths longer than 240 nm, the major absorbers in "colorless" proteins are the aromatic amino acids, phenylalanine (Phe), tyrosine (Tyr), and tryptophan (Trp), with smaller contributions from histidine (His), cystine (Cys), sulfhydryl, and the peptide bonds (Fig. 2-2). (These abbreviations for chromophoric amino acids are frequently used in the photochemical literature and may not be the same as those employed in standard texts on biochemistry.) The carboxyl and amino end groups absorb below 240 nm, and all amino acids are strong absorbers of UV radiations shorter than about 200 nm.

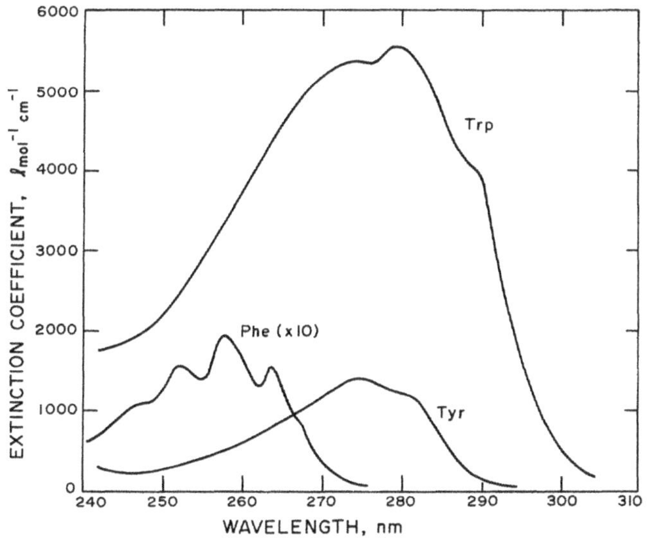

Fig. 2-2. Molar extinction coefficients of aqueous tryptophan (Trp), tyrosine (Tyr), and phenylalanine (Phe). [Adapted from *Handbook of Biochemistry,* 2nd Ed. (H. A. Sober, ed.), The Chemical Rubber Co., Cleveland (1970).] These are the only common amino acids with significant absorptions above 240 nm. The histidine absorption maximum is 211 nm. Cystine has a strong absorption at 207 nm and a weak peak near 250 nm.

2.2.1. Photochemistry of Amino Acids

Aqueous solutions of amino acids are decomposed by UV radiation, usually leading to many products. Much of the early work on amino acid photochemistry was done with 254-nm radiation from low-pressure mercury arcs. For example, irradiation of His at 254 nm gave ammonia, histamine, imidazole, acetaldehyde, and other products.[3] Some of the final products may result from the continued photolysis of earlier products, and therefore their occurrence may not provide useful information about the photochemistry of the starting amino acid. The early stages of amino acid photochemistry have been investigated in frozen systems and with flash photolysis, especially for the aromatic amino acids. There are numerous reports of protein and amino acid destruction by longer wavelength radiations, where the extent of light absorption is very low, e.g., aqueous solutions in glass vessels exposed to strong sunlight. These reactions may involve photochemical processes initiated by the weak light absorption in the long-wavelength tail of the amino acid spectra or, more likely, photosensitization by low concentrations of impurities that absorb longer wavelength light.

2.2.1.1. Initial Photochemistry of Aromatic Amino Acids[4–6]

Prior to the 1960s, it was generally assumed that bond cleavage is the major UV photolysis process in aromatic amino acids. However, early flash photolysis experiments on aqueous solutions of aromatic amino acids led to the conclusion that photoionization (electron ejection) is the most important primary act.[7] Flash photolysis transient spectra of aqueous tyrosine led to a phenoxyl-type aromatic free radical (Tyr$^{\bullet}$), formed by disruption of the phenolic OH bond, and an intense, short-lived, broad absorption band peaking in the red region. Aqueous tryptophan gave a different aromatic radical absorption (Trp$^{\bullet}$), identified with photolysis of the indole ring, and a similar red absorption band. Flash photolysis of aqueous phenylalanine led to the benzyl radical absorption ($C_6H_5CH_2^{\bullet}$) and the red absorption band. The red band was observed only at short times after the irradiation flash (<1 μs) and was absent when oxygen, hydrogen peroxide, or nitrous oxide was present. These properties characterize the "hydrated" electron, i.e., a solvated electron in water, which was first identified by Hart and Boag in "pulse radiolysis" transient spectra of water.[8] (Pulse radiolysis is similar to flash photolysis, except that the radiation is a short, intense pulse of high-energy electrons or other ionizing radiation, instead of light.) A hydrated electron is a quasi-free electron, trapped by the interactions of its electric charge with water dipoles. It is an unstable chemical species, with characteristic physical and chemical properties. Hydrated electrons are one of the strongest known reducing agents in the absence of oxygen, and they react rapidly with oxygen to form the superoxide radical anion (O_2^-), the precursor of hydrogen peroxide. The absorption spectra of these aromatic amino acid free radicals and the hydrated electron obtained by flash photolysis are reproduced in Fig. 2-3. Further study has confirmed that photoionization of aqueous aromatic amino acids and their derivatives also takes place at ordinary light intensities.[4,5]

The stable reaction products reported for the UV photolysis of aqueous tryptophan in air include tryptamine, *N*-formylkynurenine (NFK), kynurenine, aliphatic amino acids, and ammonia. The formation of NFK is especially important in protein photolysis because

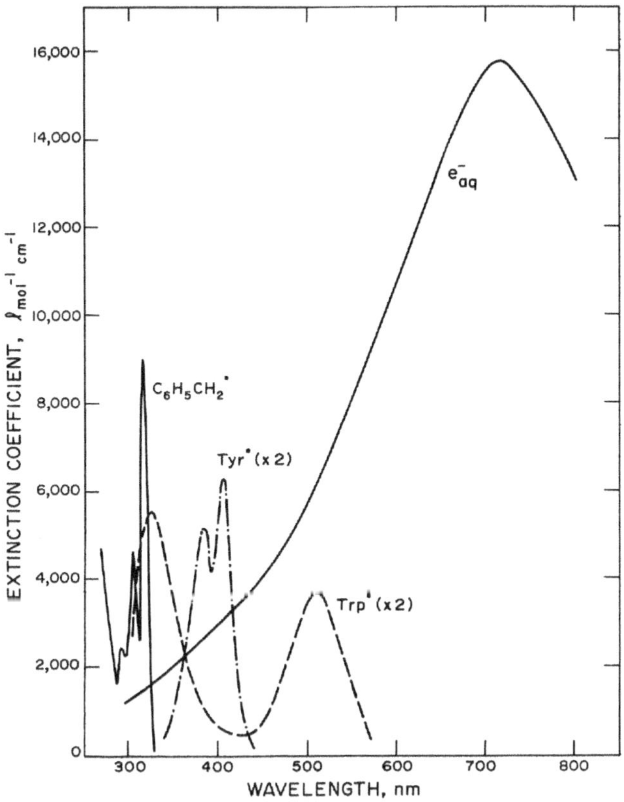

Fig. 2-3. Absorption spectra of transient species identified by flash photolysis of aqueous amino acids: half-oxidized tryptophan radical (Trp•) [adapted from J. F. Baugher and L. I. Grossweiner, Photolysis mechanism of aqueous tryptophan, *J. Phys. Chem.* **81,** 1349–1354 (1977)], half-oxidized tyrosine radical (Tyr•) [adapted from Ref. 23], benzyl radical ($C_6H_5CH_2$•) [adapted from L. J. Mittal, J. P. Mittal, and E. Hayon, Primary processes in the photochemistry of phenylalanine and derivatives in aqueous solution. Biphotonic photoionization and photodissociation reactions, J. Am. Chem. Soc. **95,** 6203–6210 (1973)], hydrated electron (e_{aq}^-) [adapted from J. K. Thomas, Methods of production of solvated electrons and their chemical and physical properties, *Radiat. Res. Revs.* **1**, 183–208 (1968)]. The estimated accuracy of the extinction coefficients is 10%.

this oxidation product of tryptophan can act as a photosensitizer of longer wavelength UV-A radiation, e.g., NFK has been implicated in the aging of the ocular lens.[(9)] UV photolysis of small tryptophan peptides in the absence of air, e.g., glycine-tryptophan-glycine, led to unusual indole derivatives with fused, eight-member lactam rings and other complex products.[(10,11)] Photolysis of aqueous tyrosine gave many products, including hydroxylated aromatics (especially dihydroxyphenylalanine or DOPA, the precursor of melanin), bityrosine, aliphatic amino acids, and ammonia. The reaction mechanism has not been worked out in detail. The permanent products reported for the continuous irradiation of aqueous phenylalanine include tyrosine and other phenolics, including DOPA, aliphatics, and ammonia.

2.2.1.2. *Initial Photochemistry of Cystine, Histidine, and Peptide Bonds*

Although cystine is a weakly absorbing amino acid, it is especially important in protein photochemistry because cystine photolysis has a high quantum efficiency and may lead to the splitting of interchain disulfide bridges. The reported permanent products of cystine photolysis include sulfhdryl, disulfide, aliphatic derivatives, and hydrogen sulfide.[(6)] The probable primary events are the splitting of S–S bonds, leading to RS•-type radicals, and the splitting of C–S bonds, leading to –S–S$^\bullet$-type radicals. Hydrated electrons react very rapidly with cystine to form a radical anion, in which an electron is temporarily trapped at the S–S bond. This reaction has been observed also with cystinyl bridges in proteins. The absorption by histidine above 240 nm is much weaker than the other protein chromophores, and it is not an important primary absorber in proteins. However, a protonated imidazole group is an electron trap, comparable to a disulfide bridge, and can serve as a temporary site for electrons photoejected from aromatic residues.[12] The peptide bonds in proteins are weak UV absorbers, but their large numbers can lead to a role in the protein photochemistry, especially in proteins without aromatic amino acids, e.g., gelatin. Peptide bonds can act as electron traps, and may provide a path for electron migration from optically excited aromatic residues to cystine and histidine sites.

2.2.2. Photochemistry of Proteins

UV irradiation of an aqueous protein solution leads to extensive changes in almost all of its properties. The "physical" effects include changes in the absorption spectrum, optical activity, apparent molecular weight, electrophoretic pattern, solubility, heat sensitivity, and other parameters. The "chemical" effects include changes in acid–base titration curves, increase of digestibility by proteolytic enzymes, modification or destruction of residues, and changes in catalytic activity of enzymes. Attempts to relate the overall damage to specific initial photochemical reactions are frustrated by the structural complexity, e.g., the "small" hormonal protein insulin contains 16 different types of amino acids, arranged on two polypeptide chains of 21 and 30 units, joined by two disulfide bridges. The UV absorption of a typical "colorless" protein has a strong "end" absorption peaking at 180–200 nm, a minimum at 240–250 nm, and the "aromatic" band maximum near 280 nm. Protein absorption spectra are similar, but not identical, to the composite spectra of mixtures of the constituent amino acids (Fig. 2-4). The differences are caused by peptide bonding and interactions between the amino acid residues. As a first approximation, it may be assumed that the fraction of light absorbed by a given amino acid in a protein can be estimated from its concentration in the protein and the molar extinction coefficient of the corresponding aqueous amino acid. However, the photochemistry of an amino acid residue in a protein may differ significantly from the aqueous amino acid. The contributing factors include the following:

1. Excitation energy may be transferred from Phe → Tyr → Trp by the Förster mechanism [Eq. (2-8)].
2. Electrons generated by the photoionization of aromatic residues [Eq. (2-12)] may migrate and react at different sites.
3. Tyrosinyl residues can by oxidized indirectly by intramolecular electron transfer from an intact tyrosine site to an oxidized tryptophan site.[12]
4. The photolytic splitting of disulfide bridges may lead to large changes in the conformation of the protein, including denaturation.
5. Initial damage to an amino acid residue in a protein may be more stable than in the free amino acid because the altered site cannot further react by diffusion.

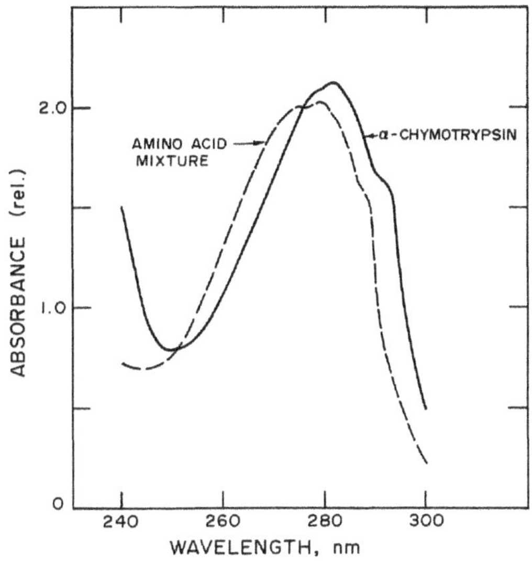

Fig. 2-4. Absorption spectra of α-chymotrypsin and a mixture of its constituent amino acids. (Adapted from Ref. 3.)

The identification of pathways leading from the initial photochemical reactions in a protein to permanent damage is a challenging problem. Most work has been done with lower molecular weight proteins, e.g., proteolytic enzymes, hormones, serum proteins, and lens proteins, especially where there is information about the primary chain sequence and the three-dimensional structure.

2.2.2.1. *Initial Photochemistry of Proteins*

Transient spectra obtained by flash photolysis of dilute, aqueous protein solutions show aromatic free radical bands, similar to those identified by the flash photolysis of tyrosine and tryptophan, and also the short-lived red absorption band of the hydrated electron, e.g., the results in Fig. 2-5 reported for the laser flash photolysis of subtilisin BPN′ at 265 nm.[13] The band maxima, identifying Tyr$^{\bullet}$ (410 nm), Trp$^{\bullet}$ (510 nm), and hydrated electrons (720 nm), have been observed in many other proteins.[2,14] Some proteins with cystinyl residues had a transient band near 420 nm, identified with the disulfide electron adduct, e.g., lysozyme, papain, and trypsin. These findings demonstrate that the dominant initial photochemical reactions in proteins are the photoionization of aromatic residues, and the temporary trapping of the ejected electrons in the water medium and in some cases at disulfide bonds. The disulfide electron adduct band was not suppressed by saturation with oxygen or nitrous oxide, which react rapidly with hydrated electrons. This result indicates that the ejected electrons did not react with disulfide bonds subsequent to their localization in the external aqueous medium as hydrated electrons. Rather, the electrons migrate directly from an optically excited aromatic residue to a disulfide bridge by a fast, intramolecular process. The absorption spectra of long-lived tryptophan and tyrosine triplet states also have been observed in the flash photolysis spectra of proteins.[14] Since triplet states may produce singlet oxygen by energy transfer,

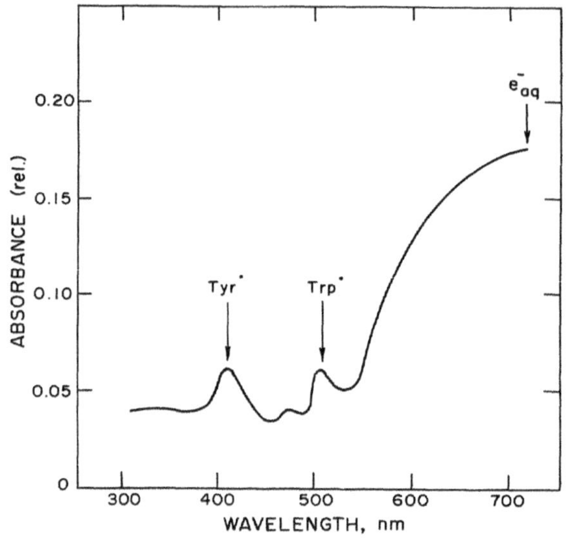

Fig. 2-5. Transient absorption spectrum of aqueous, air-free subtilisin BPN′, taken at 100-ns delay after a 17-ns laser flash at 265 nm. The arrows show the absorption maxima of half-oxidized tryptophanyl residues (Trp•), half-oxidized tyrosinyl residues (Tyr•), and hydrated electrons (e^-_{aq}) (see Fig. 2-3). This enzyme contains 3 tryptophanyl, 10 tyrosinyl, and no cystinyl residues. The observed transient bands indicate that some of the tryptophanyl and tyrosinyl residues were photoionized, and the ejected electrons were stabilized in the aqueous medium as hydrated electrons. The laser irradiation also led to loss of enzymic activity. None of the tryptophanyl or tyrosinyl residues are located near the active center. It was deduced that photolysis of a single tryptophanyl residue led to a small change in the conformation that inhibited the ability of the enzyme to bind the synthetic substrate used assay the activity. (Adapted from Ref. 13.)

which can then attack other protein sites, triplet states might be implicated in protein photolysis reactions. However, quantum yields for the UV photoinactivation of enzymes are usually independent of oxygen, which does not rule out the attack of singlet oxygen on a "nonessential" site. Carbonic anhydrase is an exception whereby photoinactivation was much faster under oxygen; evidence for the production of singlet oxygen by triplet energy transfer from NFK to oxygen was obtained in this case.[15]

2.2.2.2. *Theories of Enzyme Inactivation*[2]

Many workers have attempted to relate the quantum yields for the UV inactivation of enzymes to the amino acid composition. An early theory of Luse and McLaren is based on the hypothesis that photolysis of any chromophoric amino acid residue in an enzyme contributes additively to UV inactivation.[16] According to this approach, the inherent photosensitivity of an amino acid is related to the product of its extinction coefficient (ϵ_a) and its photodestruction quantum yield (Φ_a) in aqueous solution. The results in Table 2-1 show that cystine and tryptophan should be the most photosensitive residues at 254 nm. In the Luse and McLaren theory, the contribution of each amino acid to the photoinactivation quantum yield (Φ_{in}) for the enzyme is determined with the additive relationship:

$$\Phi_{in} = \Sigma\, n_a \epsilon_a \Phi_a / \epsilon_e \qquad (2\text{-}21)$$

where n_a is the number of type a residues in the enzyme and ϵ_e is the molar extinction coefficient of the enzyme. The application of Eq. (2-21) to common enzymes gave results in agreement with experimental data to within a factor of 2 in most cases, suggesting that the concepts of inherent residue photosensitivity and additivity are valid.[17] However, the

TABLE 2-1. Relative Photosensitivity of Amino Acids at 254 nm[a]

Compound	ϵ^{b}	Φ^{c}	$(\epsilon \times \Phi)^{d}$
Cystine (–S–S– bond)	270	0.13	35
Tryptophan	2870	0.004	12
Tyrosine	320	0.002	0.6
Phenylalanine	140	0.013	2
Histidine	0.24	<0.003	<0.007
Peptide bonds (acetylalanine)	0.2	0.05	0.01

[a]The values in this table are based on data in Ref. 3.
[b]ϵ indicates the molar extinction coefficient at 254 nm, which is the probability that light at 254 nm will be absorbed by the compound.
[c]Φ indicates the quantum yield, which is the probability that light absorbed at 254 nm will cause a chemical change.
[d]$(\epsilon \times \Phi)$ indicates a measure of the photochemical sensitivity of a particular amino acid.

earlier proposal of Setlow that Φ_{in} correlates with the cystine content in enzymes gave equally good results.[(18)] This correlation implies that conformation changes induced by the rupture of disulfide bonds are the key factor leading to the loss of enzymic activity. Augenstein and Riley attempted to rationalize the two divergent theories by proposing that cystinyl residues can be destroyed by both direct photolysis and by energy transfer from adjacent aromatic residues.[(19)] The same mechanism was extended by Dose to explain the higher yields of cystine destruction in enzymes containing aromatic residues.[(20)]

The earlier work was based on the untenable assumptions that all residues of a given type in an enzyme contribute equally to its activity and that the quantum yield for destruction of an amino acid residue in a protein is the same as the aqueous amino acid. In a recent extension of the McLaren and Luse approach, Grossweiner and coworkers replaced n_a in Eq. (2-21) by the number of essential type a residues.[(21)] In conventional biochemistry, a residue is considered essential if it is directly involved in the catalytic reaction, participates in substrate binding, or helps maintain the active enzyme conformation. In the photochemical process, the concept of "essential" may be extended to other chromophoric residues, whose photolysis leads to loss of enzymic activity. In addition, the Φ_a in Eq. (2-21) were replaced by estimates of the initial photolysis quantum yields of amino acid residues in proteins. It was found that only cystine and tryptophan need be considered as photosensitive, and that only one set of Φ_a values is required at 254 and 280 nm. The results in Table 2-2 are based on the following relationship:

$$\Phi_{in} = \Gamma_{trp} f_{trp} \Phi_{trp} + \Gamma_{cys} f_{cys} \Phi_{cys} \tag{2-22}$$

where Γ is the fraction of essential tryptophanyl or cystinyl residues, f is the fractional absorption by all tryptophanyl or cystinyl residues at the irradiation wavelength, and Φ_{trp} and Φ_{cys} are the quantum yields for destruction of tryptophan or cystine in the protein, respectively. An average value of $\Phi_{trp} = 0.05$ was used at 254 and 280 nm, based on the quantum yields for tryptophan destruction in proteins reported in the literature[(2)] and the initial yields of Trp$^{\bullet}$ formation in enzymes measured by laser flash photolysis.[(21)] The values Φ_{cys} were taken as 0.20 at 254 nm and 0.13 at 280 nm, based on the quantum yields for disulfide bond destruction in oxidized glutathione.[(22)] The results in Table 2-2 are in

TABLE 2-2. Prediction of Photoinactivation Quantum Yields[a]

Enzyme	Γ_{trp} [b]	Γ_{cys} [c]	254 nm Φ_{in}(calc)[d]	254 nm Φ_{in}(exp)[e]	280 nm Φ_{in}(calc)[d]	280 nm Φ_{in}(exp)[e]
Lysozyme	3/6	2/4	0.028	0.024	0.023	0.023
Trypsin	1/4	2/6	0.015	0.020	0.009	0.010
Papain	1/5	0/3	0.006	0.006	0.006	0.003
Carboxypeptidase A	1/6	0/1	0.005	0.005		
Subtilisin Carlsberg	0/1	0	0.000	0.007		
Ribonuclease	0	2/4	0.029	0.030	0.004	0.007

[a]The calculations in this table are based on Eq. (2-21).
[b]Γ_{trp} indicates the fraction of essential tryptophanyl residues in the enzyme based on biochemical information and the structure.
[c]Γ_{cys} indicates the fraction of essential cystinyl residues in the enzyme based on biochemical information and the structure.
[d]Φ_{in}(calc) is the calculated photoinactivation quantum yield at the indicated wavelength as described in the text.
[e]Φ_{in} is the experimental photoinactivation quantum yield at the indicated wavelength based on literature cited in Ref. 21.

good agreement with the reported data, except for subtilisin Carlsberg, which has no cystine and apparently no essential tryptophan. This model implies that energy and electron transfer between residues are not major factors in photoinactivation, and supports the concept that photosensitive residues may either be essential in the conventional biochemical sense or be located in the structure where their photochemical disruption can affect essential residues.[2,14] Examples of photosensitive tryptophanyl residues identified in flash photolysis studies include Trp 199 in trypsin, adjacent to the catalytic residue serine 198; Trp 177 in papain, in contact with the active site residue His 159; Trp 73 in carboxypeptidase A, adjacent to glutamic acid 72, which participates in the binding of essential zinc; Trp 62, Trp 63, and Trp 108 in lysozyme, which are part of the activite site. However, ribonuclease A contains no tryptophanyl residues, and none of the tyrosinyl residues are essential. Photoinactivation of ribonuclease A can be attributed entirely to the photolysis of two essential cystinyl residues (Cys 26–84 and Cys 40–95), which is favored by the higher absorption in cystine resulting from the absence of tryptophan.[23] Equation (2-22) is not applicable to the subtilisins, which contain no cystine, and none of the aromatics are located near essential residues. A laser flash photolysis study on subtilisin BPN′ led to the suggestion that Trp 113 may be photosensitive because it is hydrogen-bonded to asparagine 113, which is located at one end of a chain sequence associated with the aromatic substrate-binding sites.[13]

2.2.2.3. *Kinetics of Enzyme Inactivation*

The dependence of enzyme inactivation rates on dose exemplifies a relatively simple type of "survival curve." Photoinactivation measurements are usually carried out by exposing a dry or aqueous enzyme to monochromatic radiation, and assaying the remaining activity at periodic intervals. The actinometry is more accurate for optically dilute systems, where the enzyme absorbs a small fraction of the incident UV radiation. The analysis may be simplified by assuming that inactivation does not affect the overall light absorption by the enzyme, in which case the fraction of light absorbed by the active component is C/C_0, where C is the concentration of active enzyme after incident fluence F, and C_0 is the initial active enzyme concentration. A straightforward analysis leads to an exponential survival curve:

$$C/C_0 = \mathrm{Exp}{-}(\Phi_{in}\sigma_e F) \qquad (2\text{-}23)$$

where σ_e is the absorption cross-section of the enzyme (Section 1.5.5). Equation (2-23) predicts that a semilogarithmic plot of the surviving active enzyme fraction vs. incident fluence will be a straight line, with the slope proportional to the product of the inactivation quantum yield and the absorption cross-section. The quantity $\Phi_{in}\sigma_e$ is the "inactivation cross-section," which may be viewed as the effective cross-sectional area of the photosensitive targets on the enzyme molecule. An exponential dependence of surviving fraction on dose characterizes a "one-hit model" because it implies that the absorption of one photon by a "target" inactivates the enzyme molecule. Survival curves for cellular systems may follow Eq. (2-23) if one photochemical reaction in a single cellular target leads to the observed biological effect and this damage is not repaired during irradiation or prior to the damage assay (Chapter 4).

2.3. PHOTOCHEMISTRY OF NUCLEIC ACIDS[24–38]

Early photochemical studies on the nucleic acids dealt largely with the destructive cleavage of the pyrimidine ring, but in 1949 Sinheismer and Hastings showed that the pyrimidines could undergo photochemistry without destroying the pyrimidine ring. They reported the photochemical addition of a molecule of water across the 5,6 double bond of the pyrimidines (pyrimidine hydrate). In the 1960s, Beukers and Berends observed that UV irradiation caused the stable linkage of two adjacent thymine residues in a strand of DNA. This discovery of thymine dimer formation stimulated a resurgence of interest in UV photobiology that continues today. Unfortunately, as the importance of thymine dimers in biological inactivation and mutagenesis became apparent, there developed a tendency to give them credit for too much of UV photobiology. Many other types of photoproducts are produced in the DNA of cells, and under certain experimental conditions these photoproducts play the dominant role in UV photobiology (Chapter 4).

It is important to remember that DNA does not exhibit the same sensitivity to UV radiation under all experimental conditions. The intrinsic sensitivity of DNA to photochemical alteration can be changed by a variety of biological (e.g., growth state of cells), chemical (e.g., base analog substitution), and physical (e.g., denaturation, freezing, drying) techniques. To give an example: one photoproduct that is produced in high yield, and appears to be the major cause of death in UV-irradiated vegetative bacterial cells, is not produced to a significant extent in bacterial spores. Thus, different types of photoproducts appear to inactivate irradiated vegetative cells and spores, respectively. Simple generalizations, therefore, cannot be made as to which photoproduct in DNA is the most important to all irradiated cells under all experimental situations.

Furthermore, in addition to the intrinsic photochemical sensitivity of DNA, we must also consider the ability of a cell to repair the damage produced in DNA (Chapter 4). If one type of lesion is repaired accurately by a cell it cannot be of biological importance to that cell, while lesions that are not repaired, or are repaired inaccurately, may be of major biological importance.

2.3.1. Photochemistry of Nucleic Acid Constituents

2.3.1.1. Deoxyribose

Although carbohydrates make up about 41% by weight of the nucleic acids, they show essentially no UV absorption at wavelengths above 230 nm and therefore would not be expected to undergo direct photochemical reactions when irradiated with light of wavelengths greater than 230 nm. However, indirect effects of UV radiation on deoxyribose have been reported. For example, a cell is much more sensitive to killing by UV radiation if 5-bromouracil (BrUra) replaces the thymine in its DNA. One effect of UV radiation on BrUra in DNA is debromination, with the consequent production of a uracil radical. In the absence of another hydrogen donor, a hydrogen atom will be abstracted from the adjacent deoxyribose. This leads to the production of uracil, to the destruction of the deoxyribose, and ultimately to a chain break in the DNA.[39] Chain breaks in DNA are also produced in the presence of ketone sensitizers such as benzophenone. When certain of these sensitizers are excited by light at 313 nm they can abstract hydrogen from water to produce hydroxyl radicals, which then attack the DNA and produce chain breaks.[40] Thus, under certain conditions, chemical alterations in the carbohydrates of the nucleic acids can be brought about when the nucleic acids are exposed to UV and near-UV radiation,[41] even though the carbohydrates do not absorb light at these wavelengths.

2.3.1.2. Purines[42–44]

In pure solution, purines are approximately 10-fold more resistant to photochemical alteration than the pyrimidines. Inside cells, however, the nucleic acids are not in pure solution and, in fact, the purines (as well as the pyrimidines) participate very readily in photochemical reactions with other molecules (heteroaddition reactions).

In general, however, the photochemistry of the purines has not been as well studied as the photochemistry of the pyrimidines, but currently there is a resurgence of interest in the photochemistry of the purines. A photodimer of adenine has been identified,[45] and excisable purine photoproducts have been detected in a defined sequence of UV-irradiated human DNA.[46]

In addition, some of the photon energy absorbed by the purines can be transferred to the pyrimidines or to the sugar-phosphate backbone of DNA, and thus result in chemistry. For example, the transfer of energy from adenine to thymine has been implicated in the formation of a thymine radical and in the phosphorescence of thymine in UV-irradiated poly-dAT.

2.3.1.3. Pyrimidines

2.3.1.3a. Hydration Products[34]

When a solution of uracil (or one of its derivatives) is UV-irradiated, it loses its characteristic UV absorption peak near 260 nm, but this can be largely regenerated by heat, alkali, or acid treatment. This reversible reaction was shown to be due to the photochemical hydration of the 5,6 double bond of uracil to form 6-hydroxy-5-hydrouracil [Eq. (2-5)] (note that 5-hydroxy-6-hydrouracil is stable to heat). Photochemical hydrates of cytosine and thymine are also formed, although with less efficiency than for uracil. Pyrimidine hydrates appear to be formed through an excited precursor state in the singlet manifold, since triplet quenchers and sensitizers do not change the photochemical yield of hydrates.

The formation of a water addition photoproduct of cytosine in irradiated denatured DNA has been inferred from the appearance of a heat-reversible absorption peak around 240 nm. Dihydrocytosine derivatives (i.e., lacking the 5,6 double bond) exhibit a characteristic absorption peak at this wavelength. Irradiated native DNA, however, shows no such heat-reversible absorption peak. These and other data suggest that hydrates of cytosine are probably not formed in measurable yield in UV-irradiated, double-stranded DNA, but they are formed in single-stranded DNA. However, other photoproducts (e.g., pyrimidine dimers) can distort the DNA helix and produce *local denaturation* (i.e., single-stranded regions), and thus may permit the formation of hydrates in otherwise double-stranded DNA.

During replication and/or transcription of the DNA, there may be short regions of single-strandedness, and in these regions the formation of pyrimidine hydrates may be of importance. The possible role of pyrimidine hydrates in causing mutations has been demonstrated in an *in vitro* model system.[47] When polycytidylic acid was UV-irradiated, its coding properties in an RNA polymerase system were altered. The irradiated polymer lost its ability to code for the incorporation of guanylic acid unless adenylic acid was also added to the medium. The increase in adenylic acid incorporation in the polymer as a function of UV radiation fluence was heat-reversible under conditions known to reverse pyrimidine hydrates, and for this reason it was suggested that the code change might be the result of the formation of cytosine hydrates (i.e., the cytosine hydrate appears to base-pair with adenine rather than guanine). Therefore, the formation of pyrimidine hydrates in single-stranded regions of the DNA may play a role in the production of mutations, but this remains to be established *in vivo*.

2.3.1.3b. Thymine Glycols

Thymine glycols (Fig. 2-6) represent one of the common classes of DNA damage produced by ionizing radiation (e.g., Refs. 48 and 49), and they are also produced in low yield by UV radiation. They are thought to arise through the action of hydroxyl radicals[41,50] and are much more stable than the pyrimidine hydrates. The glycosylase activity of purified *Escherichia coli* endonuclease III shows a specificity for excising thymine glycols from UV-irradiated DNA (Chapter 4).

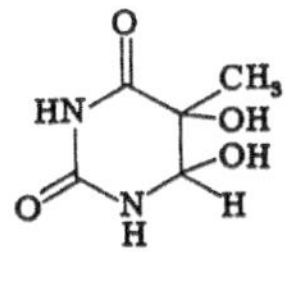

Fig. 2-6. Thymine glycol.

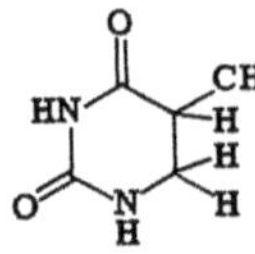

Fig. 2-7. 5,6-Dihydrothymine.

2.3.1.3c. Other Modifications of Pyrimidines

The addition of two hydrogen atoms to the 5 and 6 positions of thymine in DNA to form 5,6-dihydrothymine (Fig. 2-7) has been demonstrated *in vitro*. The biological importance of this *reducing* reaction *in vivo* is not known.

When thymine is UV-irradiated in solution, the *oxidation* products, 5-hydroxymethyl uracil and 5-formyl uracil, are formed (Fig. 2-8). Aldehydes react readily with amino groups. If 5-formyl uracil is formed *in vivo*, it has the interesting possibility of reacting with amino groups in proteins to produce DNA–protein cross-links (see Section 2.3.2.3).

2.3.1.3d. Cyclobutane-Type Dimers of Thymine, Cytosine, and Uracil(35,51)

When an aqueous solution of thymine is irradiated at 254 nm, it loses its characteristic absorption properties at a rate about one tenth that of uracil [the quantum yield (Φ) for thymine destruction is 0.4×10^{-3}]. If a solution of thymine is frozen and then UV-irradiated, it shows a greatly increased photochemical reactivity ($\Phi \simeq 0.2$), and the major product formed is a dimer. (N.B.: To be a dimer, a compound must have exactly twice the molecular weight of the monomer, otherwise it should be called an adduct; see below.) The probable effect of freezing is to bring the thymine molecules into an oriented juxtaposition favorable for a bimolecular photochemical reaction.

To form the thymine dimer, two thymine molecules are linked to each other between their respective 5 and 6 carbon atoms, thus forming a cyclobutane ring (four-carbon ring) between the two thymines [Eq. (2-2)]. There are six possible isomers of the thymine dimer, and these have been isolated from irradiated thymine oligomers (Fig. 2-9). The type I (*cis-syn*) thymine dimer is the one formed between adjacent thymines in the same strand of DNA. Certain of these isomers are stable to acid hydrolysis, while others are not.(35) Since acid hydrolysis is the usual method for liberating photoproducts from irradiated DNA, such labile photoproducts would be destroyed.

There is a wavelength dependency both for the formation and for the monomerization of the cyclobutane-type thymine dimer. After a sufficient fluence of UV radiation, a "photostationary" state is achieved, and a constant ratio of monomer to dimer is established that is characteristic of the wavelength used. At the longer wavelengths of UV radiation (near 280 nm) the formation of the dimer is favored, while at the shorter wavelengths (near 240 nm) monomer formation is favored. This response is due to differences in the absorption spectra of thymine and its dimer (Fig. 2-10), and in the

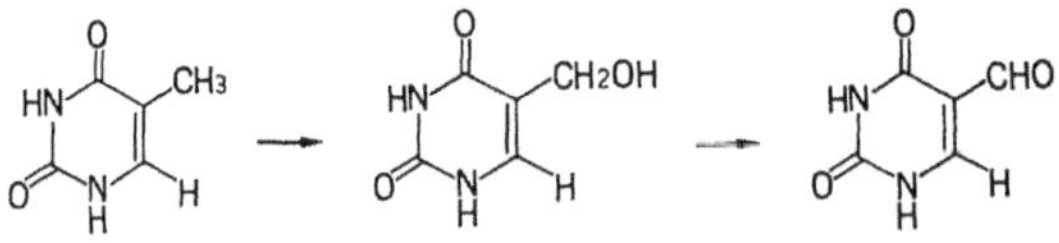

Fig. 2-8. Photochemical formation of 5-hydroxymethyl uracil and 5-formyl uracil from thymine.

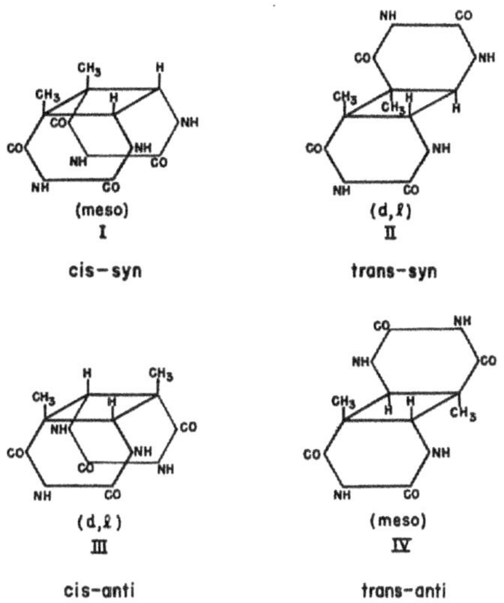

Fig. 2-9. Isomeric forms of the cyclobutane-type thymine dimers. Optical iosmers are possible for types II and III.

quantum yields for the formation and splitting of the dimer. Thus, in UV-irradiated *E. coli,* where 30–40% of the thymines could theoretically dimerize, only about 15% are dimerized even after high fluences of UV radiation at 254 nm.

Five other cyclobutane-type dimers of the natural pyrimidines are also known. These are the dimers of uracil and cytosine, and the adducts of uracil-thymine, cytosine-thymine, and uracil-cytosine. The isolation of cytosine dimers is complicated not only by the competition of the hydrate reaction but also by the fact that cytosine deaminates readily when its 5,6 double bond is saturated. Cytosine dimers are therefore readily converted to uracil dimers. It is apparent that if a cytosine dimer in the DNA of a cell were to deaminate

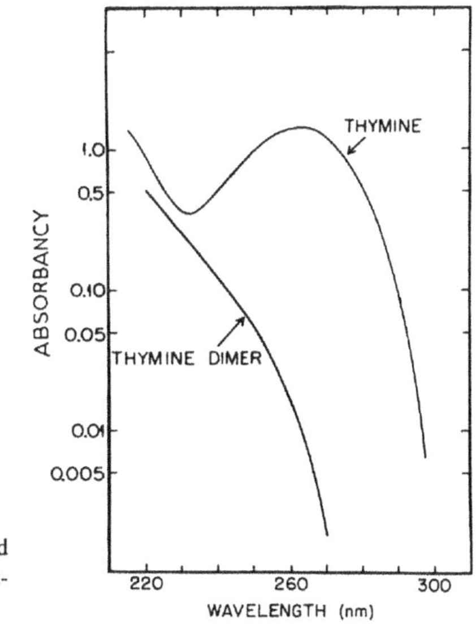

Fig. 2-10. The absorption spectra of thymine (25 μg/ml) and of purified thymine dimer (34 μg/ml) [Modifed from R. Setlow, *Biochim. Biophys. Acta* **49**, 237 (1961).]

to form a uracil dimer, and if this dimer were then split *in situ* by the photoreactivating enzyme (Chapter 4), a mutation could result since the uracil residues would base-pair with adenine rather than with guanine.

Dihydrocytosine (i.e., cytosine without its 5,6 double bond) will react with amino acids such that the amino group of dihydrocytosine is replaced by the amino group of the amino acid, resulting in a covalent link between dihydrocytosine and the amino acid.[52] Since the photohydrate of cytosine and the cyclobutane-type dimers of cytosine are analogs of dihydrocytosine, the addition of protein amino groups to these cytosine photoproducts might serve as another mechanism by which DNA (and RNA) and protein are cross-linked by UV radiation (Section 2.3.2.3).

Up to now we have considered photochemical reactions that occur as the result of the direct absorption of photons by the reacting species. Thymine dimers can be formed, however, by wavelengths of light that are not absorbed by thymine, provided that the thymine is in the presence of suitable molecules (i.e., photosensitizers) that do absorb these wavelengths. This process occurs by *triplet–triplet energy transfer* [Eq. (2-9)] and requires that the triplet state of the absorbing species (the photosensitizer) be higher in energy than the triplet state of thymine. Upon interaction, the triplet energy of the photosensitizer is transferred to the thymine, yielding thymine in its triplet state with the possibility for the subsequent formation of thymine dimers. Examples of this situation are the formation of thymine dimers by light of wavelengths >300 nm when DNA is irradiated in the presence of 10^{-2} *M* acetophenone or when bacteria are irradiated while suspended in 10% acetone. One advantage of the use of triplet state photosensitization to drive a reaction is that it can be performed at wavelengths where the reverse reaction (e.g., dimer splitting) does not occur. With this technique one should achieve essentially a quantitative conversion of all adjacent thymine residues to dimers rather than only achieving a photostationary state as obtained by the direct excitation of thymine residues (see above). Certain metal ions (e.g., Ag^+) photosensitize the formation of thymine dimers in DNA, while others (e.g., Hg^{2+}) quench this reaction.[37]

2.3.1.3e. *Spore Photoproduct*[36,51]

After high fluences of UV radiation (e.g., 2×10^4 J/m^2 at 254 nm) about 30% of the thymine in bacterial spores can be converted to what has been called the spore photoproduct (Fig. 2-11). It is a thymine-thymine *adduct,* and appears to be formed through the reaction of two different thymine radicals. The spore photoproduct does not exhibit the

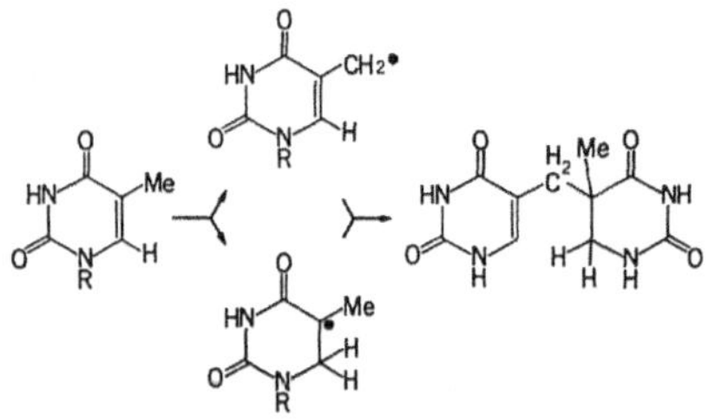

Fig. 2-11. 5-Thyminyl-5,6-dihydrothymine, a major photoproduct in UV-irradiated bacterial spores, is produced by the addition of two different radicals of thymine.

short-wavelength reversal properties of the cyclobutane-type thymine dimers. Therefore, in spores the yield of this product can approach the maximum determined by the number of thymines that are nearest neighbors in the DNA.

When spores are irradiated at various times during germination (i.e., the transition phase between the spore stage and the vegetative cell stage), the yield of spore photoproduct decreases, the yield of cyclobutane-type pyrimidine dimers increases, and the sensitivity to killing by UV radiation increases. The cyclobutane-type pyrimidine dimers appear to be about 11 times more effective in killing vegetative cells than are the spore photoproducts in killing spores. This could imply that the spore photoproduct is more efficiently repaired than the cyclobutane-type dimers.

2.3.1.3f. Pyrimidine-Pyrimidine (6-4) Adducts[25,36,51]

When a solution of DNA is UV-irradiated, or when a frozen solution of thymine is UV-irradiated and thawed, the absorption spectrum of each shows a new peak at 320 nm. This is due to the photochemical formation of (6-4) adducts (Fig. 2-12). The (6-4) adducts of TC, CC, and TT are observed in UV-irradiated DNA, but the CT adduct is not. These lesions cannot be split by reirradiation at short wavelengths, as can cyclobutane-type pyrimidine dimers. Their chief diagnostic property is their lability to hot alkali.[53] The (6-4) adducts are produced with various efficiencies at specific sites within DNA; however, on average they are produced about 100-fold less efficiently than are pyrimidine dimers. Recently, the (6-4) adducts have been shown to play a major role in UV radiation mutagenesis at specific sites in DNA (Chapter 4).

2.3.1.3g. Heteroadducts of the Pyrimidines

The pyrimidines (and purines) will also react photochemically with many other compounds (for tabulations of these reactions, see Refs. 27 and 54). For example, alcohols (R–OH) and the amino acid cysteine (R′–SH) photochemically add across the 5,6 double bond of the pyrimidines. Since some of the addition products of methanol and ethanol can be reversed by acid, they appear analogous to the water (H–OH) addition product described above, and therefore are expected to be attached to position 6. Cysteine is known to add at the 5 position and forms a stable product [Eq. (2-1)], which may be an important mechanism for the formation of DNA–protein cross-links (Section 2.3.2.3). The pyrimidines can also form photoadducts with the purines (e.g., adenine-thymine)[44] but their biological importance has yet to be determined.

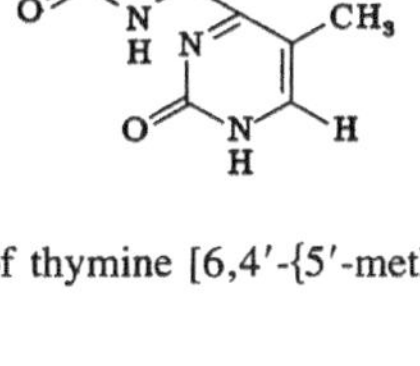

Fig. 2-12. The (6-4) adduct of thymine [6,4′-{5′-methylpyrimidin-2′-one}-thymine].

2.3.2. Photochemistry of DNA

2.3.2.1. DNA Strand Breaks

While X irradiation is very effective in producing single-strand and double-strand breaks in DNA,[48] UV irradiation is very inefficient in this respect except in the presence of sensitizers such as benzophenone (Section 2.3.1.1). However, DNA chain breaks are produced enzymatically in UV-irradiated cells (and X-irradiated cells) as necessary steps in the repair of damaged DNA (Chapter 4).

2.3.2.2. DNA–DNA Cross-Links[31]

If DNA is UV-irradiated while dry or when it is very tightly packed, as in sperm heads, DNA–DNA cross-links leading to gel formation have been observed. However, this type of reaction (i.e., gel formation) appears to be of little importance to normal wet cells.

A single interchain DNA cross-link causes the two strands of a molecule of DNA to be connected so that they can no longer be separated when the DNA is denatured with heat or formamide. Since no interchain cross-links were detected in normal phage T7 irradiated to a survival of 1%, the biological importance of this lesion seems in doubt at low fluences of UV radiation. However, this lesion may achieve a position of greater biological importance at higher fluences of UV radiation in those cells that are relatively resistant to radiation. The formation of interchain DNA cross-links has been studied *in vitro*.[31]

DNA can also be cross-lined indirectly using a photosensitizer (Chapter 3). The photosensitizer (e.g., 8-methoxypsoralen) intercalates between the stacked bases of the DNA and, when excited by the absorption of light energy, can react with pyrimidines on each strand of the DNA to produce an interchain cross-link. This photochemical reaction may be the basis for the photochemotherapy of psoriasis (Chapter 6).

2.3.2.3. DNA–Protein Cross-Links[27,54,55]

There is a progressive decrease in the amount of DNA that can be extracted free of protein from bacteria and mammalian cells following increasing fluences of UV radiation. The DNA that becomes nonextractable due to UV irradiation can be quantitatively accounted for in the precipitate containing the denatured proteins. Treatment of this precipitate with trypsin, however, yields free DNA. These results suggested that UV irradiation had caused the DNA to become cross-linked to protein. More direct proof came from experiments showing that DNA and protein could be cross-linked *in vitro* by UV irradiation.

The chemical mechanisms by which DNA and protein are cross-linked *in vivo* are still under study; however, the initial isolation of a mixed photoproduct of thymine and cysteine (5-*S*-cysteine-6-hydrothymine) [Eq. (2-1)] from the *in vitro* UV irradiation of a solution of thymine and cysteine has served as a model for the cross-linking phenomenon.

In addition to cysteine, the following amino acids add photochemically to thymine: arginine, lysine, tyrosine, tryptophan, and cystine. The following amino acids add photochemically to uracil: glycine, serine, cysteine, cystine, methionine, lysine, arginine, histidine, tryptophan, phenylalanine, and tyrosine. The other common amino acids were unreactive under the conditions tested.

Fifteen amino acids have been found to react photochemically with DNA. Cysteine, lysine, phenylalanine, tryptophan, and tyrosine were the most reactive; alanine, aspartic acid, glutamic acid, serine, and threonine were unreactive.[56]

The experiments that best demonstrate the biological importance of DNA–protein cross-linking are those using bacterial cells irradiated while frozen. *E. coli* B/r cells showed marked differences in survival after UV irradiation as a function of the temperature at which they were irradiated (Fig. 2-13). When the temperature was reduced from +21 to −79°C, an increase in sensitivity to UV radiation was shown by both a change in shoulder and a change in slope in the survival curves. At −196°C the cells were not as sensitive as at −79°C but were more sensitive than at +21°C, as evidenced by the absence of a shoulder on the survival curve. (N.B.: a shoulder on a survival curve, or a decreased slope on the straight line portion of the curve, may be taken as indirect evidence that cells have the capacity to repair DNA damage; see Section 4.1.2.)

The rate of formation of cyclobutane-type thymine dimers *decreased progressively* as the temperature of the cells during irradiation was reduced from +21 to −79°C, and then to −196°C. Therefore, there is no correlation between the production of thymine dimers and the increased killing of *E. coli* by irradiation at −79 and −196°C. This suggests that cyclobutane-type thymine dimers do not play as significant a role in the events leading to the death of irradiated frozen cells as they appear to play at room temperature. These results provide further evidence that the relative biological importance

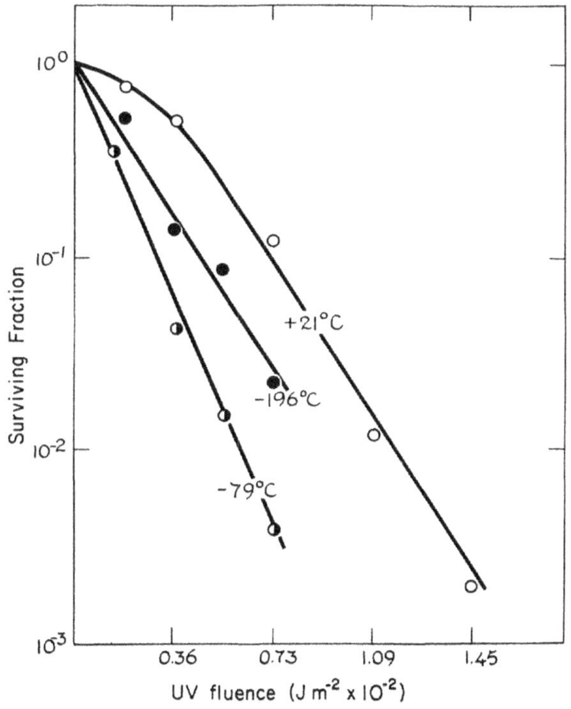

Fig. 2-13. Survival of *E. coli* B/r *thy* as a function of the fluence of UV radiation (254 nm) at different temperatures. [Modified from K. C. Smith and M. O. O'Leary, *Science* **155**, 1024 (1967).]

of a given photoproduct can change markedly, depending on growth or irradiation conditions.

In contrast, however, the amount of DNA cross-linked to protein as a function of temperature during irradiation followed the same relative pattern as the changes in survival. Thus, the photochemical event that appears to correlate with viability when cells are irradiated while frozen is the cross-linking of DNA with protein. Freezing may alter the configuration or the proximity of the protein and the DNA within the cells, so that the probability of forming DNA–protein cross-links by irradiation is greatly enhanced, thus leading to the greater lethality observed under these conditions.

The action spectrum for the killing of *Micrococcus radiodurans,* one of the most radiation-resistant organisms known, differs markedly from that for the more sensitive organism *E. coli* (Fig. 1-23) in that it shows a high sensitivity to irradiation at 280 nm as well as at 260 nm. Clasically, a response at 280 nm has indicated an involvement of protein, while a response at 260 nm has implicated nucleic acid (Section 1.5.5). It has been suggested that the resistance of this organism to UV radiation is due to its extraordinary ability to repair pyrimidine dimers, but what ultimately kills the organism is damage that involves both DNA and protein. The cross-linking of DNA and protein may constitute one type of such damage.

It is reasonable to assume that a different type of photochemistry might arise when protein and DNA are irradiated together as compared to when they are irradiated separately. Since DNA and protein do not exist in cells as pure solutions of the separate molecules but are in intimate contact with each other, it might be expected that the photochemical interaction of DNA and protein would play a significant role in the inactivation of UV-irradiated cells under certain conditions.

Proteins are not the only molecules that bind covalently to DNA due to the actions of nonionizing (i.e., UV and visible) and ionizing (i.e., X and γ) radiation, or through direct chemical attack (for tabulations of these reactions, see Refs. 27 and 54). In fact, an international symposium was held to demonstrate that the seemingly diverse fields of aging, carcinogenesis, and radiation biology are actually closely related, since the formation of DNA adducts appears to be an important initial event in each of these fields, although the types of adducts produced differ in each field.[27]

2.3.3. Photochemistry of RNA

RNA differs from DNA by virtue of having ribose instead of deoxyribose, by having uracil instead of thymine, and by having more single-stranded regions. These differences do affect the photochemistry of RNA (e.g., more pyrimidine hydrates are formed), but in general it is quite similar to that of DNA.

The photochemistry of RNA *in vivo* has not been studied as extensively as the photochemistry of DNA. This is perhaps understandable in view of the central importance of DNA to the survival of cells. Since RNA species are generally present within a cell in multiple copies, the inactivation of a few percent of these molecules would not be expected to have as adverse an effect on the survival of a cell as the photochemical alteration of even a small portion of its DNA.

Nevertheless, under special circumstances the photochemical alteration of messenger RNA and transfer RNA might be expected to have consequences for a cell. For example, in bulk tRNA in *E. coli,* about 70% of the molecules contain 4-thiouridine (S^4U) in place of uridine in position 8 of the tRNA sequence. As the consequence of the substitution of the oxygen atom at position 4 of uridine by a sulfur atom, the absorption maximum is shifted from 260 to 330 nm. Therefore, S^4U absorbs strongly in the near-UV region, whereas most other nucleic acid bases show very little absorption in this region.

The absorption of near-UV radiation by tRNA species that contain S^4U initiates a very specific photochemical reaction that results in the intramolecular cross-linking of S^4U in position 8 to the cytidine present in position 13. Such modified tRNA species show a much reduced ability to be charged with their specific amino acid; the result is a slowing down of protein synthesis.[57,58]

In some situations this slowing down of protein synthesis by near-UV irradiation can be beneficial to a cell. For example, prior near-UV irradiation protects *E. coli* B cells from a subsequent exposure to 254-nm radiation (''photoprotection'').[57,58] Photoprotection has similarities with liquid-holding recovery (Chapter 4), where macromolecular synthesis is inhibited after UV irradiation by holding cells in buffer for awhile before plating. This liquid holding results in an increase in survival in UV-irradiated (254-nm) *E. coli* cells that are deficient in postreplication repair but are proficient in excision repair. Thus, if most DNA lesions can be repaired by excision repair prior to the resumption of DNA replication, then there is less dependence by the cells on postreplication repair, which is deficient in the strains that show photoprotection and liquid-holding recovery.

Some viruses contain RNA as their genetic determinant (e.g., tobacco mosaic virus). For such viruses the photochemical alteration of their RNA leads to rapid inactivation. The biological and chemical effects of UV radiation (254 nm) on RNA have been most extensively studied with RNA viruses.[59]

2.3.4. Detection and Quantitation of Photochemical Lesions in DNA

A laboratory manual of research procedures for the study of DNA damage and repair is available.[60]

Chromatography of various types (e.g., paper, column, thin-layer) has been used to separate damaged purines and pyrimidines that have been released from UV-irradiated DNA (usually specifically labeled with radioactivity) by acid or enzymic hydrolysis.

With the new techniques of molecular biology, one can start with defined sequences of DNA (e.g., phage, plasmids, oligonucleotides), and then sequence the irradiated DNA, and assay for the nature of the damage at defined sites.

Specific antibodies to DNA damage (both conventional antisera and monoclonal antibodies) have been used effectively to detect and quantitate DNA damage.

Enzyme probes that cut the DNA strands at the sites of specific lesions are also very useful. After the DNA has been cut by the enzyme probe, the number of enzymatically induced DNA strand breaks are quantitated.

DNA strand breaks are generally quantitated by measuring the molecular weight of DNA by centrifugation in sucrose gradients. Denaturing (i.e., alkaline) gradients are used for measuring DNA single-strand breaks, and nondenaturing (i.e., neutral) gradients are used for measuring DNA double-strand breaks. For mammalian cells, the techniques of alkaline and neutral elution of DNA from filters is used to follow the formation and repair of DNA single-strand and double-strand breaks, respectively.

DNA–DNA and DNA–protein cross-links can also be measured by techniques similar to those used to assay for DNA strand breaks.

2.4. REFERENCES

1. N. J. Turro, *Modern Molecular Photochemistry,* Benjamin/Cummings, Menlo Park, Cal. (1978).
2. L. I. Grossweiner, Photochemical inactivation of enzymes, *Curr. Topics Radiat. Res. Q.* **11,** 141–199 (1976).
3. A. D. McLaren and D. Shugar, *Photochemistry of Proteins and Nucleic Acids,* Macmillan, New York (1964).
4. D. Creed, The photophysics and photochemistry of the near-uv absorbing amino acids. I. Tryptophan and its simple derivatives, *Photochem. Photobiol.* **39,** 537–562 (1984).
5. D. Creed, The photophysics and photochemistry of the near-uv absorbing amino acids. II. Tyrosine and its simple derivatives, *Photochem. Photobiol.* **39,** 563–575 (1984).
6. D. Creed, The photophysics and photochemistry of the near-uv absorbing amino acids. III. Cystine and its simple derivatives, *Photochem. Photobiol.* **39,** 577–583 (1984).
7. L. I. Grossweiner, G. W. Swenson, and E. F. Zwicker, Photochemical production of the hydrated electron, *Science* **141,** 805–806 (1963).
8. E. J. Hart and J. W. Boag, Absorption spectrum of the hydrated electron in water and aqueous solutions, *J. Am. Chem. Soc.* **84,** 4090–4095 (1962).
9. L. I. Grossweiner, Photochemistry of proteins: A review, *Curr. Eye Res.* **3,** 137–144 (1984).
10. J. Dillon, The anerobic photolysis of tryptophan containing peptides, *Photochem. Photobiol.* **32,** 37–39 (1980).
11. J. Dillon, The anaerobic photolysis of tryptophan containing peptides. II, *Photochem. Photobiol.* **33,** 137–142 (1981).
12. J. Butler, E. J. Lan, W. A. Prutz, and A. J. Swallow, Charge transfer between tryptophan and tyrosine in proteins, *Biochim. Biophys. Acta* **705,** 150–162 (1982).
13. A. Blum and L. I. Grossweiner, Laser flash photolysis and photoinactivation of subtilisin BPN′, *Photochem. Photobiol.* **36,** 617–622 (1982).
14. L. I. Grossweiner, A. G. Kaluskar, and J. F. Baugher, Flash photolysis of enzymes, *Int. J. Radiat. Biol.* **29,** 1–16 (1976).
15. P. Walrant and R. Santus, Ultraviolet and N-formyl-kynurenine-sensitized photoinactivation of bovine carbonic anhydrase: An internal photodynamic effect, *Photochem. Photobiol.* **20,** 455–460 (1974).
16. R. A. Luse and A. D. McLaren, Mechanism of enzyme inactivation by ultraviolet light and the photochemistry of amino acids (at 2537 Å), *Photochem. Photobiol.* **2,** 343–360 (1963).
17. A. D. Mclaren and O. Hidalgo-Salvatierra, Quantum yields for enzyme inactivation and the amino acid composition of proteins, *Photochem. Photobiol.* **3,** 349–352 (1964).
18. R. Setlow, A relation between cystine content and ultraviolet sensitivity of proteins, *Biochim. Biophys. Acta* **16,** 444–445 (1955).
19. L. Augenstein and P. Riley, The inactivation of enzymes by ultraviolet light. IV. The nature and involvement of cystine disruption, *Photochem. Photobiol.* **3,** 353–367 (1964).
20. K. Dose, Theoretical aspects of the u.v. inactivation of proteins containing disulfide bonds, *Photochem. Photobiol.* **6,** 437–443 (1967).
21. L. I. Grossweiner, A. Blum, and A. M. Brendzel, Primary processes in the photochemistry of proteins, in: *Trends in Photobiology* (C. Helene, M. Charlier, and Th. Montenay-Garestier, eds.), pp. 67–80, Plenum Press, New York (1982).
22. T. K. Rathinasamy and L. G. Augenstein, Photochemical yields in ribonuclease and oxidized glutathione irradiated at different wavelengths in the ultraviolet, *Biophys. J.* **8,** 1275–1287 (1968).
23. J. F. Baugher and L. I. Grossweiner, Photolysis mechanism of aqueous tyrosine and tyrosyl peptides, *Photochem. Photobiol.* **28,** 175–184 (1978).
24. S. Y. Wang (ed.), *Photochemistry and Photobiology of Nucleic Acids,* Vols. 1 and 2, Academic Press, New York (1976).

25. A. J. Varghese, Photochemistry of nucleic acids and their constituents, *Photophysiology* **7,** 207–274 (1972).
26. R. O. Rahn, Ultraviolet irradiation of DNA, in: *Concepts in Radiation Cell Biology* (G. L. Whitson, ed.), pp. 1–56, Academic Press, New York (1972).
27. K. C. Smith (ed.), *Aging, Carcinogenesis and Radiation Biology: The Role of Nucleic Acid Addition Reactions,* Plenum Press, New York (1976).
28. A. D. McLaren and D. Shugar, *Photochemistry of Proteins and Nucleic Acids,* Pergamon Press, Oxford (1964).
29. J. G. Burr, Advances in the photochemistry of nucleic acid derivatives, *Adv. Photochem.* **6,** 193–299 (1968).
30. K. C. Smith, Photochemistry of the nucleic acids, *Photophysiology* **2,** 329–388 (1964).
31. K. C. Smith, Physical and chemical changes induced in nucleic acids by ultraviolet light, *Radiat. Res. Supp.* **6,** 54–79 (1966).
32. L. Kittler and G. Löber, Photochemistry of the nucleic acids, in: *Photochemical and Photobiological Reviews,* Vol. 2 (K. C. Smith, ed.), pp. 39–131, Plenum Press, New York (1977).
33. A. Wacker, H. Dellweg, L. Träger, A. Kornhauser, E. Lodemann, G. Türck, R. Selzer, P. Chandra and M. Ishimoto, Organic photochemistry of nucleic acids, *Photochem. Photobiol.* **3,** 369–394 (1964).
34. G. J. Fisher and H. E. Johns, Pyrimidine photohydrates, in: *Photochemistry and Photobiology of Nucleic Acids,* Vol. 1 (S. Y. Wang, ed.), pp. 169–224, Academic Press, New York (1976).
35. G. J. Fisher and H. E. Johns, Pyrimidine photodimers, in: *Photochemistry and Photobiology of Nucleic Acids,* Vol. 1 (S. Y. Wang, ed.), pp. 225–294, Academic Press, New York (1976).
36. M. H. Patrick and R. O. Rahn, Photochemistry of DNA and polynucleotides: Photoproducts, in: *Photochemistry and Photobiology of Nucleic Acids,* Vol. 2 (S. Y. Wang, ed.), pp. 35–95, Academic Press, New York (1976).
37. R. O. Rahn and M. H. Patrick, Photochemistry of DNA; Secondary structure, photosensitization, base substitution and exogenous molecules, in: *Photochemistry and Photobiology of Nucleic Acids,* Vol. 2 (S. Y. Wang, ed.), pp. 97–145, Academic Press, New York (1976).
38. E. C. Friedberg, *DNA Repair,* W. H. Freeman, New York (1985).
39. F. Hutchinson, The lesions produced by ultraviolet light in DNA containing 5-bromouracil. *Q. Rev. Biophys.* **6,** 201–246 (1973).
40. R. O. Rahn, L. C. Landry, and W. L. Carrier, Formation of chain breaks and thymine dimers in DNA upon photosensitization at 313 nm with acetophenone, acetone or benzophenone, *Photochem. Photobiol.* **19,** 75–78 (1974).
41. R. Tyrrell, Damage and repair from non-ionizing radiations, in: *Repairable Lesions in Microorganisms* (A. Hurst and A. Nasim, eds.), pp. 86–124, Academic Press, New York (1984).
42. D. Elad, Photoproducts of purines, in: *Photochemistry and Photobiology of Nucleic Acids,* Vol. 1 (S. Y. Wang, ed.), pp. 357–380, Academic Press, New York (1976).
43. D. Elad, Photochemically-induced adducts of DNA, in: *Aging, Carcinogenesis and Radiation Biology* (K. C. Smith, ed.), pp. 243–260, Plenum Press, New York (1976).
44. J. Cadet, L. Voituriez, A. Grand, F. E. Hruska, P. Vigny and L.-S. Kan, Recent aspects of the photochemistry of nucleic acids and related model compounds, *Biochimie* **67,** 277–292 (1985).
45. F. P. Gasparro and J. R. Fresco, Ultraviolet-induced 8,8-adenine dehydrodimers in oligo- and polynucleotides, *Nucleic Acids Res.* **14,** 4239–4251 (1986).
46. P. E. Gallagher and N. J. Duker, Detection of UV purine photoproducts in a defined sequence of human DNA, *Molec. Cell. Biol.* **6,** 707–709 (1986).
47. G. R. Banks, D. M. Brown, D. G. Streeter, and L. Grossman, Mutagenic analogues of cytosine: RNA polymerase template and substrate studies, *J. Mol. Biol.* **60,** 425–439 (1971).
48. F. Hutchinson, Chemical changes induced in DNA by ionizing radiation, *Prog. Nucleic Acid Res. Mol. Biol.* **32,** 115–154 (1985).
49. L. H. Breimer and T. Lindahl, Thymine lesions produced by ionizing radiation in double-stranded DNA, *Biochemistry* **24,** 4018–4022 (1985).
50. M. G. Simic and S. V. Jovanovic, Free radical mechanisms of DNA base damage, in: *Mechanisms of DNA Damage and Repair* (M. G. Simic, L. Grossman, and A. C. Upton, eds.), pp. 39–49, Plenum Press, New York (1986).

51. S. Y. Wang, Pyrimidine bimolecular photoproducts, in: *Photochemistry and Photobiology of Nucleic Acids,* Vol. 1 (S. Y. Wang, ed.), pp. 295–356, Academic Press, New York (1976).
52. C. Janion and D. Shugar, Reaction of amines with dihydrocytosine analogues and formation of amino acid and peptidyl derivatives of dihydropyrimidines, *Acta Biochim. Polon.* **14,** 293–302 (1967).
53. W. A. Franklin, K. M. Lo, and W. A. Haseltine, Alkaline lability of fluorescent photoproducts produced in ultraviolet light-irradiated DNA, *J. Biol. Chem.* **257,** 13535–13543 (1982).
54. K. C. Smith, the radiation-induced addition of proteins and other molecules to nucleic acids, in: *Photochemistry and Photobiology of Nucleic Acids,* Vol. 2 (S. Y. Wang, ed.), pp. 187–218, Academic Press, New York (1976).
55. M. D. Shetlar, Cross-linking of proteins to nucleic acids by ultraviolet light, in: *Photochemical and Photobiological Reviews,* Vol. 5 (K. C. Smith, ed.), pp. 105–197, Plenum Press, New York (1980).
56. M. D. Shetlar, J. Christensen, and K. Hom, Photochemical addition of amino acids and peptides to DNA, *Photochem. Photobiol.* **39,** 125–133 (1985).
57. A. Favre, E. Hajnsdorf, K. Thiam, and A. Caldeira de Araujo, Mutagenesis and growth delay induced in *Escherichia coli* by near-ultraviolet radiations, *Biochimie* **67,** 335–342 (1985).
58. J. Jagger, *Solar-UV Actions on Living Cells,* Praeger, New York (1985).
59. M. P. Gordon, C. Huang, and J. Hurter, Photochemistry and photobiology of ribonucleic acids, ribonucleoproteins, and RNA viruses, in: *Photochemistry and Photobiology of Nucleic Acids,* Vol. 2 (S. Y. Wang, ed.), pp. 266–308, Academic Press, New York (1976).
60. E. C. Friedberg and P. C. Hanawalt (eds.), *DNA Repair: A Laboratory Manual of Research Procedures,* Marcel Dekker, New York, Vol. 1, A and B (1981), Vol. 2 (1983), Vol. 3 (1988).

3

Photosensitization

3.1. INTRODUCTION

Several types of photochemical processes occur in living organisms that are important or essential for their survival. In all of these reactions, the initial step is the absorption of photons by specialized pigments, such as chlorophyll, in the photosynthetic processes of

John D. Spikes • Department of Biology, University of Utah, Salt Lake City, Utah 84112.

plants, and carotenoid–protein complexes in animal vision. In addition to these normal photobiological processes, cells and organisms can be injured and killed by light. This occurs only if the radiation is absorbed (First Law of Photochemistry). UV-C radiation (defined in Section 1.3) and, to a lesser extent, UV-B radiation are very injurious since critical cellular components, including proteins and nucleic acids, absorb radiation very strongly in these ranges (see Chapter 2). UV-A and visible radiation typically do not cause direct damage to biological systems, since most types of biologically important molecules absorb only weakly at these wavelengths. However, in the presence of certain kinds of light-absorbing molecules termed photosensitizers, cells and organisms can be rapidly injured or killed by UV-A or visible radiation. This results from the sensitized photoalteration of critical biomolecules. Current research on photosensitized phenomena in biology includes studies of the photophysics and photochemistry of photosensitizer molecules, the mechanisms of the sensitized photodegradation of essential biomolecules, the biochemistry and physiology of photosensitized viral inactivation and cell killing, the role of endogenous sensitizers in the protection of plants against disease organisms and herbivores, the involvement of photosensitizers in plant-herbivore coevolution, and the application of sensitized reactions in practical ways such as the photosensitized treatment of diseases in humans.[1–16]

Although photosensitized effects on organisms were observed much earlier, it is generally agreed that this area of photobiology started with research performed at the end of the last century by Oscar Raab, a medical student in Munich. Raab found that very low concentrations of certain colored compounds (usually termed dyes) such as acridines and eosin, which had no effect in the dark, led to the rapid killing of the protozoan, paramecium, in the light; only those wavelengths of light absorbed by the sensitizer were effective.[1] This discovery stimulated a large amount of research on the phenomenon that was very early termed "photodynamic action" to distinguish it from simpler physical phenomena such as the photosensitization of photographic plates by dyes. Within a few years, it was shown that enzymes could be inactivated and that various types of cells as well as small animals could be killed by illumination in the presence of a number of different photosensitizers.[1] As described in the next section, a molecule must be able to do more than just absorb light to be a photosensitizer; it must, in general, be capable of being excited by light into a relatively long-lived, energy-rich form termed the triplet state (Section 1.4). It is estimated that thousands of different natural and synthetic molecules can function as photosensitizers. With most photosensitizers, biomolecules are degraded in processes that involve the uptake of oxygen; these reactions do not take place under anaerobic conditions. Many investigators confine the use of the term photodynamic action to those reactions that require oxygen.[1] There are, however, several well-studied photosensitized reactions that do not require oxygen, as will be described later.[7,17]

Research on photosensitized reactions in biology has increased sharply in recent years, in part because of the rapidly expanding use of a large variety of chemicals in agriculture, industry, medicine, and the home that sensitize humans to injury by light. Also, there is considerable interest in photochemotherapy, the use of photosensitizing drugs plus light to treat disease, as in the utilization of psoralens plus UV-A radiation for the treatment of psoriasis (PUVA), and in the use of porphyrins in the photodynamic therapy of tumors (PDT) (see Chapter 6).

3.2. PHOTOPHYSICS AND PHOTOCHEMISTRY OF PHOTOSENSITIZED REACTIONS

In photosensitized reactions, photons are absorbed by one type of molecule (the sensitizer); the resulting energy-rich state(s) of the sensitizer then undergoes reactions that ultimately result in the chemical alteration of another molecule in the system (the substrate). Photosensitized reactions are typically mediated by excited electronic states of the sensitizer. In its primary reaction, the excited sensitizer molecule can react directly with the substrate, or, alternatively, with some other molecule (frequently oxygen) in the reaction mixture, giving products that in turn can react with the substrate. Thus, as with photochemical reactions in general, photosensitized processes typically have an initial "light step" followed by one or more "dark steps" (see Chapter 2). Many photosensitized reactions are complex, and may involve several competing reaction pathways.

3.2.1. Excited States of Sensitizer Molecules[2–4,16]

Organic photosensitizer molecules in the dark (ground state = ^{0}SENS) are almost always in the singlet state; in this state the molecule has no unpaired electron spins. On the absorption of a photon, an electron in the sensitizer is promoted to a higher molecular orbital without a change in its spin; thus, the first excited state, like the ground state, is a singlet, ^{1}SENS. Very few photosensitized reactions are directly mediated by this state because of its short lifetime (~1–100 ns). Excited singlet sensitizer states can decay directly to the ground state, emitting heat or light. However, the singlet excited state of an effective photosensitizer undergoes a fast spin inversion to give a metastable triplet state, ^{3}SENS, which has two unpaired electrons (see Chapter 1 and Fig. 1-16). The triplet states of sensitizer molecules typically have much longer lifetimes than singlet states (microseconds to seconds). Thus, triplet state sensitizers will undergo large numbers of collisions with other molecules during their lifetimes and, as a result, can mediate photosensitized reactions with high efficiencies. These various processes are shown schematically below. The quantum yield of triplet formation for good photosensitizers, such as metal-free porphyrins, can be as great as 0.9.

$$^0\text{SENS} \xrightarrow{\text{LIGHT}} {}^1\text{SENS} \qquad (3\text{-}1)$$

$$^1\text{SENS} \longrightarrow {}^0\text{SENS} \qquad (3\text{-}2)$$

$$^1\text{SENS} \longrightarrow {}^3\text{SENS} \qquad (3\text{-}3)$$

$$^3\text{SENS} \longrightarrow \text{photosensitized reactions} \qquad (3\text{-}4)$$

In most photosensitized reactions, the triplet sensitizer ultimately returns to the ground state and can absorb another photon, thus acting in a somewhat catalytic fashion. In a few cases, the sensitizer is consumed stoichiometrically in the reaction. The triplet sensitizer can undergo its primary reaction with molecules in its vicinity by an electron (or hydrogen atom) transfer process (termed a type I reaction), or by an energy transfer reaction with oxygen (termed a type II reaction). In a given system, both types of processes can take place simultaneously in a competitive fashion. These processes will be described in the next two sections. Triplet sensitizers can also decay directly to the ground state, or be converted to the ground state by collisional interactions with certain types of molecules (physical quenchers, Q).

$$^3\text{SENS} \rightarrow {}^0\text{SENS} \tag{3-5}$$

$$^3\text{SENS} + \text{Q} \rightarrow {}^0\text{SENS} + \text{Q} \tag{3-6}$$

3.2.2. Photosensitization by Type I (Radical) Mechanisms

In general, a molecule in its triplet state can both abstract electrons (or H atoms) from and donate electrons (or H atoms) to other molecules more readily than when it is in the ground state. The efficiencies of these reactions depend on the chemical structures of the sensitizer and the substrate, as well as on the reaction conditions. The abstraction of an electron from a reducing substrate, SUB (the most common type of substrate in biological systems), by a sensitizer triplet gives semireduced sensitizer, $\text{SENS}^{\cdot-}$ or SENSH·, and a semioxidized form of the substrate, $\text{SUB}^{\cdot+}$ or SUB·, as follows:

$$^3\text{SENS} + \text{SUB} \rightarrow \text{SENS}^{\cdot-} + \text{SUB}^{\cdot+} \tag{3-7a}$$

$$^3\text{SENS} + \text{SUB} \rightarrow \text{SENSH}\cdot + \text{SUB}\cdot \tag{3-7b}$$

These free radical products (which have odd numbers of electrons) are very reactive chemically. In many cases, the resultant substrate radicals react with oxygen to give oxidized products of various types; often these are peroxides, which can react further to initiate free radical chain-type autooxidation processes. In many systems, a ground state sensitizer can be regenerated by the reaction of the semireduced form with ground state oxygen, $^3O_2$, to give a ground state sensitizer molecule and the superoxide radical anion, $O_2^{\cdot-}$ (or its conjugate acid, $HO_2^{\cdot}$),

$$\text{SENS}^{\cdot-} + {}^3O_2 \rightarrow {}^0\text{SENS} + O_2^{\cdot-} \tag{3-8a}$$

$$\text{SENSH}\cdot + {}^3O_2 \rightarrow {}^0\text{SENS} + HO_2^{\cdot} \tag{3-8b}$$

which in turn can react with some kinds of biomolecules.[18,19] Some semireduced sensitizers, such as flavins, undergo dismutation reactions producing one ground state and one fully reduced sensitizer molecule. This latter species may react with oxygen to give ground state sensitizer and hydrogen peroxide.

$$\text{SENSH}\cdot + \text{SENSH}\cdot \rightarrow {}^0\text{SENS} + \text{SENSH}_2 \tag{3-9}$$

$$\text{SENSH}_2 + {}^3O_2 \rightarrow {}^0\text{SENS} + H_2O_2 \tag{3-10}$$

Superoxide can also be produced with low efficiency by the transfer of an electron from triplet sensitizer to oxygen[18]:

$$^3\text{SENS} + {}^3O_2 \rightarrow \text{SENS}^{\cdot+} + O_2^{\cdot-} \tag{3-11}$$

The resulting semioxidized sensitizer, $SENS^{\cdot+}$, can react with a reducing substrate to give ground state sensitizer and semioxidized substrate,

$$SENS^{\cdot+} + SUB \rightarrow {}^0SENS + SUB^{\cdot+} \qquad (3\text{-}12)$$

which can react further as described above. Finally, in some cases, sensitizer and substrate radicals can interact to produce covalent sensitizer-substrate and substrate-substrate photoadducts:

$$SENS^{\cdot-} + SUB^{\cdot+} \rightarrow SENS\text{-}SUB \qquad (3\text{-}13)$$

$$SUB\cdot + SUB\cdot \rightarrow SUB\text{-}SUB \qquad (3\text{-}14)$$

Type I photosensitized processes are most efficient at high substrate and low oxygen concentrations, since oxygen effectively competes with substrate for interaction with triplet sensitizer. The formation of noncovalent sensitizer–substrate complexes before illumination would be expected to increase the probability of type I reactions because of the proximity of the sensitizer to the substrate molecules. Substrates most effectively photodegraded in type I reactions include those that are readily oxidized (such as amines and phenols) or reduced (such as quinones). An example of a type I reaction is the flavin-sensitized photodegradation of the amino acid methionine. On illumination, triplet flavin is generated, which abstracts one electron from the sulfur atom of the amino acid; the resulting methionine radical undergoes deamination and decarboxylation to give methional (β-methylmercaptopropionaldehyde). The flavin radical then undergoes reactions, as in Eqs. (3-9) and (3-10), resulting in the formation of ground state flavin and hydrogen peroxide.

3.2.3. Photosensitization by the Type II (Singlet Oxygen) Mechanism[2–4,20–22]

In the type II photosensitized reaction, triplet sensitizer interacts by energy transfer with ground state oxygen to give ground state sensitizer and an electronically excited singlet state of oxygen, $^1O_2$, as follows:

$$^3SENS + {}^3O_2 \rightarrow {}^0SENS + {}^1O_2 \qquad (3\text{-}15)$$

Since both the sensitizer and ground state oxygen are triplets (ground state oxygen is one of only a very few stable triplet molecules), their interaction does not require a change in spin direction and is very efficient. Oxygen can exist in two singlet excited states; the longer lived form, $^1\Delta_g$, with an excess energy of 23 kcal/mole (corresponding to a photon energy of approximately 1 eV = 1270 nm radiation) is the principal species involved in photodynamic reactions. It has a lifetime of approximately 4 μs in ordinary water and thus rapidly decays to the ground state if it does not react chemically with another molecule. Its lifetime is much longer in lipids and other nonpolar media.[20]

Ground state oxygen (a triplet) does not react very rapidly with most types of biomolecules under ordinary conditions, since organic molecules are typically in the

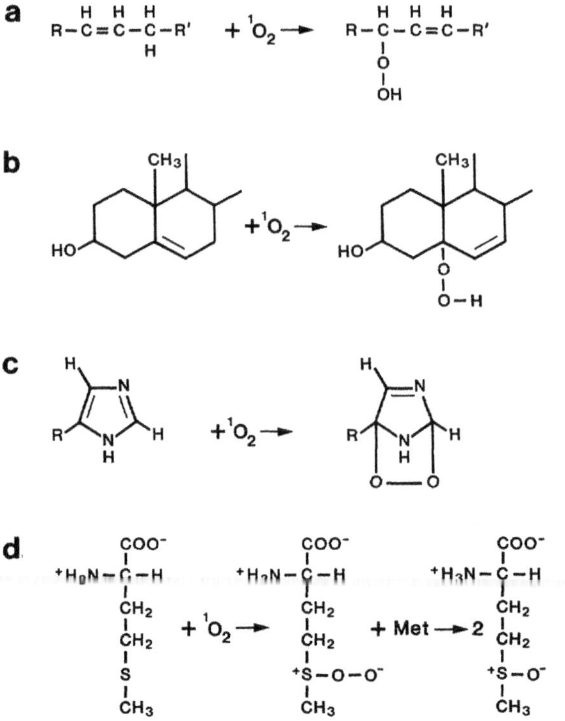

Fig. 3-1. Some of the major pathways for the interaction of singlet oxygen with biomolecules. (a) Shows the addition of singlet oxygen to an olefin with allylic hydrogen atoms (such as an unsaturated fatty acid) by an "ene"-type reaction to give the corresponding allylic hydroperoxide. (b) Shows a similar reaction in which singlet oxygen adds to cholesterol to give the 5-α-hydroperoxide, a specific singlet oxygen product. (c) Shows the addition of singlet oxygen to the diene system of a heterocyclic compound (an imidazole used as a model for histidine) to give an endoperoxide. (d) Shows the addition of singlet oxygen to an organic sulfide, the amino acid methionine (Met), to give an unstable product which in turn reacts with another molecule of methionine to give two molecules of methionine sulfoxide.

singlet state and thus the interaction with oxygen is spin-forbidden. The reaction of singlet oxygen with organic compounds, however, is not spin-forbidden. Furthermore, singlet oxygen is more electrophilic than ground state oxygen and therefore can react rapidly, albeit somewhat selectively, with electron-rich regions of many biomolecules to give oxidized species. Some of the major reaction pathways are (1) the addition of singlet oxygen to olefins with allylic hydrogen atoms (such as unsaturated fatty acids and cholesterol) to give hydroperoxides, (2) the addition of singlet oxygen to diene systems in heterocyclics (such as histidine) to give endoperoxides, and (3) the reaction of singlet oxygen with organic sulfides (such as methionine) to form the corresponding sulfoxides (see Fig. 3-1). Although the overall reaction of singlet oxygen with biomolecules sometimes appears to be simple, the process often involves several intermediate steps.

3.2.4. Anaerobic Photosensitized Reactions[7,23–27]

As mentioned earlier, a few types of light-absorbing molecules sensitize the photoalteration of biomolecules in the absence of oxygen; in some cases the reaction is even inhibited by oxygen. For example, some psoralens (furanocoumarins) intercalate between the base pairs of DNA and RNA. On subsequent irradiation under anaerobic conditions with wavelengths absorbed by the psoralen (usually UV-A), covalent photoadducts are formed between the psoralen and pyrimidine residues in the nucleic acid. These can be monoadducts (psoralen-base) or diadducts (base-psoralen-base) or both, depending on the structure of the psoralen and the illumination regime. These reactions have been studied in detail because of the extensive use of psoralens as sensitizers in the photochemotherapy of psoriasis, a disease of the skin in humans (see Chapter 6). Photochemical studies with psoralens are complicated somewhat by the fact that the illumination of these sensitizers in solution under aerobic conditions can result in the generation of singlet oxygen and the resulting oxidation of proteins, unsaturated lipids, etc. Psoralens are not the only type of anaerobic photosensitizer. For example, red blood cells are hemolyzed on illumination under anaerobic conditions in the presence of chlorpromazine (a drug used in the treatment of patients with certain mental disorders). The red blood cells are apparently lysed chemically by stable, detergentlike photoproducts resulting from the illumination of chlorpromazine.

3.2.5. Diagnostic Techniques for Determining the Mechanisms of Photosensitized Reactions[5,21,28–31]

As described above, the sensitized photoalteration of biomolecules may involve one or more of a large number of reactive species such as photoexcited states of molecules, free radicals, and reactive forms of oxygen including singlet oxygen, superoxide, hydrogen peroxide, and possibly the hydroxyl radical. Even in the simplest case of homogeneous solutions, the unambiguous establishment of reaction mechanism(s) is not easy, and in the complex environment of a living cell it becomes very difficult.

Most of the diagnostic approaches that have been used are designed to determine whether singlet oxygen is produced in the reaction and whether this species is responsible for the subsequent damage to the biological system. One technique is to examine the effects of "inhibitors" (that efficiently react with or physically quench singlet oxygen) on the rate of the photosensitized process under study. However, the use of inhibitors alone rarely gives unambiguous results. For example, if a photosensitized reaction is inhibited by low concentrations of azide (a very efficient quencher of singlet oxygen), this is often interpreted as indicating that a singlet oxygen-mediated process is involved. However, for such measurements to be useful, it must be established (as by flash techniques; see Chapter 1 and Fig. 1-7) that the inhibitor does not quench the triplet state of the sensitizer directly, since this would also decrease the reaction rate. Another kinetic technique is to compare the rate of the sensitized reaction in ordinary water (H_2O) and in heavy water (D_2O). Since the decay lifetime of singlet oxygen is over 10 times greater in D_2O than in H_2O, a significant increase of the reaction rate in D_2O suggests the involvement of singlet oxygen. Again, this technique must be used with care since it is valid only under reaction conditions where singlet oxygen deactivation by the solvent is the main rate-limiting factor. The photooxidation of "trap" compounds, such as furans and histidine, that are efficiently oxidized by singlet oxygen is sometimes taken as evidence for the production of this species in a reaction. However, many such compounds can also be oxidized to the same products by free radical processes. Cholesterol is one of the few unambiguous traps, since singlet oxygen reacts with it to form a single product, the 5-α-hydroperoxide [see Chapter 2, Eq. (2-4)], whereas interaction with radicals gives a mixture of other products. A rather new technique for the detection of singlet oxygen involves the measurement of light emission at 1270 nm, since one path of singlet oxygen decay is as follows:

$$^1O_2\ (^1\Delta_g) \rightarrow {}^3O_2 + \text{light (1270 nm)} \qquad (3\text{-}16)$$

This luminescence emission can be measured with present instrumentation, which permits the unambiguous determination of the relative yields of singlet oxygen by illuminated photosensitizers; the technique also permits measuring the rates of reaction of substrates and inhibitors with singlet oxygen. Finally, the technique of flash photolysis can be used to determine the yields and absorption spectra of triplet states of photosensitizers as well as the reaction rates of the triplets with oxygen, inhibitors, quenchers, and substrates; this technique has been of very great value in determining the mechanism(s) involved in many photosensitized processes.

3.2.6. Chemistry of Photosensitizing Compounds

Hundreds of different organic, metalloorganic, and inorganic compounds have been shown to be photosensitizers for biological systems, and there must be many, many more.(5–7,15,32,33) A large number of sensitizers occur naturally in organisms, especially in plants. These include certain alkaloids (berberine, harmaline, etc.), benzofurans, chlorophylls, flavins (riboflavin and FMN, but not FAD), polyacetylenes, a number of polycyclic polyhydroxy-quinones such as hypericin, protoporphyrin, psoralens (furanocoumarins), pyridoxal, quinones, thiophenes, etc.(33) An enormous number of synthetic compounds are photosensitizers, including acridines (acridine orange, proflavine, etc.), anthraquinones, azine dyes such as the safranines, haloaromatics, ketones, phenothiazines (methylene blue, chlorpromazine, etc.), phthalocyanines, many kinds of porphyrins, xanthenes (including eosin, rose bengal, etc.), and so on.(5–7) Inorganic photosensitizers include cadmium sulfide, cupric ions, ferric ions, ruthenium trisbipyridyl, titanium dioxide, and uranium compounds.(5–7) Many categories of light-absorbing compounds, however, are not effective photosensitizers. These include anthocyanins, azo dyes, indophenols, iron-containing porphyrins, nitro dyes, nitroso dyes, oxazine dyes, and triarylmethane dyes; presumably these compounds are not useful photosensitizers because of their low triplet yields or short triplet lifetimes. In some applications, as in the textile industry, only nonphotosensitizing dyes are used.(34) The structures of several photosensitizers commonly used in photobiological studies are shown in Fig. 3-2. All of these compounds sensitize via the type II mechanism. In addition, some of the compounds sensitize via type I mechanisms, depending on the reaction conditions. For example, with alcohol as a solvent, flavins sensitize the photooxidation of methionine to methionine sulfoxide by a singlet oxygen-mediated process; in contrast, a free radical process predominates in aqueous solvents leading to methional, as described above. The structures of several photosensitizing and nonphotosensitizing psoralens are given in Fig. 3-3.

In summary, a good photosensitizer must absorb photons efficiently (i.e., it must have a high molar extinction coefficient), it must have a high quantum yield of triplet formation, and the triplet must be long-lived. Efficient type I sensitizers must be able to donate and/or accept electrons easily, while good type II sensitizers must generate singlet oxygen with high yields. The system in which a sensitizer is used can have a large effect on its efficiency. For example, basic dyes are generally much better photosensitizers for nucleic acids than are acid dyes because they associate more closely with the nucleic acid by virtue of their charge. Some sensitizers that are very effective for substrate molecules in solution are ineffective with cells because they do not penetrate into the cell. Appropriate dark control experiments must always be done in photosensitization studies, since certain sensitizers may react directly with substrates and cells in the dark.

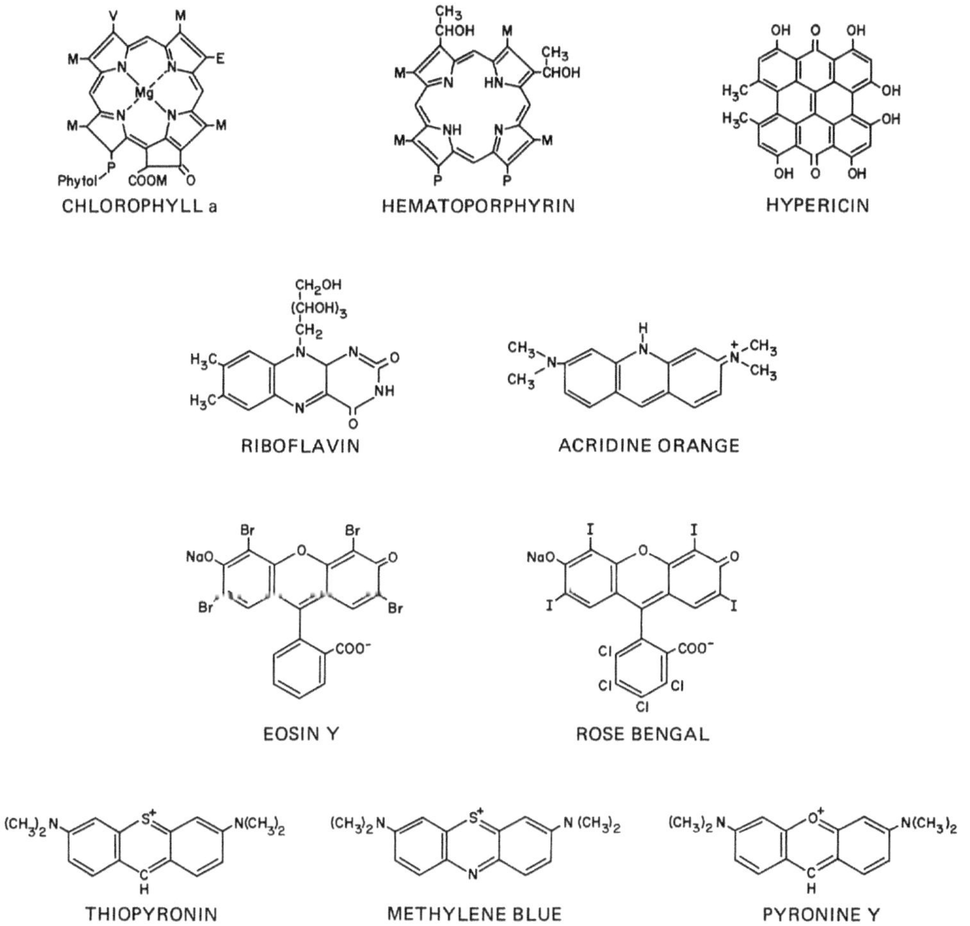

Fig. 3-2. Structures of some typical photosensitizers. E, M, P, and V in the structural formulas of chlorophyll *a* and hematoporphyrin stand for ethyl, methyl, propionic acid, and vinyl groups, respectively. Hypericin is a polycyclic polyhydroxyquinone. A number of effective sensitizers are based on fluorescein, including eosin Y (tetrabromofluorescein) and rose bengal (tetrachlorotetraiodofluorescein). Although the three compounds shown in the bottom row have rather similar structures, they differ in photosensitizing capability with thiopyronine being most effective, methylene blue less so, and pyronine Y the least effective.

A few types of cells and organisms are inherently photosensitive, since they contain naturally occurring sensitizers, as will be described later. In general, however, evolutionary pressures have ensured that most living material is well designed to minimize both direct and sensitized photodamage. For example, chlorophyll is an efficient photosensitizer, yet photosynthetic organisms are ordinarily highly resistant to chlorophyll-sensitized photodamage. In part, this results from the structural organization of photosynthetic organelles that assures the very rapid transfer of chlorophyll singlet excitation energy to regions where the electron transfer reactions of photosynthesis start. Thus, little triplet chlorophyll is generated. In addition, the organelles contain high concentrations of

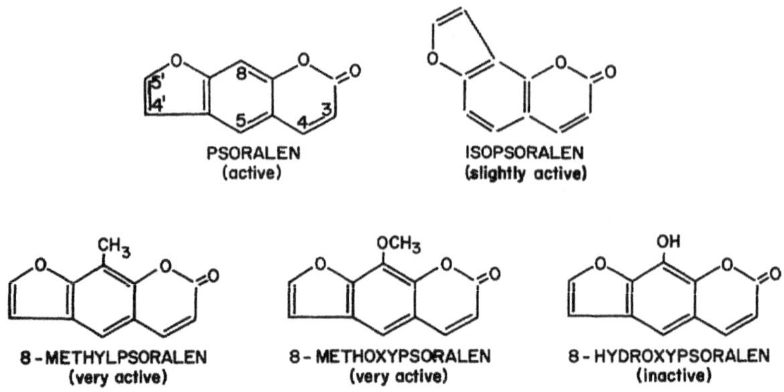

Fig. 3-3. Structures of several photochemically active and inactive psoralens. It will be noted that substitution of a methyl or methoxy group in the 8 position of psoralen increases photosensitizing activity, whereas substitution of a hydroxy group eliminates activity. A similar pattern is observed with substituents in the 5 position. Isopsoralen is also known as angelicin, while 8-methoxypsoralen is also termed methoxsalen or xanthotoxin.

carotenoids that quench both triplet chlorophyll and singlet oxygen (as well as some types of free radicals) efficiently.[35] Many other kinds of endogenous photosensitizing compounds are present in some species of plants; however, the mechanisms by which plants are protected against photodynamic damage by these compounds have not been determined.[33] Porphyrins in living organisms, such as the heme (ferric protoporphyrin) of hemoglobin and most other hemoproteins, contain a coordinated paramagnetic ferric ion; this metal ion shortens the lifetime of the porphyrin excited state to such an extent that it is no longer an effective photosensitizer.[15]

3.3. SENSITIZED PHOTOEFFECTS ON BIOMOLECULES *IN VITRO*[2,3,5]

A number of types of organic molecules characteristic of living material are susceptible to photosensitized alteration, including examples from the following categories: acids, alcohols, amines, carbohydrates, esters, growth factors, ketones, nitrogen heterocyclics, nucleic acids (and certain of their component bases), unsaturated lipids, phenols, proteins (and certain of their component amino acids), pyrroles, steroids, many vitamins, etc. The sensitivity of a particular molecule depends on its structure, on the properties of the photosensitizer, and on the reaction conditions (in particular the pH, the reactant concentrations, and the nature of the solvent). In most cases the photosensitized reactions of biomolecules turn out to be rather complex. Often the primary product will be unstable and decay very rapidly to secondary products, which may be susceptible to further sensitized photochemical changes. Reactions of a number of biologically important molecules as studied in solution are described in the next sections.

3.3.1. Lipids[2,5]

Molecules in this category that are susceptible to photosensitized oxidation include unsaturated fatty acids, unsaturated neutral fats, unsaturated phospholipids, and steroids.

A large amount of research has been done with these compounds because of their importance in the photodynamic damage to cell membranes and their involvement in the photosensitized degradation of some types of foods.[36,37] Unsaturated fatty acids can be photooxidized by both type I and type II processes. Allylic hydroperoxides are initially formed in both mechanisms, although the product isomer distributions and the reaction kinetics are different in the two cases. The type I processes are complex, giving allylic free radicals, which can react in various ways to give different hydroperoxides and, often, other products such as alcohols, epoxides, and ketones. The main products of the sensitized photooxidation of unsaturated triglycerides and phospholipids are hydroperoxides; however, dark autooxidation of the products can result in the formation of more complex mixtures of final products. In type I reactions, cholesterol is converted to a complex mixture of products including the 7-α- and 7-β-hydroperoxides. As mentioned above, the type II reaction is simpler, yielding the 5-α-hydroperoxide, with only traces of other products.

3.3.2. Purines, Pyrimidines, and Nucleic Acids

Although the photosensitized reactions of nucleic acids and their component bases have been studied extensively, these processes are still not well understood. Purines are photooxidized more readily than pyrimidines with most sensitizers; guanine and its derivatives are the most sensitive under physiological conditions. Typically, the rates of photooxidation increase with increasing pH. With guanine derivatives, the rate increases rapidly above pH 7, attaining a maximum at pH 10.5. The pH curve has an inflection at pH 9.2, corresponding to the pK for the dissociation of a proton from the N1 atom of the purine. Thus, the guanine anion is much more sensitive to photooxidation than is the unionized base. The sensitized photooxidation of purines and pyrimidines results in a complex array of reaction products. In the photooxidation of guanine and its derivatives, both rings are ruptured and a number of secondary products are produced, including guanidine, parabanic acid, urea, carbon dioxide, etc. It has been suggested that endoperoxides, hydroperoxides, and dioxetanes could be among the early photooxidation products. Different photosensitizers give different patterns of oxidation products suggesting that both type I and type II reaction pathways are involved. The mechanism of the flavin-sensitized photooxidation of purines appears to be somewhat different from that observed with other sensitizers. For example, adenine and its derivatives, which are photooxidized only very slowly with eosin and methylene blue, are photooxidized more rapidly with flavins. In the case of pyrimidines, the primary product of the photooxidation of uracil, as observed at very low temperatures, appears to be a hydroperoxide. Certain uracil derivatives, such as the 5-amino- and 5-hydroxyuracils, are much more rapidly photooxidized than the parent compound.[2,5,38]

The photodynamic treatment of nucleic acids in solution can significantly alter the chemical, physical, and biological properties of these compounds. The type of sensitizer used may have an effect on the reaction mechanism. For example, the majority of the most effective sensitizers for nucleic acids (acridines, methylene blue, thiopyronine, etc.) are positively charged and bind to the nucleic acids. This close association may increase the probability of type I reactions; the photogeneration of free radicals in the nucleic acid moiety of DNA–sensitizer complexes has been reported. In contrast, certain porphyrins sensitize the photooxidation of nucleic acids in singlet-oxygen-mediated reactions. Under physiological conditions, guanine residues in nucleic acids and synthetic polynucleotides in solutions are more rapidly photooxidized than the other bases. The destruction of guanine residues produces labile glycosylic bonds in the nucleic acid chain, which can result in single- and double-strand breaks. Other physical-chemical changes in the nucleic acids include a decrease in solution viscosity, a decrease in the melting temperature, spectral shifts, conformational alterations, and an increase in susceptibility to enzymatic degradation. Changes in the biological properties are also observed. For example, the infectivity of tobacco mosaic virus RNA as well as the transforming capability of bacterial DNA transforming principle is lost. Template, messenger, and translational activities of nucleic acids are altered as well as their antigenic properties.[5,38,39]

On exposure to light, psoralens react with nucleic acid bases in solution, especially the pyrimidines. Either the 3,4 or the 4′,5′ carbon–carbon double bond of the photoexcited psoralen adds across the 5,6 carbon–carbon double bond of the pyrimidine to give a covalent cycloadduct. In studies using 365-nm radiation, psoralens can form 4′,5′ monoadducts with the pyrimidine bases in DNA; certain of these adducts can absorb another photon and react further giving a 3,4 adduct with a pyrimidine in the complementary strand of the DNA, thus cross-linking the two strands.[38,40,41] Recently, monoclonal antibodies have been produced that specifically recognize DNA molecules modified by illumination with UV-A radiation in the presence of 8-methoxypsoralen, one photoadduct per 10^7 base pairs can be detected.[42] Some drugs, such as chlorpromazine, form covalent photoadducts with DNA.[27]

3.3.3. Amino Acids and Proteins[2,5,38]

Only five of the amino acids that occur in typical proteins are rapidly photooxidized in sensitized reactions: cysteine, histidine, methionine, tryptophan, and tyrosine, all of which have electron-rich side chains. The kinetics and mechanisms of these reactions depend on the amino acid, the sensitizer, the solvent, the pH, and the other reaction conditions. With most sensitizers, cysteine is photooxidized largely to the disulfide, cystine. With crystal violet, however, the main product is cysteic acid. The rate of photooxidation increases with increasing pH in a fashion indicating that the thiol group must be unprotonated for the reaction to occur. Histidine is efficiently photooxidized, typically by a type II process. The primary product is probably an endoperoxide; however, this is very unstable and decomposes rapidly with the formation of a variety of final products. There is a marked increase in the rate with increasing pH, showing that the unprotonated imidazole ring of the molecule is the reactive site. Methionine is photooxidized by both type I and type II processes, depending on the sensitizer, the solvent, and the reaction conditions; pH has only a small effect on the rate. The product in type I reactions is methional, while the sulfoxide is produced by type II reactions, as described above. A number of products result from the photooxidation of tryptophan; both free radical and singlet oxygen pathways appear to be involved, and the array of reaction products resulting can be complex. One well-established stable product is *N*-formylkynurenine, which itself is an efficient photosensitizer. The detailed mechanisms of tyrosine photooxidation are not known, but both type I and type II processes are probably involved. The reaction products are not well established, although rupture of the tyrosine ring occurs. The reaction rate increases at high pH indicating that the phenolate anion is the most reactive form of the molecule.[2,5]

Almost all proteins are rapidly photooxidized on illumination in the presence of appropriate photosensitizers and oxygen. All classes of enzymes have been studied as well as blood plasma proteins, hemoglobin, ovalbumin, bacterial toxins, protein hormones, the structural protein collagen, etc. In most cases, the sites of damage are cysteinyl, histidyl, methionyl, tryptophyl, and tyrosyl residues in the macromolecule. Photooxidizable residues exposed at the surface of the protein molecule are degraded more rapidly than residues located in the interior of the molecule. Photosensitized reactions can be used to probe the three-dimensional structures of protein molecules by using sensitizers or protective agents that bind at specific sites on the protein. The pattern of amino acid residue photooxidation depends on the protein, the sensitizer, and the reaction conditions. For example, histidyl residues are not photooxidized at low pH but are very sensitive at high pH, as is the case for the free amino acid. Marked changes in the physical-chemical properties of photodynamically treated proteins are often observed. These include spectral changes, increased susceptibility to proteolytic enzymes and to denaturation by urea, and changes in conformation, electrophoretic mobility, extent of cross-linking, solubility, ability to bind cofactors, etc.[5]

In most cases, the normal biological function of a protein is changed or destroyed by sensitized photooxidation. With the exception of horseradish peroxidase and superoxide dismutase, all enzymes are rapidly inactivated; in a very few cases, enzymes show an initial increase in activity on photooxidation, followed by inactivation. Peptide hormones such as glucagon, insulin, etc., lose their biological function, and protein toxins from bacteria, plants, and snake venoms are inactivated. Proteins lose their antigenicity and ability to react with specific antibodies. Some photosensitizers form covalent photoadducts with proteins, e.g., riboflavin photobinds to the enzyme lysozyme. In some cases, covalent protein–protein cross-links are formed on illumination of proteins in the presence of photodynamic sensitizers. For example, illumination of hemoglobin subunits in the presence of methylene blue produces covalent dimers and higher aggregates. Sensitized cross-linking of proteins to nucleic acids and various other molecules can occur. Some proteins contain chromophores which act as photosensitizers. For example, bacterial glutamate decarboxylases, which require pyridoxal-5′-phosphate for activity, are inactivated on illumination in the presence of the cofactor.[5]

Psoralens interact photochemically with proteins via mechanisms that depend on the reaction conditions. UV-A irradiation of solutions containing psoralen-protein mixtures under anaerobic conditions results in the slow formation of covalent psoralen-protein adducts. Under aerobic conditions, psoralens sensitize the photoinactivation of enzymes by a singlet oxygen-mediated process. Finally, if psoralen solutions are illuminated under aerobic conditions and then mixed with bovine serum albumin in the dark, covalent psoralen-albumin adducts form; this suggests that reactive products are formed during illumination that can subsequently react with proteins.[25]

3.3.4. Other Biomolecules

Alcohols have been studied as model compounds for the photosensitized oxidation of carbohydrates. Anthraquinone and ketonic sensitizers, on illumination, abstract a hydrogen atom from the α carbon of the alcohol giving an alcohol radical; this in turn reacts with oxygen to yield aldehydes or carboxylic acids for primary alcohols, and ketones for secondary alcohols. Alcohols as well as simple sugars, polysaccharides such as cellulose, and complex carbohydrates (glycosaminoglycans) are photooxidized with rather low efficiencies. Hexitols such as sorbitol are oxidized to the corresponding hexose, and then to the hexonic acid. Much research has been carried out on cellulose because of its importance in the textile industry. Mechanistically, as for alcohols, cellulose is photooxidized by a type I process, giving fibers with decreased mechanical strength. Thus, dyes used with textiles should not be efficient photodynamic sensitizers, as mentioned earlier. Complex polysaccharides such as hyaluronic acid, heparin, and sodium alginate are photooxidized by free radical processes; these reactions result in chain scission giving depolymerized products. A number of important molecules that occur in only small amounts in living material are efficiently photooxidized, including ascorbic acid, biotin, glutathione, folic acid, lipoic acids, tocopherols, and plant growth hormones such as indoleacetic acid.[2,5,6]

3.4. SENSITIZED PHOTOEFFECTS ON VIRUSES[6]

Many viruses are efficiently inactivated by photosensitized treatment; this would be expected since they are made up of nucleic acids and proteins, and, in some cases, lipids. A large amount of research has been done in this area for a number of reasons, e.g., to elucidate the mechanisms of viral inactivation, to use photosensitized reactions as probes of the molecular structure of viruses, to examine the use of photodynamically inactivated viruses in the preparation of vaccines and immunodiagnostic reagents, and to explore the use of photosensitized reactions in the therapy of superficial viral infections in laboratory animals and humans.

Sensitized photoeffects on bacterial viruses (bacteriophages) have been studied extensively.[6] Both nucleic acids and proteins in bacteriophage can be altered by photodynamic treatment; in many cases nucleic acid strands are broken, DNA–protein crosslinks are formed, and mutant forms of the bacteriophage are sometimes produced. In some cases, photodynamically damaged bacteriophage can be repaired in appropriate host bacteria. Rates of photodynamic inactivation vary widely with different strains of bacteriophage because of differences in the permeability of the protein head membrane to sensitizers. As might be expected with a substrate as complicated as bacteriophages, photosensitized effects can be complex. For example, the inactivation of bacteriophage with *N*-formylkynurenine as sensitizer results from singlet oxygen-mediated damage to the purines of the DNA and to the tail proteins; inactivation also involves the direct interaction of excited sensitizer with the bacteriophage. With basic sensitizers such as acridine, acridine orange, proflavine, etc., the targets for bacteriophage P22 photoinactivation are the injection proteins; these appear to be photosensitized by dye molecules stacked in a region of the bacteriophage DNA located close to these proteins.[43]

Plant viruses, too, are inactivated by photodynamic treatment. For example, tobacco mosaic virus is inactivated with a number of different sensitizers. The isolated infectious RNA from tobacco mosaic virus is inactivated more rapidly and more completely than the virus itself; this suggests that the protein coat of the virus can have a protective effect, as in the case of some bacteriophages. The main site of photodynamic damage in tobacco mosaic virus appears to be the guanine residues of the RNA. As with bacteriophage, photodynamic treatment can produce mutations in plant viruses.[6]

Animal viruses from a number of different families can be inactivated by photosensitized treatment including adeno, herpes, orthomyxo (influenza), papova (SV-40), paramyxo (measles, mumps, etc.), picorna (polio), pox (vaccinia), reo, toga (equine encephalitis), etc.[44] Some viruses, such as polio, are resistant to photosensitized inactivation since photosensitizers do not penetrate the protein coat. However, if the virus is grown in the presence of a sensitizer (which results in its incorporation into the viral structure) or if it is treated with sensitizer at a high pH (which increases the permeability of the protein coat) it becomes light-sensitive. As with bacteriophages and plant viruses, mutants of animal viruses have been produced by photodynamic treatment.[6]

A dye-light procedure for the treatment of herpes simplex infections in humans has been developed. In this, a solution of a photodynamic sensitizer, such as neutral red or proflavine, is applied to the lesions; the area is then exposed to light.[45] However, there is considerable concern over the risk/benefit ratio of this type of photochemotherapy since it is reported that such treatment can transform mammalian cells (Chapter 6).[46,47] The most common sensitizers used for photoinactivating animal viruses are positively charged dyes (methylene blue, neutral red, and acridines such as proflavine and acridine orange) which bind to the viral nucleic acids; psoralens, which efficiently sensitize the photoinactivation of almost all kinds of viruses, intercalate into the viral nucleic acid. Thus, on illumination in the presence of these sensitizers, the major effect is probably on the viral nucleic acid. The use of negatively charged sensitizers, such as porphyrins, may result in viral inactivation by effects on proteins, thus avoiding the production of virus mutants that might induce tumors.[48] Also, in the case of those viruses containing lipids, the use of lipid-soluble photosensitizers might avoid effects on nucleic acids and thus decrease the possibility of producing oncogenic changes in the virus.[49]

3.5. SENSITIZED PHOTOEFFECTS ON CELLS

Phototoxicity is a general term for the damaging or killing of cells by photosensitized reactions. Mechanistic studies on phototoxicity are difficult because of the extreme complexity of cells. For example, the cell membrane acts as a differential barrier to the penetration of photosensitizers into the cell because of its charge and hydrophobic core. Also, cells are not merely tiny bags containing simple solutions but are composed of a very large number of microregions. Because of the many kinds of molecules making up cells, this provides microenvironments with a wide array of physical-chemical properties. As a result, photosensitizer molecules penetrating into cells will not ordinarily be in simple aqueous solution but, depending on their size, charge, and solubility properties, will be bound noncovalently at various subcellular sites, be associated with the more hydrophobic regions of membraneous organelles, etc. Thus, the microscopic pattern of the cellular localization of sensitizers, and therefore the pattern of subcellular damage resulting from illumination, can vary with the sensitizer, the cell type, the physiological condition of the cell, etc. Furthermore, the mechanism, and thus the end product of a photosensitized reaction, might be altered by the binding of the sensitizer to a cellular component or by the dielectric constant, oxygen solubility properties, etc., of the microenvironment.[50]

3.5.1. Model Cell Systems[51]

One preliminary approach to the elucidation of reaction mechanisms in biological systems is to use so-called model systems, systems that simulate in some ways, albeit much oversimplified, the properties of the real system. For example, changing the dielectric properties (and solvent properties) of aqueous solutions by the addition of a miscible organic solvent such as methanol can have significant effects on the efficiencies and mechanisms of sensitized photooxidations.[52] In some cases, photosensitizers show an increased efficiency when bound to nucleic acids or proteins. Bound dye can sensitize photodamage to the molecule to which it is bound as well as to other molecules in the reaction system. Sensitizer binding may also favor type I photoprocesses.[5,50] In aqueous systems, surfactant (detergent) micelles have a charged, hydrophilic exterior and a hydrophobic interior, and thus crudely resemble biomembranes. Hydrophobic sensitizers localize in the inner part of the micelle, and hydrophilic sensitizers can associate with the surface region; in both cases, on illumination, singlet oxygen can be generated with high efficiency and oxidize substrates both inside and outside of the micelle.[50] Similar types of studies have been carried out with liposomes (microscopic, bilayered phospholipid vesicles), which structurally are even more like cell membranes.[51,53] As additional model studies are carried out, especially with the more complex organized assemblies, the resulting information will further our understanding of the mechanistic aspects of the photosensitized killing of cells and of photodynamic damage to tissues.[54] It should be stressed that the actual mechanism of cell killing will depend on the sensitizer, the reaction conditions, and the type of cell involved.

3.5.2. Microorganisms[6,55–57]

All types of procaryotic microorganisms (rickettsia, mycoplasmas, blue-green algae, and bacteria) can be killed on illumination in the presence of appropriate photosensitizers; most of the studies in this area have been carried out with bacteria. Many categories of sensitizers are effective, including psoralens. Photosensitized damage can be mediated by effects on nucleic acids, proteins, or unsaturated lipids. It appears that most sensitizers have to bind to or penetrate into bacterial cells to be effective. However, *Escherichia coli* cells are killed on illumination in the presence of rose bengal bound covalently to water-insoluble polystyrene beads. This sensitizer remains outside of the cells, so killing probably results from the diffusion of singlet oxygen into the cells. With some bacteria, the cell envelope of treated cells is more easily lysed. The swimming behavior of flagellated bacteria, such as *Salmonella typhimurium,* is altered by photosensitized reactions; longer treatment results in the complete loss of motility. A number of the metabolic activities of bacteria can be inhibited by photodynamic treatment including glycolysis, protein synthesis, respiration, etc.; in some cases, particular enzymes are inactivated. Carotenoids, which occur in high concentrations in some kinds of bacteria, protect the organisms against photosensitized damage. Some photosensitization studies have been carried out with isolated bacterial organelles. For example, photodynamic treatment inactivates the respiratory system located in cell membranes isolated from *Sarcina lutea.* Also, isolated *E. coli* ribosomes are damaged by photosensitized treatment; both ribosomal RNA and protein are altered, and with some sensitizers, covalent RNA–protein cross-links are formed. The photodynamic treatment of intact bacterial cells can produce DNA–protein cross-links. Furthermore, mutations are produced in various strains of *E. coli, S. typhimurium,* etc., using acridine dyes, methylene blue, phenothiazines, psoralens, and the like as photosensitizers. Some strains of bacteria are able to repair certain kinds of sensitized photodamage. *E. coli* strains carrying various combinations of genes that control sensitivity to near-UV and far-UV radiation have been used to analyze the mechanisms of photosensitized cell inactivation and mutagenicity of a number of sensitizers derived from plants.(58)

Many kinds of photosensitization studies have been performed using eucaryotic microorganisms. As indicated earlier, modern studies on biological photosensitization began with the work of Oscar Raab, who used a ciliated protozoan, paramecium, as his test organism. Since Raab's pioneering work other studies have been carried out with paramecia; the effects of a number of different types of photosensitizers have been examined, including carcinogenic hydrocarbons. Xanthene dyes sensitize paramecia to photobehavioral responses, e.g., positive or negative phobophototactic behavior, depending on the sensitizer concentration, light intensity, etc. The photodynamic treatment of paramecium also causes loss of motility, a delay in cell division, and induces a subsequent increase in sensitivity to heat, Another ciliate, *Tetrahymena pyriformis,* has also been used in a number of photodynamic studies. Trypanosomes, which are flagellated protozooans, and amebas have also been utilized. The green algae *Chlorella* and *Chlamydomonas,* as well as the dinoflagellate *Gymnodinium,* are inactivated by photodynamic treatment. A large amount of work has been done with yeasts using well-defined genetic strains. Little is known of the genetic basis of sensitivity to photodynamic killing in eucaryotic microorganisms, although both sensitive and resistant mutants of yeast have been isolated. Different mechanisms can be involved in the photosensitized damage to yeast cells depending on the sensitizer, reaction conditions, etc. Some porphyrins, for example, do not penetrate into yeast cells, and thus probably sensitize only membrane damage. Other sensitizers, such as eosin and rose bengal, penetrate into the cytoplasm and sensitize damage there, while some basic dyes, including acridine orange, penetrate into the nucleus where they sensitize damage. Mutations of various types are readily produced in yeast cells with a number of sensitizers. Many structurally different psoralens have been studied as photosensitizers for yeast cells. Difunctional psoralen derivatives efficiently sensitize both cell killing and the induction of nuclear mutations, but are rather ineffective in the production of "petite" cytoplasmic mutations. In contrast, monofunctional derivatives are active in the photoproduction of petite mutations, but are poor sensitizers for cell killing and the induction of nuclear mutations.(6,55,57,59)

3.5.3. Cells from Multicellular Organisms(6,7)

Cells from multicellular organisms have been used extensively for photosensitization studies. In the case of higher plants, cell division in isolated leaf mesophyll cells from the plant *Calystegia* (a morning glory) is inhibited by illumination in the presence of 8-methoxypsoralen.(60) Cercosporin, a complex quinone-type photosensitizer produced by certain types of plant pathogenic fungi, sensitizes the rapid photodynamic killing of protoplasts isolated from tobacco cells grown in culture. Damage appears to be mediated by the photooxidative destruction of membrane lipids that results in the bursting of the protoplasts.[61] Protoplasts isolated from bean leaf cells show membrane blebbing and lysis on illumination in the presence of hematoporphyrin derivative as sensitizer.[62] Some work has been done with cells from invertebrate animals.[6] For example, the treatment of the sperm of marine invertebrates with photosensitizers causes a loss of motility and a decrease in the fertilizing capability. Treatment of marine invertebrate eggs can stimulate parthenogenetic cell division; longer treatment causes loss of fertilizability and blocks the formation of the fertilization membrane. Extensive studies have been carried out on the effects of photodynamic treatment on the axons of the nerve cells from the lobster, snail, octopus, and squid. Large light doses in the presence of photosensitizers that penetrate into the membrane block sodium channels, depolarize the membranes, and thus stop nerve impulse propagation. Frog muscle is stimulated to contract by light in the presence of appropriate photosensitizers.[59]

Mammalian erythrocytes have been a favorite experimental material ever since it was shown in 1908 that rabbit red blood cells hemolyzed on illumination in the presence of hematoporphyrin or chlorophyll preparations.[1] The mature mammalian red blood cell is the simplest biological model for cell membrane studies, since the cells do not contain nuclei or much in the way of the usual cytoplasmic organelles. Red blood cells can be processed to give resealed "ghosts" in which essentially only the plasma membrane remains. Photodynamic treatment produces an increase in the passive permeability of the cell membrane to small cations that results in the colloid osmotic lysis of the cells. Membrane components damaged include unsaturated fatty acids, cholesterol, and enzymes; extensive cross-linking of membrane structural proteins also occurs.[55,57,59,63–65]

An enormous amount of research has been done in recent years on photosensitized effects using normal and neoplastic cells from humans and other mammals in culture.[7,55,57,59,63] The kinds of lesions expected in photosensitized cells can be predicted. For example, cell membranes contain enzymes, structural proteins, and unsaturated lipids, all of which can be photodegraded. Thus, sensitizers located in the membrane would cause alterations in catalytic and transport processes and changes in membrane permeability and mechanical properties. Sensitizers located in various regions of the cytoplasm could mediate the photodegradation of enzymes, tRNA, and components of the various cell organelles. Finally, with sensitizers that localize in the nuclear region, illumination would result in guanine destruction in the DNA, nucleic acid strand breaks, and the formation of DNA-protein adducts; these photochemical changes could in turn interfere with cell division and produce mutations. All of these changes, which in many cases result in cell death, have been observed, depending on the cell type, sensitizer, and reaction conditions. Cell survival curves (see Chapter 2 and Fig. 2-14) typically have an initial curved region (shoulder) followed by a linear section corresponding to the exponential killing of the cells. The width of the shoulder, its shape, and the slope of the linear region depend on the sensitizer, the reaction conditions, and the cell type and physiological status; some cell lines, as studied in synchronous cultures, show a marked difference in photodynamic sensitivity during different phases of the cell division cycle.[7]

With mammalian cells in culture, more is probably known about sensitized effects on the plasma membrane than on other organelles and components.[7,64,66] Effects include inhibition of membrane transport processes, leakage of potassium and other components out of the cells, increases in the passive permeability of the membrane to many kinds of molecules, changes in the mechanical properties of the membrane, alterations in the binding and antigenic characteristics of the membrane, and changes in membrane morphology (bleb and hole formation). With many kinds of mammalian cells and with moderately hydrophobic sensitizers such as protoporphyrin, deuteroporphyrin, etc., the membrane is the principal site of damage. A number of kinds of damage are observed with photosensitizers that penetrate into the cytoplasm of mammalian cells. In some cases, the mitochondria show the first ultrastructural changes observed in photodynamically treated cells. Mitochondrial, lysosomal, and cytosol enzymes are inactivated. Typically, both protein and RNA synthesis are markedly decreased on the photodynamic treatment of cells. A number of more detailed photodynamic studies have been performed using organelles isolated from the cytoplasm of mammalian cells. For example, with isolated mitochondria, a number of changes occur including swelling, structural disruption, uncoupling and inhibition of oxidative phosphorylation, and inactivation of various enzymes.[63,67] Photodynamic treatment alters the membranes of isolated lysosomes permitting lysosomal enzymes to leak out; with isolated microsomes, the mixed-function oxidase system is inactivated and lipids are peroxidized. Finally, a number of characteristic effects are observed with sensitizers that penetrate into the nucleus. The nuclear membrane can be damaged, alkali-labile lesions are produced in the DNA, and chromosome breaks and sister chromatid exchanges occur. Guanine residues in the DNA strands of intact cells are altered as demonstrated by immunofluorescence techniques. DNA strand breaks in mammalian cells are usually repaired rapidly.[55] There are very few reports of mutation production in mammalian cells using oxygen-requiring photosensitizers. Psoralens appear to be more photomutagenic.[7] As this section clearly indicates, photodynamically treated mammalian cells can be injured at one or more of numerous sites, and by a variety of reaction mechanisms, depending on the cell type, sensitizer, reaction conditions, etc.[7,55,57,59,63–67]

3.6. SENSITIZED PHOTOEFFECTS ON MULTICELLULAR ORGANISMS[6,8]

3.6.1. Plants

Under some conditions plants can be injured or killed by reactions photosensitized by endogenous chlorophyll. For example, if photosynthesis in leaves is blocked by heat treatment or carbon dioxide deprivation, illumination kills the photosynthetic tissues; no photodamage is observed in the absence of oxygen.[6] Chloroplasts contain high concentrations of carotenoids, which protect against photodynamic damage under ordinary conditions by quenching triplet chlorophyll and singlet oxygen. Superoxide dismutase, which is also present in photosynthetic structures, appears to protect leaves against superoxide-mediated photodamage.[68] As mentioned in Section 3.2.6, many species of plants contain endogenous photosensitizing compounds in addition to chlorophylls. It would be of interest to understand the mechanisms by which such plants are protected against photodamage, since some of these compounds are potent photosensitizers for plant tissues. For example, berberine, a quinoline alkaloid synthesized by several plant species, strongly inhibits the growth of onion root tips on illumination.[69] Illumination of pea plant tissue in the presence of psoralens, which occur in many plant species, increases the rate of synthesis of certain proteins in the tissue; this suggests a possible growth photoregulatory function for such compounds. Leaves, stems, and roots of plants are injured on illumination after treatment with exogenous sensitizers.[6] For example, illumination of pea leaf tissue after exposure to eosin inhibits photosynthesis, inactivates certain chloroplast enzymes such as ribulose biphosphate carboxylase, bleaches chlorophyll, and disrupts chloroplast structure. The inhibition of photosynthesis results from an interference with electron transport processes in Photosystem II (see Chapter 12).[70] Roots sensitized with dyes such as erythrosin or rose bengal show a positive phototropism on unilateral illumination due to growth inhibition on the illuminated side; illumination in the presence of riboflavin decreases the rate of the hormonally controlled growth of pea stem sections. Protoplasmic streaming in plant cells and cyclosis in barley root hairs is initially stimulated and then inhibited by photodynamic treatment. Chromosome breaks in barley and bean plant tissues are produced as a result of illumination in the presence of basic photosensitizing dyes such as acridine orange and methylene blue.[6]

3.6.2. Nonmammalian Animals[6]

Relatively little work has been done on the responses of lower invertebrates to photodynamic treatment, although hydra, several kinds of worms, and rotifers have been used to a small extent as experimental material, primarily in photokilling studies.[1,6] The nematode worm *Aphelenchus avenae* is killed efficiently on illumination in the presence of dithiophene sensitizers and oxygen.[71] The aquatic larval forms of the blood fluke *Schistosoma mansoni* are rapidly killed on illumination in the presence of low concentrations of acridine orange and methylene blue. Light causes contraction of the adductor muscle of larval mussels in the presence of some sensitizers.[6] Nauplii of the brine shrimp *Artemia salina* lose motility on illumination in the presence of a number of polycyclic aromatic hydrocarbons; the relative photosensitizing efficiencies of the compounds used correlates well with their relative carcinogenic activities.[72] More recently, a number of studies have been done with insects. The aquatic larvae of several different kinds of mosquitoes are efficiently killed when illuminated in water containing low concentrations of photosensitizers. A number of other types of insects including the fruit fly (*Drosophila*), the housefly, the face fly, the imported fire ant, and the boll weevil can be photokilled with many different types of sensitizers, including several xanthene dyes (erythrosin B, rose bengal, etc.), acridine red, substituted butadienes, thiophenes, etc.[73] The use of photosensitizing compounds in the control of insect pests and the role of naturally occurring sensitizers in the resistance of plants to insects will be discussed in Sections 3.8.2 and 3.9.

Only a few photosensitization studies have been reported with the nonmammalian vertebrates. Tadpoles, newts, and small fish can be killed by photodynamic treatment, and frogs have been used for some studies on the effects of photodynamic treatment of muscle tissue and blood vessels.[1,6] Chickens, ducks, geese, and turkeys become light-sensitive as a result of eating psoralen-containing seeds from several different plant species. Ducks, which appear to be the most sensitive, suffer phototoxic damage to the beak, foot webs, periorbital region, and retina; this can result in severe deformation of the beak and feet, and blindness.[74,75]

3.6.3. Mammals[6,8]

Light responses in sensitized humans and some other mammals can involve two different mechanisms, phototoxicity and photoallergy, depending on the photosensitizer and the organism involved. Phototoxic reactions are immediate and will probably occur in all mammals if sufficiently high sensitizer and light doses are used.[76–79] The photoallergic response is delayed, and appears to result from the photosensitized formation of protein "photoantigens," which in turn produce an immunological type of reaction (see Chapter 6).[78] The phototoxic response in white mice, for example, involves a characteristic array of symptoms on illumination including hyperactivity, skin itching, skin injury (reddening, edema, necrosis, and ulceration), and retinal damage. In addition, intestinal hemorrhage, decreased blood pressure, and circulatory collapse occur in severe phototoxicity; these responses probably result from the photodynamic formation of toxic materials in the skin that are then transported to all parts of the body in the bloodstream.[1] Other mammals, including rats, guinea pigs, rabbits, and dogs, show similar responses.[77,79] With most sensitizers, such as porphyrins, oxygen is necessary for phototoxic responses. Psoralens can sensitize phototoxic skin reactions that are typically expressed as erythema followed by significant skin darkening.[26] Some hereditary or drug-induced diseases in mammals result in the accumulation of abnormal amounts of iron-free porphyrins in the organism. These conditions, termed porphyrias, often make the animal highly photosensitive.[80] For example, mice fed with large doses of the antibiotic griseofulvin accumulate excessive amounts of protoporphyrin, a phototoxic porphyrin, in the liver and red blood cells, with smaller amounts in the blood serum. On illumination, these mice show acute phototoxic responses in the ears, tail, and depilated skin.[15] Some chemicals protect mammals against photosensitized injury. For example, β-carotene, which is an efficient quencher of triplet photosensitizers, free radicals, and singlet oxygen, partially protects mice against photosensitization with hematoporphyrin.[81] Thiols and α-tocopherol also have some protective action.[6]

Cows, sheep, and other grazing mammals sometimes show a type of photosensitivity resulting from a disturbance in liver function. The photosensitizer involved is phylloerythrin, a phototoxic derivative of chlorophyll produced in the animal gut by bacterial action. This material is normally excreted in the bile; however, in a number of types of liver dysfunction it accumulates in the body to serious photosensitizing levels. These conditions can be produced by chemicals termed hepatotoxins that occur in several kinds of plants; in addition, some mycotoxins produced by fungi growing on plants cause liver damage. In most cases, the chemical nature of the toxin has not been established. Some viral diseases in animals damage the liver and thus induce a secondary photosensitization. Finally, some breeds of sheep have a hereditary abnormality in the phylloerythrin-excreting mechanism of the liver. Ligation of the bile duct in sheep can result in photodynamic death if the animal is fed green plants; if the animal is protected from light, or if it is fed grain or some other chlorophyll-free food, photosensitivity does not develop. A hereditary type of porphyria sometimes occurs in cattle, which results in the accumulation of phototoxic porphyrins including protoporphyrin and uroporphyrin in the body. Cattle with this condition (termed "pink tooth," inherited as a simple recessive) show a reddish brown pigmentation of the bones, teeth, and urine; hairless and lightly pigmented regions of the skin become highly photosensitive.[75,82–84]

As mentioned above, many species of plants contain potent photosensitizers other than chlorophylls. If grazing animals eat such plants they can become photosensitive; this is a serious problem in certain areas of the world. Phototoxic effects are greatest in unpigmented areas of the skin; historically, white sheep and white horses could not be raised in some geographic regions. Plants of the genus *Hypericum* are especially troublesome since many species contain a potent photosensitizer, hypericin (or related compounds); the structure of this sensitizer is shown in Fig. 3-2. Many *Fagopyrum* species (buckwheat) also produce pigments (fagopyrins) that photosensitize grazing animals and swine. Because of this, buckwheat leaves are not used as fodder; fortunately, the buckwheat grain contains little photosensitizer, so that eating buckwheat cakes should not be hazardous to your health! Finally, members of several plant genera (including *Ammi* and *Cymopterus*) contain psoralens, which are very effective photosensitizers as described above. In some areas these plants pose a significant threat to sheep, for example.(75,82,85)

3.6.4. Humans

Photosensitization in humans will be considered only very briefly since this area will be covered in Chapter 6 on photomedicine. Light sensitivity in humans accompanied by an excretion of abnormal porphyrins in the urine was observed in the last century. Experimentally, photosensitization in humans was first studied in 1913 by Meyer-Betz, a physician, who injected himself with 200 mg of hematoporphyrin, and then exposed parts of his body to sunlight in controlled ways. Exposure resulted in an immediate burning sensation in the exposed skin and in a few minutes erythema, edema, and some pain started to develop; the severity of the reaction increased with increasing light dose. Meyer-Betz's skin remained light-sensitive for two months.(8) As in the case of other mammals, humans show both phototoxic and photoallergic responses.(78) Photosensitization in humans is not uncommon and can result from ingestion of a sensitizer, from skin contact with a sensitizer, from an injection of a photosensitizing drug, or from certain diseases including hereditary and drug-induced porphyrias, pellagra, and lupus erythematosus. Many individuals are photosensitized by commonly used drugs including certain antibiotics, antihistamines, antiseptics, barbiturates, diuretics, sulfonamides, phenothiazine tranquilizers such as chlorpromazine, etc. Also, some chemicals used in agriculture, industry, the home, and research are phototoxic.(76,78) For this reason, compounds being considered for commercial use should be routinely screened for possible photosensitizing activity.(79) Furthermore, adequate precautions should be taken in medical practice when using drugs that might photosensitize patients. Individuals with the hereditary disease erythropoietic protoporphyria (inherited as a simple dominant) have unusually high levels of protoporphyrin in their erythrocytes, feces, and serum. On illumination with light that is absorbed by the porphyrin, many of these people have itching and burning sensations in the exposed skin areas, followed by the development of erythema and edema; over a period of time more severe skin damage can develop. Persons with porphyria cutanea tarda (a type of drug-induced porphyria) accumulate abnormal levels of uroporphyrin in the body; in many cases this results in high sensitivity to light.(86,87) Some success has been obtained in using large doses of β-carotene for the treatment of patients with photosensitizing porphyrias, especially those with erythropoietic protoporphyria.(81)

3.6.5. Sensitized Photocarcinogenesis

It is well known that chronic exposure to UV-B radiation induces skin cancers in laboratory animals (and presumably in humans).(88,89) Cancers can also be induced by visible or UV-A radiation in animals treated with certain photosensitizers. For example, it has been reported that laboratory animals treated with hematoporphyrin, proflavine, or neutral red develop cancers on chronic exposure to light. None of these compounds act as chemical carcinogens in the dark.[90] Animals develop cancers when exposed to UV-A radiation following treatment with photosensitizing psoralens.[26] The mechanisms involved in photosensitized carcinogenesis are not known. However, the phenomenon is perhaps not unexpected since photosensitized treatment can produce chemical alterations in DNA, chromosome breaks, and sister chromatid exchanges in mammalian cells. Such treatment also transforms mammalian cells, and cells transformed by PUVA treatment can produce tumors when implanted in animals. Many photosensitized reactions produce hydrogen peroxide and singlet oxygen; it has been suggested that both of these species can act as chemical carcinogens.[7,26]

3.7. NATURAL PHOTOSENSITIVITY

Some kinds of cells are killed or injured on illumination with cool white or daylight-type fluorescent lamps (which emit visible light and some UV-A radiation). Cellular damage in this case may be mediated by light absorbed by the flavin and/or heme components of certain proteins. For example, light treatment of isolated rat hepatocytes severely damages catalase (a heme-containing enzyme), microsomes, and mitochondria; the cell membranes are not affected. Microsomes and mitochondria isolated from the cells are also damaged by light. Photodamage is decreased by antioxidants such as ascorbic acid and α-tocopherol. In many mammals, the retinal cells are very light-sensitive; recent studies suggest that the damage is mediated by singlet oxygen generated by illuminated retinal in the cells. Many kinds of mammalian cells exhibit light sensitivity when incubated in tissue culture media that contain flavins. Cell killing, production of chromosome breaks, formation of cross-links in DNA, and neoplastic transformation have been reported. These effects may be caused by hydrogen peroxide produced in the flavin-sensitized photodynamic degradation of tryptophan or tyrosine in the culture medium. The most effective wavelengths for both the endogenous and medium-sensitized effects are in the UV-A and blue regions; thus, mammalian cell cultures should probably be handled and stored under lights emitting only at longer wavelengths. Natural photosensitivity, presumably sensitized by endogenous chromophores, has been observed with many kinds of microorganisms. Effects such as cell killing and mutagenesis commonly occur. Even a few multicellular animals are injured or killed by visible light. For example, the small aquatic annelid *Tubifex* is killed within a few minutes on illumination in the presence of oxygen; this organism is normally found in habitats where both the light intensity and the oxygen concentration are low.[6,7,15]

3.8. BIOLOGICAL APPLICATIONS OF PHOTOSENSITIZED REACTIONS

In general, photosensitized reactions are harmful to living organisms. They are also harmful in other ways. For example, some colored textiles are photodegraded in reactions sensitized by the dyes used. Light degrades many kinds of beverages and foods via sensitized reactions. Beer rapidly develops an unpleasant flavor on exposure to radiation in the UV-A and blue regions of the spectrum. This results from riboflavin-sensitized photooxidation reactions that produce certain kinds of organosulfur compounds. For this reason, the glass (typically brown) used in the manufacture of beer bottles is designed to absorb radiation in the UV-A and blue regions. Milk also develops an off-flavor on exposure to light, again by flavin-sensitized reactions. Many foods, in particular those containing high concentrations of unsaturated lipids, are light-sensitive, including olive oil, cooking oils, and snack foods such as potato and corn chips.[36,37]

Although we tend to think of photosensitized reactions as being detrimental, a number of attempts have been made to utilize such reactions in the solution of practical problems. Some of these are described below.

3.8.1. The Health Sciences[91]

The most successful applications of photosensitized reactions in the health sciences have been in the treatment of certain diseases (see Chapter 6). Light can be used in therapy in two ways. In phototherapy, the light involved is absorbed by endogenous molecules in the organism. The best known example of this is the use of blue light to treat premature infants with hyperbilirubinemia.[92] In this condition, which is common in premature infants, bilirubin (a normal breakdown product of hemoglobin) can accumulate in the blood of an infant to levels that produce irreversible brain damage. Illumination of hyperbilirubinemic infants with high-intensity light of wavelengths absorbed by bilirubin causes photochemical reactions that result in an increased excretion of bilirubin and thus brain damage is prevented. Very large numbers of infants are treated with this technique each year.

In photochemotherapy, an exogenous photosensitizing drug is first administered to the patient; then after an appropriate time interval, the patient is illuminated with light of wavelengths absorbed by the sensitizer. The most successful application of this technique has been in the treatment of plaque-type psoriasis, an inflammatory disease of unknown etiology in which increased epidermal proliferation results in the formation of raised, scaly, red patches on the skin. Patients are typically given an oral dose of 8-methoxypsoralen, and then, after a short wait, the psoriatic areas are irradiated using UV-A lamps. This type of treatment is usually termed PUVA therapy. In some cases, the sensitizer is applied topically to the psoriatic areas of the skin before irradiation. Many thousands of patients with psoriasis are treated each year with this form of therapy, and with a high degree of success. Other skin diseases can also be treated with PUVA therapy.[10,93]

More recently, a photochemotherapeutic technique has been applied to the treatment of tumors. It has been known for a number of years that certain porphyrins are selectively retained in solid tumors in laboratory animals and humans. Since the porphyrin retained in tumors fluoresces bright red on illumination with blue or UV-A radiation, this permits the visual localization of tumors, especially small metastases. Illumination, usually with a laser, of a tumor with retained photosensitizing porphyrins can destroy the tumor with relatively little effect on the adjacent normal tissue. Several thousand patients have now been treated by this technique, termed photodynamic therapy (PDT), with generally encouraging results. It appears to be especially promising for endobronchial tumors and superficial bladder tumors.[11–15,94,95] Much research needs to be done in this area to develop more specific site-selective photosensitizing drugs. One approach that looks very promising is to covalently attach a photosensitizer, such as a porphyrin, to monoclonal antibodies directed toward particular types of tumor cells.[96]

Photosensitized reactions have been used in other ways in the health sciences. For example, they have been used to inactivate viruses and snake venom toxins for use in the preparation of vaccines and antivenins. In the public health area, it has been proposed that photosensitized reactions might be an economical tool for water purification. In particular, it has been shown that the treatment of effluents from sewage disposal plants with photosensitizers such as methylene blue followed by exposure to artificial light or sunlight efficiently destroys both bacteria and viruses. Chemical pollutants, such as phenols in industrial wastewater, can also be destroyed by photosensitized reactions. Only rapidly degraded sensitizers should be used in these applications to avoid building up detrimental levels of photosensitizing molecules in the environment.[91]

3.8.2. The Agricultural Sciences[73]

There is considerable interest at present in the use of photodynamic reactions to control insect pests. Although many kinds of conventional insecticides have been successfully used, they often show undesirable side-effects, including toxicity to nontarget organisms such as fish and mammals, including humans; the induction of mutations and cancers; and long-term persistence in the environment. A number of synthetic sensitizing dyes, including acridine red, methylene blue, and xanthenes (especially halogenated fluoresceins such as erythrosin, phloxine, and rose bengal), photosensitize the injury and killing of the larvae and/or adults of some economically important insects. Susceptible organisms include the mosquito, codling moth, house fly, cockroach, mealworm, imported fire ant, boll weevil, face fly, cabbage butterfly, cabbage looper, and corn earworm. In preliminary field tests, it appears possible to control both mosquitoes and house flies under certain conditions. Some sensitizers such as erythrosin B are inexpensive, appear to be relatively nontoxic to mammals at the concentrations used, and are rapidly photodegraded in the environment. However, more detailed toxicological studies need to be carried out. Hopefully, synthetic photoinsecticides will have useful applications, especially in the control of insect pests around poultry-producing facilities, barns, feedlots, manure storage areas, etc.[73,97,98]

Plants synthesize a large number of different phototoxic compounds.[33,99,100] Recently, in addition to the work with synthetic sensitizers, considerable interest has developed in the use of some of these naturally occurring materials as photoinsecticides. For example, studies have been made using α-terthienyl, a compound synthesized by many plants of the aster family. When applied to experimental ponds at appropriate concentrations, this compound sensitizes the photokilling of the larvae of the mosquito *Aedes intrudens* with relatively little effect on nontarget organisms including the caddisfly, shrimp, snail, and trout.[101] The furanocoumarins, 8-methoxypsoralen, and sphondin, which are synthesized by some plants, efficiently sensitize the delayed photokilling of early instar larvae of the mosquito *Aedes aegypti;* later instar larvae are less sensitive.[102] Although questions must be answered about the environmental safety of these types of photosensitizers, the preliminary work is very encouraging. Also, there are many additional types of naturally occurring photosensitizers to be tested.[33]

A novel development in herbicides has occurred recently, based on knowledge of the chemical pathways involved in the biosynthesis of chlorophylls in plants. Application of a tetrapyrrole precursor, such as aminolevulinic acid, to plants leads to the synthesis of large amounts of the tetrapyrrole intermediates formed in the chlorophyll pathways. These accumulate and then sensitize the photodynamic killing of the plant; this occurs within only a few hours for susceptible species. Considerable selectivity of killing is observed, depending on the plant species. For example, monocotyledenous plants such as barley, corn, and wheat show low sensitivity, while dicotyledenous weeds including lambs-quarter, purslane, and mustard are highly sensitive. These different responses appear to depend on the tetrapyrrole biosynthetic pathways and the rates of turnover of tetrapyrroles in the different plant species.(103)

3.9. ROLE OF PHOTOTOXIC PHYTOCHEMICALS IN THE ECOLOGY AND COEVOLUTION OF PLANTS AND HERBIVORES

Photosynthetic plants produce and accumulate an enormous array of so-called secondary substances or metabolites (allelochemicals), most of which have no known metabolic, nutritional, or physiological function in the plant itself. Large numbers of these phytochemicals are toxic to herbivores, such as mammals and insects, as well as to bacteria and fungi. Thus, they appear to play an important ecological role in the defense of plants against other organisms, especially herbivores. These materials are highly diverse chemically, including alkaloids, cyanogenic and triterpenoid glycosides, flavanoids, non-protein amino acids, phenols, etc.(104) A large number of these secondary plant metabolites are potent photosensitizers, typically with absorption peaks in the UV-A region of the spectrum. Thus, in the presence of sunlight, phototoxic reactions may play a significant ecological role in the defense of plants against predators.(99,105–109) Some of these plant phototoxins, such as the thiophenes and furanocoumarins, are being studied as potentially commercial photoinsecticides, as described in the preceding section.

It has been suggested that secondary plant substances are involved importantly in herbivorous insect–plant coevolution.(110) In this type of evolutionary process it is envisioned that, by mutation and recombination, photosynthetic plants would evolve capable of producing compounds that, by chance, decrease their suitability as food; such plants would suffer less predation than their ancestors. However, insects, by mutation, could evolve mechanisms to resist the new phytochemicals.(111)

Recent experimental studies on the associations observed between furanocoumarin-producing plants and certain insect species furnish good supporting evidence for the process of coevolution.(111) Furanocoumarins occur in hundreds of species of higher plants, almost all of which are in two plant families, the Umbelliferae (carrot-parsley family) and the Rutaceae (citrus family). Furanocoumarins occur in two structural forms, linear (psoralens) and angular (isopsoralens), as shown in Fig. 3-3. The angular forms occur in lower concentrations in plants than the linear forms. Caterpillars of the black swallowtail butterfly (*Papilio polyxenes*) feed exclusively on Umbelliferae that contain high levels of linear furanocoumarins; in spite of this, they do not show phototoxic responses. The caterpillars avoid plants that contain appreciable quantities of angular furanocoumarins, which, in contrast to the linear compounds, are phototoxic to them. In this species, tolerance to linear furanocoumarins results from the rapid and almost complete detoxification of the sensitizer by microsomal mixed-function oxidases located in both the midgut and body cells. Angular psoralens are metabolized more slowly and thus can apparently accumulate to photosensitizing levels.[112,113] In contrast to the resistance of the black swallowtail caterpillars, many insects, such as the armyworm, which feed on a variety of plants are lethally photosensitized by furanocoumarins; experiments show that the armyworm detoxifies furanocoumarins only very slowly.[113] There is evidence that the production of angular furanocoumarins by plants has evolved more recently than that of the linear compounds. It has been suggested that this may be a coevolutionary response to the selective pressure of herbivores, such as the black swallowtail, that can detoxify the linear furanocoumarins.[114] In conclusion, it should be mentioned that some insects avoid the phototoxic effects of plant photosensitizers by other methods. For example, dark pigmentation in the cuticle of the insects could decrease light sensitivity. Also behavioral responses are involved. Some insect species feed only at night and thus avoid high light intensities. Others (leaf rollers) avoid light by wrapping themselves in leaves, while some bore into the stems of plants where they are protected.[111,115]

3.10. CONCLUSIONS

Basic studies of photosensitized processes as described in this chapter are quite justifiable in their own right in terms of increasing our understanding of this important and rather commonly occurring photobiological phenomenon. In addition, from the examples given, it can be seen that photosensitized reactions have become useful tools in some areas including medicine and agriculture. Also, phototoxic reactions involving natural sensitizers in plants are contributing to our understanding of some ecological patterns as well as coevolutionary mechanisms.

We now understand the initial processes involved in photosensitized reactions reasonably well, including the production and properties of the excited states of sensitizers, the immediate reactions of triplet sensitizers by free radical and energy transfer (singlet oxygen) pathways, and the chemical behavior of the various reactive oxygen species that can be produced. Our understanding of the mechanistic organic chemistry involved in the sensitized photodegradation of biomolecules is less complete. Finally, although much excellent work has been done in recent years, our understanding of events at the cellular and organismal levels is still very incomplete. Clearly, there is much research yet to be done on photosensitized reactions in biological systems.

3.11. REFERENCES

1. H. F. Blum, *Photodynamic Action and Diseases Caused by Light*, Reinhold, New York (1941). (Reprinted in 1964 with an updated appendix by Hafner Publ., New York.)
2. C. S. Foote, Photooxidation of biological model compounds, in: *Oxygen and Oxy-Radicals in Chemistry and Biology* (M. A. J. Rodgers and E. L. Powers, eds.), pp. 425–439, Academic Press, New York (1981).
3. C. S. Foote, Mechanisms of photooxygenation, in: *Porphyrin Localization and Treatment of Tumors* (D. R. Doiron and C. J. Gomer, eds.), pp. 3–18, Alan R. Liss, New York (1984).
4. G. Laustriat, Molecular mechanisms of photosensitization, *Biochimie* **68**, 771–778 (1986).

5. R. C. Straight and J. D. Spikes, Photosensitized oxidation of biomolecules, in: *Singlet O_2*, Vol. 4 (A. A. Frimer, ed.). pp. 91–143, CRC Press, Boca Raton (1985).
6. J. D. Spikes, Photodynamic reactions in photomedicine, in: *The Science of Photomedicine* (J. D. Regan and J. A. Parrish, eds.), pp. 113–144, Plenum Press, New York (1982). (This paper lists a large number of earlier reviews on photosensitization.)
7. J. D. Spikes, Photosensitization in mammalian cells, in: *Photoimmunology* (J. A. Parrish, M. L. Kripke, and W. Morison, eds.), pp. 23–49, Plenum Press, New York (1983).
8. J. D. Spikes, Photosensitized reactions in mammals, in: *Experimental and Clinical Photoimmunology*, Vol. I (R. A. Daynes and J. D. Spikes, eds.), pp. 70–80, CRC Press, Boca Raton (1983).
9. J. D. Regan and J. A. Parrish (eds.), *The Science of Photomedicine*, Plenum Press, New York (1982).
10. W. L. Morison, *Phototherapy and Photochemotherapy of Skin Diseases*, Praeger, Westport (1983).
11. D. Kessel and T. J. Dougherty (eds.), *Porphyrin Photosensitzation*, Plenum Press, New York (1983).
12. D. R. Doiron and C. J. Gomer (eds.), *Porphyrin Localization and Treatment of Tumors*, Alan R. Liss, New York (1984).
13. A. Andreoni and R. Cubeddu (eds.), *Porphyrins in Tumor Phototherapy*, Plenum Press, New York (1984).
14. G. Jori and C. Perria (eds.), *Photodynamic Therapy of Tumors and Other Diseases*, Libreria Progetto, Padova (1985).
15. G. Jori and J. D. Spikes, Photobiochemistry of porphyrins, in: *Topics in Photomedicine* (K. C. Smith, ed.), Plenum Press, New York (1984).
16. R. V. Bensasson, G. Jori, E. J. Land, and T. G. Truscott (eds.), *Primary Photo-Processes in Biology and Medicine*, Plenum Press, New York (1985).
17. T. B. Fitzpatrick and M. A. Pathak, Historical aspects of methoxsalen and other furocoumarins, *J. Invest Dermatol.* **32,** 299–312 (1959).
18. M. A. J. Rodgers and E. L. Powers (eds.), *Oxygen and Oxy-Radicals in Chemistry and Biology*, Academic Press, New York (1981).
19. W. Bors, M. Saran, and D. Tait (eds.), *Oxygen Radicals in Chemistry and Biology*, Walter de Gruyter, Berlin (1984).
20. M. A. J. Rodgers, Activated oxygen, in: *Primary Photo-Processes in Biology and Medicine* (R. V. Bensasson, G. Jori, E. J. Land, and T. G. Truscott, eds.), pp. 181–195, Plenum Press, New York (1985).
21. N. I. Krinsky, Biology and photobiology of singlet oxygen, in: *Oxygen Radicals in Chemistry and Biology* (W. Bors, M. Saran, and D. Tait, eds.), pp. 453–463, Walter de Gruyter, Berlin (1984).
22. A. A. Frimer (ed.), *Singlet O_2*, 4 vols., CRC Press, Boca Raton (1985).
23. P.-S. Song, Photochemistry and photobiology of psoralens, *Photochem. Photobiol.* **29,** 1177–1197 (1979).
24. G. Rodighiero, F. Dall'Acqua, and M. A. Pathak, Photobiological properties of monofunctional furocoumarin derivatives, in: *Topics in Photomedicine* (K. C. Smith, ed.), pp. 319–398, Plenum Press, New York (1984).
25. F. Dall'Acqua, S. Caffieri, and G. Rodighiero, Photoreactions of furocoumarins (psoralens and angelicins), in: *Primary Photo-Processes in Biology and Medicine* (R. V. Bensasson, G. Jori, E. J. Land, and T. G. Truscott, eds.), pp. 259–272, Plenum Press, New York (1985).
26. F. Dall'Acqua and G. Rodighiero, Biological and medical aspects of furocoumarins (psoralens and angelicins), in: *Primary Photo-Processes in Biology and Medicine* (R. V. Bensasson, G. Jori, E. J. Land, and T. G. Truscott, eds.), pp. 277–293, Plenum Press, New York (1985).
27. I. E. Kochevar, Phototoxicity mechanisms: Chlorpromazine-photosensitized damage to DNA and cell membranes, *J. Invest. Dermatol.* **76,** 59–64 (1981).
28. C. S. Foote, F. C. Shook, and R. B. Abakerli, Characterization of singlet oxygen, in: *Methods in Enzymology, Vol. 105, Oxygen Radicals in Biological Systems* (L. Packer, ed.), pp. 36–47, Academic Press, New York (1984).
29. J. B. Verlhac, A. Gaudemer, and I. Kraljić, Water soluble porphyrins and metalloporphyrins as photosensitizers in aerated aqueous solutions. I. Detection and determination of quantum yields of formation of singlet oxygen, *J. Chimie* **8,** 401–406 (1984).
30. R. V. Bensasson, E. J. Land, and T. G. Truscott, *Flash Photolysis and Pulse Radiolysis: Contributions to the Chemistry of Biology and Medicine*, Pergamon Press, Oxford (1983).

31. M. A. J. Rodgers, Instrumentation for the generation and detection of transient species, in: *Primary Photo-Processes in Biology and Medicine* (R. V. Bensasson, G. Jori, E. J. Land, and T. G. Truscott, eds.), pp. 1–23, Plenum, New York (1985).
32. L. Santamaria and G. Prino, List of the photodynamic substances, in: *Research Progress in Organic, Biological and Medicinal Chemistry,* Vol. 3, Part 1 (U. Gallo and L. Santamaria, eds.), pp. XI–XXXV, North-Holland, Amsterdam (1972).
33. G. H. N. Towers, Photosensitizers from plants and their photodynamic action, *Progr. Phytochem.* **6,** 183–202 (1980).
34. H. E. A. Kramer, The lightfastness or the non-photochemistry of dyes, *Chimia* **40,** 160–169 (1986).
35. N. I. Krinsky, Photobiology of carotenoid protection, in: *The Science of Photomedicine* (J. D. Regan and J. A. Parrish, eds.), pp. 397–407, Plenum Press, New York (1982).
36. J. D. Spikes, Photodegradation of foods and beverages, in: *Photochemical and Photobiological Reviews,* Vol. 6 (K. C. Smith, ed.), pp. 39–85, Plenum Press, New York (1981).
37. I. Rosenthal, Photooxidation of foods, in: *Singlet O_2,* Vol. 4 (A. A. Frimer, ed.), pp. 145–163, CRC Press, Boca Raton (1985).
38. J. Amagasa, Dye binding and photodynamic action, *Photochem. Photobiol.* **33,** 947–955 (1981).
39. S. Kawanishi, S. Inoue, and S. Sano, Photodynamic guanine modification by hematoporphyrin is specific for single-stranded DNA with singlet oxygen as a mediator, *J. Biol. Chem.* **261,** 6090–6095 (1986).
40. G. D. Cimino, H. G. Gamper, S. T. Isaacs, and J. E. Hearst, Psoralens as photoactive probes of nucleic acid structure and function: Organic chemistry, photochemistry, and biochemistry, *Annu. Rev. Biochem.* **54,** 1151–1193 (1985).
41. J. Cadet, M. Berger, C. Decarroz, J. R. Wagner, J. E. Van Lier, Y. M. Ginot, and P. Vigny, Photosensitized reactions of nucleic acids, *Biochimie* **68,** 813–834 (1986).
42. R. M. Santella, N. Dharmaraja, F. P. Gasparro, and R. L. Edelson, Monoclonal antibodies to DNA modified by 8-methoxypsoralen and ultraviolet A light, *Nucleic Acid Res.* **13,** 2533–2544 (1985).
43. J. L. Bryant, Jr., and J. King, DNA injection proteins are targets of acridine-sensitized photoinactivation of bacteriophage P22, *J. Mol. Biol.* **180,** 837–863 (1984).
44. C. V. Hanson, Inactivation of viruses for use as vaccines and immunodiagnostic reagents, in: *Medical Virology,* Vol. 2, Proc. Int. Symp. 1982, pp. 45–79, Elsevier, Amsterdam (1983).
45. J. L. Melnick and C. Wallis, Photodynamic inactivation of herpesvirus, in: *The Science of Photomedicine* (J. D. Regan and J. A. Parrish, eds.), pp. 545–559, Plenum Press, New York (1982).
46. F. Rapp and J.-L. H. Li, Transforming activity of viruses after dye-light inactivation, in: *The Science of Photomedicine* (J. D. Regan and J. A. Parrish, eds.), pp. 571–594, Plenum Press, New York (1982).
47. L. E. Bockstahler, T. P. Coohill, K. B. Hellman, C. D. Lytle, and J. E. Roberts, Critical review and risk evaluation of photodynamic therapy for herpes simplex, in: *Int. Encycl. Pharmacol. Ther. (Viral Chemother.),* Vol. 1 (D. Shugar, ed.), pp. 479–509, Pergamon Press, Oxford (1984).
48. L. E. Schnipper, A. E. Lewin, M. Swartz, and C. S. Crumpacker, Mechanisms of photodynamic inactivation of herpes simplex viruses: Comparison between methylene blue, light plus electricity, and hematoporphyrin plus light, *J. Clin. Invest.* **65,** 432–438 (1980).
49. W. Snipes, G. Keller, J. Woog, T. Vickroy, R. Deering, and A. Keith, Inactivation of lipid-containing viruses by hydrophobic photosensitizers and near-ultraviolet radiation, *Photochem. Photobiol.* **29,** 785–790 (1979).
50. G. Jori and J. D. Spikes, Photosensitized oxidations in complex biological structures, in: *Oxygen and Oxy-Radicals in Chemistry and Biology* (M. A. J. Rodgers and E. L. Powers, eds.), pp. 441–459, Academic Press, New York (1981).
51. J. D. Spikes, Photosensitizing properties of porphyrins in model cell systems, in: *Porphyrins in Tumor Phototherapy* (A. Andreoni and R. Cubeddu, eds.), pp. 51–60, Plenum Press, New York (1984).
52. E. Reddi, M. A. J. Rodgers, J. D. Spikes, and G. Jori, The effect of medium polarity on the hematoporphyrin-sensitized photooxidation of L-tryptophan, *Photochem. Photobiol.* **40,** 415–421 (1984).
53. L. I. Grossweiner and G. Goyal, Photosensitization of liposomes by porphyrins, *J. Photochem.* **25,** 253–265 (1984).
54. J. D. Spikes and R. C. Straight, Photodynamic behavior of porphyrins in model cell, tissue and tumor systems, in: *Photodynamic Therapy of Tumors and Other Diseases* (G. Jori and C. Perria, eds.), pp. 45–53, Libreria Progetto, Padova (1985).

55. T. Ito, Photodynamic agents as tools for cell biology, *Photochem. Photobiol. Rev.* **7,** 141–186 (1983).
56. G. Bertoloni, A. Viel, A. Grossato, and G. Jori, The photosensitizing activity of haematoporphyrin on mollicutes, *J. Gen. Microbiol.* **131,** 2217–2223 (1985).
57. J. Moan, Porphyrin photosensitization and phototherapy, *Photochem. Photobiol.* **43,** 681–690 (1986).
58. R. W. Tuveson, M. R. Berenbaum, and E. E. Heininger, Inactivation and mutagenesis by phototoxins using *Escherichia coli* strains differing in sensitivity to near- and far-ultraviolet light, *J. Chem. Ecol.* **12,** 933–948 (1986).
59. J. P. Pooler and D. P. Valenzeno, Dye-sensitized photodynamic inactivation of cells, *Med. Phys.* **8,** 614–628 (1981).
60. P. Ries and R. Beiderbeck, Photosensibilization of isolated mesophyll cells of *Calystegia sepium* L., *Z. Naturforsch.* **34c,** 1215–1217 (1979).
61. M. E. Daub, Peroxidation of tobacco membrane lipids by the photosensitizing toxin, cercosporin, *Plant. Physiol.* **69,** 1361–1364 (1982).
62. B. Kjeldstad, M. L. Olsten, T. Christensen, and A. Johnsson, Photodynamic damage to protoplasts with a hematoporphyrin derivative, *Plant Sci.* **44,** 139–143 (1986).
63. G. Moreno, Photosensitization of mammalian cells by psoralens and porphyrins, *Biochimie* **68,** 869–873 (1986).
64. D. P. Valenzeno, Photomodification of biological membranes with emphasis on singlet oxygen mechanisms, *Photochem. Photobiol.* **46,** 147–160 (1987).
65. A. W. Girotti, J. D. Thomas, and J. E. Jordan, Prooxidant and antioxidant effects of ascorbate on photosensitized peroxidation of lipids in erythrocyte membranes, *Photochem. Photobiol.* **41** 267–276 (1985).
66. R. Santus and J. P. Reyftmann, Photosensitization of membrane components, *Biochimie* **68,** 813–818 (1986).
67. T. M. A. R. Dubbelman, Porphyrin-photosensitized modification of subcellular structures, in: *Photodynamic Therapy of Tumors and Other Diseases* (G. Jori and C. Perria, eds.), pp. 93–99, Libreria Progetto, Padova (1985).
68. H. D. Rabinowitch and I. Fridovich, Superoxide radicals, superoxide dismutases and oxygen toxicity in plants, *Photochem. Photobiol.* **37,** 679–690 (1983).
69. M. L. Molero, M. J. Hazen, and J. C. Stockert, Photodynamic effects of berberine sulfate on the growth rate of *Allium cepa* roots, *J. Plant Physiol.* **120,** 91–94 (1985).
70. J. P. Knox and A. D. Dodge, The photodynamic action of eosin, a singlet oxygen generator. The inhibition of photosynthetic electron transport, *Planta* **164,** 30–34 (1985).
71. F. J. Gommers, J. Bakker, and H. Wynberg, Dithiophenes as singlet oxygen sensitizers, *Photochem. Photobiol.* **35,** 615–619 (1982).
72. D. D. Morgan and D. Warshawsky, The photodynamic immobilization of *Artemia salina* nauplii by polycyclic aromatic hydrocarbons and its relationship to carcinogenic activity, *Photochem. Photobiol.* **25,** 39–46 (1977).
73. J. R. Heitz and K. R. Downum (eds.), *Light-Activated Pesticides,* ACS Symp. Ser. 339, American Chemical Society, Washington, D.C. (1987).
74. A. Shlosberg and M. N. Egyed, Photosensitization in ducklings induced by seeds of *Cymopterus watsonii* and *C. longpipes, Avian Diseases* **22,** 576–583 (1978).
75. G. W. Ivie, Chemical and biochemical aspects of photosensitization in livestock and poultry, *J. Natl. Cancer Inst.* **69,** 259–262 (1982).
76. J. H. Epstein, Chemical phototoxicity in humans, *J. Natl. Cancer Inst.* **69,** 265–268 (1982).
77. K. L. Gabriel, P. D. Forbes, and R. E. Davies, Phototoxicity, *J. Toxicol., Cutaneous Ocul. Toxicol.* **4,** 185–191 (1984).
78. L. C. Harber, I. E. Kochevar, and A. R. Shalita, Mechanisms of photosensitization to drugs in humans, in: *The Science of Photomedicine* (J. D. Regan and J. A. Parrish, eds.), pp. 323–347, Plenum Press, New York (1982).
79. D. P. Valenzeno and J. P. Pooler, Phototoxicity, the neglected factor, *J. Am. Med. Assoc.* **242,** 453–454 (1979).
80. J. T. Hindmarsh, The porphyrias: Recent advances, *Clin. Chem.* (Winston-Salem, N.C.) **32,** 1255–1263 (1986).

81. M. M. Mathews-Roth, Beta-carotene therapy for erythropoietic protoporphyria and other photosensitivity diseases, *Biochimie* **68,** 875–884 (1986).
82. S. J. Galitzer and F. W. Oehme, Photosensitization: A literature review, *Vet. Sci. Commun.* **2,** 217–230 (1978).
83. A. E. Johnson, Toxicologic aspects of photosensitization in livestock, *J. Natl. Cancer Inst.* **69,** 253–258 (1982).
84. G. R. Ruth, S. Schwartz, and B. Stephenson, Bovine protoporphyria: The first nonhuman model of this hereditary photosensitizing disease, *Science* **198,** 199–202 (1977).
85. A. C. Giese, Hypericism, *Photochem. Photobiol. Revs.* **5,** 229–255 (1980).
86. I. A. Magnus (ed.), *Dermatological Photobiology: Clinical and Experimental Aspects,* Blackwell, Oxford (1976).
87. H. Ippen, Photodermatoses, in: *The Science of Photomedicine* (J. D. Regan and J. A. Parrish, eds.), pp. 349-394, Plenum Press, New York (1982).
88. F. Urbach, Ultraviolet radiation and skin cancer, in: *Topics in Photomedicine* (K. C. Smith, ed.), pp. 39–142, Plenum Press, New York (1984).
89. P. D. Forbes, R. E. Davies, F. Urbach, and C. Cole, Photocarcinogenesis, *J. Toxicol. Cutaneous Ocul. Toxicol.* **4,** 219–236 (1985).
90. L. Santamaria, A. Arnaboldi, L. Andreoni, A. Bianchi, and P. Daffara, Photocarcinogenesis by methoxypsoralen, neutral red, proflavine and long UV radiation, *Cancer Detect. Prev.* **8,** 447–454 (1985).
91. J. D. Spikes, The historical development of ideas on applications of photosensitized reactions in the health sciences, in: *Primary Photo-Processes in Biology and Medicine* (R. V. Bensasson, G. Jori, E. J. Land, and T. G. Truscott, eds.), pp. 209–227, Plenum Press, New York (1985).
92. T. R. C. Sisson and T. P. Vogel, Phototherapy of hyperbilirubinemia, in: *The Science of Photomedicine* (J. D. Regan and J. A. Parrish, eds.), pp. 477–509, Plenum Press, New York (1982).
93. J. A. Parrish, R. S. Stern, M. A. Pathak, and T. B. Fitzpatrick, Photochemotherapy of skin diseases, in: *The Science of Photomedicine* (J. D. Regan and J. A. Parrish, eds.), pp. 595–623, Plenum Press, New York (1982).
94. D. Kessel (ed.), *Methods in Porphyrin Photosensitization,* Plenum Press, New York (1985).
95. T. J. Dougherty, Photosensitizers: Therapy and detection of malignant tumors, *Photochem. Photobiol.* **45,** 879–889 (1987).
96. A. R. Oseroff, G. Ara, D. Ohuoha, J. Aprille, J. C. Bommer, M. L. Yarmush, J. Foley, and L. Cincotta, Strategies for selective cancer photochemotherapy: Antibody-targeted and selective carcinoma cell photolysis, *Photochem. Photobiol.* **46,** 83–96 (1987).
97. J. R. Heitz, Xanthene dyes as pesticides, in: *Insecticide Mode of Action* (J. R. Coates, ed.), pp. 429–457, Academic Press, New York, (1982).
98. J. R. Robinson, Photodynamic insecticides: A review of studies on photosensitizing dyes as insect control agents, their practical application, hazards and residues, *Residue Rev.* **88,** 69–100 (1983).
99. J. T. Arnason, G. H. N. Towers, B. J. R. Philogène, and J. D. H. Lambert, The role of natural photosensitizers in plant resistance to insects, *ACS Symp. Ser. 208* (Plant Resist. Insects), 139–151 (1983).
100. K. R. Downum, Photoactivated biocides from higher plants, *ACS Symp. Ser. 296* (Nat. Resist. Plants Pests), 197–205 (1986).
101. J. T. Arnason, B. J. R. Philogène, P. Morand, J. C. Scaiano, N. Werstiuk, and J. Lam, Thiophenes and acetylenes: Phototoxic agents to herbivorous and blood-feeding insects, *ACS Symp. Ser. 339* (Light-Activated Pesticides), 255–264 (1987).
102. J. Kagan, P. Szczepanski, V. Bindokas, W. D. Wulff, and J. S. McCallum, Delayed phototoxic effects of 8-methoxypsoralen, khellin and sphondin in *Aedes aegypti, J. Chem. Ecol.* **12,** 899–914 (1986).
103. C. A. Rebeiz, A. Montazer-Zouhoor, J. M. Mayasich, B. C. Tripathy, S. M. Wu, and C. C. Rebeiz, Photodynamic herbicides and chlorophyll biosynthesis modulators, *ACS Symp. Ser. 339* (Light-Activated Pesticides), 295–328 (1987).
104. G. A. Rosenthal and D. H. Janzen (eds.), *Herbivores: Their Interaction with Secondary Plant Metabolites,* Academic Press, New York (1979).
105. D. Ashkenzay, Y. Kashman, A. Nyska, and J. Friedman, Furocoumarins in shoots of *Pituranthos triradiatus* (Umbelliferae) as protectants against grazing by Hyrax (Procaviidea: *Procavia capensis syriaca*), *J. Chem. Ecol.* **11,** 231–239 (1985).

106. G. H. N. Towers, Significance of phototoxic phytochemicals in insect herbivory, *J. Chem. Ecol.* **12,** 813–821 (1986).
107. K. R. Downum and E. Rodriguez, Toxicological action and ecological importance of plant photosensitizers, *J. Chem. Ecol.* **12,** 823–834, (1986).
108. D. E. Champagne, J. T. Arnason, B. J. R. Philogène, P. Morand, and J. Lam, Light-mediated allelochemical effects of naturally occurring polyacetylenes from Asteraceae on herbivorous insects, *J. Chem. Ecol.* **12,** 835–858 (1986).
109. M. R. Berenbaum, Charge of the light brigade: phototoxicity as a defense against insects, *ACS Symp. Ser. 339* (Light-Activated Pesticides), 206–216 (1987).
110. J. B. Harborne (ed.), *Biochemical Aspects of Plant and Animal Coevolution,* Academic Press, New York (1978).
111. M. Berenbaum, Coumarins and caterpillars: A case for coevolution, *Evolution* **37,** 163–179 (1983).
112. G. W. Ivie, Biological actions and metabolic transformations of furanocoumarins, *ACS Symp. Ser. 339* (Light-Activated Pesticides), 217–230 (1987).
113. D. L. Bull, G. W. Ivie, R. C. Beier, and N. W. Pryor, In vitro metabolism of a linear furanocoumarin (8-methoxypsoralen, xanthotoxin) by mixed-function oxidases of larvae of black swallowtail butterfly and fall armyworm, *J. Chem. Ecol.* **12,** 885–892 (1986).
114. M. R. Berenbaum and P. Feeny, Toxicity of angular furanocoumarins to swallowtails: Escalation in the coevolutionary arms race, *Science* **212,** 927–929 (1981).
115. R. A. Larson, Insect defenses against phototoxic plant chemicals, *J. Chem. Ecol.* **12,** 859–870 (1986).

4

UV Radiation Effects

DNA Repair and Mutagenesis

4.1. REPAIR OF PHOTOCHEMICAL DAMAGE TO DNA

4.1.1. Introduction

The first indication that cells might have the capacity to recover from radiation damage was the observation that minor modifications in the postirradiation treatment of the cells

Kendric C. Smith • Department of Radiation Oncology, Stanford University School of Medicine, Stanford, California 94305.

(e.g., growth media, temperature, etc.) had a marked effect upon the ultimate viability of irradiated cells. Thus in 1937, Hollaender and Claus found that higher survival levels of UV-irradiated fungal spores could be obtained if they were allowed to remain in water or salt solution for a period of time before assaying for survival. Roberts and Aldous[1] extended these observations by showing that the shapes of the UV radiation survival curves for *Escherichia coli* could be changed drastically by holding the irradiated cells in media devoid of an energy source for various times before plating on nutrient agar to assay for survival (Fig. 4-1). The molecular basis of this phenomenon, known as liquid-holding recovery, will be discussed along with other "recovery phenomena" in Section 4.1.3.

The isolation of a radiation-sensitive mutant of *E. coli* (B_{s-1}) by Ruth Hill in 1958 inaugurated the era of the genetic approach to the study of DNA repair. Genetic studies have been responsible for most of the discoveries about the molecular basis of DNA repair. For example, one can measure the initial yield of a particular type of radiation-induced lesion in the DNA of a wild-type strain (i.e., radiation-resistant) and a mutant strain (i.e., radiation-sensitive), and then follow the kinetics and extent of repair of that lesion, and thereby determine if the mutation under investigation has an effect on the repair of that type of DNA lesion. Then by constructing strains that carry two mutations that affect DNA repair, one can determine through survival and DNA repair studies whether the two mutations act in the same or in different biochemical pathways of repair.

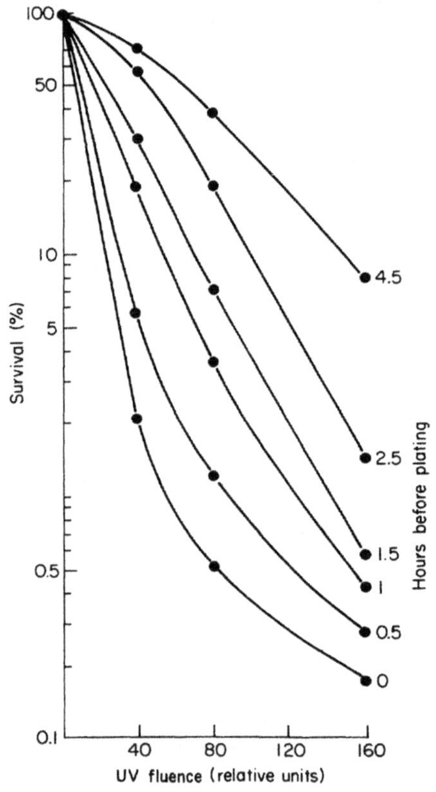

Fig. 4-1. UV radiation survival curves of *E. coli* B. After irradiation the cells were suspended in a liquid medium devoid of an energy source for the times indicated before being plated on nutrient agar. It is evident that both the slopes and the shapes of the survival curves can be altered by the postirradiation treatment of the cells. These data, from 1949, provided evidence that cells could recover from radiation damage. (Modified from Ref. 1.)

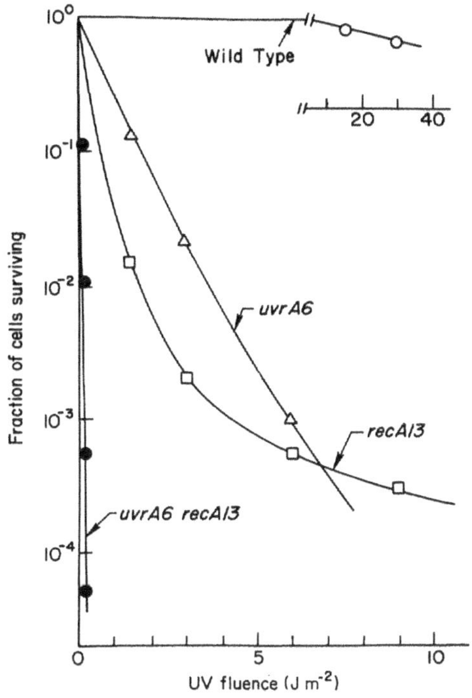

Fig. 4-2. UV radiation survival curves of DNA repair-deficient mutants of *E. coli* K-12. The mutations are defined in Table 4-1. Note that the double mutant is much more radiation-sensitive than either of the single mutants, suggesting that the *uvrA* gene and the *recA* gene act in different pathways of DNA repair. It is now known that the *uvrA* mutation blocks excision repair, and the *recA* mutation blocks postreplication repair (Section 4.1.5) and inducible processes (Section 4.1.4). [Modifed from P. Howard-Flanders and R. P. Boyce, *Radiat. Res. Suppl.* **6**, 156 (1966).]

For example, since the *uvrA recA* double mutant is much more UV radiation-sensitive than are either of the singly mutant strains (Fig. 4-2), it suggests that the *uvrA* and *recA* genes function in separate pathways of DNA repair. As will be seen later (Section 4.1.5), a *uvrA* mutation blocks excision repair, and a *recA* mutation blocks postreplication repair and inducible repair functions (Section 4.1.4).

The discovery of excision repair (Section 4.1.5.2) by Howard-Flanders and coworkers and by Setlow and coworkers came from the observation that, after UV irradiation, the wild-type strain broke down some of its DNA into acid-soluble material in a radiation dose-dependent manner, but the radiation-sensitive strain did not. It did not seem logical at first that DNA degradation could be beneficial to survival, but subsequently it was shown that this DNA degradation was not random. UV radiation-induced lesions (i.e., thymine dimers) were selectively cut out of the DNA in the wild-type strain, but not in the sensitive strain. It is this ability to make comparative measurements between the wild-type strain and different radiation-sensitive strains that has allowed the rapid progress in our understanding of DNA repair in *E. coli*. The paucity of radiation-sensitizing mutations in mammalian cells has slowed progress in understanding DNA repair in mammalian cells.[2–4] Many mutations that affect radiation sensitivity are known for yeast,[5] but the understanding of their biochemical basis is far behind that for *E. coli*. This explains why the bulk of the information for this chapter comes from work on *E. coli*.

A textbook on DNA repair,[2] the proceedings of an international meeting on this subject,[6] and a laboratory manual of research procedures for studying DNA repair[7] have recently been published.

4.1.2. Survival Curves

A radiation survival curve is a plot of the logarithm of the surviving fraction of a population of cells vs. the dose of radiation (a semilogarithmic plot). For ionizing radiation (e.g., X rays) the dose is generally the *absorbed* dose (100 rad = 1 gray), but for nonionizing radiation (e.g., UV radiation) it is generally the *incident* dose (fluence) (joules/m^2). This difference (i.e., absorbed vs. incident dose) simply reflects the types of dosimetry in common use for these two types of radiation.

Before it was discovered that cells could recover from radiation-induced damage, survival curves were interpreted in terms of *target theory,* i.e., in terms of physical *targets* and inactivating *hits* within the targets.(8,9) If the survival curve is a straight line on a semilogarithmic plot, it suggests that the cells were killed when one target within the cell was hit once by the radiation. More complicated survival curves (e.g., shouldered curves such as those labeled 2.5 and 4.5 in Fig. 4-1) were interpreted as being due to the presence of multiple targets and/or the requirement for multiple hits within the targets.

In a form similar to the Lambert–Bouguer–Beer law (Section 1.5.1), one can express the simplest case of cellular inactivation (i.e., a straight line on a semilogarithmic plot) by the relationship $N/N_o = e^{-\sigma D}$, where N_o is the number of live cells initially present and N is the number of cells surviving after a dose D. The constant σ is called the inactivation cross-section and is the product of the quantum yield (Φ) for inactivation (i.e., Φ = number of cells killed/number of photons absorbed) and the absorption cross-section. The expression σD is the average number of lethal hits per target. [The determination of σ over a range of wavelengths would constitute an action spectrum (Section 1.5.5) for inactivation.] The inactivation cross-section can be determined from the slope of the survival curve and is the reciprocal of the dose to give 37% survival (N.B.: e^{-1} = 0.37). At this dose (mean lethal dose), designated as D_{37} (dose) or F_{37} (fluence), there is an average of one hit per sensitive target. A similar term for shouldered survival curves is D_0 and is calculated from the linear portion of the curve. A description of the difference between D_0 and D_{37} and the definition of other parameters that can be derived from survival curves are shown in Fig. 4-3.

Most survival curves are not straight lines but come in a variety of shapes (e.g., Fig. 4-1). The mathematics of the curve fitting becomes much more complicated for such curves,(8,9) and attempts to explain the mechanism(s) of cell lethality on the basis of such curve fitting become even more difficult. A recent development (1972) in curve fitting is the linear-quadratic model for inactivation. Many (but not all!) X-ray survival curves for mammalian cells can be fit by the expression $N/N_o = e^{-(\alpha D + \beta D^2)}$. This suggests that part of the targets are inactivated by single-hit kinetics (i.e., a linear function of dose D) and some by two-hit kinetics (i.e., dose squared, D^2). Alpha and beta are constants.

Another interpretation of the diverse shapes of survival curves has tried to incorporate the concept of DNA repair.(e.g., 10) The simplest interpretation is that a shoulder on a survival curve indicates the capacity to repair and a linear survival curve indicates that the cells have no repair capacity. At high doses when a shouldered survival curve becomes exponential (i.e., linear on a semilogarithmic plot), it is presumed that either the repair systems themselves have been inactivated or they have been "swamped" by the large amount of damage produced by such high doses. These interpretations are oversimplified, however, since a linear survival curve does not necessarily mean that cells have no DNA repair capacity.

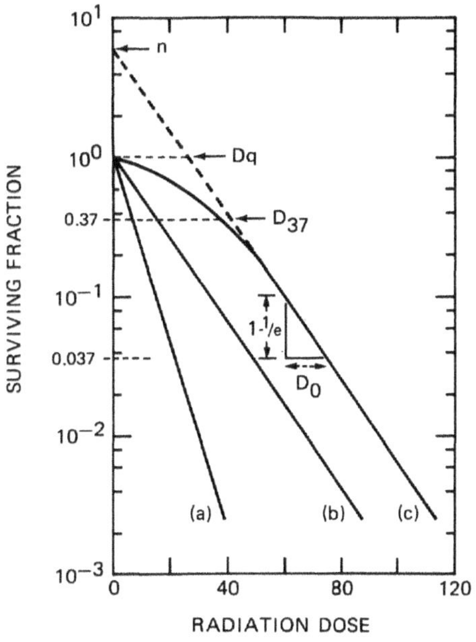

Fig. 4-3. Examples of survival curves and their associated parameters. Straight-line survival curves (a and b) can be characterized by a single parameter, the D_{37} (i.e., the dose to yield a survival of 37%). For a shouldered survival curve (c), three parameters apply. The extrapolation number (n) is one measure of the width of the shoulder, and is the point where an extrapolation of the linear portion of the survival curve crosses the ordinate. D_q (quasi-threshold dose) is also a measure of the width of the shoulder, and is the dose at which the extrapolation line, described above, intersects the 100% survival line. The D_0 is the same concept as the D_{37}, but it is determined on the linear portion of the survival curve. While $D_0 = D_{37}$ on an exponential survival curve, these two parameters differ greatly on a shouldered survival curve. Note that the D_0 for curve c is equal to the D_{37} of curve b, i.e., the lines are parallel. These parameters were largely developed for studies using ionizing radiation and the term dose (i.e., absorbed dose). Most frequently for UV radiation studies, fluence (incident dose) is used rather than dose, so the comparable parameters would be F_0, F_{37}, and F_q.

Such curve fitting has not greatly advanced our understanding of the molecular basis of radiation biology. However, "mathematical analysis does compel one, at least, to make explicit the assumptions and ideas used to describe the phenomena."[11] Since such mathematical treatments have played such a large part in the history of radiation biology, a student of radiation biology (UV or ionizing radiation) should become knowledgeable in this area, but should also be aware of its limitations.

4.1.3. Recovery Phenomena

4.1.3.1. Liquid-Holding Recovery

Recovery phenomena have been an important part of radiation biology.[12] The recovery phenomenon now known as *liquid-holding recovery* (LHR) provided the first evidence that cells could recover from radiation-induced DNA damage (Section 4.1.1). The initial observation was that when *E. coli* cells were held in the absence of nutrients for a period of time after UV irradiation, not only did the shape of their survival curve change with time, but the cells also exhibited a much higher survival than if they were plated on growth medium immediately after UV irradiation (Fig. 4-1).

Genetic studies have shown that there are two major requirements for observing LHR.[13] First, the cell must be proficient in nucleotide excision repair (Section 4.1.5.2), and second, it must be at least partially deficient in postreplication repair (Section 4.1.5.3). In fact, cells completely deficient in postreplication repair (i.e., *recA* cells) show the greatest amount of LHR.

In a fully repair-proficient cell (i.e., wild-type), if a DNA lesion is not repaired by the excision repair process it can usually be repaired by the postreplication repair process. However, nucleotide excision repair can occur in buffer that is devoid of nutrients, but postreplication repair requires the presence of complete growth medium (i.e., macromolecular synthesis is required). Therefore, by holding cells in buffer without essential nutrients, both DNA replication and postreplication repair are prevented, and only excision repair can proceed. Such liquid holding is beneficial to the survival of cells deficient in postreplication repair because the more lesions that are repaired by the excision repair process before the cells are placed in growth medium where DNA replication can proceed, the fewer lesions that will be left to cause the formation of DNA daughter strand gaps, which are the substrates for a repair process in which the cells are deficient. LHR has also been studied in yeast.

In mammalian cells, a similar phenomenon after X irradiation is called *plateau phase recovery,* or the repair of *potentially lethal damage.*(14) In this process, cells that have entered stationary phase (i.e., the cells are confluent and nutrients in the growth medium are exhausted) show a higher survival after X irradiation if they are kept in stationary phase for several hours rather than being immediately subcultured and plated in fresh medium for the determination of survival.

4.1.3.2. Minimal Medium Recovery

UV-irradiated, excision repair-deficient cells of *E. coli* exhibit *minimal medium recovery* (MMR), although a better name for this process would be rich medium lethality (see below). The observation was that excision repair-deficient cells show a higher survival after UV irradiation if they are plated on minimal medium rather than on rich medium, regardless of the type of medium in which they were grown before irradiation. Since this process was described in excision repair-deficient cells, the presumption was that some aspect of postreplication repair was inhibited by components of the rich growth medium. This has proven to be the case. The major processes of postreplication repair (i.e., the repair of DNA daughter strand gaps and the repair of DNA double-strand breaks) are partially inhibited by rich growth medium. This inhibitory effect of rich medium can be reproduced by just adding the common amino acids to minimal growth medium. A mutation, *mmrA,* has been isolated that blocks this effect of rich medium on survival and repair, but the *mmrA* gene product is not known,(15) nor is the mechanism known by which excess amino acids inhibit postreplication repair.(16)

4.1.3.3. Medium-Dependent Resistance

E. coli cells grown to logarithmic phase in rich medium and plated on rich medium are about three times (*dose modification factor,* i.e., ratio of doses to produce equal survival) more resistant to X radiation and about 1.5 times more resistant to UV radiation than are cells grown in and plated on minimal medium, i.e., the cells show *medium dependent resistance* (MDR).(17) Therefore, MDR reflects a selective effect on the repair of X-ray-induced DNA damage. Cells that exhibit MDR also show a much greater capacity for postirradiation protein synthesis and for the repair of DNA double-strand breaks. MDR is an inducible process that is absent in *recA, lexA,* and *recBC* cells, and is partially inhibited by other mutations.(18)

Such recovery phenomena (especially LHR) were particularly useful for gaining insights about DNA repair in the period before mutations affecting DNA repair were available. Now in combination with genetic studies, certain of these recovery phenomena offer the possibility of studying the physiological control of DNA repair (e.g., MMR, MDR).

4.1.4. Inducible Responses

Since the isolation of the first radiation-sensitive mutant of *E. coli* in 1958, numerous mutations that affect radiation sensitivity have been isolated, and the biochemical deficiency of many of these mutants has been determined (Table 4-1). Through the characterization and comparison of these mutants have come most of the discoveries of the multiple processes of DNA repair.

Another important discovery from such genetic studies was that some of the repair systems are constitutive and some are inducible. Furthermore, certain of the constitutive repair proteins (e.g., UvrA and UvrB proteins) can be induced to higher concentrations in irradiated cells.

4.1.4.1. SOS System[2,19–21]

In 1974, Radman postulated that DNA damage, or the physiological consequence of such damage (e.g., the temporary blockage of DNA synthesis), causes the induction of repair processes in *E. coli.* He called this phenomenon "SOS" repair (N.B.: SOS is the international distress signal). (reviewed in Ref. 2) Since UV radiation mutagenesis was only observed under conditions of functional SOS repair, it was presumed that this mutagenesis

TABLE 4-1. Some of the More Important Genes of Escherichia coli That Affect DNA Repair

Gene symbol	Map position (min)[a]	Gene product and/or function[b]
ada	48	Methyltransferase, adaptive response
alkA	43	3-Methyladenine-DNA glycosylase II (inducible)
lexA	92	"SOS" gene repressor, inducible repair
lig	52	DNA ligase, all "dark" repair
phr	16	DNA photolyase, photoreactivation
polA	87	DNA polymerase I, ER, PRR
polB	2	DNA polymerase II, unknown
polC	4	DNA polymerase III, DNA replication
recA	58	SOS repair, recombination, ER, PRR
recB,C,D	61	Exonuclease V, recombination, ER, PRR
recF	83	Recombination, ER, PRR
ssb	92	Single-strand DNA-binding protein
tag	72	3-Methyladenine DNA-glycosylase I (noninducible)
umuC,D	26	UV radiation mutagenesis, PRR
uvrA,B,C	92,18,42	Excision nuclease, ER
uvrD	85	Helicase II, ER, PRR

[a]From B. J. Bachmann, Linkage map of *Escherichia coli* K-12, ed. 7, *Microbiol. Rev.*, *47*, 180–230 (1983).
[b]ER = nucleotide excision repair; PRR = postreplication repair.

was due to error-prone DNA repair. The SOS phenomenon is under the control of the *recA* and *lexA* genes. DNA-damaging agents other than UV radiation can also induce the SOS system, e.g., ionizing radiation, and chemicals such as mitomycin C. Since the SOS system is inducible, inhibitors of RNA and protein synthesis prevent its expression.

One candidate for the "signal" that causes the SOS system to be induced is the sudden increase in the amount of single-stranded DNA that occurs in a cell due to the distortion of the DNA helix by lesions in the DNA and/or due to the stalling of DNA replication at the site of a lesion. In support of this idea, single-stranded DNA activates the protease function of the RecA protein (product of the *recA* gene) *in vitro*. The LexA protein (product of the *lexA* gene) is the repressor of the SOS genes. The activated RecA protein catalyzes the cleavage of the LexA protein, and when the concentration of LexA protein in the cell falls to a low level the SOS genes become depressed, including the *recA* and *lexA* genes themselves. Other inducible SOS genes include the *uvrA*, *uvrB*, and *uvrD* genes, which are involved in excision repair; the *recN* and *ruv* genes, which are involved in recombination; a number of *din* (*d*amage-*in*ducible) genes, some of which are still uncharacterized; and the *umuC* and *umuD* genes, which play a key role in UV radiation mutagenesis but play a lesser role in X-ray mutagenesis (Section 4.2.4). (N.B.: Recent evidence indicates that the activated RecA protein also catalyzes the cleavage of the UmuD protein.) Thus, a large set of genes that perform diverse functions are induced in the SOS process to optimize the cell's chances for survival.

It is still not totally resolved as to whether or not mammalian cells show an SOS response.[e.g., 22,23]

4.1.4.2. *Adaptive Response*[2,20,21]

Other inducible DNA repair processes are known that are independent of the SOS system. The best characterized of these is the *adaptive response*, the induced resistance to alkylating agents. In 1977, Samson and Cairns reported that cells of *E. coli* that had been treated with low levels of the methylating agent *N*-methyl-*N'*-nitro-*N*-nitrosoguanidine (MNNG) developed resistance to the lethal and mutagenic effects of a subsequent challenge with a higher dose of MNNG. Two types of processes are induced in the adaptive response. One of these is related to base excision repair, and one is an entirely new mechanism involving an enzyme called a methyltransferase (see below).

The product of the *alkA* gene is 3-methyladenine-DNA glycosylase II, and its concentration increases approximately 20-fold as a consequence of the adaptive response. This enzyme has a broad range of activities for removing a number of methylated purines and pyrimidines from DNA, as the first step in base excision repair (see Section 4.1.5.2). The product of the *tag* gene is 3-methyladenine-DNA glycosylase I. It is not inducible and shows a high specificity for removing 3-methyladenine from DNA.

The product of the *ada* gene not only plays a role in regulating the adaptive response by a mechanism not yet understood but also participates in the repair process as a methyltransferase. Methyltransferase is a new kind of enzyme, a *suicide enzyme*. The classical definition of an enzyme (an organic catalyst) is that it speeds up the reaction but is not consumed in the reaction. The methyltransferases do not follow this rule. They transfer a methyl group, most frequently from O^6-methylguanine, to a specific cysteine residue in their structure, and thereby inactivate themselves.

4.1.4.3. Heat-Shock System[20]

When cells are shifted to lethal temperatures for a sublethal period of time (i.e., they are given a heat shock), they induce a number of heat-shock proteins (detectable by gel electrophoresis) and become resistant to a subsequent exposure to high temperatures. The definition of high temperature differs for different species, but for mammalian cells it is around 43°C and for *E. coli* it is around 50°C. The gene that controls this response in *E. coli* is the *htpR* gene. The molecular basis of this heat-shock response (i.e., is it due to damage to DNA or to membranes, or both) is under active investigation in many types of cells. This is especially true for human cells, since hyperthermia is now being used in the treatment of certain types of human tumors.[24]

There is some overlap in these three inducible systems, e.g., MNNG can induce *some* of the SOS proteins, and SOS-inducing agents (e.g., UV radiation, nalidixic acid, etc.) can induce *some* of the heat-shock proteins.

4.1.4.4. Oxidation System

Evidence is accumulating for the existence of a fourth inducible system, which is specific for oxidative damage to DNA, i.e., the prior treatment of cells with a low concentration of H_20_2 induces resistance to subsequent exposure to H_20_2 but not to MNNG. Thus, this response appears to differ in specificity from the adaptive response for alkylating agents[25] but shows some overlap with the heat-shock response.[26] A similar overlap exists between H_20_2 and near-UV radiation.[27] This result is consistent with the fact that many of the photochemical reactions involving near-UV radiation also require oxygen (see Chapter 3).

4.1.5. Mechanisms of DNA Repair

4.1.5.1. Photoreactivation

Under certain conditions, the damaged part of a DNA molecule can be restored to its functional state *in situ* (i.e., without breaking the sugar-phosphate backbone of the DNA). This restoration can be either by chemical or enzymatic means. *Chemical restoration* can result from the spontaneous "decay" of the damage to the original structure, e.g., the dehydration of pyrimidine photohydrates (Section 2.3.1.3) or by the recombination of radicals to yield a restored molecule (e.g., $R^{\bullet} + H^{\bullet} \rightarrow RH$).

Photoreactivation, the enzymatic splitting of cyclobutane-type pyrimidine dimers *in situ,* is mediated by exposure to visible light, and was the first repair system to be elucidated at the molecular level.[28] In 1949, Kelner observed that the survival of UV-irradiated bacteria could be greatly enhanced if the cells were subsequently exposed to an intense source of blue light. In 1958, Rupert and coworkers demonstrated the existence of a photoreactivating enzyme (DNA photolyase) and established its basic properties. In the dark, the enzyme binds tightly to a cyclobutane-type pyrimidine dimer in UV-irradiated DNA to form an enzyme–substrate complex. The absorption of light between 300 and 450 nm activates this complex, the substrate is converted from a cyclobutane-type pyrimidine dimer to two monomeric pyrimidines, and the enzyme is released. This reaction is shown schematically in Fig. 4-4.

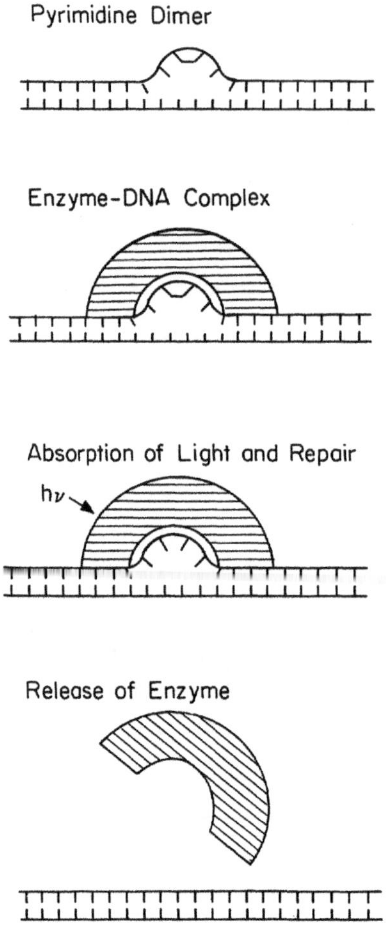

Fig. 4-4. A general model for photoreactivation. The photoreactivating enzyme (DNA photolyase) binds to a cyclobutane-type pyrimidine dimer in UV-irradiated DNA to form an enzyme–substrate complex. The absorption of light between 300 and 450 nm activates the complex, the pyrimidine dimer is converted to monomeric pyrimidines, and the enzyme is released.

As much as 90% of the lethal damage induced in bacteria by low fluences of UV radiation at 254 nm can be photoreactivated, thus indicating the importance of pyrimidine dimers as lethal lesions. The gene for the photoreactivating enzyme has been mapped in *E. coli*(29,30) and shown to be a flavoprotein.(31)

Photoreactivating enzymes have been found in a wide range of species from the simplest living cells, the mycoplasmas, to the skin and white blood cells of man.(32)

4.1.5.2. Excision Repair

Two types of excision repair are known, nucleotide excision repair and base excision repair. The former system constituted the first discovery (in 1964) of a mechanism of *dark repair,* i.e., "dark", to distinguish it from photoreactivation that is mediated by visible light (Section 4.1.5.1). Base excision repair is a more recent discovery.

In *nucleotide excision repair,*(2) the damaged section of a DNA strand is removed

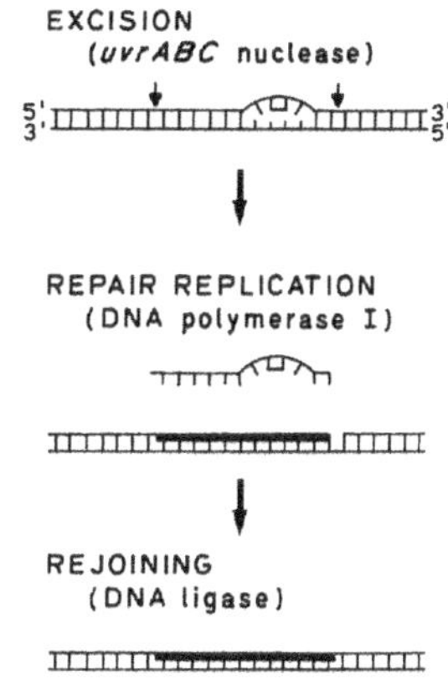

Fig. 4-5. A model for the major pathway of nucleotide-excision repair in *E. coli*. The UvrABC nuclease recognizes the lesion, shown here as a cyclobutyl-pyrimidine dimer, and cuts on both sides of the lesion, producing a gapped structure that is about 12 nucleotides long. Repair replication (heavy line) fills this gap, using the opposite strand of DNA as the template. Finally, the break in the repaired DNA strand is sealed by DNA ligase. See the text for more details. [Modified from A. Sancar and W. D. Rupp, *Cell* **33**, 249–260 (1983).]

and replaced with undamaged nucleotides to restore the normal function of the DNA. A schematic representation of the probable steps involved in the *major* pathway of this process is shown in Fig. 4-5.

The first step in nucleotide excision repair is the recognition of the damage in the DNA. The UvrABC nuclease appears to recognize the distortion in the DNA helix due to the presence of the damage rather than recognizing the damage itself. This conclusion is based on the observation that this excision repair system functions on both radiation and chemical damage whose molecular structures are very different. This complex enzyme, coded for by the *uvrA*, *uvrB*, and *uvrC* genes (Table 4-1), produces breaks in the DNA chain at the eighth phosphodiester bond on the 5′ side and the fourth or fifth phosphodiester bond on the 3′ side of a pyrimidine dimer. The *uvrD* gene product, helicase II, then appears to cause the release of the UvrC protein (through some type of protein–protein interaction rather than through its helicase activity), and DNA polymerase I (product of the *polA* gene) then displaces the DNA fragment containing the lesion and fills the gap using the DNA strand opposite the gap as the template.[33] When all of the nucleotides have been replaced, DNA ligase (product of the *lig* gene) seals the two ends of the chain as the final step in the repair process.

These results with the *uvrD* gene product were from *in vitro* studies. Studies *in vivo*, however, give a somewhat different picture. Since a *uvrD* mutation sensitizes a *recA* strain to UV radiation only very slightly, it suggests that the *uvrD* gene plays very little role in *recA*-independent excision repair. However, a *uvrD* mutation causes a marked sensitization of a *uvrB* strain, suggesting that the *uvrD* gene plays a significant role in postreplication repair. This apparent contradiction between the *in vivo* and *in vitro* results on the role of the *uvrD* protein in excision repair remains to be resolved.

Nucleotide excision repair has been divided into two pathways (Fig. 4-6). The major pathway for the repair of excision gaps requires DNA polymerase I (*polA*), and repair can proceed by the mechanism described in Fig. 4-5, even when the cells are in buffer. The minor pathway (in terms of the number of lesions repaired) is dependent on a functional *recA* gene and requires that the cells be in complete growth medium (see below).

Another feature that distinguishes these two pathways is the size of the repair

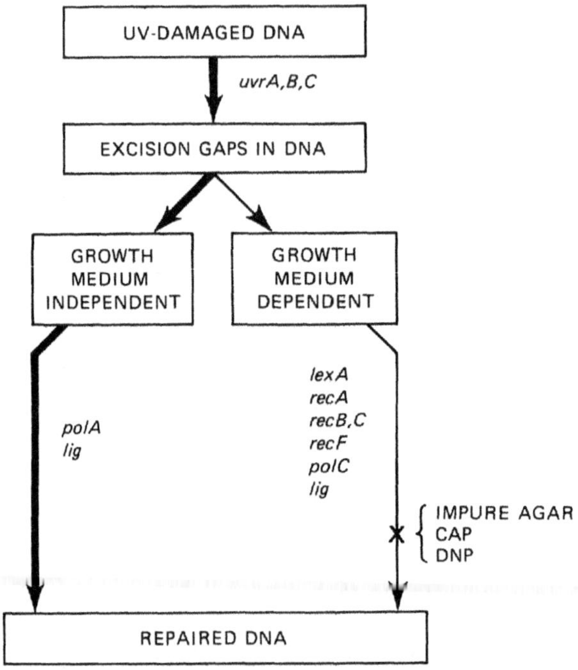

Fig. 4-6. The genetic and physiological control of the different pathways of nucleotide excision repair in *E. coli*. Most of the excision gaps are repaired by the growth medium-independent, *polA*-dependent pathway by the mechanism shown in Fig. 4-5. A small number of gaps are repaired by the growth medium-dependent, *recA*-dependent pathway. The latter pathway appears to function only in the replicated portion of the chromosome (see text and Fig. 4-7). The mutations are defined in Table 4-1. Chloramphenicol (CAP), dinitrophenol (DNP), and impure agar inhibit the growth medium-dependent pathway. [Modified from D. A. Youngs, E. Van der Schueren and K. C. Smith, *J. Bacteriol.* **117**, 717–725 (1974).]

"patches." Hanawalt and coworkers have shown that short patches (20–30 nucleotides long) are produced in the *polA*-dependent process, while long patches (200–1500 nucleotides long) are produced in the *recA*-dependent process.[34]

Based on recent knowledge about the multiple pathways of postreplication repair (Section 4.1.5.3), a model for *recA*-dependent nucleotide excision repair has been proposed.[35] Consistent with this model, *recA*-dependent excision repair only occurs in the replicated portion of the chromosome where sister duplexes exist, and where intrachromosomal recombination can occur (Fig. 4-7); the repair of excision gaps that occur in cells with unreplicated chromosomes is *recA*-independent. The majority of this *recA*-dependent excision repair is *recF*-dependent, and a small portion is *recB*-dependent. The *recF* and *recB* gene products are suggested to function in *recA*-dependent excision repair in a manner similar to their function in postreplication repair (see below), i.e., in the portion of the chromosome replicated prior to UV irradiation, the RecF pathway repairs excision gaps by a recombinational process that results in the formation of long repair patches (Fig. 4-7C-E), and the RecB pathway repairs the DNA double-strand breaks that arise at unrepaired gaps (Fig. 4-7F). This model for *recA*-dependent excision repair will become more clear after reading Section 4.1.5.3 on postreplication repair.

A study of excision repair in mammalian cells has led to the observation of an apparent correlation between carcinogenesis and the defective repair of DNA.[36] Fibroblasts from patients with the hereditary disease xeroderma pigmentosum exhibit much-reduced levels of excision repair after UV irradiation. It has been suggested that this deficiency in excision repair is related to the induction of fatal skin cancers in these patients after exposure to sunlight. In general, excision repair appears to be more accurate

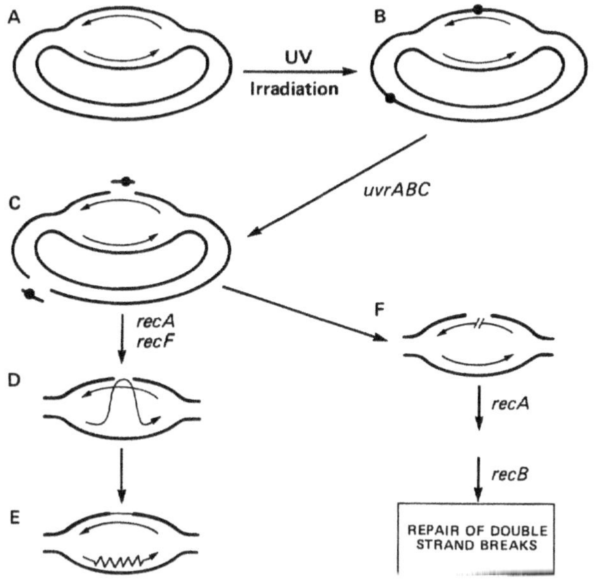

Fig. 4.7. The *recA*-dependent repair of excision gaps in UV-irradiated *E. coli.* Lesions can be produced in both the replicated and nonreplicated regions of the chromosome (A, B), but only the excision gaps produced in the replicated region of the chromosome (C) can be repaired by an intrachromosomal recombinational process that is *recF*-dependent (D). This process leaves a gap in the homologous sister duplex that can be filled by long-patch repair replication (wavy line), using the parental strand opposite the gap as a template (E). If the daughter strand opposite the excision gap is cut (–/ /–; F), the resulting double-strand break is repaired primarily by a *recB*-dependent process (see text for further discussion). (From Ref. 35.)

than postreplication repair. Therefore, if an individual is deficient in "error-free" repair, then all DNA lesions must be repaired by the remaining "error-prone" system. Error-prone DNA repair is the cause of UV radiation mutagenesis (Section 4.2.4), and mutations are believed to be an early step in some mechanisms of carcinogenesis. The DNA repair capacities of cells from patients with various genetic diseases that predispose them to cancer have been reviewed,(2,3) as well as the immunological defects that are manifested in these diseases.(4)

In *base excision repair,*(2,37) altered bases (e.g., thymine glycol, 3-methyladenine, uracil, hydroxymethyluracil, hypoxanthine, urea) are recognized by a class of enzymes called *DNA glycosylases,* which are specific to varying degrees for particular types of altered bases. These enzymes split off the altered base at the *N*-glycosylic bond that links the base to the sugar in the DNA backbone. The site of the base loss is called an *AP site* (for *ap*urinic or *ap*yrimidinic). These steps are shown schematically in Fig. 4-8. AP sites are substrates for *AP endonucleases,* of which there are two types. One type cuts on the 5′ side and the other type cuts on the 3′ side of an AP site. If the two AP endonucleases act in concert, then in theory a gap can be formed in the DNA that is only one nucleotide long. However, if after an initial cut by one of the AP endonucleases, an exonuclease degrades the broken strand, the gaps could be much larger. These gaps are presumably repaired by the same processes described above for the repair of the gaps formed during nucleotide excision repair.

4.1.5.3. Postreplication Repair(2,38)

The first indication that excision repair was not the only mechanism by which cells could repair radiation damage to their DNA in the dark was the observation that bacterial

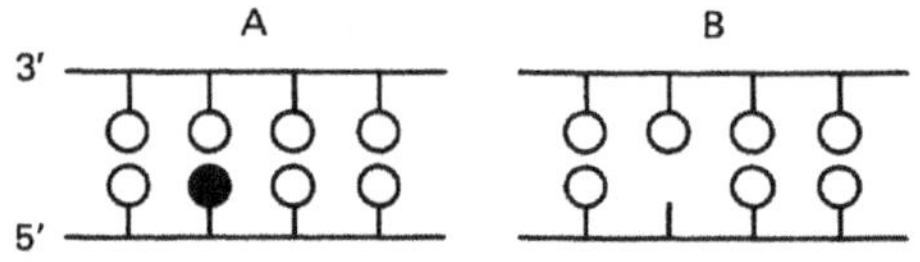

Fig. 4.8. Initial steps in base excision repair. An altered base (dark circle in A) is recognized by a specific enzyme called a DNA glycosylase. The base is removed from the DNA by hydrolyzing the *N*-glycosylic bond that links the base to the sugar in the DNA backbone. The site of the base loss (in B) is called an AP site (for *ap*urinic or *ap*yrimidinic). In subsequent steps (not shown), an AP endonuclease cuts the DNA backbone at the AP site, and the resulting DNA gap is presumably repaired by the same processes described for the repair of the gaps formed during nucleotide excision repair (Section 4.1.5.2).

cells deficient both in excision repair (*uvrA*) and in genetic recombination (*recA*) are much more sensitive to killing by UV radiation than cells carrying either mutation alone (Fig. 4-2). These genetic studies suggested that the *uvrA* and *recA* genes act in different biochemical pathways of repair and that certain steps in genetic recombination must be important in the repair of radiation damage. Postreplication repair was discovered in 1968 by Rupp and Howard-Flanders using excision repair-deficient cells.

The observations that led to the model for postreplication repair were that (1) the DNA that is synthesized in *E. coli* immediately after UV irradiation has discontinuities when assayed on alkaline sucrose gradients and (2) the mean length of the newly synthesized DNA approximates the average distance between pyrimidine dimers in the parental strand. With further incubation of the cells, these discontinuities (i.e., DNA daughter strand gaps) disappear, and the DNA approximates the molecular size of that from unirradiated control cells. This process is shown schematically in Fig. 4-9.

This model is an oversimplification because it shows that the genetic transfers (Fig. 4-9c–d) result in lesion-free daughter strand DNA. However, in *E. coli* about half of these genetic transfers result in the covalent joining of daughter strand DNA to parental DNA that contains pyrimidine dimers.[38] Therefore, several cycles of DNA replication and postreplication repair are needed before all of the dimers are "diluted" out of the daughter DNA. The transfer of dimers from parental DNA to daughter DNA appears to be due to the random resolution of the Holliday junction, an intermediate in recombination; if resolved one way it yields dimer-free daughter strands, and if resolved in a second way it transfers pyrimidine dimers to the daughter strands.[39] For mammalian cells, however, very few pyrimidine dimers are detected in daughter strand DNA after postreplication repair.[40] It is not clear whether this means that Holliday junctions are only resolved in one way in mammalian cells (i.e., to yield dimer-free daughter strands) or that daughter strand gaps are repaired in mammalian cells by a mechanism that differs from that found in *E. coli*.

Studies on the genetic control of postreplication repair indicate that there are two distinct processes; both require a functional *recA* gene.[41,42] One process is for the repair of DNA daughter strand gaps. About half of the DNA daughter strand gaps are repaired by a process that is *recF*-dependent, and about half are repaired by a process that is *recF*- and *recB*-independent and is constitutive.[43] The genetic control of this latter pathway is not established, although the *umuC* gene appears to play a small role, and may define an error-prone (i.e., mutagenic; see Section 4.2.4) pathway of postreplication repair. The *polA* gene (especially the $5' \rightarrow 3'$ exonuclease activity of DNA polymerase I) also plays a role in this *recF*-independent pathway for the repair of daughter strand gaps.

The second process of postreplication repair requires a functional *recB* gene, and repairs the DNA double-strand breaks that arise at unrepaired daughter strand gaps. The

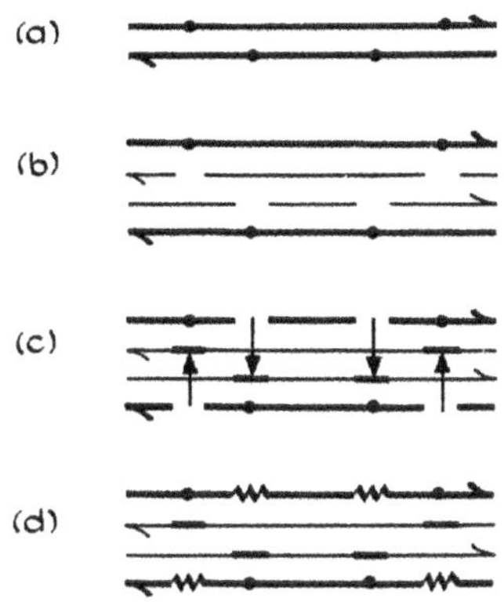

Fig. 4-9. The postreplication repair of DNA daughter strand gaps in UV-irradiated *E. coli*. (a) Dots indicate photochemical lesions produced in the two strands of DNA. (b) DNA synthesis proceeds up to and then skips past the lesions in the parental strands, leaving gaps in the daughter strands. (c) Filling of the gaps in the daughter strands with material from the parental strands by a recombinational process that also requires a functional *recA* gene. About half of the time, however, lesions are also transferred from the parental strands to the daughter strands (not shown).[39] (d) Gaps in the parental strands are repaired by repair replication. [Modified from K. C. Smith, *Photophysiology* **6**, 209–278 (1971).]

recB pathway for the repair of double-strand breaks is about as important to the survival of UV-irradiated *E. coli* as the *recF* pathway for the repair of daughter strand gaps, as judged by the fact that, in the *uvrB* background, a *recF* or a *recB* mutation sensitizes cells to killing to about the same extent. These pathways of postreplication repair are shown schematically in Fig. 4-10.

The recombination deficiency and radiation sensitivity of *recBC* strains are suppressed by an additional mutation in the *sbcB* gene, which is the structural gene for exonuclease I, a single-strand specific 3′ → 5′ DNA exonuclease. The presence of an *sbcB* mutation also restores the proficiency of *recBC* cells to repair DNA double-strand breaks, and this repair is dependent on the *recF* and *recJ* genes. Since the RecBCD enzyme (Exo V) has a DNA helicase activity that requires blunt or nearly blunt ends of DNA duplexes (i.e., it will not unwind DNA that has a long single-stranded tail), it suggests that the double-strand breaks that are formed at daughter strand gaps are normally processed to blunt ends by Exo I (*sbcB*) and Exo V (*recBCD*) before being repaired by the *recBCD*-dependent recombination process. When daughter strand gaps containing single-stranded tails are not degraded by Exo I and Exo V (i.e., in *sbcB recBC* cells), they become substrates for a *recF*-dependent recombination process. Therefore, depending on the structure of a DNA double-strand break and the genetic background of the cell, a double-strand break may be repaired by the RecBCD pathway (the primary pathway in a wild-type cell) or by the RecF pathway, or both.

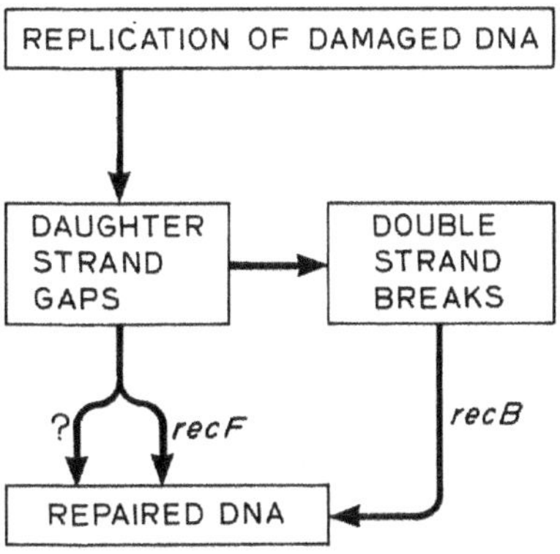

Fig. 4-10. Schematic diagram of the interaction of the multiple pathways of postreplication repair in *E. coli*. During DNA replication, noncoding DNA lesions are skipped, producing DNA daughter strand gaps (Fig. 4-9). These gaps are repaired by two processes, one of which requires a functional *recF* gene; the one marked with "?" is a pathway that functions in the absence of the *recF* and *recBC* genes. If these gaps are not repaired, they can be converted to DNA double-strand breaks, which are repaired primarily by a process that requires a functional *recB* gene.

The repair of DNA double-strand breaks in UV-irradiated *E. coli* is very complex and poorly understood. Its complexity is best exemplified by the number of genes that appear to control the repair of double-strand breaks. With the exception of the *umuC* gene, practically all of the genes that have been implicated in postreplication repair are also involved, to varying degrees, in the repair of double-strand breaks. While the *recF* and *recJ* mutations appear to specifically affect the repair of double-strand breaks in UV-irradiated *uvrA recBC sbcB* cells, mutations in *recA, recBC, recN, radB, ssb, uvrD, lexA,* and *polA* all produce a deficiency in the repair of double-strand breaks in *sbcB*$^{+}$ cells. Presumably, some of these genes are preferentially involved in either the RecF or the RecBCD pathway for the repair of double-strand breaks, and some may be involved in both pathways. A similar set of genes has also been observed to control the repair of X-ray-induced DNA double-strand breaks; however, in this case there is a requirement for the *recF* and *recJ* gene products in *sbcB*$^{+}$ cells. This suggests that additional types of double-strand breaks are formed by X irradiation.

The postreplicational formation and repair of DNA double-strand breaks after UV irradiation has also been observed in cultured human cells. Therefore, this aspect of postreplication repair appears to be similar for both *E. coli* and human cells.

4.1.5.4. Conclusions

One cannot help but be impressed with the multitude and sophistication of the repair systems possessed by cells. A significant percentage of the energy of a cell is spent in synthesizing enzymes to repair and maintain the integrity of the genetic information coded in its DNA. There is evidence, however, to suggest that these repair systems have not evolved just to repair damage produced in DNA by environmental agents, but rather they appear to have a necessary function in the everyday life of a cell. This conclusion is based on the observation that while certain mutants deficient in one DNA repair protein grow normally in the absence of radiation, certain mutants deficient in two repair proteins (e.g., *polA recA*) are not viable. Thus, normal growth processes appear to produce lesions in DNA that must be repaired in order to maintain viability. Such normal metabolic damage to DNA has also been shown to contribute to spontaneous mutagenesis (Section 4.3).

With the recent availability of a number of highly purified DNA repair proteins, e.g., RecA, RecBC, UvrA, UvrB, UvrC, UvrD, Phr, PolA, and Lig, and of DNA-sequencing gels and similar techniques, biochemical studies on DNA repair *in vitro* are under way in a number of laboratories. As a result, major advances in our understanding of the biochemistry of DNA repair and of recombination should be forthcoming.

4.2. UV RADIATION MUTAGENESIS

4.2.1. Introduction

The action spectrum for UV radiation mutagenesis in bacteria mimics the absorption spectrum for nucleic acid (DNA). Therefore, to understand the molecular basis of UV radiation mutagenesis one needs to understand the biochemistry of DNA in terms of both replication and repair.

4.2.1.1. DNA Replication

A UV-irradiated cell may suffer a mutation because a DNA lesion produces a coding error during the replication of DNA. It has been mentioned (Section 2.3.1.3) that the photohydrate of cytosine codes for adenine rather than for guanine in an *in vitro* replication system. Such miscoding lesions, which lead directly to mutations via DNA replication, must be rare events *in vivo* after UV irradiation, since mutations are not induced by UV irradiation in the repair-deficient *recA* and lexA strains, where mutations can only be produced by replication errors (see below). In support of this concept, mutations can be induced in these strains by treatment with certain chemicals (e.g., ethylmethanesulfonate) that produce altered DNA bases with different coding properties. The fact that a mutational event is *recA*-independent has been regarded as diagnostic of the involvement of miscoding lesions and of DNA replication in the mutagenic process.

4.2.1.2. DNA Repair

Noncoding lesions in DNA elicit a repair response. The second mechanism for mutagenesis requires the induction of an error-prone DNA repair system that produces changes in the base sequence of the DNA in the course of restoring viability. Evidence for this mechanism of mutagenesis followed from the initial observation that certain DNA repair-deficient strains of *E. coli* (i.e., *recA* and *lexA*) cannot be mutated by UV irradiation, suggesting that error-prone DNA repair plays the major role in UV radiation mutagenesis.[19] In general, postreplication repair is considered to be more error-prone than excision repair.

4.2.1.3. Types of Mutations

One type of DNA alteration that leads to mutations is the substitution of one base pair by another. *Base substitution* mutations can be of two kinds. In *transitions* a purine is replaced by another purine (e.g., G → A), or a pyrimidine is replaced by another pyrimidine (e.g., C → T). In *transversions* a purine is replaced by a pyrimidine or vice versa [i.e., (G or A) → (C or T)]. If the base substitution causes a change of one amino acid in the protein and the protein exhibits an altered function, this is called a *missense* mutation. If the base substitution produces a chain-terminating codon (i.e., a triplet of nucleotides that does not code for any amino acid), it is called a *nonsense* mutation, and an incomplete protein is synthesized.

The second type of mutagenic change in DNA is the deletion or insertion of a base pair (or a larger unit) to yield a *frameshift* mutation. The addition or removal of a base pair would shift the reading frame of the genetic code such that a string of incorrect codons and incorrect amino acids would result.

4.2.2. Mutation Assays

Bacteria have been popular subjects for the study of the molecular basis of mutagenesis because they grow rapidly and can easily be obtained in large numbers. Most bacterial studies have followed the acquisition of resistance to killing by bacteriophage or by antibiotics, and the reversion to prototrophy (i.e., an *auxotrophic* bacterium may require an amino acid for growth, but a *prototrophic* revertant would not).

The reversion to prototrophy has been the most popular assay because low frequencies of mutations can easily be detected and the same medium (partially enriched medium containing ~1 μg/ml of the required amino acid) can be used to score both mutations and survival. When plating for survival at high dilution, auxotrophs form small (i.e., they run out of the required amino acid) but visible colonies. When plated at high cell densities for scoring for mutations, the auxotrophic population forms a limited "lawn" of cells against which prototrophic mutants form large colonies. The irradiated cells need a small amount of the required amino acid in order to "fix" the mutation through error-prone DNA repair before they become prototrophs.

With the new DNA-sequencing techniques, one no longer needs to depend on earlier indirect techniques to classify the type of mutation (i.e., missense, nonsense, frameshift) that is produced (e.g., based on the types of chemicals that are needed to revert the mutation). Now the exact base change due to the mutation can be determined. Through such DNA-sequencing procedures it was shown that the (6–4) pyrimidine adduct (Section 2.3.1.3) was the mutagenic lesion at certain "hot spots" in the *lacI* gene, as opposed to the previous conclusion that cyclobutane-type pyrimidine dimers were responsible for all UV radiation mutagenesis.[44]

Mutation studies in mammalian cells are not as easy to perform, and there are relatively few assays (i.e., loci) from which to choose. Recently, a new technique has been introduced that overcomes many of these problems, and the mutations can be readily assayed at the DNA sequence level. In one example of this procedure, the bacterial *lacI* gene, carried on a *shuttle vector* (i.e., an autonomously replicating plasmid), is introduced into mammalian cells and allowed to replicate in the cell nucleus. The cells are then treated with the mutagen, and one or two days later the vector DNA is recovered and transfected back into *E. coli* for the analysis of mutations in the *lacI* gene.[45] Other shuttle vector techniques are also under development for the study of mutagenesis in mammalian cells.[46]

4.2.3. Kinetics of Induction

Classically, UV radiation mutagenesis was categorized as being dependent on the interaction of two radiation-induced events, i.e., it was a two-hit process, while X-ray mutagenesis was categorized as being a one-hit process. That is, if one plots mutant frequency against dose on a log-log plot, the slopes of the straight lines would be 2 and 1, respectively. With the observation that UV radiation mutagenesis was inducible, there was an attempt to explain the two-hit response as one hit in the gene to be mutated, and one hit somewhere else in the cell to induce error-prone DNA repair. There is some support for this hypothesis after low fluences of UV radiation.[19]

Like most generalities, the apparent distinction between X-ray mutagenesis and UV radiation mutagenesis on the basis of the hit number is not strictly correct. Both radiations show one-hit and two-hit mutagenesis. For UV radiation mutagenesis, the one-hit mode has been correlated with the production of *back mutants* (i.e., the direct correction of the initial genetic error), while the two-hit process has been correlated with the formation of *suppressor mutations* [i.e., a second mutation in the cell, usually a mutation affecting the anticodon of a transfer RNA (tRNA), suppresses the effect of the first mutation].[47] Is there some unique structural feature (i.e., base sequence) in tRNA genes that requires (or predisposes them to) two-hit kinetics for UV radiation mutagenesis? For example, it would require two hits to produce an overlapping DNA daughter strand gap. Very many fundamental questions about UV radiation mutagenesis, such as this, remain to be answered.

4.2.4. Genetic Control of Mutagenesis

The observation that the *recA* and *lexA* strains of *E. coli* could not be mutated by UV irradiation led to the discovery that UV radiation mutagenesis is not a passive process but is due to inducible, error-prone DNA repair.[19] Because the *recA* and *lexA* genes seemed to be involved in the regulation of the complex SOS response, it was thought that there might be other gene products that participate more directly in error-prone DNA repair. To this end, the *umuC* and *umuD* mutations were isolated, which block all of targeted UV radiation mutagenesis.[reviewed in 20] The *umuC* and *umuD* genes are part of the SOS system but their mechanism of action is not known. A *umuC* mutation has a small effect on the repair of DNA daughter strand gaps after UV irradiation, and it has been suggested that the *umuC* gene may be involved in the repair of overlapping DNA daughter strand gaps,[48] since such lesions cannot be repaired by the usual mechanism of postreplication repair. However, if one of the gaps could be repaired by some unique process such as *translesion synthesis* (see below), then the other gap could be repaired by classical postreplication repair.

Actually, translesion synthesis was offered some years ago as a model to explain all of UV radiation mutagenesis.[19] In this model, an inhibitor of the editing function of DNA polymerase is induced, thus causing the polymerase to synthesize past noncoding lesions, yielding modified DNA. Recently, the RecA protein has been found to bind to UV-irradiated DNA and to inhibit the editing subunit of DNA polymerase III holoenzyme.[49] In addition, a model has been presented that suggests that while nucleotides can be inserted opposite noncoding lesions in UV-irradiated cells, the UmuDC proteins are required for the resumption of replication past the lesion.[50]

The *mucA* and *mucB* genes in the plasmid pKM101 are analogs of the *umuD* and *umuC* genes in *E. coli*. A nonmutable *umuC* strain can be made mutable by infection with pKM101, and a *umuC*$^+$ strain can be made more mutable by infection with pKM101.[20] The Ames tester strains, which are used for screening environmental mutagens, often carry the plasmid pKM101 to increase their sensitivities to mutagens.

While a *umuC* mutation knocks out all of targeted UV radiation mutagenesis, it only knocks out part of X-ray mutagenesis.[51] Furthermore, it is the "oxygen effect" of X-ray mutagenesis that is blocked by a *umuC* mutation [i.e., while wild-type cells show enhanced mutagenesis if they are X-irradiated in the presence of oxygen (vs. the absence of oxygen), a *umuC* strain shows no such oxygen effect]. A small portion of oxygen-independent X-ray mutagenesis is also *umuC*-dependent and leads to (AT or GC) → TA transversion mutations. The *umuC*-independent, oxygen-independent mutagenesis results in GC → AT and AT → GC transitions.

The fact that the *umuDC* genes only control a portion of X-ray mutagenesis raises the possibility that other genes control the *umuDC*-independent portion of X-ray mutagenesis.

Recently, three mutations (*xmu*, *X*-ray *mu*tability) that block *umuC*-independent X-ray mutagenesis have been isolated in this laboratory and await further study.

4.3. SPONTANEOUS MUTAGENESIS[2]

4.3.1. DNA Replication

Until very recently, it was generally believed that spontaneous mutagenesis was due entirely to errors made during DNA replication, as a result of the inherent inaccuracy of the replication system.

One theory for spontaneous mutagenesis is that when a hydrogen atom on a DNA base shifts from one position to another, the base assumes a different configuration (i.e., undergoes a *tautomeric shift*) such that it can form an abnormal base pair. The occurrence of such tautomeric shifts at the moment of DNA synthesis would result in mutations if not subsequently repaired.

If a DNA polymerase does put in an incorrect nucleotide, it has an editing function that can cut out the incorrect nucleotide, and then it can try again to insert the correct one. However, DNA polymerase does occasionally make errors that it fails to correct.

The second line of defense against replication errors is *mismatch repair*.[2,20,53] When a mismatched base pair is recognized shortly after replication, the incorrect base is cut out and replaced with the correct one. In *E. coli* the DNA that is newly synthesized is not yet methylated, whereas the parental strand is methylated at the adenine residues in GATC sequences. This pattern of methylation is what tells the mismatch repair system which strand of DNA to cut in order to remove the incorrect base. Mutations that inhibit mismatch repair (e.g., *mutH*, *mutL*, *mutS*) greatly increase the level of spontaneous mutagenesis.

In a mutant deficient in DNA adenine methylase (*dam*), which methylates the adenines of newly synthesized DNA, the cell doesn't know which strand to cut for mismatch repair, and has a 50:50 chance of cutting the correct one. Therefore, the spontaneous mutation level is increased in *dam* mutants. Furthermore, since neither strand is methylated, the cell sometimes cuts both strands, producing a DNA double-strand break. The production of such double-strand breaks in a *dam* strain can be prevented by an additional *mutL* or *mutS* mutation that blocks mismatch repair.[54]

4.3.2. DNA Repair

About half of spontaneous mutagenesis in wild-type *E. coli* is due to error-prone DNA repair. This conclusion is based in part on the observations that about half of spontaneous base substitution mutagenesis is under the control of the *recA* gene, and that mutations in DNA repair genes that either reduce UV radiation mutagenesis (e.g., *lexA*, *umuC*) or enhance UV radiation mutagenesis (e.g., *uvrA*, *uvrB*) have a similar effect on spontaneous mutagenesis. This latter result has also been interpreted to mean that some of the DNA lesions that lead to spontaneous mutagenesis are "UV-like" in their requirements for repair.

What is the source of this UV-like damage? Excited state intermediates are formed enzymatically by certain oxidases. For example, the action of horseradish peroxidase on isobutyraldehyde results in the production of acetone in its excited state. This enzymatically generated, excited state acetone can transfer its energy to DNA and produce damage *in vitro*. Recently, the metabolism of phenylalanine has been shown to produce UV-like lesions in DNA (based on genetic studies) that result in mutations in *E. coli*. Phenylalanine is only weakly mutagenic, which is a good thing since we can't live without it. (N.B.: Oxygen is also mutagenic, and also necessary for life.) In contrast, the presence of excess cystine, arginine, or threonine appears to be antimutagenic in *E. coli*.

DNA repair is not only important in spontaneous mutagenesis for *E. coli*, but also for yeast, and for all other organisms where data are available on this topic.[52] Since excision repair-deficient strains of *E. coli* show an enhanced level of spontaneous mutagenesis, it allows one to speculate that humans that are deficient in excision repair (i.e., patients with xeroderma pigmentosum) should show an abnormal amount or type of internal organ cancer. There are some data to support this hypothesis.[e.g., 55]

It is important to remember that the spontaneous mutation rate of a cell culture or of an organism is the net result of a large number of both mutagenic and antimutagenic processes that involve both DNA replication and DNA repair.

4.4. REFERENCES

1. R. B. Roberts and E. Aldous, Recovery from ultraviolet irradiation in *Escherichia coli*, *J. Bacteriol.* **57**, 363–375 (1949).
2. E. C. Friedberg, *DNA Repair*, W. H. Freeman, New York (1985).
3. E. C. Friedberg, U. K. Ehmann, and J. I. Williams, Human diseases associated with defective DNA repair, *Adv. Radiat. Biol.* **8**, 85–174 (1979).
4. K. B. Hellman, G. B. Schuller, and W. R. Lewis, Immunological defects in human radiation sensitive disorders, in: *Topics in Photomedicine* (K. C. Smith, ed.), pp. 143–182, Plenum Press, New York (1984).
5. R. H. Haynes and B. A. Kunz, DNA repair and mutagenesis in yeast, in: *Molecular Biology of the Yeast Saccharomyces: Life Cycle and Inheritance*, Monograph 11A, pp. 371–414, Cold Spring Harbor Laboratory, Cold Spring Harbor, New York (1981).
6. E. C. Friedberg and B. A. Bridges (eds.), *Cellular Responses to DNA Damage*, Alan R. Liss, New York (1983).
7. E. C. Friedberg and P. C. Hanawalt (eds.), *DNA Repair: A Laboratory Manual of Research Procedures*, Vol. 1, Marcel Dekker, New York, (1981), Vol. 2 (1983), Vol. 3 (1988).
8. D. E. Lea, *Actions of Radiation on Living Cells*, Cambridge University Press, London (1955) (Reprinted 1962).
9. W. Harm, *Biological Effects of Ultraviolet Radiation*, Cambridge University Press, Cambridge (1980).
10. R. H. Haynes, The influence of repair processes on radiobiological survival curves, in: *Cell Survival After Low Doses of Radiation: Theoretical and Clinical Implications* (T. Alper, ed.), pp. 197–208, Wiley, New York (1975).
11. R. H. Haynes, F. Eckardt, and B. A. Kunz, The DNA damage-repair hypothesis in radiation biology: Comparison with classical hit theory. *Br. J. Cancer* **49**, suppl. VI, 81–90 (1984).
12. P. A. Swenson, Physiological responses of *Escherichia coli* to far-ultraviolet radiation, in: *Photochemical and Photobiological Reviews*, Vol. 1 (K. C. Smith, ed.), pp. 269–387, Plenum Press, New York (1976).
13. M. Tang and K. C. Smith, The expression of liquid holding recovery in ultraviolet-irradiated *Escherichia coli* requires a deficiency in growth medium-dependent DNA repair, *Photochem. Photobiol.* **32**, 763–769 (1980).
14. M. Bertrand and D. F. Deen, Factors influencing the recovery from potentially lethal damage (PLD) in mammalian cells *in vitro* and *in vivo*, *Cancer Treat. Rev.* **7**, 1–15 (1980).
15. R. C. Sharma, N. J. Sargentini, and K. C. Smith, New mutation (*mmrA1*) in *Escherichia coli* K-12 that affects minimal medium recovery and postreplication repair after UV irradiation, *J. Bacteriol.* **154**, 743–747 (1983).

16. R. C. Sharma and K. C. Smith, A mechanism for rich-medium inhibition of the repair of daughter-strand gaps in the deoxyribonucleic acid of UV-irradiated *Escherichia coli* K12 *uvrA*, *Mutation Res.* **146,** 177–183 (1985).
17. N. J. Sargentini, W. P. Diver, and K. C. Smith, The effect of growth conditions on inducible, *recA*-dependent resistance to X-rays in *Escherichia coli*. *Radiat. Res.* **93,** 364–380 (1983).
18. N. J. Sargentini and K. C. Smith, Quantitation of the involvement of the *recA*, *recB*, *recC*, *recF*, *recJ*, *recN*, *lexA*, *radA*, *radB*, *uvrD*, and *umuC* genes in the repair of X-ray-induced DNA double-strand breaks in *Escherichia coli*, *Radiat. Res.* **107,** 58–72 (1986).
19. E. M. Witkin, Ultraviolet mutagenesis and inducible DNA repair in *Escherichia coli*, *Bacteriol. Rev.* **40,** 869–907 (1976).
20. G. C. Walker, Mutagenesis and inducible responses to deoxyribonucleic acid damage in *Escherichia coli*, *Microbiol. Rev.* **48,** 60–93 (1984).
21. G. C. Walker, Inducible DNA repair systems, *Annu. Rev. Biochem.* **54,** 425–437 (1985).
22. M. Radman, Is there SOS induction in mammalian cells? *Photochem. Photobiol.* **32,** 823–830 (1980).
23. T. G. Rossman and C. B. Klein, Mammalian SOS system: A case of misplaced analogies, *Cancer Invest.* **3,** 175–187 (1985).
24. G. M. Hahn, *Hyperthermia and Cancer*, Plenum Press, New York (1982).
25. B. Demple and J. Halbrook, Inducible repair of oxidative DNA damage in *Escherichia coli*, *Nature* **304,** 466–468 (1983).
26. P. C. Lee, B. R. Bochner, and B. N. Ames, AppppA, heat-shock stress, and cell oxidation, *Proc. Natl. Acad. Sci. USA* **80,** 7496–7500 (1983).
27. R. M. Tyrrell, A common pathway for protection of bacteria against damage by solar UVA (334 nm, 365 nm) and an oxidising agent (H_20_2), *Mutation Res.*. **145,** 129–136 (1985).
28. W. Harm, C. S. Rupert, and H. Harm, The study of photoenzymatic repair of UV lesions in DNA by flash photolysis, *Photophysiology* **6,** 279–324 (1971).
29. D. A. Youngs and K. C. Smith, Genetic location of the *phr* gene of *Escherichia coli* K-12, *Mutation Res.* **51,** 133–137 (1978).
30. A. Sancar and C. S. Rupert, Correction of the map location for the *phr* gene in *Escherichia coli* K-12, *Mutation Res.* **51,** 139–143 (1978).
31. A. Sancar and G. B. Sancar, *Escherichia coli* DNA photolyase is a flavoprotein, *J. Mol. Biol.* **172,** 223–227 (1984).
32. E. C. Friedberg, K. H. Cook, J. Duncan, and K. Mortelmans, DNA repair enzymes in mammalian cells, in: *Photochemical and Photobiological Reviews*, Vol. 2 (K. C. Smith, ed.), pp. 263–322, Plenum Press, New York (1977).
33. I. Husain, B. van Houten, D. C. Thomas, M. Abdel-Monem, and A. Sancar, Effect of DNA polymerase I and DNA helicase II on the turnover rate of UvrABC excision nuclease, *Proc. Natl. Acad. Sci. USA* **82,** 6774–6778 (1985).
34. P. C. Hanawalt, P. K. Cooper, A. K. Ganesan, R. S. Lloyd, C. A. Smith, and M. E. Zolan, Repair responses to DNA damage: Enzymatic pathways in *E. coli* and human cells, *J. Cellular Biochem.* **18,** 271–283 (1982).
35. K. C. Smith and R. C. Sharma, A model for the *recA*-dependent repair of excision gaps in UV-irradiated *Escherichia coli*, *Mutation Res.* **183,** 1–9 (1987).
36. J. E. Cleaver, Repair processes for photochemical damage in mammalian cells, *Adv. Radiat. Biol.* **4,** 1–75 (1974).
37. T. Lindahl, DNA repair enzymes, *Annu. Rev. Biochem.* **51,** 61–87 (1982).
38. K. C. Smith, T. V. Wang, and R. C. Sharma, *recA*-Dependent DNA repair in UV-irradiated *Escherichia coli*, *J. Photochem. Photobiol., B:Biology* **1,** 1–11 (1987).
39. A. K. Ganesan, Persistance of pyrimidine dimers during post-replication repair in ultraviolet light-irradiated *Escherichia coli*, *J. Mol. Biol.* **87,** 103–119 (1974).
40. A. J. Fornace, Recombination of parent and daughter strand DNA after UV-irradiation in mammalian cells, *Nature* **304,** 552–554 (1983).
41. R. H. Rothman, T. Kato, and A. J. Clark, The beginning of an investigation of the role of *recF* in the pathways of metabolism of ultraviolet-irradiated DNA in *Escherichia coli*, in: *Molecular Mechanisms for Repair of DNA* (P. C. Hanawalt and R. B. Setlow, eds.), pp. 283–291, Plenum Press, New York (1975).
42. T. V. Wang and K. C. Smith, Mechanisms for *recF*-dependent and *recB*-dependent pathways of postreplication repair in UV-irradiated *Escherichia coli uvrB*, *J. Bacteriol.* **156,** 1093–1098 (1983).

43. R. C. Sharma and K. C. Smith, A minor pathway of postreplication repair in *Escherichia coli* is independent of the *recB, recC* and *recF* genes, *Mutation Res.* **146,** 169–176 (1985).
44. W. A. Franklin and W. A. Haseltine, The role of the (6-4) photoproduct in ultraviolet light-induced transition mutants in *E. coli, Mutation Res.* **165,** 1–7 (1986).
45. J. S. Lebkowski, S. Clancy, J. H. Miller, and M. P. Calos, The *lacI* shuttle: Rapid analysis of the mutagenic specificity of ultraviolet light in human cells, *Proc. Natl. Acad. Sci. USA* **82,** 8606–8610 (1985).
46. M. P. Calos, Mutation of autonomously replicating plasmids, in: *Gene Transfer* (R. Kucherlapati, ed.), pp. 243–261, Plenum Press, New York (1986).
47. N. J. Sargentini, R. C. Bockrath, and K. C. Smith, Three mechanisms for ultraviolet radiation mutagenesis in *Escherichia coli* K-12 *uvrB5:* Specificity for the production of back and suppressor mutants, *Mutation Res.* **106,** 217–224 (1982).
48. T. V. Wang and K. C. Smith, Role of the *umuC* gene in postreplication repair in UV-irradiated *Escherichia coli* K-12 *uvrB, Mutation Res.* **145,** 107–112 (1985).
49. C. Lu, R. H. Scheuermann, and H. Echols, Capacity of RecA protein to bind preferentially to UV lesions and inhibit the editing subunit (ϵ) of DNA polymerase III: A possible mechanism for SOS-induced targeted mutagenesis, *Proc. Natl. Acad. Sci. USA* **83,** 619–623 (1986).
50. B. A. Bridges, R. Woodgate, M. Ruiz-Rubio, F. Sharif, S. G. Sedgwick, and U. Hubscher, Current understanding of UV-induced base pair substitution mutation in *E. coli* with particular reference to the DNA polymerase III complex, *Mutation Res.* **181,** 219–226 (1987).
51. N. J. Sargentini and K. C. Smith, *umuC*-dependent and *umuC*-independent γ- and UV-radiation mutagenesis in *Escherichia coli, Mutation Res.* **128,** 1–9 (1984).
52. N. J. Sargentini and K. C. Smith, Spontaneous mutagenesis: The roles of DNA repair, replication, and recombination, *Mutation Res.* **154,** 1–27 (1985).
53. M. Radman and R. Wagner, Mismatch repair in *Escherichia coli, Annu. Rev. Genet.* **20,** 523–528 (1986).
54. T. V. Wang and K. C. Smith, Inviability of *dam recA* and *dam recB* cells of *Escherichia coli* is correlated with their inability to repair DNA double-strand breaks produced by mismatch repair, *J. Bacteriol.* **165,** 1023–1025 (1986).
55. K. H. Kraemer, M. M. Lee, and J. Scotto, DNA repair protects against cutaneous and internal neoplasia: Evidence from xeroderma pigmentosum, *Carcinogenesis* **5,** 511–514 (1984).

5

Environmental Photobiology

5.1. INTRODUCTION

The abiotic and biotic components that make up the surroundings or environment of an organism can exert considerable influence on the effects of light-mediated processes within the organism. Environmental factors such as temperature, the availability of water and nutrients, and interactions with other organisms modify the effects of light-mediated processes within an organism and thereby ultimately affect its growth and survival in the ecosystem. The examination of individual processes involving light allows us to understand the mechanism of these processes within an organism. How radiation is absorbed, what wavelengths are utilized, and what action or effect the radiation elicits in biological systems are questions of primary concern for photobiologists. For example, studies on the absorption of light by chlorophyll and the photochemical reactions of the photosynthetic apparatus are important for understanding the fundamental mechanism of photosynthesis. Important considerations in this mechanism would include the absorption spectrum for chlorophyll and the number of photons required to provide sufficient energy for the light reaction of photosynthesis. Although knowledge of how the photosynthetic mechanism

Ronald Robberecht • Department of Range Resources, University of Idaho, Moscow, Idaho 83843.

works is essential, this alone does not allow us to predict how the whole organism would respond in nature. This is because the actual capacity of a plant for carbon assimilation and biomass production is determined by the complex interplay of plant genetics and physiology, and the environment. The fundamental nature of processes involving light are not changed, but rather their effect is modified by the way in which the organism interacts with its environment. It is therefore important to consider light-mediated processes in the context of the whole organism and its interaction with the environment. *Environmental photobiology* thus provides a bridge between the understanding of fundamental processes involving light within an organism and the effects of these processes on the whole organism in the ecosystem.

The spectral distribution and irradiance of incoming solar radiation can be significantly altered in terrestrial ecosystems by the vegetation canopy. Further selective attenuation will occur as radiation penetrates the leaf. Therefore, the position of a plant in a forest canopy or leaves within an individual plant canopy will to a large extent determine the wavelength quality and irradiance available for light-mediated processes. The situation is similar for aquatic ecosystems, where the selective attenuation of radiation by water occurs. Water clarity and the location of aquatic plants and microorganisms are critical variables in this ecosystem.

Environmental photobiology thus involves an understanding of how the general environment alters the radiation regime for organisms as well as how the general environment influences the behavior of the organism in the ecosystem after light-mediated processes have produced an effect. The field of environmental photobiology encompasses a wide variety of topics, ranging from the more simple case of how one environmental factor affects one organism, to the most complex level of the ecosystem where the interaction among species and multiple environmental factors must be considered. An environmental factor that has recently caused great scientific and public concern is the potential intensification of UV radiation on earth due to a partial depletion of stratospheric ozone. Because of the effectiveness of UV radiation to cause damage in biological systems and because this environmental problem is of global concern, it will be used to illustrate many of the aspects of environmental photobiology in this chapter.

5.2. SOLAR RADIATION

5.2.1. The Solar Spectrum

The solar energy incident on the earth's atmosphere is relatively constant at approximately 1.39 kW/m^2 and has a spectral distribution that approximates a 6000°K blackbody radiation curve (Fig. 5-1).[1,2] Because of reflection, molecular and particulate scattering, and absorption of radiation in the atmosphere, solar irradiance is reduced as it penetrates the atmosphere, so that on average only approximately one-half of the radiation incident at the top of the atmosphere reaches the ground. The amount of solar radiation incident at the surface of the earth is highly dependent on cloud cover and the clarity of the atmosphere, as well as solar angle. In addition to reduced irradiance at the earth's surface, the change in spectral distribution, or wavelength quality, as radiation penetrates the atmosphere, vegetation canopies, or water is particularly significant. Since the action of

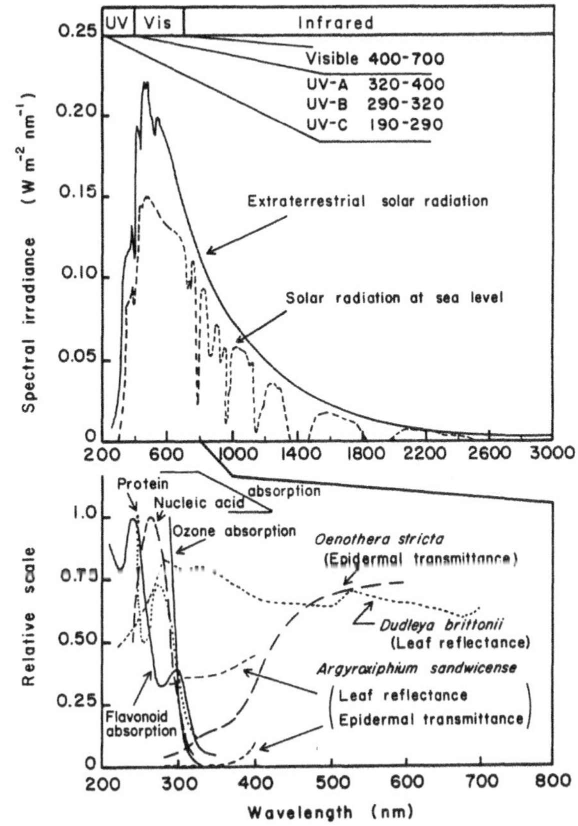

Fig. 5-1. Solar radiation at the top of the atmosphere and at sea level (top). The solar spectrum at the earth's surface is truncated at approximately 295 nm by stratospheric ozone. The absorption bands in the infrared region are due to the absorption of solar radiation by water, oxygen, carbon dioxide, and ozone molecules in the atmosphere. (Adapted from Ref. 1.)

The absorption, transmittance, and reflectance characteristics for various chromophores and plant tissues (bottom). The leaf epidermis is a highly selective filter of UV-B radiation as shown by the UV transmittance spectra for *Argyroxiphium sandwicense* (silversword) and *Oenothera stricta* (evening primrose). (Adapted from Refs. 23 and 26.) The highly pubescent leaf surface of *A. sandwicense* and the highly glaucous or wax-covered leaf surface of *Dudleya brittonii* (live-forever) show that these specialized leaf morphologies can result in high levels of UV and visible reflectance. (Adapted from Refs. 23 and 24.) The relatively low level of UV-B transmittance for glabrous leaves such as those of *O. stricta* is primarily due to the UV absorption characteristics of flavonoid and related phenolic compounds (shown is the spectrum for 7-hydroxyisoflavone, adapted from Ref. 31). The ozone absorption coefficient, normalized to 290 nm, shows that extremely effective absorption occurs at wavelengths below 300 nm. (Adapted from Ref. 7.) Although the absorption of UV radiation by nucleic acids and proteins is maximal in the UV-C waveband, these chromophores do absorb UV-B. Partial depletion of stratospheric ozone is predicted to result in increased UV-B irradiance as well as a shift in the solar spectrum at the ground surface toward shorter wavelengths. There would thus be a greater degree of overlap between the UV-B waveband and the absorption spectra of nucleic acids and proteins. (Adapted from Ref. 32.)

radiation on biological processes is highly wavelength-dependent, the spectral distribution of radiation in the environment of the organism, and how the organism itself alters the radiation through, e.g., pigmentation, will greatly influence the ultimate effect of radiation on physiological processes in the organism.

The solar radiation curve shown in Fig. 5-1 indicates that the solar spectrum at the ground surface can be partitioned according to the biological effectiveness of the major wavebands. On earth, the solar spectrum is truncated in the UV waveband at approximately 295 nm by the absorption of stratospheric ozone. In this region of the solar spectrum, the UV-B waveband (290–320 nm) is of particular interest because of its potential to cause damage to organisms.[3] The biological effectiveness of UV radiation increases logarithmically with decreasing wavelength. This increase is primarily due to the increased overlap between the UV-B waveband and the absorption spectra of nucleic acids and proteins as the wavelength decreases.[4] The visible radiation waveband at 400–700 nm, also referred to as *p*hotosynthetically *a*ctive *r*adiation (PAR), is primarily significant for its role in photosynthesis and plant photomorphogenesis, although it also has significant thermal and photodestructive effects.[5] The infrared (IR) region extends from 700 to approximately 4000 nm, and is primarily significant in regard to thermal effects on organisms. References 1–7 are suggested for further reading on the quantitative aspects of solar radiation.

5.2.2. Stratospheric Ozone and the Evolution of Terrestrial Organisms

Ozone forms a thin layer in the stratosphere with a maximum concentration between 20 and 26 km above sea level. It absorbs solar UV-B radiation with increasing effectiveness at shorter wavelengths so that essentially no radiation below 295 nm penetrates through the atmosphere to the ground.[2] Ozone (O_3) is constantly formed in the upper atmosphere through the combination of molecular oxygen (O_2) and atomic oxygen (O). The latter is formed from the photodissociation of O_2 by short-wavelength UV-C radiation (<240 nm).[8] Other photochemical processes in the upper atmosphere, involving reactions catalyzed by oxides of nitrogen and chlorine (e.g., NO_x and Cl), result in the breakdown of ozone. This natural formation and breakdown of ozone is in balance and results in an effective shield against injurious UV radiation at the ground. Ozone photochemistry in the stratosphere is limited by oxygen availability and may be affected by natural perturbations such as stratospheric temperature variations, solar proton events, changes in solar irradiance, and chemical inputs from volcanic eruptions.[9]

Most of the ozone is produced above the equator where solar irradiance is maximal. Ozone formed at these latitudes subsequently diffuses in the stratosphere toward the poles. Thus, while ozone concentrations in the stratosphere above the equator may average 0.29 cm (the thickness of a vertical column of stratospheric ozone condensed to standard temperature and pressure), ozone concentrations above the poles may exceed 0.40 cm at the end of the winter season.[8] Considerable seasonal and daily fluctuations in ozone concentration may occur, however. Concentrations of stratospheric ozone tend to be highest in late winter and early spring, and lowest in late summer and early fall. Because the smallest seasonal variation in solar irradiance occurs above the equator, stratospheric ozone at this latitude also shows the smallest seasonal variation. At temperate latitudes the seasonal variation in stratospheric ozone concentration appears to be small, and at first glance not too biologically relevant. Ozone measured at Arosa, Switzerland, for example, varied from a mean monthly high of 0.27 cm in April to a low of 0.205 cm in November (Fig. 5-2).[2] The relatively small monthly fluctuations in ozone shown for this location, however, can result in large variations in UV-B irradiance. This is because the absorption of UV radiation by ozone increases exponentially with ozone thickness.

The significance of the stratospheric ozone layer and its seasonal variation becomes more apparent when viewed in regard to the biological effectiveness of UV-B radiation. Biological effectiveness refers to the capacity of UV-B radiation to cause damage to living organisms and is based on the biological weighting factor used to quantify UV-B response. The damage induced at each wavelength of the UV-B waveband to DNA molecules, photosynthesis, or the whole organism has been used to develop biological weighting factors. These weighting factors, also known as action spectra, are highly wavelength-

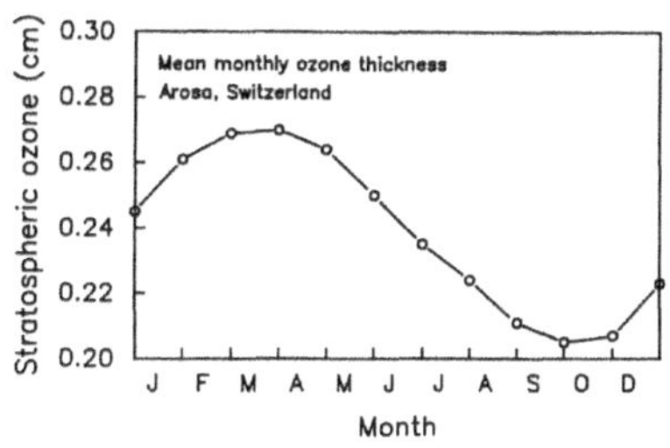

Fig. 5-2. Mean monthly stratospheric ozone concentration at standard temperature and pressure for Arosa, Switzerland. Because UV-B irradiance varies exponentially with ozone concentration, the apparently small monthly changes in ozone at this temperate latitude location result in large changes in UV-B irradiance. (Adapted from Ref. 2.)

dependent in the UV region and tend to increase exponentially with decreasing wavelength. An equation that integrates the weighting factor and solar irradiance over the UV-B waveband is:

$$\text{Biologically effective UV-B} = \int_{280}^{320} I(\lambda) \cdot E(\lambda) \cdot d\lambda$$

where $I(\lambda)$ is the spectral irradiance at wavelength λ, and $E(\lambda)$ is the biological weighting factor.(6) The weighting factor for DNA damage in response to UV-B irradiation, for example, increases more than four orders of magnitude as the wavelength decreases from 320 to 280 nm. This indicates that the potential for UV radiation-induced damage increases dramatically with small decreases in wavelength. When both the biological effectiveness of UV radiation and the wavelength dependency of the absorption of UV radiation by ozone are considered in respect to the natural latitudinal gradient of ozone thickness, along with increased solar angles at higher latitudes, a pronounced gradient in biologically effective UV-B radiation results.(6) When the weighting factor for UV-B radiation-induced damage to DNA is used in the equation above, the biologically effective solar UV-B irradiance at low latitudes can exceed that at high latitudes by an order of magnitude.

Prior to the development of the stratospheric ozone layer, the earth's atmosphere was transparent to short-wavelength UV radiation (<290 nm), wavelengths of sufficient energy and overlap with the absorption spectra of nucleic acids and proteins to cause lethal damage to organisms. The processes that led to the development of an ozone layer about 600 million years ago remain unclear. It has been hypothesized that oxygen in the earth's atmosphere increased slowly, at first through the photodissociation of water and later through photosynthesis.(10) Since oxygen is essential for the photochemical production of ozone, a threshold concentration of atmospheric oxygen was necessary before ozone could be formed in significant amounts. An ozone concentration of 1% of present atmospheric levels may have been reached about 600 million years ago. This primeval atmosphere, still rich in highly photoreactive wavelengths, precluded terrestrial organisms.(10) Plant and animal life at that time was confined to aquatic environments where the penetration of UV radiation was restricted. Early aquatic organisms were confronted with the dilemma of occupying aquatic habitats deep enough to confer protection from harmful short-wavelength UV radiation, yet having sufficient visible radiation available for photosynthesis. It was not until approximately 400 million years ago when ozone concentrations are believed to have reached 10% of present levels that adaptation to the terrestrial environment was possible. The earliest fossil records of land plants and animals correlates well with this theory of ozone development. Although considerable debate exists on the exact processes that led to the development of the ozone layer and over what time scale this occurred, it is clear that UV-B irradiance in the early atmosphere was more intense than present levels. Furthermore, this radiation environment would have presented considerable selective pressure for adaptation in terrestrial organisms under this intense UV-B regime.

Mechanisms that attenuated UV-B radiation before damage occurred in nucleic acids would have been of protective value and of great adaptive significance for survival in the intense UV-B radiation environment. The more effectively an organism was protected from UV-B radiation, the less likely that the rate of damage would exceed the capacity of enzymatic mechanisms to repair UV radiation-induced damage. Early land plants adapted to the relatively high UV-B radiation environment through the development of a pigment system that provided effective protection from UV-B-induced injury.[10,11] This system consists of flavonoid and related phenolic compounds, which are colorless to light yellow pigments with a strong absorption of UV radiation at wavelengths less than 320 nm (Fig. 5-1). These compounds are present in most tissues of higher land plants, and form an effective UV-B radiation filter in the outer tissue layers of leaves, stems, and flower parts. With a few exceptions, flavonoid compounds are not present in algae.[12]

Flavonoid and related phenolic compounds are synthesized in plants in response to UV-B irradiation. Their presence in leaf epidermal tissue and around chloroplasts in the mesophyll reduces the potential for radiation-induced damage to sensitive targets in the plant. Although the present stratospheric ozone layer provides an effective filter for short-wave UV radiation and presently limits this waveband to 295 nm (Fig. 5-1), flavonoid and phenolic compounds in leaves and other plant tissues still appear to have adaptive significance for plant survival at present levels of UV-B irradiance. These pigments as well as stratospheric ozone are not only effective filters of the more actinic wavelengths below 295 nm, but are transparent to visible radiation.

5.2.3. The Solar Radiation Environment in Aquatic Ecosystems

Water clarity in the environment of aquatic organisms directly affects the degree of penetration of solar radiation into water.[13] Figure 5-3 illustrates the degree of penetration and the spectral distribution of solar radiation to the 1-m water depth for distilled water and for lakes varying in water clarity. In the IR waveband, at wavelengths above 700 nm, water is relatively opaque to radiation.[14] This strong absorption of IR radiation is a physical property of water and is not particularly affected by the level of impurities in the water. Impurities in water such as dissolved organic compounds do, however, exert a highly significant influence on the penetration of visible and UV radiation. The effect of these impurities increases with decreasing wavelength, so that while distilled water transmits approximately 98% of the incident solar radiation below 550 nm to a depth of 1 m, lakes with very high concentrations of dissolved organic compounds absorb essentially all radiation below 550 nm in the first meter of water. Thus, the radiation environment for aquatic organisms can be one of low solar irradiance and of a spectral distribution greatly different from that found at the surface. Aquatic organisms in oligotrophic lakes, rela-

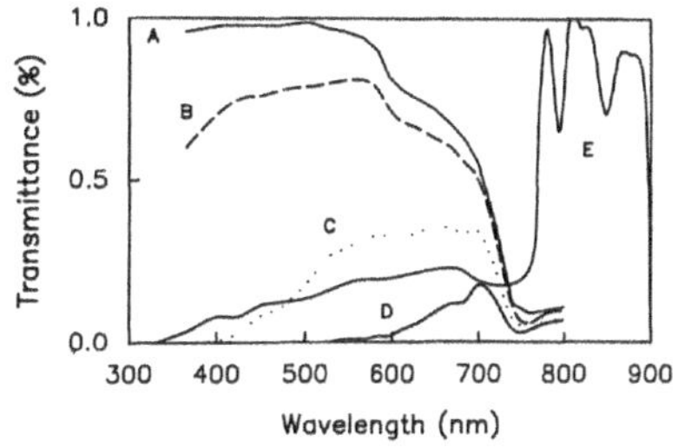

Fig. 5-3. The penetration of solar radiation into water and vegetation canopies. Curves A–D illustrate the level of transmittance of solar radiation into 1-m-deep water of different degrees of clarity (curve A is for distilled water, and curves B–D are for lakes with increasing levels of dissolved organic compounds that tend to attenuate UV wavelengths more effectively than longer wavelengths). (Adapted from Ref. 14.) Curve E shows that vegetation canopies tend to selectively absorb the visible or photosynthetically active wavelengths (400–700 nm) and are highly transparent to near-infrared radiation (700–2000 nm). The radiation environment beneath a canopy is therefore shifted to the infrared region. (Adapted from Ref. 15.)

tively clear and nutrient-poor waters, would be exposed to high levels of solar UV-B and visible irradiance. Those in eutrophic estuaries or other coastal waters, which are aquatic environments rich in dissolved organic matter and nutrients, would exist in an environment much reduced in UV and visible wavelengths.

The reduced penetration of photosynthetically useful wavelengths into eutrophic estuaries, lakes, or coastal waters would be expected to reduce the primary productivity of these aquatic ecosystems.[13] However, significant environmental and behavioral factors can ameliorate the effects of a reduced solar radiation regime. The movement of water, either as upwelling along coastal regions, stream flow, or the seasonal turnover of thermal strata in lakes, is a major factor in the productivity of aquatic ecosystems. This movement will not only influence the distribution of nutrients and dissolved organic matter in the water, but also may influence the location and movement of planktonic communities. Nonmotile phytoplankton may benefit from periodic water movements if that movement results in transport to the water surface layers where exposure to solar radiation would be higher. In contrast, motile and phototactic algae may overcome the effects of water movement by their ability to migrate to the upper water layers to increase the daily reception of solar radiation. Because photosynthetically useful wavelengths are absorbed in the upper 1-m layer of relatively eutrophic ecosystems (Fig. 5-3), aquatic macrophytic algae that are attached to rocks or sediments in lakes or in coastal waters must rely on efficient mechanisms for the absorption and utilization of radiation deficient in these wavelengths. This may involve, for example, a greater reliance on accessory pigments for photosynthesis (Chapter 12). Thus, the reduced radiation regime of eutrophic ecosystems could be compensated for by organisms that are adapted for low-light conditions, or with mechanisms involving periodic migration to regions of higher solar irradiation. Reference 13 is recommended for a more detailed discussion on light-mediated processes and aquatic ecosystems.

5.2.4. The Solar Radiation Environment beneath Vegetation Canopies

The radiation environment beneath plant canopies is also quite different in intensity and spectral distribution from the solar radiation regime at the top of the vegetation layer. Absorption of solar radiation by successive layers of leaves within an individual plant canopy or within a stand of vegetation, however, results in a radiation environment distinctly different in spectral distribution from that in aquatic ecosystems (Fig. 5-3). The mechanisms involved in the alteration of solar radiation by vegetation is similar for canopies of individual plants and for the complex canopy structure of whole plant communities.

Wavelength selectivity of the leaf is the most significant factor determining the level and quality of irradiance underneath a vegetation canopy. Since leaves absorb essentially all UV-B radiation, the lower leaves in a plant canopy tend to be well protected from exposure to this photochemically active waveband. Coincident with the absorption spectra of photosynthetic chromophores such as chlorophyll *a* and *b*, and with the spectra of accessory pigments such as carotenoids and xanthophylls, approximately 90% of the visible radiation incident on the leaf surface is absorbed. With this high level of absorption at each leaf layer in the canopy, the radiation available to leaves in the lower strata of a plant canopy, or the understory of a plant community, is rapidly depleted of wavelengths that are photosynthetically useful. Because leaves are highly transparent to near-IR radiation, the selective absorption of UV-B and visible radiation and the selective transmittance of near-IR radiation results in a radiation regime for the understory of a plant community that is depleted of PAR and rich in near-IR radiation (see Fig. 5-3 and Ref. 15).

The progressive spectral shift toward the IR with increasing depth into the plant canopy can significantly affect plant growth and development. Understory plants have adapted to the PAR-poor and IR-rich radiation environment through leaf structural and functional modifications. Relative to leaves developed in full sunlight, attributes of these shade-adapted leaves include the tendency toward larger leaves, reduced mesophyll thickness and chlorophyll *a*/*b* ratios, and increased density of chloroplast membrane systems. In addition, the accessory pigments such as xanthophylls tend to be more important for shade-adapted plants. Functionally, the shade-adapted leaves have greatly reduced rates of respiration, photosynthesis, and transpiration. In obligate shade-adapted plants, these characteristics result in efficient utilization of irradiance, but also tend to preclude the capacity to acclimate to full insolation.

Radiation penetration into plant canopies is often described simply by a modified version of the Lambert–Beer extinction law:

$$I = I_o e^{-k\mathrm{LAI}}$$

where I represents irradiance at some depth in the canopy, I_o is solar irradiance incident at the top of the canopy, k is the extinction coefficient for a particular vegetation stand, and LAI is the ratio of total leaf area to ground area (m^2/m^2) above the depth at which the final irradiance is estimated. While this relationship is instructive for viewing general aspects of radiation attenuation in different types of vegetation stands and is useful for comparing the attenuating capacity of different types of plant communities, accurate descriptions of such attenuation are considerably more complex. It is of particular importance to note that this equation does *not* describe the shift in spectral distribution to the IR waveband as radiation penetration into the canopy occurs. Of particular importance is the fact that characterization of radiation penetration into plant stands involves the integration of all aspects of vegetation architecture and solar radiation.[16] Some of the more significant considerations include leaf optical properties (reflectance, absorptance, and transmittance), leaf orientation in relation to the sun and to other leaves, the arrangement of foliage strata in the canopy, changes in solar angle throughout the day and year, and the affect of changes in the direct and diffuse components of solar radiation. All of these components would have seasonal variability, especially in plant communities dominated by deciduous trees, and daily variability as solar angle and radiation penetration through the atmosphere changes. The integration of all these aspects into mathematical models forms a considerably greater task than utilizing the Lambert–Beer law, and necessitates the use of computer modeling. The end result of differences in vegetation architecture is that the potential for photosynthesis within the canopy will differ and thereby significantly affect the productivity of the plant stand. References 16–21 are suggested for further reading on the relationship between canopy architecture and solar radiation.

Competition for photosynthetically active radiation in plant stands is perhaps the fullest expression of how light-mediated processes are linked to the functional aspects of the organism in its environment. The competitive capacity of an organism relative to neighboring plants can have profound effects on its growth, survival, and productivity. While it is not possible to truly separate competition among plants for soil nutrients or water from competition for light, plants that out-compete their neighbors for critical resources in light-limited environments may be able to occupy a higher position in the canopy where irradiance is higher. This would be achieved through relatively faster stem and leaf development. Interception of radiation by plants in the higher strata of the canopy would thus inhibit the growth of plants beneath them. The ability to increase plant height faster than neighboring plants and the efficient display of foliage for the interception of visible solar radiation are essential characteristics for competitiveness in plant communities.(21)

5.3. HOW ORGANISMS PROTECT THEMSELVES FROM SOLAR UV RADIATION

Solar radiation can affect organisms only if the radiation is absorbed. And since the effect on light-mediated processes is highly wavelength-dependent, environmental factors or aspects of the organism itself that alter the spectral distribution of radiation before it is absorbed may significantly influence how well the organism functions in its environment. The potential damage to nucleic acids caused by UV-B exposure, and the subsequent effects on growth, reproduction, and survival, can exert a considerable influence on the natural selection for plants and animals with effective protective mechanisms against UV-B radiation damage. Mechanisms that prevent damage, or at least reduce it to levels within the organism's capacity to repair damage, would therefore have adaptive significance for the organism.

5.3.1. Terrestrial Plants

Higher plants are confronted with the problem of balancing the need to maximize the interception of visible radiation for photosynthesis and to minimize the interception of UV-B radiation and its possible damaging effects. Foliage display for maximal interception of visible radiation may be optimum for photosynthesis, but also exposes leaves to possible high rates of damage by UV-B radiation. Avoidance of UV-B radiation through changes in leaf orientation relative to the sun is a possible adaptation, but would simultaneously reduce the interception of visible radiation by the leaf. Adaptations that involve changes in leaf anatomy, such as leaf surface pubescence or thick cuticular waxes, tend to decrease UV-B penetration into the leaf primarily by increasing the reflectivity of the leaf surface. However, these anatomical adaptations are generally not wavelength-selective and attenuate UV and visible radiation to the same extent. This often results in relatively low rates of photosynthesis, growth, and annual productivity for such species.

Although leaf pubescence or thick cuticular wax layers can increase the reflectance of UV-B radiation from the top of these layers, cuticular waxes and cell wall constituents are themselves relatively transparent to UV-B and visible radiation. These plant tissues, therefore, do not substantially contribute to the attenuation capacity of the epidermis.(22) The geometry of cell walls, including the amount of air space within the leaf, may increase multiple reflection of radiation within the leaf, and thereby reduce to some extent the UV-B radiation incident on sensitive targets within the leaf.

The highly pubescent leaves of *Argyroxiphium sandwicense* DC. (silversword) or the thick wax-covered leaves of *Dudleya brittonii* Johansen (live-forever) are notable examples of species occupying relatively high irradiance environments.(23,24) The leaf surfaces of these plants are highly reflective in the UV-B and visible wavebands. Additional absorption of radiation by flavonoid compounds in the epidermal layer results in essentially no UV-B penetration to the mesophyll. However, these highly reflective leaf surfaces also greatly reduce the penetration of visible radiation for photosynthesis, which may be a significant disadvantage. Because this type of protective system requires a relatively permanent modification of leaf anatomy, it is not responsive to the daily or seasonal fluctuations in UV-B irradiance. A lack of flexibility and wavelength selectivity limits UV-B acclimation potential and the efficacy of such leaf surface modifications as a UV-B protective mechanism for plants. Furthermore, species with highly pubescent or wax-covered leaves tend to be relatively rare and, in general, the green glabrous or smooth leaves are the dominant leaf type found in nature.

A UV-B protective mechanism that is highly wavelength-selective for this waveband and the capacity to respond and acclimate to changes in the UV-B radiation environment would provide a great adaptive advantage for terrestrial plants. The presence of flavonoid and related phenolic compounds in the upper epidermis of leaves may be an essential part of such a protective mechanism. The absorption characteristics of this class of compounds produce an effective attenuation of UV-B and shorter wavelength radiation (Fig. 5-1) and no absorption of visible radiation (with the exception of anthocyanins, which strongly absorb radiation between 520 and 560 nm). Several studies have clearly shown the relationship between UV-B irradiation and the increased synthesis of these compounds. The link between UV-B irradiation and flavonoid induction has been indicated both at the biochemical level, as increased levels of the enzyme phenylalanine ammonia-lyase that is involved in the production of flavonoids, and at the whole-plant level, as increased UV-B absorbance in the epidermis and mesophyll.(25,26) This direct link between UV-B irradiation and increased flavonoid concentrations forms a protective mechanism that is potentially responsive to seasonal and daily changes in the UV-B radiation environment.

Absorption characteristics of leaves indicate that the epidermis functions as a highly selective filter of solar radiation. Reflectance of UV-B from green glabrous leaves is relatively low at 3–5%, and increases in the middle part of the visible waveband to approximately 15–20%. The outer layers of the leaf can attenuate over 90% of the UV-B radiation incident on the leaf surface. With additional UV-B absorption in the mesophyll, essentially no UV-B radiation penetrates through the entire leaf, and so leaves are opaque to this waveband. This substantial attenuation of UV-B radiation in leaf tissue is due to flavonoid and related phenolic compounds. With increasing wavelength into the UV-A and visible regions, the epidermis becomes increasingly more transparent (Fig. 5-1). The epidermis has maximal transmittance of more than 80% in the visible waveband; mesophyll absorption is maximal in this region. In the near-IR region the entire leaf is both highly reflective and transparent, with the combined reflectance and transmittance of near-IR radiation as high as 95%. This characteristic has adaptive significance in reducing the thermal energy load for the leaf. The combined optical characteristics of the leaf epidermis and mesophyll thus produce a highly wavelength-selective filter well suited to protect the leaf from injurious UV-B radiation and IR radiation heat stress, and to maximize the penetration of photosynthetically active visible radiation.

The reproductive parts of plants may also be sensitive targets for UV-B radiation. However, the effect of this radiation on pollen and anthers, and the protective role of the surrounding flower petals, has been difficult to study. This is related both to the small size of these reproductive structures and to the difficulty in predicting the length of time that pollen is exposed to UV-B irradiation in nature. Flower petals and anther walls are largely opaque to UV-B radiation.[27] Pollen is therefore completely protected from UV-B exposure while the flower is closed and, after the flower opens, remains protected by the UV-opaque anther wall. When exposed to direct UV-B irradiance, pollen is rather sensitive to injury, as shown by significant reductions in the germination percentages of the exposed pollen. The length of time that pollen is exposed to UV-B irradiation in nature, that period between dehiscence of pollen from the anther and the penetration into the stigma, is the critical factor. This potential exposure period is relatively short, however, and current studies suggest that under natural field conditions pollen of temperate latitude plant species is not adversely affected by current levels of UV-B irradiance.

5.3.2. Animals and Aquatic Organisms

Unlike plants, animals do not require sunlight to synthesize their food and are mobile. Continuous exposure to solar radiation throughout an animal's lifetime is not required, as in the case of plants. Rather, exposure is regulated by behavior of the animal. When exposed to solar radiation, however, animals are subject to the damaging effects of UV-B radiation. The adaptations that have evolved in animals to cope with damaging UV-B radiation are in some ways similar to those in plants, i.e., they are protective in that they minimize UV radiation-induced damage. Two major protective mechanisms for reducing the potential for UV-B-induced damage involve optical properties of the outer skin layers and surface coverings. Several chromophores in the epidermal layer of human skin will absorb UV-B radiation. These include aromatic amino acids, urocanic acid, nucleic acids, and melanin. The latter chromophore gives human skin a dark-pigmented appearance, and people with reduced skin pigmentation are generally more susceptible to sunburn and UV-B related skin cancers than are highly pigmented people. This is especially the case for albinos.[9] Thus, as in the case for the protective role of UV-absorbing flavonoid compounds in the plant leaf epidermis, the UV-absorbing chromophores in the human skin epidermis provide a degree of protection from damaging UV-B radiation. This protective mechanism must be highly effective because only about 5% of the UV radiation incident on human skin is reflected. Thus, the major portion of this incident radiation penetrates the skin. Surface coverings, such as hair or fur, can substantially reduce UV radiation incident on the skin. For human populations, clothing provides a suitable covering to reduce the exposure of skin to sunlight. However, the recent rise in UV-related skin cancers for light-skinned people in, for example, the United States of America, Europe, and Australia highlight the need to modify recreational or work activities that involve extensive exposure of skin to sunlight. In the past, exposure of the skin to sunlight was restricted by the use of more clothing. The function of fur is primarily related to its thermal properties, but it can also protect against physical damage to the skin and against insects. In addition, the color or pigmentation of fur often functions as camouflage. While these may be the primary functions of fur, a skin covering of fur can reduce the penetration of UV radiation to the skin, and thereby provide protection from damaging UV radiation. The white fur of polar bears and baby harp seals is an interesting exception to this.(28) The fur of these animals is actually composed of hollow, transparent hairs that are entirely lacking in pigmentation. The rough inside surface of the hollow hairs reflects visible radiation and thus appears white. Furthermore, only UV radiation is "funneled" to the skin through the hollow-cored hairs.(28) The result is that the white fur of these animals appears black when photographed with UV-sensitive film. Because of the low solar angles and the relatively high concentrations of stratospheric ozone at arctic and antarctic latitudes (Section 5.2.2), UV-B irradiance is relatively low in polar habitats. Therefore, the present or predicted enhanced levels of UV-B radiation will probably not be deleterious for arctic populations of polar bears and harp seals.

The mobility of animals is another effective mechanism by which these organisms can reduce exposure to solar radiation. It is a mechanism of avoidance, and primarily involves behavioral control over the duration of exposure and the time of day that an individual is exposed. For humans, the desire for outdoor recreational activities has outweighed caution against too much exposure. Avoidance of UV-B exposure by animals is an indirect benefit of behaviors that minimize exposure to the midday sun and high temperatures, such as nocturnal or early morning and evening feeding habits. The combined affects of skin pigmentation, hair and fur, mobility, and behavior thus form an effective defense against damaging UV radiation.

The primary form of protection from UV-B radiation for aquatic organisms involves mobility, movements that change the position of the organism in the vertical profile of the water column. As discussed above in Section 5.2.3, water circulation patterns or turbulent mixing of water surface layers can transport planktonic communities to different locations in the water. Also, motile and phototactic algae have the ability to migrate nearer to or further from the water surface. These movements will directly determine the UV-B exposure time for these organisms. Since with few exceptions flavonoid compounds are not present in algae, these organisms lack the protective pigment system of higher vascular plants. Regulation of UV-B exposure time through movement in the vertical profile of their water habitat is thus significant for aquatic organisms. Because the water medium itself acts as a filter of UV-B radiation (Fig. 5-3), aquatic organisms, particularly those in estuaries or zones of upwelling where mixing of dissolved organic compounds and other impurities may occur, are to some extent sheltered from the full intensity of UV-B radiation.

5.4. BIOLOGICAL CONSEQUENCES OF STRATOSPHERIC OZONE DEPLETION

A significant environmental problem of global magnitude is the prediction that atmospheric pollutants will result in a partial depletion of the stratospheric ozone layer. The release of pollutants such as chlorofluorocarbons (CFCs) and nitrous oxide (N_2O) into the atmosphere is predicted to reduce the equilibrium ozone column thickness at all latitudes in the coming decades, i.e., the increased concentration of these pollutants will increase the rate of ozone destruction relative to the rate at which ozone is naturally produced.(8) Ozone is destroyed in chemical reactions involving oxides of nitrogen (e.g., NO_2), which are chemically active molecules formed from the photooxidation of nitrous oxide (N_2O). The latter compound is produced naturally by bacteria in the soil and water. Additional sources of N_2O related to human activities include high-altitude aircraft exhaust, human and animal waste, and industrially produced nitrogen fertilizers. These additional new sources of N_2O have significantly increased the atmospheric concentration of this molecule by 2.7% from 1964 to 1981, a trend that is expected to continue into the future.(9) Chlorofluorocarbons (e.g., CF_2Cl_2 and $CFCl_3$) are a commercially produced class of compounds found in refrigeration systems and propellants. When these compounds are decomposed by solar radiation in the upper atmosphere, atomic chlorine reacts with ozone in chemical reactions similar to those involving nitrous oxides.(8) Other potential sources of chlorine atoms include methyl chloroform (CH_3CCl_3), an industrial solvent, methyl chloride (CH_3Cl), and carbon tetrachloride (CCl_4).

Computer simulation models of atmospheric chemistry represent our primary source for the prediction of ozone depletion rates and tend to be very complex. The degree of ozone depletion predicted varies with input variables such as present and future pollutant production rates, their rate of diffusion to the stratosphere, the length of time these compounds are resident in the stratosphere, the photochemical reactions and their rates, and the behavior of the stratosphere. Because of the complex nature of these interacting variables as well as incomplete knowledge of atmospheric processes, computer simulation models often vary in their predictions for the expected depletion rate of ozone. This has resulted in predictions of stratospheric ozone depletion ranging from 5 to 20% for the effect of CFCs alone. Refinements in the models have improved the estimates of ozone depletion and, as reported by the National Research Council of the United States National Academy of Sciences, the degree of ozone depletion from present levels is expected to be 5–9% due to CFCs alone and 10–16% due to a doubling of N_2O.(9) Such reductions in the ozone layer will not only increase UV-B irradiance reaching the earth's surface but should also shift the terrestrial solar spectrum slightly toward shorter, more photobiologically effective wavelengths. Since this spectral shift would increase the overlap of UV-B wavelengths with the absorption spectra of nucleic acids and proteins, the increase in UV-B irradiance is expected to have biologically significant effects. When the combined effects of a 15% decrease in ozone at temperate latitudes, the spectral shift toward shorter wavelength, and increased UV-B irradiance weighted for DNA damage are considered, the potential future radiation environment could be 44% greater in effective UV-B radiation (see equation in Section 5.2.2).(6)

The discovery in 1985 of an ozone depletion zone over the continent of Antarctica, an ozone "hole" that appears every antarctic spring, has renewed concern over the potential for ozone destruction and the intensification of UV-B irradiance. Observations over Antarctica indicate that ozone concentrations during the spring season have declined by up to 40% over the past decade.[29] While the cause of this substantial ozone depletion zone above Antarctica is not yet fully understood, it is believed to involve a combination of stratospheric cooling and circulation patterns and a higher abundance of chlorinated molecules in this polar region. Public concern over this global environmental problem has resulted in an international agreement, signed in 1987 by 23 nations and sponsored by the United Nations Environment Program, for a 50% reduction in CFC production by 1999.

5.4.1. The Response of Individual Plant Species and the Consequences for Ecosystem Productivity

The survival of organisms in the predicted enhanced UV-B radiation environment may depend greatly on the effectiveness of their protective mechanisms and the efficiency of mechanisms that repair UV-B induced damage. Effective protection from UV radiation-induced damage is particularly important for plants, since these organisms depend on solar radiation for photosynthesis and cannot avoid exposure to UV-B radiation. In an enhanced UV-B radiation environment, the various mechanisms that repair UV-B-induced damage to nucleic acids (Chapter 4) may not be able to keep pace with the rate at which damage occurs. Therefore, organisms with the capacity to increase their attenuation of UV-B radiation, such as plants with a relatively opaque leaf epidermis, may sufficiently reduce the rate of damage below the point at which repair mechanisms are saturated.

An increase in the attenuation capacity of the epidermis, through the synthesis of increased amounts of flavonoid and related phenolic compounds, does occur under artificially enhanced UV-B environments. For example, flavonoids in the epidermal tissue of *Oenothera stricta* Ledeb. (evening primrose) increased by up to 100% after exposure to an enhanced UV-B environment.[26] This radiation regime also significantly reduced UV-B transmittance through intact epidermal tissue by as much as 33%. Transmittance of visible radiation through the epidermis was not affected. However, once the daily UV-B exposure exceeded the highest levels normally found in this species' field habitat, photosynthesis was significantly decreased. This suggests that an upper threshold of exposure exists above which the capacity of protective mechanisms to reduce injury is exceeded. *Oenothera stricta* is native to South America at low-elevation, temperate latitude sites, but has in recent years invaded high irradiance habitats in the Hawaiian Islands. The ability of this species to colonize such habitats is evidence that it can acclimate to current high UV-B irradiance levels. Whether sufficient acclimation capacity exists within this species, or others like it, to tolerate the enhanced UV-B environment of the future is still unclear.

The capacity for acclimation to higher UV-B irradiance levels varies among plant species, and agronomic plants tend to be more sensitive to enhanced UV-B irradiation than wildland plants. These conclusions, however, are based on studies in which filtered fluorescent lamps were used to produce an enhanced UV-B environment. The lamp/filter systems used do not perfectly simulate the natural spectral distribution of UV radiation in sunlight, and plants studied in greenhouse or controlled environment chambers tend to be more sensitive than field-grown plants to UV irradiation. This greater UV-B sensitivity

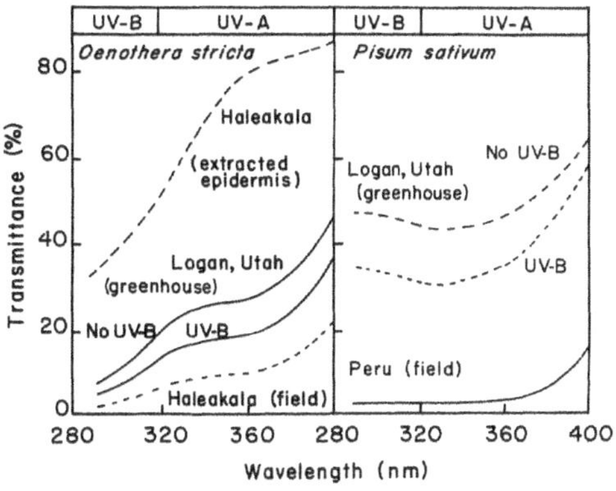

Fig. 5-4. The degree of UV transmittance through leaf epidermal tissue of *Pisum sativum* "Alderman" (pea) and *Oenothera stricta* (evening primrose). The latter species has colonized high-irradiance habitats on the Haleakala Crater of the Hawaiian island of Maui, and *P. sativum* is a common agronomic species. For plants examined under greenhouse conditions at Logan, Utah, the reduction in epidermal transmittance for *O. stricta* was evident after 11 days of a biologically effective UV-B exposure approximately 20% less than that experienced on cloudless days in the Hawaiian habitat. Field-grown plants exhibited the lowest level of epidermal UV-B transmittance, and thus the greatest protection from UV-induced damage. After flavonoid and related phenolic compounds were extracted from the epidermis, the capacity of this tissue to attenuate UV radiation was substantially reduced. A similar response to UV-B irradiation was evident for *P. sativum*, a plant species relatively sensitive to enhanced UV-B radiation. Exposure to UV-B radiation in controlled experiments reduced epidermal transmittance but, as with *O. stricta*, the greatest attenuation of UV radiation occurred in field-grown plants. The high degree of plasticity in epidermal transmittance exhibited by *P. sativum* may explain why this temperate latitude cultivar is successfully cultivated in high-elevation fields of the Peruvian Andes, where biologically effective UV-B radiation is up to 30% higher than that experienced by the greenhouse plants. (Adapted from Refs. 23 and 26.)

may in part be attributed to the lower level of UV-A and visible radiation available in controlled environments for UV repair mechanisms requiring light (i.e., photoreactivation; see Chapter 4). Controlled experiments can, however, be used to detect differences in UV-B radiation sensitivity and their capacity to acclimate to higher UV-B levels among species. Based on reductions in photosynthesis and biomass after UV-B irradiation in such experiments, *Rumex patientia* L. (dock), *Pisum sativum* L. (pea), *Cucurbita pepo* L. (squash), and *Glycine max* (L.) Merr (soybean) are among the more UV-sensitive species.[30] Some species under field conditions exhibit considerably greater capacity for acclimation to UV-B radiation than when examined under experimental conditions. As illustrated for *O. stricta* and *P. sativum* "Alderman" in Fig. 5-4, the degree of UV-B transmittance is reduced in response to intensified UV-B irradiation under experimental conditions. The reduction in epidermal UV-B transmittance, however, is substantially greater in the plants grown in their high UV-B irradiance field habitats.

The consequences of an enhanced UV-B environment may be expressed in more subtle ways than what has been demonstrated for individual plants cultivated under the artificially enhanced UV-B regimes. Rarely in nature are organisms exposed to only one environmental stress factor at a time. Usually several factors act on an organism to affect its growth, survival, and reproductive capacity in the ecosystem. How the whole organism is affected depends, however, to a great extent on factors secondary to the initial influence of radiation on light-mediated processes. Growing season temperatures, solar radiation, water availability, and nutrient supply are among the more significant environmental factors that influence organisms. Particularly important is the interplay between these factors and the developmental stage of the organism. The influence of interacting factors such as UV-B irradiance, water and nutrient supply, and air temperature on a plant, for example, would most likely be different during the seedling stage than at the mature plant or flowering stage. In addition to the effects of environmental factors, the availability of resources in the environment that are critical to the organism as well as competition among organisms for these resources should have a significant influence on the ultimate response of organisms in an enhanced UV-B radiation environment. Obtaining answers to how plants in nature will respond to an intensified UV-B environment requires multifactor experiments. However, executing such multifactor experiments so that reliable predictions can be made about plant behavior under a variety of situations has been quite difficult.

Ecosystems, because of their large scale and complexity, render extrapolations from experiments on individual plants to their behavior in whole ecosystems tenuous. Although predictions about ecosystem level responses may be based on the relative sensitivity of individual plant species, e.g., reduced photosynthesis and growth in some sensitive species and acclimation in others, these predictions would have a large level of uncertainty. Differences in UV-B sensitivity among species may indicate that overall primary productivity of natural terrestrial ecosystems will not be significantly reduced under an intensified UV-B environment. Rather, this future radiation environment may result in more subtle changes in relative species composition and abundance. Growth and productivity of species with moderate to high UV-B sensitivity may be reduced so that they are not as competitive with species more resistant to UV radiation-induced damage. Species in this latter category could thus become more dominant and abundant in an ecosystem. The actual situation is considerably more complex since various stages of plant growth can be more sensitive to UV-B radiation than others, and will also respond differently to environmental factors at each of these stages. Agricultural ecosystems are composed of monocultures, and are thus generally less complex than natural ecosystems. Although differences in soils, water and nutrient supply, species composition, and genetic variation are minimized in agricultural systems, experimentation difficulties encountered with these systems are similar to those with plants in natural ecosystems. Simulation of actual field conditions in experiments has been difficult to achieve with the artificial UV-B radiation systems that have been used. However, the yields of some economically important crops may be reduced in an intensified UV-B environment, an assessment based largely on the response of agronomic plants in greenhouse and controlled environment chambers. In such experiments, *G. max, P. sativum, Vigna unguiculata* (L.) Walp. (cowpea), *Phaseolus vulgaris* L. (bean), and *Brassica oleraceae* L. "acephala" (collard) are among the crops that have exhibited reductions in photosynthesis, leaf area, or productivity.[(30)] Species that have been found to be relatively more resistant to UV-B radiation injury include such important crops as *Hordeum vulgare* L. (barley), *Oryza sativa* L. (rice), *Triticum aestivum* L. (wheat), and *Zea mays* L. (corn). Of all the species examined in numerous studies, approximately 20% have been found to be very sensitive to UV-B irradiation, 50% exhibited moderate sensitivity, and 30% were relatively resistant to UV-B radiation injury.[(30)] There are substantive inconsistencies among various studies in regard to the sensitivity of species to UV-B radiation, however. The frequent contradictory findings among studies may stem from the lack of uniform experimental techniques, variations in the UV radiation lamp/filter systems used to simulate an enhanced UV-B radiation regime, and differences in UV-B radiation exposure, plant cultivation conditions, and cultivars. Considerable caution should thus be used in the interpretation of the results from studies on the UV-B radiation sensitivity of crop species. However, there is wide agreement that species and varieties exhibit substantial differences in their response to UV-B radiation. Important sources of variation include genetic differences among species and among cultivars that determine the capacity for UV-B attenuation in the epidermis and the efficacy of mechanisms for the repair of UV-B radiation injury. The phenological stage at which the stress of enhanced UV-B radiation is initiated may also be a significant factor in the apparent sensitivity of a species to UV-B radiation. Current research sponsored by the United States Environmental Protection Agency on the response of representative crops plants (e.g., wheat, corn, and soybeans) to enhanced UV-B radiation should significantly increase our understanding of how agricultural ecosystems will respond under the predicted intensified UV-B radiation environment of the future.

5.4.2. Effects on Terrestrial and Aquatic Animal Populations

The consequences of an intensified UV-B environment for animal populations in terrestrial ecosystems is difficult to assess since this aspect has received little study. Because of the protective pigments and fur, and particularly because of the behavioral mechanisms that reduce exposure to damaging UV radiation, animals in natural terrestrial ecosystems may be less affected by increased UV-B radiation than plants. The information available for domesticated animals suggests that the food yield from domestic livestock will not be significantly reduced even at the most extreme predicted levels of ozone depletion.[9] Also, the degree of UV radiation-induced damage found in experiments with cattle, where only one breed has been found susceptible to UV-B-caused cancer eye,[9] has not warranted significant concern about the deleterious consequences of increased UV-B radiation for livestock.

There has been considerable concern, however, over the possible impact of UV-B intensification on humans. Noncancerous and acute effects of UV-B radiation exposure include sunburn and inflammation of the cornea, and in some individuals UV-B exposure may alter aspects of the immune system. Long-term exposure to UV-B radiation, especially in light-skinned Caucasians, is either the direct cause of nonmelanoma skin cancers (e.g., basal cell and squamous cell cancers) or a contributing factor in the case of skin melanoma.[9] The incidence of both skin melanoma and other UV-B-induced skin cancers is correlated with the increase in solar irradiance that occurs from northern to equatorial latitudes. This suggests that UV-B radiation is a contributing causal factor. Also, people whose occupations require day-long exposure to sunlight tend to exhibit higher rates of UV-B related skin cancers on exposed body areas such as the head and neck. Although the incidence of skin cancer is predicted to increase significantly in an enhanced UV-B environment, it is unclear how changes in human behavior that reduce exposure time will influence the incidence of such skin diseases.

In aquatic ecosystems the effect of higher levels of UV-B radiation will to a large extent depend on the degree of UV-B attenuation by water and the exposure time for the organism. Ecosystem level research is difficult and predictions for the effects on ecosystem productivity currently rely on studies performed under controlled conditions. Such studies show that even current levels of UV-B radiation are sufficient to depress primary productivity near the water surface of marine ecosystems.[9] Because phytoplankton form the base of food chains in aquatic ecosystems, any significant reduction in the productivity of phytoplankton is expected to profoundly affect organisms at higher levels in the food chain, such as populations of anchovy, herring, shellfish, and crustaceans. The dynamic nature of aquatic ecosystems, where UV-B attenuation by water (Fig. 5-3) is highly variable and where variations in the vertical position of planktonic organisms can considerably affect their level of UV-B exposure, results in large uncertainties for predictions of ecosystem response to enhanced UV-B radiation.

5.5. CONCLUSIONS

The effects of light-mediated processes on the growth and development of an organism are highly influenced by environmental factors and how the organism responds to environmental stress. In nature, organisms are simultaneously exposed to several interacting environmental factors, e.g., solar radiation, water and nutrient availability, and temperature, all of which can exhibit considerable daily and seasonal variation. The response of an organism to these environmental factors also changes during its lifetime. Therefore, experiments in environmental photobiology must consider both the dynamic nature of the environment and the plasticity of an organism's response to these factors. Although experiments that closely reflect field conditions are desirable, they tend to increase in complexity as the number of environmental factors considered increases. This complexity presents the environmental photobiologist with formidable problems in experimental design. Moreover, as the systems studied become progressively more complex from the cellular level, to the whole organism, to the ecosystem, the experiments also tend to increase in difficulty and expense. The degree of uncertainty in the results may also increase as the number of uncontrolled variables increases at, say, the ecosystem level.

Because of the increased complexity of experiments involving multiple factors, studies on the how the effects of light-mediated processes are influenced by the environmental factors with which an organism interacts have tended to be single-factor experiments. The effects of temperature, nutrient or water stress, and radiation on photosynthesis, for example, have generally been examined one factor at a time. This single-factor approach has also been the primary experimental procedure used to examine the response of organisms to enhanced UV-B radiation. While such experimental designs do indeed increase our understanding of the mechanisms of processes involving light, and how the effect of these processes are affected by a particular environmental factor, the results often do not allow extrapolation to field conditions. Reliable extrapolations to field situations are particularly crucial for anticipating the consequences of enhanced UV-B radiation for natural and agricultural ecosystems. Therefore, the design of experiments that permit predictions on how organisms would be affected under field conditions, where many environmental factors interact and vary, poses a significant challenge to the field of environmental photobiology.

5.6. REFERENCES

1. P. R. Gast, A. S. Jursa, J. Castelli, S. Basu, and J. Aarons, Solar electromagnetic radiation, in: *Handbook of Geophysics and Space Environments* (S. L. Valley, ed.), pp. 16-1–16-38, U.S. Air Force Cambridge Research Laboratories, McGraw-Hill, New York (1965).

2. L. R. Koller, *Ultraviolet Radiation,* Wiley, New York (1965).
3. J. Jagger, *Solar-UV Actions on Living Cells,* Praeger, New York (1985).
4. R. B. Setlow, The wavelengths in sunlight effective in producing skin cancer: A theoretical analysis, *Proc. Natl. Acad. Sci. USA* **71,** 3363–3366 (1974).
5. J. Ross, Radiative transfer in plant communities, in: *Vegetation and the Atmosphere,* Vol. 1 (J. L. Monteith, ed.), pp. 13–55, Academic Press, London (1975).
6. M. M. Caldwell, R. Robberecht, and W. D. Billings, A steep latitudinal gradient of solar ultraviolet-B radiation in the arctic-alpine life zone, *Ecology* **61,** 600–611 (1980).
7. M. Iqbal, *An Introduction to Solar Radiation,* Academic Press, New York (1983).
8. G. Brasseur and A. De Rudder, Agents and effects of ozone trends in the atmosphere, in: *Stratospheric Ozone Reduction, Solar Ultraviolet Radiation and Plant Life* (R. C. Worrest and M. M. Caldwell, eds.), pp. 1–28, Springer-Verlag, Berlin (1986).
9. National Research Council, National Academy of Sciences, *Causes and Effects of Stratospheric Ozone Reduction: An Update,* National Academy Press, Washington, D.C. (1982).
10. B. Lowry, D. Lee, and C. Hébant, The origin of land plants: A new look at an old problem, *Taxon* **29,** 183–197 (1980).
11. J. W. McClure, Physiology and functions of flavonoids, in: *The Flavonoids* (J. B. Harborne, T. J. Mabry, and H. Mabry, eds.), pp. 970–1055, Academic Press, New York (1975).
12. T. Swain, Evolution of flavonoid compounds, in: *The Flavonoids* (J. B. Harborne, T. J. Mabry, and H. Mabry, eds.), pp. 1096–1129, Academic Press, New York (1975).
13. R. G. Wetzel, *Limnology,* 2nd ed., Saunders, Philadelphia (1983).
14. H. R. James and E. A. Birge, A laboratory study of the absorption of light by lake waters, *Trans. Wis. Acad. Sci. Arts Lett.* **31,** 1–154 (1938).
15. D. M. Gates, H. J. Keegan, J. C. Schleter, and V. R. Weidner, Spectral properties of plants, *Appl. Opt.* **4,** 11–20 (1965).
16. J. Ross, *The Radiation Regime and Architecture of Plant Stands,* Dr. W. Junk, The Hague (1981).
17. M. M. Caldwell, T. J. Dean, R. S. Nowak, R. S. Dzurec, and J. H. Richards, Bunchgrass architecture, light interception, and water-use efficiency: Assessment by fiber optic point quadrat and gas exchange, *Oecologia* **59,** 178–184 (1983).
18. R. L. McCown and W. A. Williams, Competition for nutrients and light between the annual grassland species *Bromus mollis* and *Erodium botrys, Ecology* **49,** 981–990 (1968).
19. H. M. Rawson, R. L. Dunstone, M. J. Long, and J. E. Begg, Canopy development, light interception and seed production in sunflower as influenced by temperature and radiation, *Aust. J. Plant Physiol.* **11,** 255–265 (1984).
20. J. Warren Wilson, Analysis of growth, photosynthesis and light interception for single plants and stands, *Ann. Bot.* **48,** 507–512 (1981).
21. M. M. Caldwell, Plant architecture and resource competition, in: *Potentials and Limitations of Ecosystem Analysis* (E. -D. Schulze and H. Zwölfer, eds.), pp. 164–179, Springer-Verlag, Berlin (1987).
22. A. Frey-Wyssling, *The Plant Cell Wall,* Gebrüder Borntraeger, Berlin (1976).
23. R. Robberecht, M. M. Caldwell, and W. D. Billings, Leaf ultraviolet optical properties along a latitudinal gradient in the arctic-alpine zone, *Ecology* **61,** 612–619 (1980).
24. T. W. Mulroy, Spectral properties of heavily glaucous and non-glaucous leaves of a succulent rossette-plant, *Oecologia* **38,** 349–357 (1979).
25. E. Wellmann, UV dose-dependent induction of enzymes related to flavonoid biosynthesis in cell suspension cultures of parsley, *FEBS Lett.* **51,** 105–107 (1975).
26. R. Robberecht and M. M. Caldwell, Protective mechanisms and acclimation to solar ultraviolet-B radiation in *Oenothera stricta, Plant, Cell Environ.* **6,** 477–485 (1983).
27. S. D. Flint and M. M. Caldwell, Influence of floral optical properties on the ultraviolet radiation environment of pollen, *Am. J. Bot.* **70,** 1416–1419 (1983).
28. R. E. Grojean, J. A. Sousa, and M. C. Henry, Utilization of solar radiation by polar animals: An optical model for pelts, *Appl. Opt.* **19,** 339–346 (1980).
29. R. S. Stolarski, The Antarctic ozone hole, *Sci. Am.* **258,** 30–36 (1988).
30. A. H. Teramura, Effects of ultraviolet-B radiation on the growth and yield of crop plants, *Physiol. Plant.* **58,** 415–427 (1983).

31. T. J. Mabry, K. R. Markham, and M. B. Thomas, *The Systematic Identification of Flavonoids,* Springer-Verlag, Berlin (1970).
32. M. M. Caldwell, Plant response to solar ultraviolet radiation, in: *Plant Physiological Ecology. I. Encyclopedia of Plant Physiology,* Vol. 12A (O. L. Lange, P. S. Nobel, and C. B. Osmond, eds.), pp. 169–197, Springer-Verlag, Berlin (1981).

6

Photomedicine

John H. Epstein • Department of Dermatology, University of California School of Medicine, San Francisco, California 94143-0536.

6.1. INTRODUCTION

The medical aspects of the science of photobiology are among the most important parts of this discipline, since they are concerned primarily with direct effects of nonionizing radiation on humans. Two organ systems are directly affected by sunlight: the eye and the skin. The eye is discussed in Chapter 9. Therefore, the present discussion will be concerned primarily with cutaneous reactions to the sun.[1–6]

The solar radiation that reaches the earth's surface ranges from around 290 nm (or perhaps 288 nm) in the UV spectrum through the visible, infrared, and well beyond. Almost all of the photobiological reactions that occur in the skin are induced by radiation between 290 (or 288) and 320 nm (UV-B). These are the wavelengths that inhibit mitosis, DNA, RNA, and protein synthesis, make vitamin D, induce skin cancer, stimulate pigment formation, and produce the erythema response that we call "sunburn." These wavelengths comprise what is generally termed the "sunburn spectrum." The longer wavelength UV radiation between 320 and 400 nm (UV-A) does cause a few minor photobiological effects such as immediate pigment darkening (IPD) and the immediate erythema response. Major biological effects of UV-A rays include their accentuation of the acute injury induced by, and enhancement of, the carcinogenic potential of the UV-B spectrum. Furthermore, UV-A is responsible for the vast majority of the reactions produced by exogenous photosensitizers. An excellent review of UV-A effects is presented in Ref. 7. The rest of the sun's spectrum has little specific effect on the skin, though infrared radiation can produce a thermal burn and can enhance photobiological responses initiated by UV radiation.

6.2. OPTICAL PROPERTIES OF THE SKIN[1,6:pp.147–194]

For photobiological reactions to occur, three components are necessary: the biological system, the radiation, and a radiation absorber in the biological system. The biological system with which we are concerned is the skin (Figs. 6-1, 6-2). The radiation emanates from the sun. The radiation-absorbing molecules in skin generally remain to be determined; however, at least one important molecule would appear to be DNA.

If a photoactive molecule is present in or on the dead cells of the stratum corneum, it will most likely act as a protective filter. If the molecule is in the Malpighian or germinative layers of the epidermis, it may initiate a photoreaction after abosrbing radiation

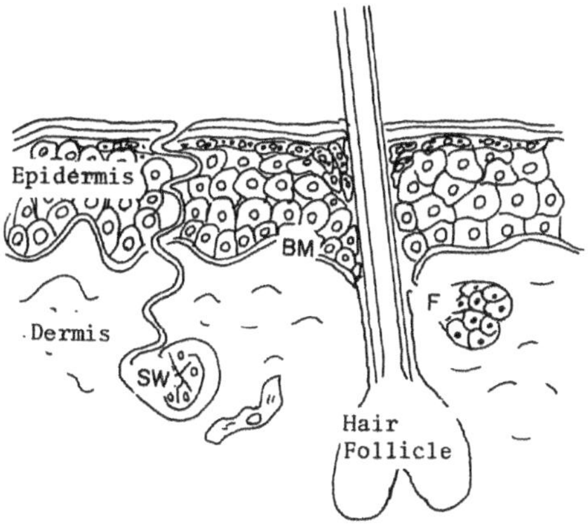

Fig. 6-1. Structure of skin. The skin consists of three main layers. The epidermis is the outermost layer, which is separated from the dermis by a basement membrane (BM). The dermis consists primarily of connective tissue, mostly collagen, and some elastic tissue produced by fibrocytes. The dermis also contains blood and lymph vessels, hair follicles, sweat glands (SW), sebaceous glands (F), muscles, and nerves. Below this is a fat layer not shown on the diagram.

of the proper wavelength. It is clear that the radiation must penetrate to the site of the absorbing molecule for the reaction to occur. The penetration of radiation is governed, in part, by the optical properties of the skin

Skin reflects a large amount of incident visible and near-infrared radiation. White skin reflects these radiations, as well as UV-A radiation, much more effectively than black skin. Very little reflection of UV radiation shorter than 320 nm occurs with either type of pigmentation.

Scattering and absorption play a greater role in the attenuation of sunburn radiation penetration than does reflection. The flattened cells of the stratum corneum, melanosomes, nucleic acids, proteins, lipids, histidine, urocanic acid, peptides, cholesterol, and the like limit penetration of this UV radiation by both scattering and absorption.

Because of the importance of the shorter wavelength UV radiation in the production of photobiological reactions, extensive transmittance studies have been conducted at these wavelengths. Although these investigations have produced a number of conflicting re-

Fig. 6-2. Structure of epidermis. The epidermis consists of four layers. The basal layer (B) is composed of a single or double layer of germinative cells where cell division occurs for the replacement of the epidermis. The Malpighian layer (M) consists of multiple squamous or prickle cells (keratinocytes) with intercellular bridges. These cells emanate from the basal layer, and eventuate in the outer dead stratum corneum layer. The Malpighian cells become the granular layer (G) where specific proteins are formed, presumably necessary for the formation of keratin. The cells then lose their nuclei, contain large amounts of keratin, and form the protective stratum corneum (C). They are then shed into the environment. The melanocytes (P) are dendritic cells located in the basal layer. Their dendritic (armlike) processes extend up into the Malpighian layer. These cells form the pigment melanin, which they pass into the Malpighian cells through their dendrites. The melanocytes and associated Malpighian cells compose the epidermal melanin unit.

sults, it is generally agreed that the bulk of incident radiation in the sunburn or UV-B spectrum (290–320 nm) is absorbed in the stratum corneum. In Caucasian skin, at least 20% of this radiation reaches the Malpighian cells, and probably 10% penetrates to the upper dermis. The stratum corneum of black skin absorbs a greater amount of this energy due to the presence of melanin. A large proportion of the longer wavelength UV radiation (UV-A) and of visible radiation penetrates into, and can be absorbed by, photosensitizers in the dermis.

With the absorption of this energy, an excited state is induced in the chromophore. The excited state may be deactivated by a number of methods including thermal decay or emission of light energy, usually of a longer wavelength than that absorbed (Chapter 1). Cutaneous photosensitivity reactions are brought about either by direct alteration in the absorbing molecule or by the transference of this energy to other molecules or cellular components such as membranes, mitochondria, nucleic acids, and the like.

6.3. ACUTE EFFECTS OF SUNLIGHT ON THE SKIN

6.3.1. Sunburn[1,3,6:pp.219–260]

The sunburn reaction in the skin is by far the most common adverse effect produced by sunlight. The cutaneous changes induced by the erythemogenic radiation depend on the amount of radiation, the degree of melanin pigmentation, and the thickness of the stratum corneum. As would be expected, the morphological and microscopic responses are less intense with less radiation and/or more protection. However, sunburn will occur even in pigmented skin if exposed to enough radiation of the proper wavelength.

Erythema (reddening of the skin) is the most visually prominent aspect of the sunburn response. It appears in a biphasic pattern consisting of an immediate faint erythema that occurs during exposure and disappears shortly thereafter. The delayed response appears 2–4 h later, reaches a peak in 14–20 h, and persists for 24–48 h. The injury is then followed by desquamation of the dead epidermal cells.

The action spectrum for sunlight-induced erythema is confined primarily to radiation between 288 or 290 and 320 nm (Fig. 6-3). The standard erythemal curve described by Coblentz and Stair in 1934 showed a maximum efficiency at 297 nm, and minimum efficiency at 280 and 320 nm. Subsequent studies have provided various results with the peak erythemogenic potential ranging between 290 and 294 nm. In addition, using parameters of efficiency and available energy, the most erythemogenic wavelengths in the sun's spectrum lie between 305 and 308 nm.[1:pp.184–187] A mild erythema can be produced by radiation between 320 and 400 nm, but this requires a 1000 times greater exposure. However, UV-A markedly enhances the erythema induced by UV-B.[4:pp.131–141,8]

Radiation at wavelengths shorter than the UV-B spectrum, produced by artificial light sources, are also erythemogenic. In truth, UV-C radiation at 254 nm is more efficient in producing erythema than is the UV-B spectrum. Though the various studies have not produced exactly consistent curves, there is agreement that radiation at 254 nm is most efficient with a rapid reduction in erythemal efficacy, especially at 270–280 nm. On this point there is some discrepancy, some investigators noting a depression and others a plateau at 270–280 nm, relative to 290 nm. The curve continues toward but not to zero at

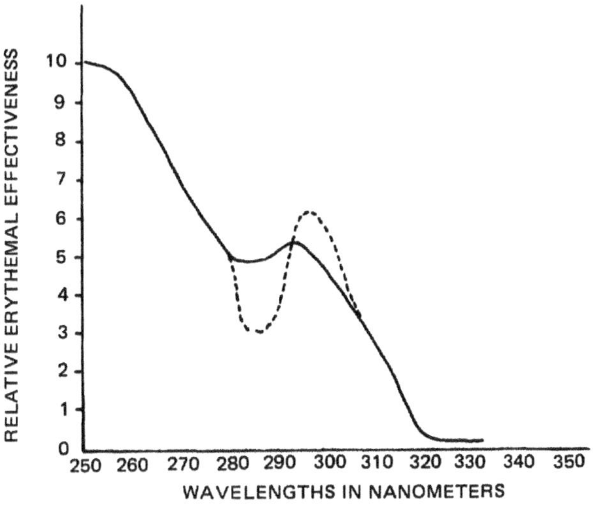

Fig. 6-3. The erythemal action spectrum. This spectrum has not been completely agreed upon, and this figure represents the different concepts noted in Ref. 1. It is not clear whether there is a dip (dotted line) or a plateau (solid line) in the effectiveness of wavelengths around 280 nm.

320 nm. The erythema produced by radiation at wavelengths shorter than 290 nm differs from the UV-B response in appearance and latent period, i.e., it is a pink rather than a deep red color, and it reaches a peak at 8 h rather than 14–20 h. Therefore, it may well represent a different phenomenon.

The term "sunburn" stands for a number of complicated cutaneous responses to injury induced primarily by wavelengths between 288 or 290 and 320 nm. Obviously, many changes occur simultaneously. In addition, it is even difficult to differentiate primary and secondary effects. The nature of the chromophore that absorbs the light energy in order to initiate the primary photochemical responses is not known. Proteins containing aromatic amino acids have been considered likely candidates because of their absorption spectra and the profound effects of UV radiation on these molecules. However, a number of other substances, including nucleoproteins (DNA and RNA), urocanic acid, melanin, and even unsaturated fatty acids of phospholipids, may play a role in the initial absorption of light energy.[4:pp.117–130,6:pp.225] Lipid peroxidation has been demonstrated both *in vitro* and *in vivo*. Since lysosomal membranes are preferentially damaged by lipid peroxides, their formation may play an important part in UV radiation-induced injury. It seems quite likely that multiple chromophores are involved in this complex process.

The pathogenesis of the sunburn erythema has perhaps received the most attention. However, it has not been established conclusively whether this is a primary or secondary phenomenon. Though the latter has been considered more probable, direct UV damage to the upper dermal blood vessels has been demonstrated and some studies suggest that there is a broad-action spectrum from 250 nm into the long-wavelength UV region that causes a direct effect on blood vessels.[1:pp.182,183] In addition, there is a superimposed sharp peak around 300 nm that induces the formation of a diffusable vasodilating substance.

This brings up the question as to the character of the chemical mediator of the sunburn response.[4:pp.143–146,6:pp.221–224] The accumulated evidence indicates that one of the most prevalent vasoactive substances in mammalian tissue, histamine, is not responsible for this process. In addition, though kinins and possibly serotonin are released following exposure to sunburn energy, they are not likely to be related to the characteristic delayed erythema response. More recent evidence suggests that the vasoactive mediator is a prostaglandin.[9] However, further work is needed to establish the validity of this concept since the information to date is primarily circumstantial.

6.3.2. Microscopic Anatomy

6.3.2.1. Light Microscopy

Histologically very little is seen in human skin for several hours after exposure to UV radiation. In fair-skinned people the first recognizable change consists of focci of dyskeratotic cells (sunburn cells) in the epidermal Malpighian layer noted by 24 h. Within 48 h, damage is noted throughout the epidermis [e.g., abnormal staining (dyskeratotic) and vacuolated cells and shrunken nuclei]. The dendrites of the melanocytes also become quite prominent at this time. Regeneration is usually in progress, and the damaged cells form a desquamating (sloughing) layer by 72 h. Hyperplasia (thickening) of the epidermis is quite striking by this time (Fig. 6-4). At 96 h there is a great increase in dihydroxyphenylalanine (DOPA)-positive melanocytes, with marked arborization of their dendrites. These melanocytic changes occur in pigmented skin as well, but the dyskeratotic changes are not nearly as notable and cellular disorganization is much less prominent.

Histochemical and biochemical techniques have demonstrated a number of other acute epidermal changes following irradiation. [1,4:pp.117–130,147–156,6:pp.219–260] These include lysosomal membrane alterations and rupture, lipid peroxide formation, inhibition of the activity of a number of enzymes, increases in both sulfhydryl and disulfide groups, and the transformation of urocanic acid from a *trans* to a *cis* isomer. This last noted change may well play an important protective role in dissipating sunburn energy. Also, glycogen accumulates in the epidermal basal cells 12 h postirradiation, presumably to supply energy for the subsequent burst of mitotic activity.

Dermal changes generally are not prominent. Vasodilatation occurs, and in rodents an inflammatory infiltrate consisting of polymorphonuclear leucocytes can by seen as early as 40 min postirradiation, which becomes maximal between 8 and 24 h. This cellular reponse is minimal, if present at all, in human skin.

The action spectrum for the microscopic changes is found primarily in the UV-B spectrum. However, as with the erythema response, the long-wavelength UV-A radiation accentuates the microscopic damage initiated by the shorter wavelengths (UV-B).[8]

6.3.2.2. Electron Microscopy[4:pp.195–215,6:pp.219–260]

Electron microscopic studies have shown that structural changes occur well before they are demonstratable by light microscopy. The earliest changes noted to date occur in melanocytes associated with the immediate pigment-darkening phenomenon, with changes in microtubules, melanosome patterns, and thick filaments. Within 2–3 h postirradiation, marked changes are also noted in the cytoplasm, nucleus, and nucleolus of the epidermal cells, including vacuole formation, filament aggregation, and apparent loss of lysosomes.[10,11] Thus, it appears that a number of structural changes occur shortly after UV exposure, which seem likely to result from injury to several sites simultaneously rather than representing only a sequence of events.

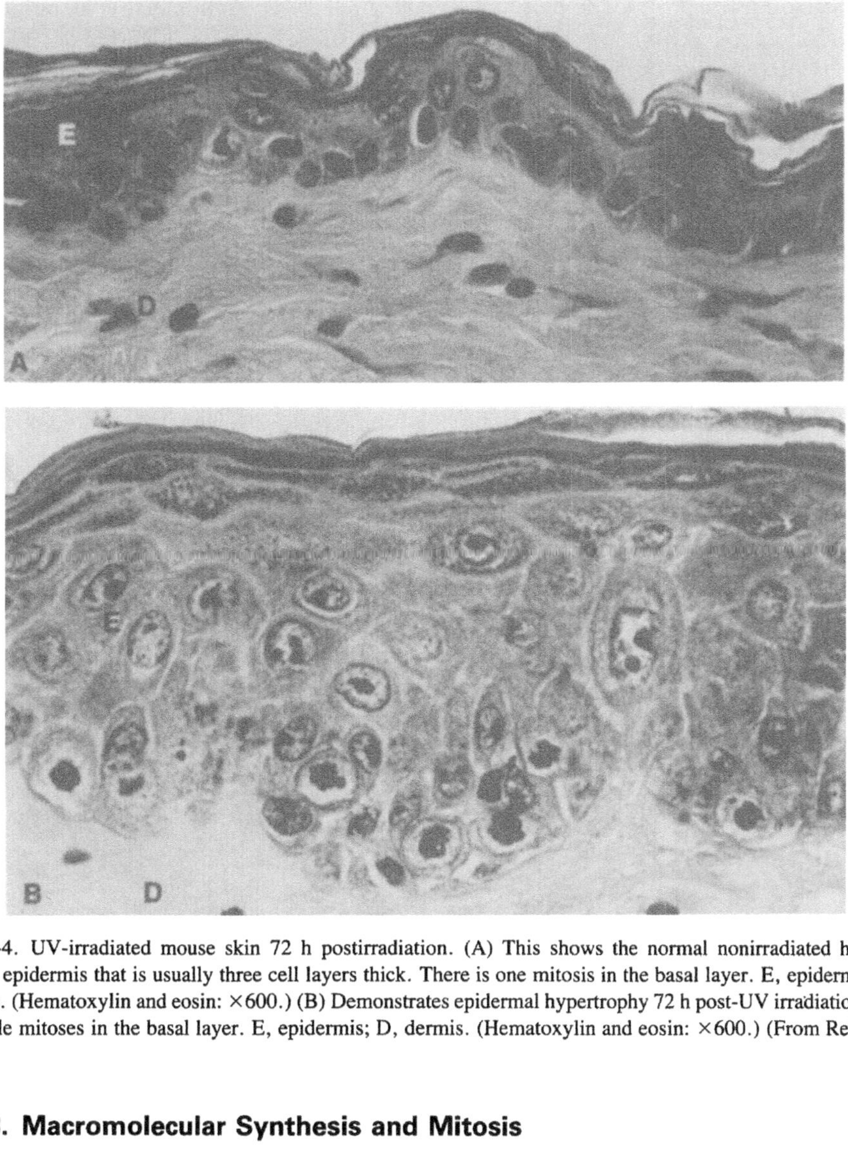

Fig. 6-4. UV-irradiated mouse skin 72 h postirradiation. (A) This shows the normal nonirradiated hairless mouse epidermis that is usually three cell layers thick. There is one mitosis in the basal layer. E, epidermis; D, dermis. (Hematoxylin and eosin: ×600.) (B) Demonstrates epidermal hypertrophy 72 h post-UV irradiation with multiple mitoses in the basal layer. E, epidermis; D, dermis. (Hematoxylin and eosin: ×600.) (From Ref. 13.)

6.3.3. Macromolecular Synthesis and Mitosis

The sunburn spectrum has a profound influence on DNA, RNA, and protein synthesis, and mitosis in mammalian skin *in vivo*.(12,13) Using autoradiographic, biochemical, and metaphase-arresting techniques, it has been demonstrated that synthesis of these macromolecules and mitosis are inhibited within the first hour postirradiation with a moderate erythema dose. This inhibition persists for at least 6 h. By 24 h, recovery occurs followed by acceleration of these functions, reaching a peak at 48–72 h; the activity then gradually subsides. However, increased mitotic activity in mouse skin has been demonstrated up to 40 days after a single exposure.(14) The stage of increased mitotic activity is associated with epidermal hyperplasia, which also gradually subsides over the next 2 months. The mechanism of this hyperplasia and increased mitoses is not established.

Recent findings relating to the presence of epidermal mitotic inhibitors (chalone) and the actions of cyclic AMP and GMP present interesting but as yet unproven possibilities.[15,16] In addition, a mitotic stimulating substance has been identified postirradiation, suggesting that this hyperplasia is due to a combination of the removal of inhibition and the stimulation of growth.

6.3.4. DNA Repair[4:pp.67–77,299–315,6:pp.91–112,17,18]

The inhibition of epidermal cellular DNA synthesis is one of the earliest post-UV irradiation events that occurs in mammalian skin *in vivo,* similar to responses noted in cultured mammalian cells (Fig. 6-5). Among the number of DNA lesions that could occur, pyrimidine dimer formation, primarily between thymine bases, has received an extraordinary amount of attention. Currently, three repair mechanisms for DNA base damage in bacteria have been described; photoreactivation, excision repair, and postreplication repair (Chapter 4). In mammalian cells, the excision repair system appears to be the primary mechanism.[17,18] Autoradiographic (Fig. 6-5), density gradient and chromatographic techniques have been used to demonstrate that thymine dimerization can be induced by sunlight as well as by artificial UV radiation. The DNA can be repaired by

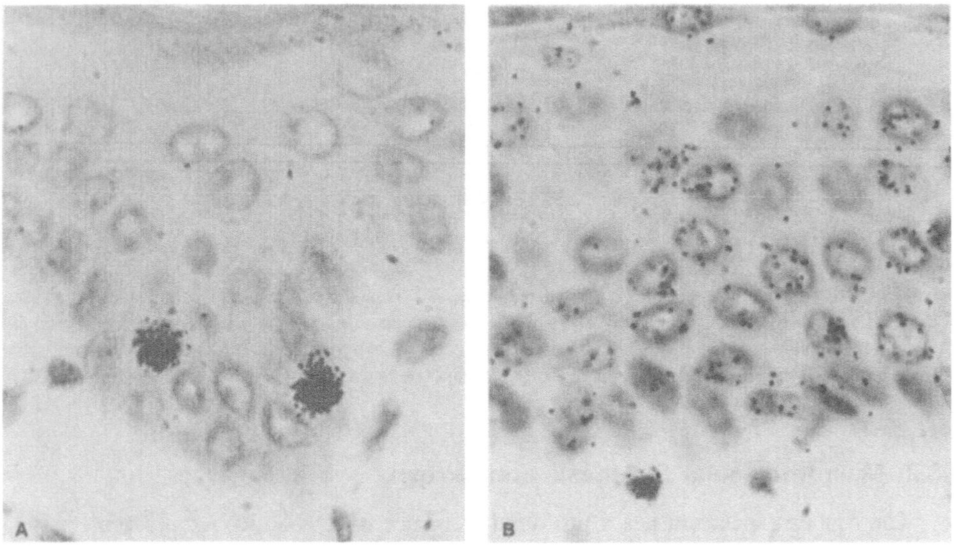

Fig. 6-5. DNA repair 15 min postirradiation. (A) Autoradiographs utilizing tritiated thymidine (^{3}H-TdR) as the radioactive tracer. This figure shows a non-UV-irradiated human epidermis with the labeling (black dots) concentrated in two of the basal cells in a dense labeling pattern, which represents premitotic semiconservative DNA synthesis. (Hematoxylin and eosin: ×720.) (B) Autoradiograph of human skin labeled 15 min post-UV irradiation using ^{3}H-TdR as the radioactive tracer. A sparse labeling pattern is seen throughout the epidermal cell nuclei. This sparse labeling represents unscheduled DNA synthesis or repair replication that is absent or markedly reduced in the skin of patients with xerodema pigmentosum. In normal human skin postirradiation, the upper dermal fibrocytes and vascular endothelial cells show this unscheduled DNA synthesis, giving visual proof of the penetration of the sunburn rays into the upper dermis. (Hematoxylin and eosin: ×720.)

excision of the dimers and their replacement by normal bases in the proper sequence by the dark repair enzyme system. This damage and repair occurs in the epidermis *in vivo* and *in vitro,* as well as in cell culture systems. The discovery that the human photosensitive disease xeroderma pigmentosum (XP) is characterized by a defect in this repair system has supplied a model for the evaluation of the importance of this repair system in human skin.[4:pp.299–315] It is of interest that bacteria with a defect in excision repair not only are more sensitive to the killing effects of UV radiation, but also form increased numbers of mutations postirradiation. A similar situation exists for cells from patients with XP. The primary photosensitive problem in patients with XP is their inordinate sensitivity to the carcinogenic effects of sunlight. Whether this represents cellular mutation or not remains to be determined.

A postreplication repair process has been described in mouse L cells in culture, and is apparently dependent on *de novo* DNA synthesis, and not recombination mechanisms.[4:pp.91–106,19] Recent studies by Lehman *et al.*[20] indicate that this repair mechanism is defective in cultured cells from four patients with xeroderma pigmentosum (XP variant).

6.3.5. Melanogenesis[1,4:pp.165–215,6:pp.219–260]

Melanins are black or brownish yellow pigments that appear to be complex polymers of DOPA-quinone, 5,6-dihydroxyindole, and 5,6-dihydroxyindole-2-carboxylic acid at various oxidation levels, held together by a variety of bond types. This pigment in mammalian skin is produced from tyrosine and DOPA through enzymatic action in specialized cells, i.e., melanocytes. This activity occurs on or in the cytoplasmic melanosome system. Light energy can both induce the immediate pigment-darkening (IPD) phenomenon and stimulate new pigment formation.

IPD is characterized morphologically by a darkening of the skin that occurs essentially immediately upon exposure to radiation between 320 and 700 nm. It is presumably due to the oxidation of reduced or partially reduced melanin already present in epidermal melanocytes (Fig. 6-2). It reaches a maximum at 1–2 h and decreases between 3 and 24 h postirradiation. Electron microscopic studies suggest that IPD is related to a distal distribution of the melanosomes toward the tips of the dendrites elicited by the active motor force of microtubules and microfilaments, and an apparent transfer to keratinocytes (Malpighian cells) without an obvious effect on the character of the melanosome. There is no significant change in the melanogenic organelles necessary for the synthesis of new melanosomes.

New melanin formation (delayed tanning) is characterized morphologically by a brown pigmentation of the skin starting by 48–72 h postirradiation. It reaches a peak by 13–21 days and then gradually subsides over the next several months. The action spectrum for this response falls primarily in the sunburn range (290–320 nm). However, long-wavelength UV and visible radiation, at least as long as 700 nm, can initiate new melanin formation if enough energy is applied. UV radiation from 250 to 280 nm is also distinctly less effective in inducing new melanin production. Light and electron microscopy correlated with biochemical studies indicate that this new pigment formation occurs in or on cytoplasmic organelles called melanosomes, which are formed in the epidermal melanocytes. Subsequently, the pigmented melanosomes are transferred to epidermal ker-

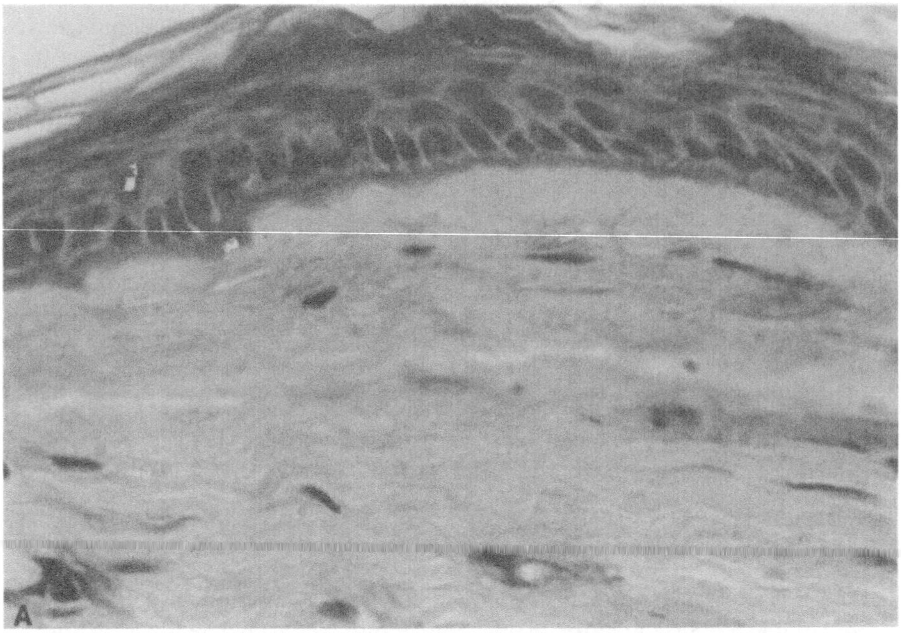
A

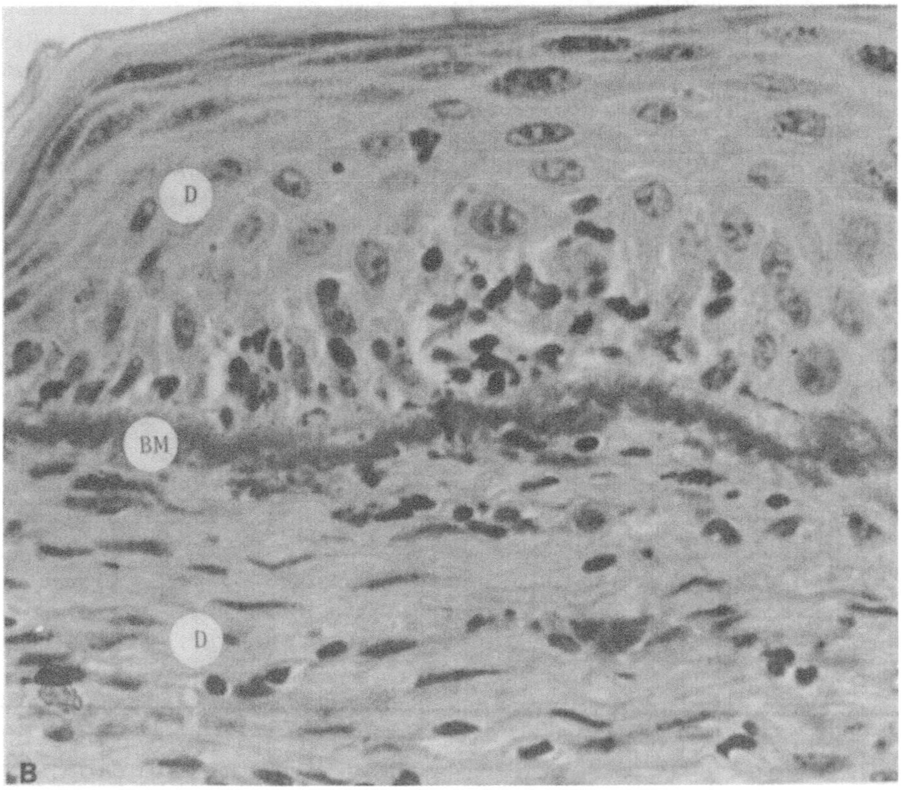
D
BM
D
B

atinocytes, which together with the melanocytes make up the *epidermal melanin unit* (Fig. 6-2). The pigment is then carried to the stratum corneum where it provides the most distinctive visually dark or hyperpigmented color effect. This new pigmentation provides the most significant available natural barrier to sunlight-induced skin damage.

Hyperpigmentation of the skin in "delayed tanning" is associated with (1) proliferation of melanocytes and activation of dormant melanocytes, (2) melanocytic hypertrophy and increased dendritic arborization, (3) increased melanosomal synthesis, (4) increased rate of melanization of melanosomes, (5) increased transfer of melanosomes to keratinocytes that is related to an increase in turnover of keratinocytes, (6) increased size of melanosome complexes (more notable in Caucasoids and Mongoloids), and (7) tyrosinase activation by a direct effect on tyrosinase-inhibiting sulfhydryl compounds in the epidermis.[4:pp.165–194]

6.4. CHRONIC EFFECTS OF SUNLIGHT ON THE SKIN[1,3,6:pp.261–292]

6.4.1. Solar (Actinic) Degeneration

Repeated exposures of Caucasoid skin over a number of years result in gross changes, including wrinkling, atrophy, hyper- and hypopigmented spots, dilated blood vessels, yellow raised areas, and actinic keratoses. A distinctive furrowed leathery change may be noted on the back of the neck, especially of very fair-skinned people who have received extensive sun exposure, such as sailors and farmers.

Histologically a shortening or flattening of the rete ridges (epidermal downward projections, Fig. 6-1), a possible thinning of the epidermis, and the presence of many abnormal cells in disorderly arrangement are seen in chronically sun-damaged human epidermis. The melanocytes show great variations in size, distribution, and tyrosinase content, although their absolute number generally remains unaltered. In addition, there are greater differences in size and melanin-forming activity among the basal melanocytes as compared to normal skin. Pigment transfer from melanocytes to keratinocytes also appears to be impaired in areas of chronic solar damage.[21]

More dramatic changes occur after repeated exposure of hairless mouse skin under controlled conditions.[22] These include a striking thickening of all of the layers of the epidermis and the epidermal-dermal basement membrane (Fig. 6-6). In addition, there is an increase in the number of basal cells in DNA synthesis and mitosis, and a decrease in the transit time of cells through the epidermis.

In contrast to the relatively limited demonstrable epidermal changes, profound dermal alterations occur associated with solar radiation-induced degeneration in human skin. There is a progressive degeneration in the upper portion of the dermis. Specific changes

Fig. 6-6. Effects of chronic UV radiation exposures to mouse skin. (A) Hairless mouse skin after repeated UV radiation exposures for 1 month showing a benign hyperplasia with regular mild epidermal hypertrophy, and moderate but regular thickening of the basement membrane (BM), and thickening of the dermis. E, epidermis; D, dermis. (Hematoxylin and eosin: ×600.) (B) Hairless mouse skin after 4 months of repeated UV irradiation showing a stage of premalignant to malignant change. It is characterized by abnormalities of size, shape, polarity, and chromatin patterns of the basal cells, and a marked thickening and clumping of the basement membrane, which has become discontinuous at several points. (Hematoxylin and eosin: ×600.) (From Ref. 26.)

include the development of dilated blood vessels, the accumulation of acid mucopolysaccharides and abnormal-appearing fibrocytes, the loss of mature collagen, but an increase in the soluble component, and marked increase and degeneration of elastic tissue referred to as actinic elastosis. Actinic elastosis, the most prominent and obvious connective tissue alteration due to chronic solar damage, is a dynamic progressive process that has been detected as early as the first decade of life.[4:pp.157–163] The earliest change appears to be a simple increase in number of elastic fibers. Subsequent alterations include thickening, curling, and increased branching of the fibers with eventual replacement of the dermis and disorganization of the connective tissue into amorphous masses. Although some questions may still exist as to the origin of the fibers that stain like elastic tissue, biochemical and electron microscopic studies have confirmed that actinic elastosis is due to the accumulation of elastic tissue.

The action spectrum for the induction of elastosis in experimental animals falls in the sunburn range. It has been postulated that this connective tissue change is the result of photochemically induced alterations in fibroblast function by UV radiation rather than degradation of connective tissue elements.[1] In support of this concept, the direct injury of human dermal connective tissue cellular DNA *in vivo* has been observed within a few minutes after irradiation with wavelengths shorter than 320 nm.[23] Therefore, the radiation that produces sunburn most likely also plays an important role in chronic solar damage to dermal connective tissue. Whether the longer wavelengths also have a significant influence must remain a matter of speculation at this time. However, recent studies have shown that chronic UV-A and infrared radiation can produce connective tissue changes in mouse skin.[24]

6.4.2. Carcinogenesis[3,4:pp.259–283,22,25–27]

6.4.2.1. *Human Skin Cancer Formation*[4:pp.266–280,27:pp.488–489]

Examination of the sun's role in the production of human skin cancers does not lend itself to direct experimentation. However, extensive astute observations have strongly suggested the etiological significance of light energy in the induction of these tumors. Skin cancers in Caucasians are generally most prevalent in geographic areas of the greatest insolation (i.e., receiving the greatest amount of solar radiation) and among people who receive the most exposure (i.e., people who work outdoors). They are rare in Negroes and other deeply pigmented individuals who have the greatest protection against UV-induced injury. Furthermore, the lightest complexioned individuals, such as those of Scottish and Irish descent, appear to be most susceptible to skin cancer formation when they live in geographic areas of high-UV fluence. When skin cancers do occur in the darkly pigmented races, they are not distributed primarily in the sun-exposed areas, as they are in light-skinned people. The tumors in these pigmented individuals are most commonly stimulated by other forms of trauma such as chronic leg ulcers, irritation due to not wearing shoes, the use of a *kangeri* (an earthenware pot that is filled with burning charcoal and strapped to the abdomen for warmth), the wearing of a *dhoti* (loin cloth), and so on. In contrast, the distribution of skin cancer in the Bantu albino and in patients with xeroderma pigmentosum follows sun exposure patterns. The arguments supporting the role of sunlight in human skin cancer formation may be summarized as follows: (1) skin cancers do occur predominantly on the sun-exposed parts of the body, (2) they are more common in regions of the earth that receive the most sunlight, and (3) pigmented races are much less susceptible to skin cancer formation than Caucasians.

Though these arguments do not constitute absolute proof and are themselves not wholly established, there is a considerable body of circumstantial evidence supporting the role of sunlight in at least three types of skin cancers: basal cell epitheliomas, squamous cell carcinomas, and melanomas.

The formation of the most common skin cancers (and therefore the most common human malignancies), the basal cell epithelioma and squamous cell carcinoma, is influenced by four major factors.[4:pp.259–283]

1. *Total lifetime of sunlight exposure.* Exposure to sunlight for prolonged periods, such as occurs in Texas and Australia, appears to be an important factor in the development of skin cancer. However, a comparison of skin cancer formation in Galway, Ireland, and Philadelphia indicates that hours of exposure does not constitute the complete story. Cancers occurred in Philadelphia with much less total exposure.
2. *Intensity and duration of the UV component in sunlight.* Recent studies indicate that skin cancer formation can be correlated with the duration of exposure to the erythemally effective radiation (290–320 nm) more accurately than the duration of total sun exposure. Using this criteria, the discrepancy between cancer formation occurring in rural Galway, with much more total sunlight exposure, and urban Philadelphia becomes easier to understand. The maximum exposure to the sunburn spectrum possible in Galway would be 1.33 months/year and in Philadelphia it would be 4 months/year (if people were out of doors continuously during daylight hours). On this basis, the "effective" exposure times of the Philadelphia and Galway skin cancer groups become more similar.
3. *Genetic predisposition.* The condition xeroderma pigmentosum is the most striking example of genetic predisposition to skin cancer formation.[4:pp.299–315] Those individuals who lack the ability to repair UV-damaged DNA generally develop all varieties of sunlight-induced malignancies within the first two decades of life. However, genetic predisposition also appears to play an important role in the formation of basal and squamous cell cancer in the general population. A number of studies have shown a distinct association between these cancers and light eye color, fair complexion, light hair color, poor ability to tan, ease of sunburning, and a history of repeated sunburn reactions. In addition, individuals of Celtic origin are more prone to this type of carcinogenesis than other light-complexioned individuals.
4. *Factors unrelated to sunlight.* Exposure to sunlight, primarily to erythemally effective wavelengths, appears to be the primary factor in the formation of these skin malignancies. The morphological distribution of the most common cutaneous cancer, the basal cell epithelioma, indicates that other factors must also play a role. Squamous cell carcinomas appear to be directly related to sun exposure. In contrast, about one-third of basal cell epitheliomas occur on areas of the skin receiving minimal exposure to sunlight. Thus, although sunlight appears to be a dominant factor in the induction of this type of tumor, other as yet undetermined influences must also participate.

Melanomas represent the most dreaded type of skin cancer. Unlike basal cell and squamous cell cancers, these lesions metastasize readily to other organs in the body and have a comparatively high mortality rate. Fortunately, they are relatively uncommon. Unlike the two common malignancies, the influence of sunlight on melanoma formation is not as well established.[26] Melanomas are not found primarily on sun-exposed areas, and the protective effect of melanin pigment is not so obvious as it is in other forms of skin cancer. However, a number of surveys suggest that sunlight does influence the development of at least some of these malignancies.[28] Caucasians with melanomas statistically tend to have light eyes, light hair, fair complexions, and spend more time outdoors when compared to a control group of patients without melanomas. Surveys of the geographic distribution of this cancer, even in genetically similar populations, have demonstrated a much greater melanoma prevalence associated with high insolation (i.e., exposure to sunlight), as compared to low insolation. The occurrence of solar radiation-induced melanomas in patients with xeroderma pigmentosum, and the production of melanomas from benign pigmented lesions by chronic exposure to UV radiation in experimental animals, further confirms the potential of sunlight to stimulate the production of these tumors under proper circumstances. In addition, the distribution of melanomas developing from circumscribed precancerous melanosis strongly suggests that the sun is at least responsible in part for such lesions. However, the anatomical and geographic distribution of melanomas indicates that factors other than sunlight must play an important role in the etiology and increasing incidence of these tumors.[29]

6.4.2.2. *Experimental Carcinogenesis*

6.4.2.2a. *Quantitiative Investigations*

Traditionally, the skin of the ear of the albino mouse and, to a lesser extent, the albino rat was used for the experimental production of tumors by UV radiation because this area lacks the three main natural inhibiting factors: pigment, a thick stratum corneum, and thick hair growth. The type of tumor produced in this tissue was primarily a sarcoma that was quite adequate for the monumental quantitative studies accomplished by Blum and coworkers.[25] Using the ears of an inbred, albino, haired mouse strain and a UV radiation source primarily emitting radiation shorter than 320 nm, they demonstrated that skin cancer development follows the law of reciprocity as far as dose–rate dependence is concerned. Increasing the dose or shortening the intervals between exposures accelerated tumor formation but did not alter the shape of the incidence curve.

A factor in the production of skin cancer by UV radiation that has apparently not been explored is the influence of protraction of the daily dose on reciprocity. Various experimenters have used exposures varying in length from a few seconds to several hours, while exploring such variables as influence of chemicals, action spectrum, etc. There are apparently no published data on whether a radiation source has an equally good chance of producing skin cancer if a given fluence is delivered in a few seconds or is attenuated over several hours. However, recent studies by Forbes *et al.*[19] demonstrated that the delivery of the same dose in divided doses (i.e., five exposures a week) was more effective in producing tumors than when delivered once a week. The reason for this lack of the reciprocity effect is not clear.

Under controlled circumstances, skin cancers will develop if enough energy is delivered for a sufficient period of time.[25] However, tumors will not appear no matter what energy is used if it is not applied long enough. Blum[25] surmised that UV radiation-induced cancer formation is a continuous process that begins with the initial exposure. The appearance of tumors within the lifetime of the animal depends on sufficient acceleration of the growth process. In support of the concept that growth acceleration is an important component of carcinogenesis, the production of squamous cell carcinomas in hairless mouse skin with one exposure to UV radiation followed by croton oil promotion has been reported.[26]

The mechanism of tumor growth acceleration remains to be established. The production of mitosis-stimulating substances or, perhaps more importantly, the removal of mitosis-inhibiting materials ("chalones") may well play a role in this process.

6.4.2.2b. *Qualitative Investigations*[4:pp.259–283,22,26]

The evaluation of progressive changes occurring in tissue during UV radiation-induced epidermal cancer development was examined in the skin of the hairless mouse; these tumors are identical to primary growths induced with sunlight radiation in human skin. In addition, these animals proved useful in studying the effect of UV radiation on melanoma formation.[26]

A number of investigators have considered dermal influence to be of prime importance in the development of epidermal skin cancers. Although traditionally actinic elastosis was considered to play an important role in this respect, several studies have shown that such changes are not essential to the formation of epidermal malignancies in humans or experimental animals.[22,26] However, other changes, including dissolution of elastic tissue and collagen, proliferation of young collagen, accumulation of acid mucopolysaccharides, mast cells, and fibroblasts, formation of new elastic tissue, and alterations in dermal vasculature, have been reported.[24,26] Some or all of these changes may be associated with significant dermal influences on the development of epidermal tumors.

The progressive development from benign hyperplasia through stages of actinic keratosislike lesions to frank invasive malignancy was evaluated in the hairless mouse system, using histochemical and radioactive tracer (^{3}H-thymidine) techniques to examine anatomic changes and cellular kinetics.[22,26] These studies demonstrated an accumulation of acid mucopolysaccharides, a loss of insoluble collagen in the upper dermis, and the proliferation of mast cells and fibrocytes. The most striking response occurred in the epidermal-dermal basement membrane (BM) (Fig. 6-6). A progressive thickening of this structure was noted with the development of epidermal hyperplasia. As irregular and abnormal epidermal cell proliferation occurred, the BM became thicker, irregular, clumped, and frayed in appearance, and with frank invasive malignancy, it disappeared altogether. Similar breaks in the BM have been noted with the invasion of human skin cancers and chemically induced experimental tumors. Electron microscopic studies have suggested that the microprojection of cells through the basal lamina may represent the earliest stage of tumor invasion.

Epidermal cell kinetic studies demonstrated a progressive increase in the number of germinative basal cells synthesizing DNA and dividing, with a shortening of the DNA synthesis time and G2 period associated with the development of epidermal changes.[22]

In addition, there was a progressive reduction in the cell transit time through the epidermis despite its increasing thickness. With frank malignancy, the presence of a germinative cell layer was lost, and mitoses, many abnormal in appearance, were present throughout the tumor. Thus, the process of carcinogenesis was characterized by the acceleration of cell formation, maturation, and turnover. Furthermore, the epidermal germinative basal cell appeared to be the primary or initial site of abnormal proliferation.

Experimentally, the transformation of benign pigmented growths to malignant melanomas in the skin of hairless mice has been induced with radiation shorter than 320 nm.[27] Repeated exposures over several months resulted in the production of large invasive melanocytic tumors with histologic, autoradiographic, and electron microscopic characteristics of malignancy. In addition, several of these tumors metastasized to the regional lymph nodes. These results indicate that benign pigmented lesions can be transformed to malignant growths in experimental animals by UV radiation. Whether a similar process occurs in human skin remains to be established.

The enhancing effect of heat on the degree of cutaneous injury and the intensity of erythema response to UV radiation has been well documented.[27] Furthermore, it has been demonstrated that increased temperatures present at the time of exposure to UV radiation accelerate tumor production. Clinical experience also suggests that heat does in fact aggravate UV radiation-induced cancer formation in human skin.

The presence of carcinogenic and tumor-promoting chemicals in our environment has made the experimental evaluation of chemical influences on UV radiation-induced carcinogenesis of practical importance.[26] Recent studies have demonstrated that UV radiation and chemical carcinogenic stimuli are additive in nature. In addition, repeated applications of the noncarcinogen croton oil results in significant cancer formation following a single UV exposure. These and other experimental findings suggest that environmental chemicals may play a role in the development of skin tumors.[27]

The relationship of immunological factors to the control of cancer formation, growth, invasion, and rejection presents another important aspect of carcinogenesis, which is being examined experimentally.[30–32] Recent studies have demonstrated that subcarcinogenic amounts of UV-B radiation will induce a specific population of suppressor T lymphocytes, which prevent rejection of skin cancers that have been produced by carcinogenic amounts of UV-B radiation in experimental animals. Whether such tumor tolerance occurs in humans is not known at this time.

Perhaps the most intriguing aspect of UV radiation-induced carcinogenesis concerns the relationship of the acute responses to the eventual cancer formation.[3,17,25–27] Light-induced carcinogenesis occurs only when acute phototoxic erythema is produced as well. The croton oil experiments also indicate that the process of UV radiation-induced cancer formation begins with the initial exposure.[27] However, the early changes leading to cancer formation remain undetermined. A number of studies have demonstrated a similar but not identical effect of UV radiation and chemical carcinogens on certain vital functions in mammalian epidermis, including DNA and RNA synthesis, cell turnover, and mitotic rate.[13] The most intriguing findings to date relating acute photoinjury to cancer formation have developed from the study of DNA repair systems and of the genetic disease xeroderma pigmentosum (XP). As noted in Section 6.3.4, this is a rare genetic disease in which photosensitivity is expressed as an inordinate susceptibility to the development of sunlight-induced skin cancers (Fig. 6-7).[4:pp 299–315,20,33] The cells of these

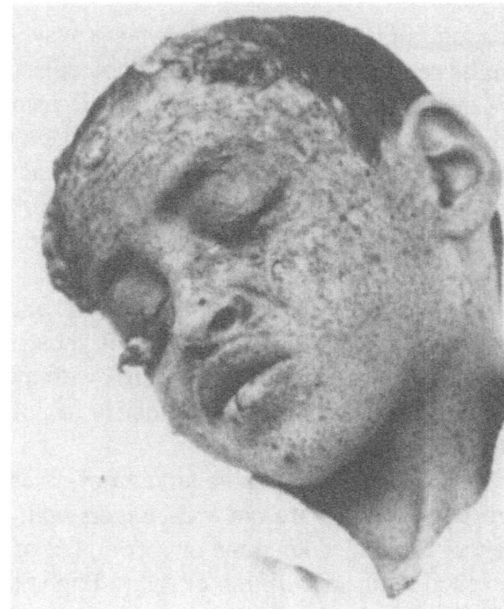

Fig. 6-7. Clinical picture of a patient with xeroderma pigmentosum showing actinically damaged skin, multiple actinic keratoses, and tumors. (Courtesy of William Spencer, M.D.)

patients are defective in their ability to repair UV radiation-damaged DNA.[20,27,33] Cultured XP cells do have a higher mutation rate than normal human cells following UV irradiation.[34] It has been postulated that this defect may lead to a high mutation rate following sunlight irradiation, and thus result in cancer formation. It should be noted that even if this theory proves correct for the cancer susceptibility of XP patients, it may not relate to sunlight-induced cancers in the general population.

6.5. ADVERSE CUTANEOUS REACTIONS TO SUNLIGHT [2–4,6,35]

Most of the cutaneous effects of sunlight are injurious. The following part of this discussion is designed to define the pathogenesis of some of the adverse effects of this energy. The explanations will of necessity be somewhat crude due to the limitations of our knowledge at this time.

6.5.1. Photosensitivity[3]

Photosensitivity is the broad term used to describe adverse reactions to sunlight or artificial light energy. Two types of photosensitivity reactions may occur. The reactions may be phototoxic or photoallergic in nature.

6.5.1.1. Phototoxicity[35]

Light-induced damage in the skin that is not dependent on an allergic mechanism may be considered to be phototoxic. Theoretically, these reactions could occur in everybody if the skin were exposed to enough light energy of the proper wavelengths, and enough molecules that absorb these wavelengths were present. Of course, the radiation must penetrate to the absorbing molecules for the reaction to occur. The sunburn reaction is the classic example of a phototoxic response. Although all types of skin will sunburn, much more energy is required to produce this response in deeply pigmented skin than in very light skin. Clinically, phototoxic reactions usually are characterized by erythema and at times by edema (swelling) that occurs within a few minutes to several hours after exposure, followed by hyperpigmentation and desquamation (peeling) confined to the exposed areas (Fig. 6-8).

Histologically, epidermal cell degeneration may be prominent when the photosensitizer is in the epidermis. This is most notable after the application of an exogenous photosensitizer to the skin. Edema with a mild to moderate inflammatory cell infiltration into the dermis consisting primarily of polymorphonuclear leukocytes may be seen (Fig. 6-9).

Despite a remarkable amount of investigation, the mechanisms by which phototoxic responses occur are not well understood. In the case of an exogenous photosensitizer, either the molecule alone or a complex of the chemical and cellular organelles becomes excited by the absorption of light. Triplet states and/or free radicals may thus be formed, and the dissipation of this energy may result in a number of changes including peroxide

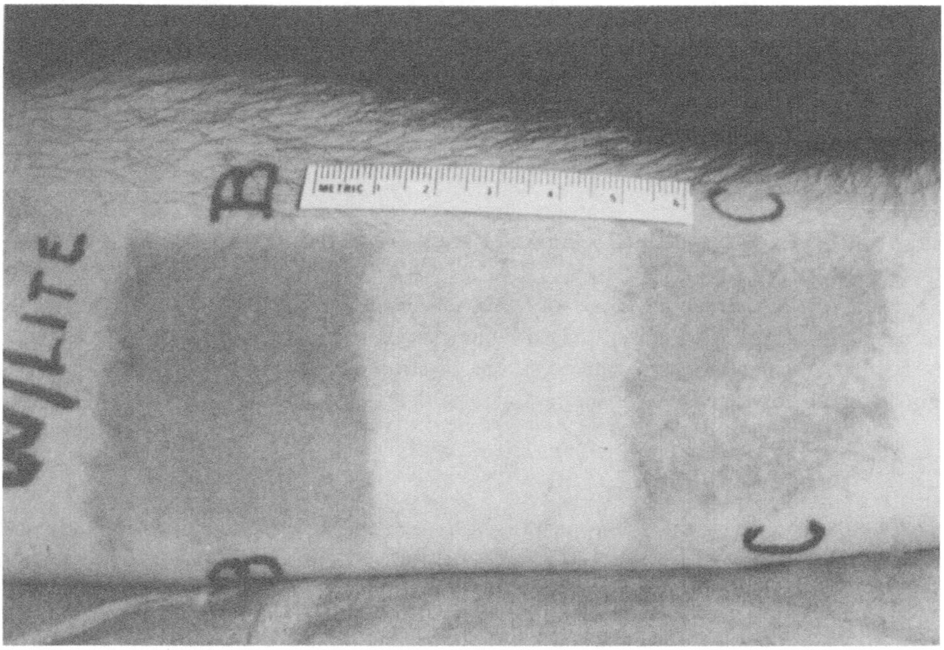

Fig. 6-8. Phototoxic reactions. Demonstrates a phototoxic reaction in two areas on the forearm, B and C, due to the topical application of a psoralen compound followed by exposure to the action spectrum for this chemical (UV radiation between 320 and 400 nm). It is characterized by erythema and edema at this point. The area between B and C received the UV exposure without prior application of the psoralen compound. (Courtesy of H. I. Maibach, M.D. and F. N. Marzulli, M.D.)

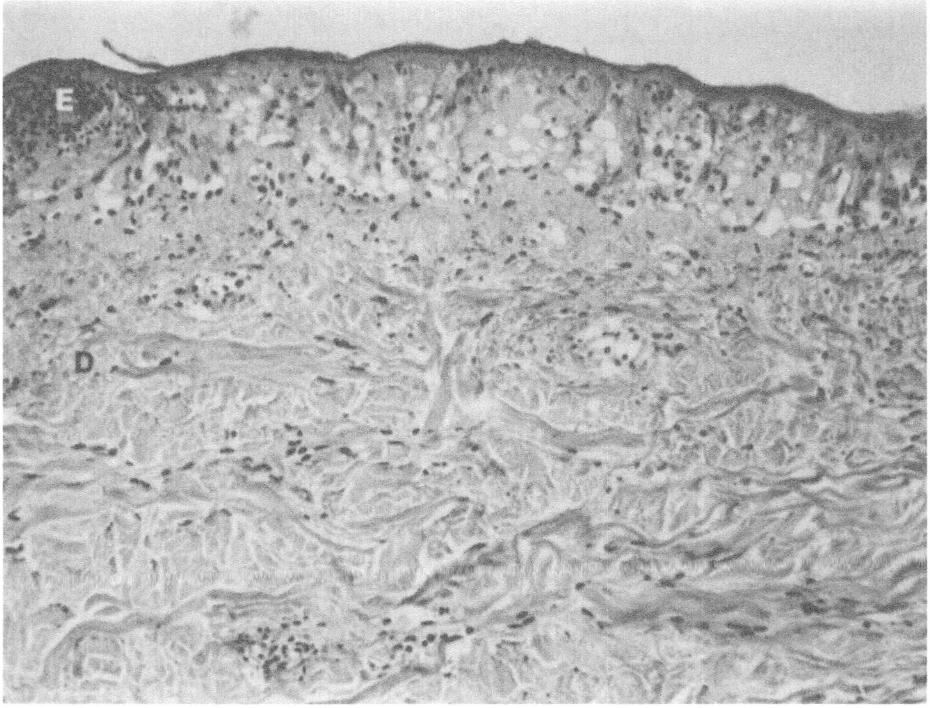

Fig. 6-9. Histology of the phototoxic reaction. Biopsy of the phototoxic reaction shown in Fig. 6-8 demonstrating a marked destruction of the epidermis with intracellular and extracellular edema (swelling), vacuole formation, and cell death. The dermis shows very little cellular infiltrate. E, epidermis; D, dermis. (Hematoxylin and eosin: ×50.) (From Ref. 3.)

formation, cell membrane or lysosomal membrane damage, and nuclear and/or mitochondrial injury.

It is likely that mechanisms vary with the photosensitizer.[36] In support of this idea, there is evidence that the phototoxicity induced by the furocoumarins is associated with the formation of an adduct with DNA. Chlorpromazine appears to form similar adducts, primarily with RNA. Other chemical photosensitizers require the presence of oxygen. This last type of phototoxicity has been termed "photodynamic action" (Chapter 3). Certain dyes and chemicals such as methylene blue, acriflavine, rose bengal, and porphyrins produce phototoxic effects on living and nonliving substrates only in the presence of oxygen. The photodynamically active substance becomes excited and forms a triplet state or a free radical. Singlet oxygen formation may play a role. The excited chemical also may form peroxides and then oxidize the substrate. Other possibilities include passing the energy from the excited chemical to the biological substrate, which then becomes oxidized, or the activated chemical may be able to accept electrons resulting in oxidation of the substrate. After excitation, the photosensitizing molecules return to the ground state and are structurally unchanged. Thus, they may partake in the induction of photodynamic reactions as long as they are associated with the substrate. Recent studies have also indicated that complement, mast cells, and prostaglandins may also play an important role in mediating some phototoxic responses.[36:p.47,37,38]

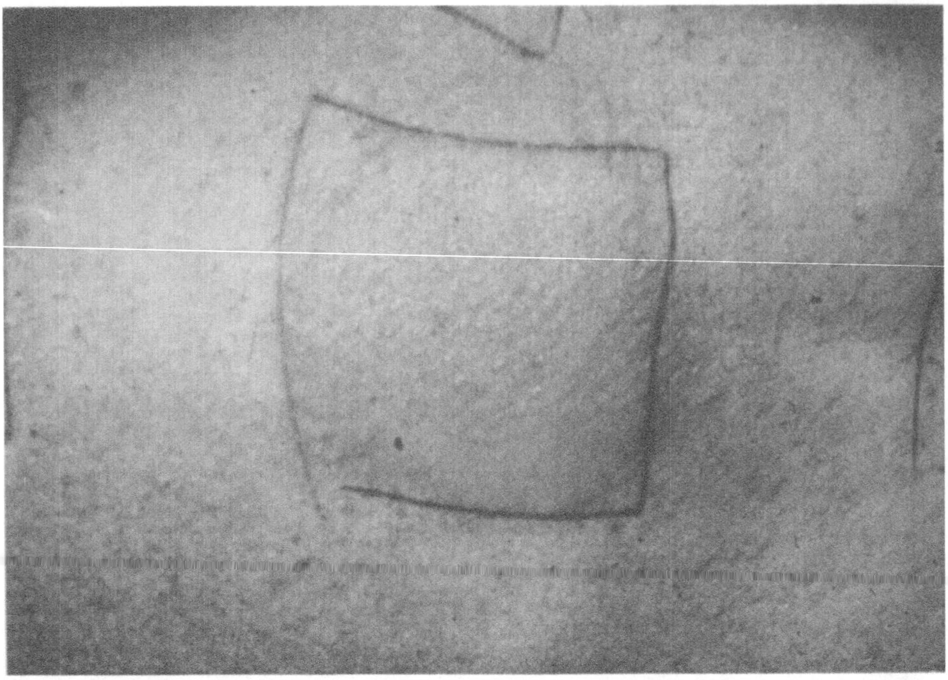

Fig. 6-10. Solar urticaria. Clinical picture of urticarial wheal (hive) occurring shortly after exposure to UV radiation and subsiding within 30 min. (From Ref. 3.)

6.5.1.2. Photoallergy[35,39]

Photoallergy can be defined as an acquired, altered capacity of the skin to react to light energy alone or in the presence of a photosensitizer that is presumably dependent on the development of a circulating antibody or a cell-mediated immune response. These reactions are generally uncommon, and the clinical patterns range from an immediate hive response (Fig. 6-10) to delayed itching, scaling, or weaping lesions (Fig. 6-11). The eruption frequently extends beyond the exposed areas, and eruptions in distant, previously involved sites may occur. Under controlled conditions, the general characteristics of allergic responses can be identified, including detection of an incubation period, spontaneous flare, and transfer of the process to normal subjects. Usually, less radiation exposure is required to induce a photoallergic response than a phototoxic response, when both are dependent on the same action spectrum.

Photoallergy also differs histologically from phototoxicity. The immediate urticarial (hive) lesions show very little microscopically other than some edema and blood vessel dilatation. The delayed reactions present a dense round-cell infiltrate around the blood vessels in the dermis that is characteristic, though not diagnostic, of these responses (Fig. 6-12). Similar to phototoxicity, photoallergy can be produced with or without the presence of known exogenous photosensitizers.

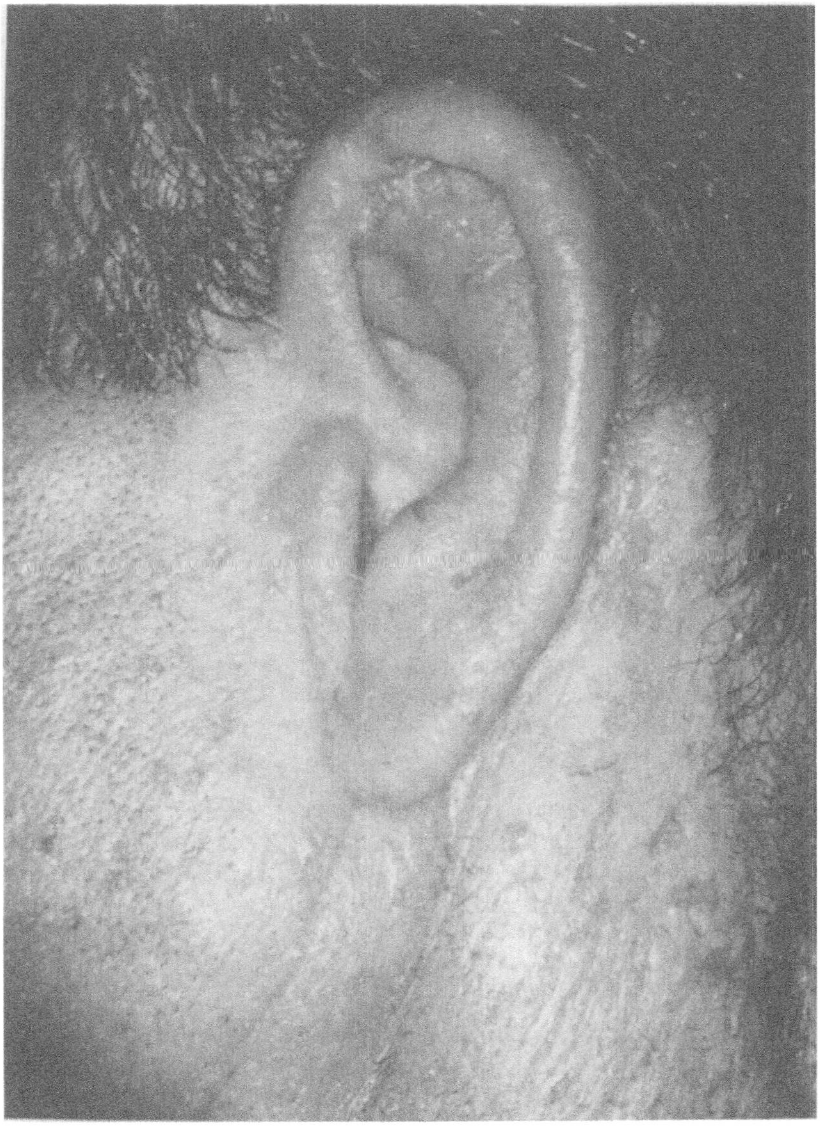

Fig. 6-11. Allergic photocontact dermatitis. Clinical picture of an eczematous photoallergic contact dermatitis induced by utilization of a soap containing tribromosalicylanilide and exposure to the action spectrum (UV radiation between 320 and 400 nm).

6.5.2. Clinical Problems

It is not appropriate in this chapter to dwell on the clinical responses that fall primarily in the field of individuals with dermatological training. However, an outline of these reactions hopefully may serve a useful purpose in putting medical effects of light

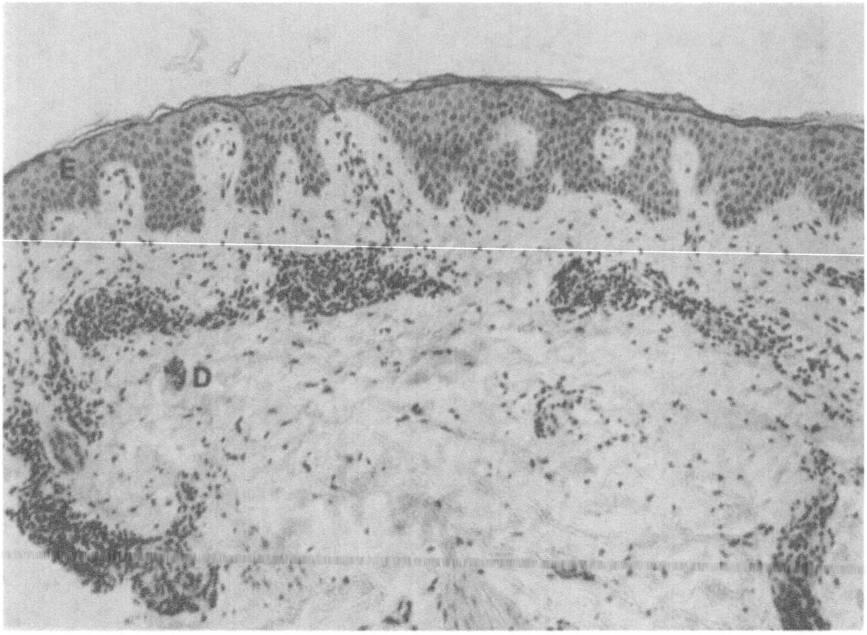

Fig. 6-12. Histology of the allergic photocontact dermatitis. Biopsy of an allergic photocontact reaction showing a dense perivascular round cell (collection of black dots in the dermis) infiltrate in the dermis characteristic of a delayed hypersensitivity response. E, epidermis; D, dermis. (Hematoxylin and eosin: ×80.) (From Ref. 3.)

energy in perspective. The adverse clinical effects may be divided into three large categories: (1) responses due to lack or loss of protection, (2) those due to the presence of a photosensitizing chemical, and (3) those not due to a deficiency in protection or to the presence of a known photosensitizer.

6.5.2.1. *Lack or Loss of Protection*[2,3]

Melanin pigment is the primary natural protective agent in the skin. Therefore, almost all of the clinical problems in this category relate to a melanin deficiency. This group includes patients with albinism who have a defect in the enzymatic tyrosinase system and are therefore unable to produce sufficient melanin; vitiligo, in which there is a loss of pigment cells and thus a loss of melanin; phenylketonuria (PKU), in which pigment dilution occurs due to an inherited error in phenylalanine metabolism; Chediak–Higashi syndrome, a genetic disorder with pigment dilution, apparently due to a structural abnormality in the melanosomes; and, the most common of all photosensitive conditions, i.e., fair complexion with light hair and eye color. This category also includes patients with xeroderma pigmentosum, in whom the lack of melanin does not play a role (Fig. 6-7). As noted in Section 6.3.4, the defect in this disease relates to an inability to repair UV radiation-damaged DNA.

The clinical reactions in this category consist of acute sunburning and, with chronic repeated injury, skin degeneration, actinic keratoses (Fig. 6-13), and skin cancer forma-

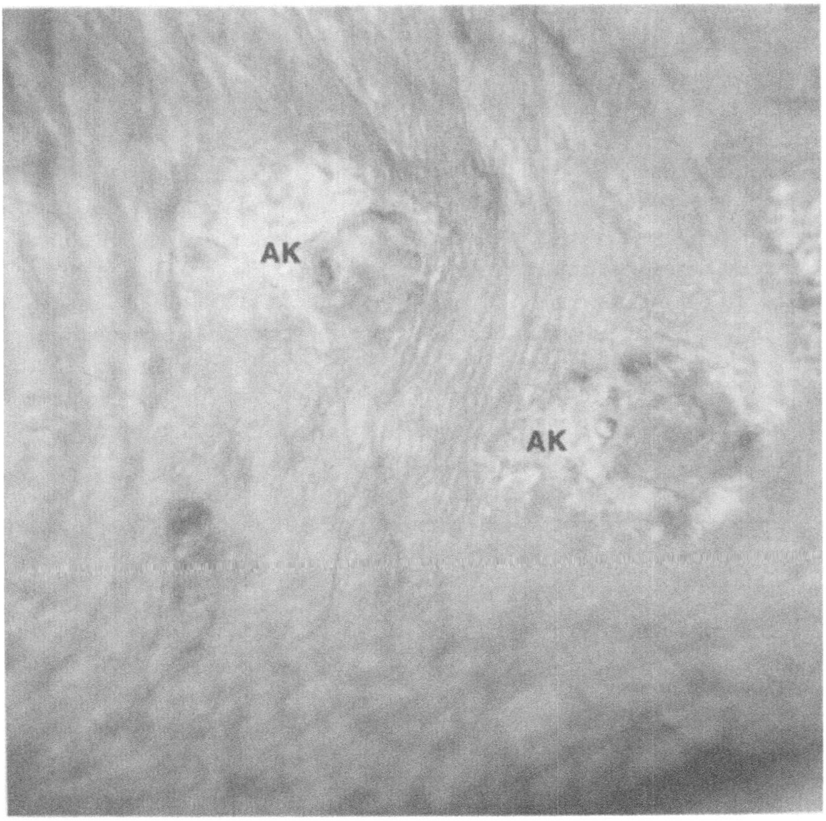

Fig. 6-13. Chronic sun (actinic) damage. Chronic actinic damage of the back of a hand with actinic keratoses (AK).

tion. The responses are phototoxic in nature, and the action spectrum falls in the sunburn range (290–320 nm).

6.5.2.2. Photosensitized Reactions[2,3,36]

A photosensitizer in this discussion will be considered to be a substance that, as the result of absorption of the sun's energy, induces an adverse response in the skin. The photosensitizer may be endogenous or exogenous in origin.

The porphyrins are the only well-established photosensitizers made by the human body. These are pyrrole ring structures that are essential to cellular metabolism throughout the body. Among other things, the heme part of hemoglobin is formed by these molecules. However, in the group of diseases named the porphyrias, large amounts of porphyrins are made in the erythropoietic (bone marrow) and/or the hepatic (liver) tissues. In these states, the porphyrins are not associated with iron and are potent photosensitizers. As such, they induce cutaneous changes ranging from marked photodestruction of the skin and underlying tissues to simple hyperpigmentation, depending on sun exposure and the amount of porphyrin molecules available. The cutaneous changes are due to photodynamic phototoxic responses.

Exogenous photosensitizers may get to the skin by topical application (contact) or through the bloodstream (systemic).

(1) Topical chemicals may cause photosensitivity reactions advertently, as occurs with tars and psoralen compounds in the treatment of psoriasis and vitiligo, or inadvertently, as occurs with the psoralen molecules in plants and perfumes, halogenated salicylanilides in deodorant soaps, sulfonamide and phenothiazine medications, and so on.[2,17] The reactions are usually phototoxic, though occasionally they are photoallergic in type. The action spectrum usually falls in the range of 320–400 nm.

(2) Systemically administered chemicals also may induce photosensitivity reactions advertently, as occurs with the use of psoralen compounds in the treatment of vitiligo and psoriasis, or inadvertently, which is the usual case. These inadvertent photoreactions are generally induced by commonly used medications including antibacterial sulfonamides, thiazide diuretics, sulfonylurea antidiabetic drugs, phenothiazines, and the broad-spectrum antibiotic demethylchlortetracycline.[36] The vast majority are phototoxic in nature, though occasionally delayed hypersensitivity responses may occur. The action spectrum for both phototoxic and photoallergic reactions usually includes long-wavelength UV radiation.

6.5.2.3. *Protection Normal and No Known Photosensitizer*

Photoreactions not due to deficient protection or the presence of known photosensitizers comprise a catch basket of conditions, including allergic responses such as the immediate antibody-mediated solar urticaria (hive response) and the apparently cell-mediated polymorphous light eruption (PMLE), certain autoimmune diseases, vitamin deficiencies, and a number of genetic problems. In general, these are uncommon diseases, except for PMLE. The action spectrum and mechanisms vary with the conditions. Because of the complicated and primarily clinical nature of this category, it will not be discussed further.[2,3]

6.6. BENEFICIAL EFFECTS OF SUN AND/OR ARTIFICIAL LIGHT ENERGY

Although adverse effects are much more common than beneficial responses, UV radiation does have therapeutic uses.

6.6.1. Vitamin D Formation[1,2,4:pp.247–258,6:pp.195–218,40]

Perhaps the only completely established beneficial effect of UV radiation on normal human skin is the formation of vitamin D. Vitamin D or, better stated, the D vitamins affect the transport of calcium from the gut, and also affect bone metabolism. These vitamins prevent rickets (softening of bones) in children, and osteomalacia (loss of bone substance) in adults by ensuring proper bone calcification.

Vitamin D_3 (calciferol) is made in the skin, most likely in the Malpighian cells (Fig.

6-2), through conversion of 7-dehydrocholesterol by the action of UV radiation. Further hydroxylation to 25-OH calciferol provides a more efficient compound. It has been suggested that skin color, and therefore the amount of protection against UV radiation, has been geographically regulated to prevent naturally occurring vitamin D intoxication.[41] Thus, the light skin of northern Europe was needed because of the limited amount of natural exposure to UV radiation, and the dark skin of peoples who live in the tropics is a necessary protection against producing too much vitamin D. However, as yet no natural vitamin D intoxication has been reported in light-complexioned individuals who have migrated southward. Thus, vitamin D formation by UV irradiation remains primarily a beneficial effect.

6.6.2. Treatment of Skin Diseases[6:pp.511–532]

Niels Finsen, an outstanding physician and an early pioneer in photobiology, received the Nobel prize in 1903 for his use of UV radiation in the treatment of cutaneous tuberculosis. Though this form of treatment is no longer necessary due to medical advances, the therapeutic value of light energy has been described for a large number of conditions including acne, eczema, pityriasis rosea, vitiligo, psoriasis, and herpes simplex. Perhaps we have the most information as to such effects in the last three diseases.

6.6.2.1. Vitiligo

Vitiligo is a cutaneous disorder characterized by a loss of pigment cells and thus pigment, the etiology of which is speculative at present (Fig. 6-14). Phototherapy represents almost the only available treatment, and success is limited at best. The procedure consists of the topical or systemic administration of psoralen compounds followed by exposure to radiation that includes the action spectrum of these chemicals, i.e., from 320 to 380 nm.[4:pp.335–368,783–791] It should be noted that the action spectra for the stimulation of pigmentation and the induction of erythema are quite similar, but they may not be identical. Until recently it was felt that a significant phototoxic erythema effect was necessary for the induction of repigmentation. However, certain studies indicate that the pigment response may be separate from the phototoxic injury (as defined by the erythema and visible damage).[42] The basic mechanism for the induction of hyperpigmentation induced by psoralen compounds and long wavelength UV radiation has not been established. However, the following observations have been noted following application of 4,5′,8-trimethylpsoralen (TMP) and exposure to these wavelengths of radiation: (1) an increased number of functioning epidermal melanocytes occurs due to proliferation and/or activation of inactive cells (mitotic activity has been detected between 48 and 72 h after TMP plus UV-A irradiation), (2) melanocyte hypertrophy and increased extension of dendrites around the keratinocytes, (3) increased number of melanosomes in melanocytes and keratinocytes, (4) increased tyrosinase activity in melanocytes, (5) increased rate of transfer of melanosomes to keratinocytes, and (6) dispersion of melanosomes in keratinocytes in nonaggregated distribution. (This last effect has only been described in Caucasoid skin.) This dispersed distribution of melanosomes may persist for more than 9 months, suggesting that some long-term gene depression may occur, perhaps related to the photoaddition of psoralens to DNA.

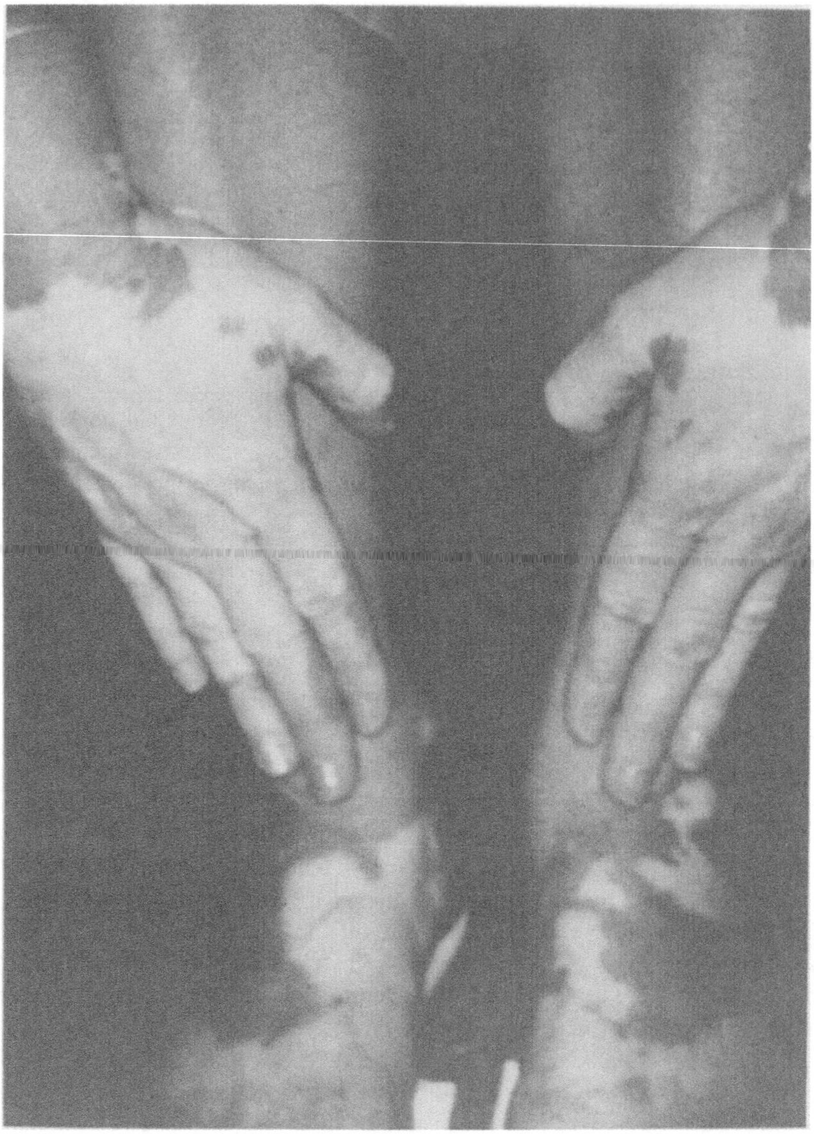

Fig. 6-14. Clinical picture of vitiligo showing depigmentation over the hands and feet areas caused by loss of pigment cells.

6.6.2.2. Psoriasis

Psoriasis is a common dermatological disease of unknown etiology that affects 1–3% of the world's population, and between 2 and 8 million people in the United States. The clinical picture varies from localized scaling plaques to generalized exfoliation (Fig. 6-15). The most notable pathological characteristic of this disorder is a marked increase in epidermal cell proliferation with rapid turnover of the germinative cells.

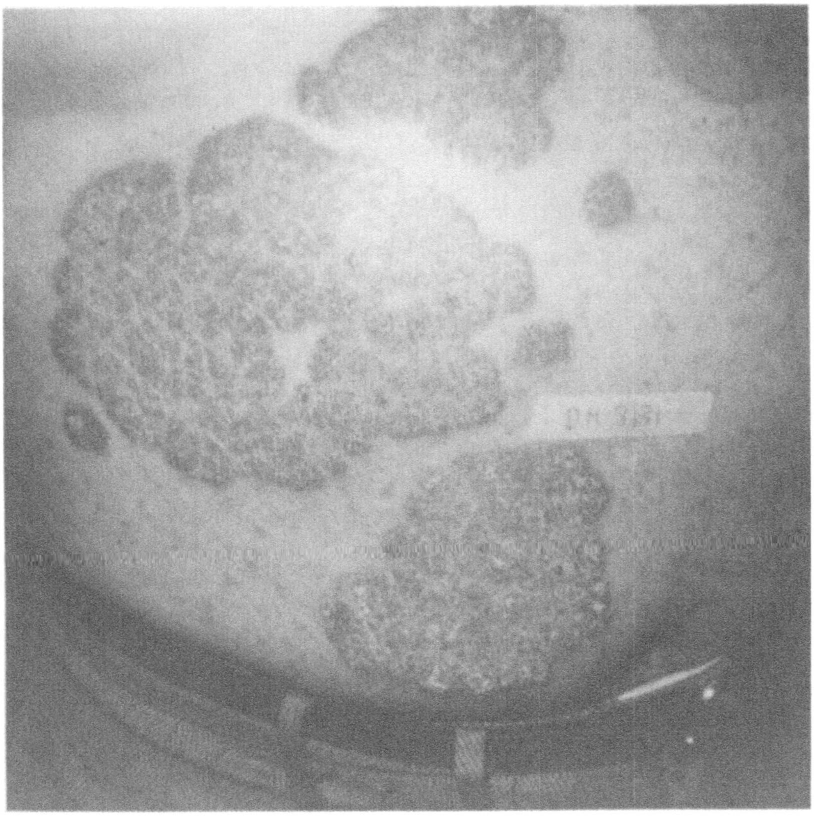

Fig. 6-15. Clinical picture of psoriasis characterized by large, erythematous, scaling plaques.

Treatment of psoriasis has revolved around attempts at the inhibition of this increased cellular proliferation, and phototherapy represents one of the most time-honored, effective methods in this respect[4:pp.793–796,6:pp.511–532] Astute clinical observations and subsequent controlled studies have established that radiation in the UV-B region (290–320 nm) is therapeutically effective in this disease. In 1925, Goeckerman described increased benefits from the topical application of crude coal tar to the sunburn exposure regimen. Since the action spectrum of the coal tar photosensitization is in the UV-A range, the increased effects had to be additive in nature, i.e., the tar and UV-B acted independently. One of the disadvantages of the Goeckerman regimen was the messy nature of the tar, which generally had to be used in a hospital setting. Within the past few years, regression of lesions following topical application of psoralen compounds and UV-A radiation has been reported. Most recently, an apparently extremely effective treatment has been described using systemic 8-methoxypsoralen and UV-A radiation emitted from a newly developed high-intensity system (PUVA).[43]

Though the mechanism of the effect of phototherapy on psoriasis has not been completely established, a reduction in cellular proliferation most likely plays a part. UV radiation shorter than 320 nm has a profound influence on epidermal cellular DNA synthesis and mitosis, which are inhibited shortly after irradiation and remain so for several hours.[11] Similarly, psoralens plus long-wavelength UV radiation significantly inhibit these functions, presumably due to photoaddition between the chemical and thymine bases of the DNA. It is also possible that phototherapy influences the dermal vasculature either directly or indirectly, and thus alters the course of the psoriatic lesions. Further study is obviously needed to clarify these concepts.

6.6.2.3. *Herpes Simplex Infection*[6:pp.545–559,571–594,44]

The herpes simplex virus is a common human pathogen that causes primary and recurrent skin and mucosal infections, and occasionally may involve other organs. Recently, a possible relationship to human cancer formation has been described. By far the most common clinical problem is the recurrent herpes simplex infection that is characterized by localized vesicular lesions involving the face, especially the lips, and the genitalia. Two varieties of closely related viruses are responsible for the lesions: those that occur above the waist are usually due to the type 1 virus, and those that appear below the waist are generally due to the type 2 variety. The eruption may occur as frequently as at monthly intervals or less, and thus presents a most disturbing and at times debilitating problem.

A group of investigators have advocated the use of photoinactivating procedures utilizing photodynamically active dyes, such as proflavine and neutral red, and visible light to inactivate the virus in these lesions. This concept resulted from the fact that photoinactivation of the viruses could be accomplished in tissue culture by this procedure.

At present the clinical efficacy and safety of this herpes simplex virus inactivation has not been established. The clinical results are as yet not convincing due to the great difficulty in obtaining enough patients to have adequate controls. In addition, *in vitro* studies suggest that there may be a risk of inducing cancer by this procedure.[6:pp.571–594,44] Thus, this treatment should be considered with some reservation until further information is available.

6.6.2.4. *Photochemotherapy of Tumors*[6:pp.625–638,45]

Photochemotherapy of tumors using exogenous chemicals and photosensitization of tumor tissue apparently was initially attempted in the early 1900s.[46] In the 1970s the concept was examined further by a number of investigators.[6] Most of these reactions *in vivo* are mediated through a transfer of energy from the triplet excited state of the sensitizer to molecular oxygen producing a highly reactive singlet oxygen which is cytotoxic.[6:pp.113–144] The chemical in wide use at this time is named hematoporphyrin derivative (HpD). This chemical localizes to tumor cells apparently due primarily to one of its components, di-hematoporphyrin ether (DHE).[47]

Though the major absorption peak for HpD is at the Soret band around 400 nm, the 630-nm band is the most important for therapy because tissue absorption at 630 nm is minimal, which allows greater penetration into the tissues.[48] Filtered xenon and tungsten lamps have been used to treat cutaneous and subcutaneous lesions. However, the argon pumped dye laser at 630 nm and optical fiber delivery systems can be used in all body cavities.[45]

This photochemotherapy has been and is being used to treat lung, esophageal, bladder, occular, head and neck, neurological, and gynecological tumors with partial to complete response in 85% of the patients.[6:p.89] However, more cutaneous and subcutaneous lesions have been treated to date. These include metastatic breast cancers, melanomas, and basal cell epitheliomas. The results have been most encouraging.

It should be noted that the HpD is also being used for tumor detection because it does localize in cancer and dysplastic tissue. Fluorescence of the HpD is usually stimulated by light rays around 400 nm from a filtered mercury-arc lamp or a krypton ion laser.[45:p.87]

Neither HpD nor DHE are toxic in the dose utilized in the absence of the light energy. However, they are retained in the skin and do induce a photosensitivity that may last 3–4 weeks.

6.6.3. Treatment of Systemic Diseases (Neonatal Jaundice)

Historically, phototherapy has been prescribed for a variety of diseases. As noted previously, there is evidence that UV radiation is beneficial for diseases of calcium metabolism, such as rickets and osteomalacia, due to the photochemical formation of vitamin D in the skin. Perhaps the one other clinical condition in which there is good evidence for the benefits of phototherapy is neonatal jaundice.[4:pp.231–258,6:pp.477–510,49]

Bilirubin ($C_{33}H_{33}O_6N_4$), the chief pigment in human bile, is formed from heme when hemoglobin from red blood cells is broken down primarily by the Kupffer cells of the liver. It appears to result from the opening of the porphyrin ring. In the neonatal period, increased amounts of unconjugated bilirubin are present in the circulation. This is due to the shorter survival of red blood cells, and to the functional immaturity of the neonatal liver, such that it is limited in its capacity to convert bilirubin from a lipid-soluble to a water-soluble compound (i.e., by conjugation) that can be excreted in the urine. In addition, increased permeability of the blood–brain barrier makes the newborn infant more susceptible to central nervous system damage by the deposition of bilirubin in brain cells (kernicterus). The danger is most notable in the premature infant. At present, therapy is designed to keep unconjugated bilirubin levels below 10–15 mg%. One of the mechanisms for the reduction in circulating unconjugated bilirubin, which has received considerable attention recently, is phototherapy.

In vitro studies indicate that light energy primarily between 450 and 500 nm will cause photodegradation of bilirubin. The exposure of experimental animals (Gunn rats) and infants to this energy has established that there is some alteration of bilirubin by the light energy. In addition, there is a striking reduction in circulating unconjugated bilirubin. *In vitro* studies have demonstrated that wavelengths in the blue part of the visible spectrum (primarily 450–500 nm) will photooxidize bilirubin. These *in vitro* photooxidation products lack the toxicity of bilirubin. The *in vivo* derivatives are less lipophilic and are more readily excreted without requiring conjugation. Therefore they are less likely to accumulate in the central nervous system tissue where the damage occurs in kernicterus. The mechanism of the beneficial effects of this phototherapy does not depend on photodegradation of bilirubin as originally believed. In infants exposed to blue light the major components of the excreted products are isomers of bilirubin. Thus, phototherapy enhances the excretion of bilirubin predominantly by photoisomerization.[49]

As with all therapies, the potential disadvantages must be considered. Fortunately, the photoproducts produced appear to be either nontoxic or so readily excreted that they do not accumulate in the central nervous system. Photodynamic injury to cutaneous cells must be considered as a possible detrimental effect, though no such response has been observed as yet. In addition, a number of injurious effects on bacteria and purified DNA have been recorded. Whether such findings relate to clinical responses remains to be determined. Perhaps the most potentially disturbing aspect of this phototherapy relates to possible retinal damage, which has been induced in newborn piglets exposed to blue fluorescent lights. No such damage has been noted in human infants to date despite extensive exposure. This is most likely due to the extensive use of eye protection.

6.7. PHOTOPROTECTION

The concept of cutaneous photoprotection encompasses protection against all of the adverse effects of sun on the skin. These range from acute phototoxic, photoallergic, idiosyncratic, and specific disease-oriented responses to chronic damage, and carcinogenesis. Obviously, the available sunscreens provide excellent methods for such protection, and recent data indicate that these screens also will inhibit UV radiation-induced cancer formation in experimental animals. Also, there are a number of non-sunscreen photoprotective measures that can be used to reduce the incidence and severity of acute adverse reactions, and thus perhaps reduce the chronic changes. These range from physical avoidance of sun exposure to the use of certain systemic agents.

6.7.1. Sunscreens[50]

Sunscreens protect the viable cells of the skin against actinic damage by absorbing and reflecting the radiation impinging on the skin. Most sunscreens are chemicals that absorb UV radiation and are incorporated in a cream, lotion, or gel vehicle. They are usually effective as an invisible thin film and are cosmetically acceptable to most people. The most widely used chemical sunscreens contain *para*-aminobenzoic acid (PABA), PABA esters (glyceryl PABA, amyldimethyl PABA, and octyldimethyl PABA), benzophenones (oxybenzone and dixoybenzone), cinnamates (octylmethoxycinnamate and cinoxate), salicylates (homomenthyl salicylate), and anthranilates. Physical sunblocks (zinc oxide and titanium dioxide) are usually opaque formulations that reflect and scatter radiation. They are cosmetically unacceptable to many individuals but may be essential for those who are unusually sensitive to UV as well as visible radiation, and for protection of limited areas such as the nose or lips, which receive a great deal of sun exposure.

Most chemical sunscreens (e.g., PABA and PABA esters) absorb and filter out the sunburn-producing radiation (UV-B, 290–320 nm). Only benzophenones and, to a limited extent, anthranilates exhibit sufficient additional absorption in the UV-A (320–400 nm) range to label them as broad-spectrum sunscreens.

The topical sunscreens containing PABA or PABA esters that are effective in the prevention of erythema reactions to sunburn radiation (290–320 nm) have not proved to be highly effective in the prevention of phototoxic and photoallergic reactions induced by near-UV and visible radiations. A combination of two sunscreens is usually recommended: (1) one that absorbs UV-B radiation and (2) one that also absorbs UV-A radiation. The topical psoralen-induced phototoxic reaction (8-methoxypsoralen and 4,5′,8-

trimethylpsoralen) caused by 320–380 nm radiation can be reduced by certain benzophenone derivatives.

The topical formulations in oil base with emollients, containing two or more light-absorbing ingredients such as 4% ethylhexyl-*p*-methoxycinnamate + 3% 2-hydroxy-4-methoxybenzophenone + 5.5% 2-phenylbenzimidazole sulfonic acid appear to be very effective formulations in the protection of human skin against the effects of sunburn UV radiation. Such a formulation is also effective at high elevations during skiing or mountain climbing, and against drug-induced phototoxic reactions.

Adverse effects from topical sunscreens include yellow discoloration of clothes by PABA, especially when the clothing is exposed to the sun. A few patients develop contact-type eczematous dermatitis from PABA, PABA esters, benzophenones, or other sun-screening ingredients. Patients allergic to benzocaine, procaine, *para*-phenylenediamine, and sulfanilamide may cross-react with PABA and exhibit allergic reactions. Impurities in the form of *ortho-* and *meta*-esters of PABA may cause contact and photocontact dermatits. Certain sunscreens cause selective burning and smarting reactions on the face while sunbathing.

6.7.2. Physical Factors[50]

Clothing provides an optical filter against radiation penetration. Hats are essential for preventing damage to sparsely haired scalps, and broad brims protect ears and noses and the like. Dark, especially black, clothes absorb the light and heat, and thus may be quite uncomfortable. Tighter weaves tend to let less radiation through to the skin. Also one should avoid using synthetic fabrics that allow the penetration of sunburning radiation.

The time of day also plays a role. The UV-B rays reach the earth's surface from the sun predominantly in the middle of the day. They are most intense between 11:30 A.M. and 12:30 P.M. Avoidance of exposure between 10:00 A.M. and 2:00 P.M. will reduce the danger of UV-B radiation exposure by 50–60%.

Melanin pigment provides the most effective physiological protection produced by our skins. It provides an optical filter and is a stable free radical that can quench other free radicals, which may be generated during photoreactions. Dark-complexioned people require a great deal more energy to develop sunburns or other phototoxic responses than do light-complexioned individuals. Also dark skins are much more resistant to photocarcinogenesis.

The stratum corneum also provides a fair to good optical barrier for the passage of UV radiation. Thickening of the stratum corneum does protect the lower vulnerable layers of the skin.

Through the *cis-trans* isomerization process, urocanic acid provides some protection against the damaging effects of UV-B radiation.

6.7.3. Systemic Chemical Agents

6.7.3.1. Antimalarials[51]

Chloroquine (Aralen), dihydroxychloroquine (Plaquenil), and quinacrine (Atabrine) do not alter UV-B radiation-induced sunburn or other phototoxic responses. However, they are effective in certain photosensitive diseases such as polymorphous light eruptions (PMLE), lupus erythematosus (LE), and solar urticaria. The mechanisms responsible for these effects are not clear, though their antiinflammatory properties most likely play a role. Regular ophthalmologic examinations are necessary, however, because of the possible retinal damage that might be induced by chloroquine or Plaquenil. However, recent evaluations indicate that the dosage schedules in use are unlikely to produce such effects. Complete blood count (CBC) evaluations are indicated since rare hematopoietic toxic effects may occur. Other toxic effects are noted in the references.(50,51)

6.7.3.2. *β-Carotene*(52–54)

β-Carotene and phytone, a colorless precursor of β-carotene, have been shown to reduce the intensity of acute UV-B radiation-induced phototoxic reactions, and β-carotene has been shown to inhibit UV-B radiation-induced carcinogenesis experimentally. However, these effects are quite limited and may have little, if any, clinical significance. β-Carotene does appear to be at least partially effective in the treatment of certain of the porphyrias, such as erythropoietic protoporphyria (EPP). Reports on its beneficial effects on other photosensitive diseases such as solar urticaria and hydroa aestivale require more thorough evaluation. Recent data on its effect on PMLE are encouraging.(55) β-Carotene appears to produce its beneficial effects through quenching singlet oxygen

6.7.3.3. *Systemic Psoralens*(8,10,56,57)

Systemic 8-methoxypsoralen (8-MOP) and 4,5′,8-trimethylpsoralen (trisoralen) with UV-A radiation have been used to increase pigment production, and thus lead to greater natural protection. In general, this is not a suggested approach except under unusual circumstances, because of the potential damage to the skin and eyes. However, 8-MOP plus UV-A radiation has proved to be an effective therapy for patients with PMLE, actinic reticuloid, and solar urticaria. The mechanisms of these responses, at least in the former two diseases, may relate in part to eliminating reactive cells in the dermis. Protection and/or stabilization of mast cells may play a role in its effects on solar urticaria.

6.7.4. Topical Psoralens(11,12,58,59)

5-Methoxypsoralen (5-MOP) has been incorporated into sunscreens in small amounts to accelerate tanning. Experimental studies indicate that this ingredient is phototoxic and photomutagenic. In addition, it is photocarcinogenic experimentally in the concentrations present in these sunscreens. Thus, it does not appear to be an appropriate approach to protection.

6.7.5. Quick-Tanning Preparations(50)

Topical agents containing 3–5% dihydroxyacetone and oral preparations containing β-carotene and canthoxanthine can be used to produce color in the skin. This approach appears to be safe, though the long-term effects of canthoxanthine have not been completely evaluated. These agents do not produce any photoprotective effect from sunburn damage; thus the color is not photoprotective.

6.8. PHOTOIMMUNOLOGY(6:pp.293–322,39,60–65)

Photoimmunology could be considered the overall term for all of the light-influenced immunological phenomena, including the cutaneous photoallergic reactions noted in Section 6.5.1.2. However, for the purpose of this discussion, the term will represent the effects of nonionizing radiation on immune phenomena excluding these photoallergic reactions.

6.8.1. Immediate Hypersensitivity(6:pp.298–303,39:pp.636–637)

Immediate hypersensitivity reactions are mediated through antibody formation. Antibodies are produced by B lymphocytes and plasma cells following stimulation by antigenic substances, and T-helper and suppressor lymphocytes modulate this antibody formation. Large amounts of UV-C radiation *in vitro* have been shown to inhibit various antibody activities, i.e., agglutination, complement fixation, etc. However, the amount of energy needed makes the clinical relevance of these findings questionable. Though surface markers and viability of B and T lymphocytes can be altered by psoralen and UV-A photosensitization (PUVA) and UV radiation *in vitro*, the effect of UV radiation or PUVA on antibody formation *in vivo* has not been clarified as yet.

Mast cells are an integral part of IgE-mediated immediate hypersensitivity responses. These cells can be altered by UV radiation. *In vitro* studies have shown that UV-B and UV-A radiation, in small enough doses to be potentially clinically relevant, can disrupt these cells and release their granules. In contrast, erythemogenic exposures to UV-B radiation over 4–32 days will markedly increase the number of mast cells in a mouse ear. A mild increase was noted with chronic UV-B radiation and none with chronic erythrogenic amounts of UV-A exposures. Perhaps from a clinically relevant point of view the response of these cells to PUVA therapy *in vivo* is of the most significance. PUVA treatment has been shown to stabilize mast cell membranes to prevent degranulation.[66] This relationship requires further evaluation. PUVA therapy has been found to be of value in urticaria pigmentosum, a disease characterized by the accumulation of mast cells in the skin, and in solar urticaria, which at least in some cases appears to be an IgE-mediated immediate hypersensitivity response.

6.8.2. Delayed Hypersensitivity (DHS) Cell-Mediated Immune Response (CMI)(6:pp.303–305,39:p.637)

There is an increasing body of evidence that UV radiation and PUVA have a profound influence on CMI responses and DHS. Clinically, chronic sun-exposed skin is less reactive to allergens than nonexposed skin. UV-B and PUVA irradiation has been shown to locally and systemically suppress contact DHS in animals that had been sensitized to dinitrochlorobenzene (DNCB). Also, localized reduction in both the initiation or sensitization and elicitation phase of DHS has been reported in PUVA-treated sites in experimental animals. In addition, PUVA treatment of psoriasis has been reported to suppress DHS to DNCB in humans. UV-B as well as PUVA have been shown to suppress contact sensitization to nitrogen mustard. The mechanism and specificity of this suppression was not determined.

Recent studies in mice have demonstrated the inhibition of the afferent phase of DHS, which is associated with the formation of antigen specific T-suppressor cells by both UV-B and PUVA irradiation.[61] There is some evidence that this affect relates to antigen presentation to splenic macrophages in the UV-B-treated mice.[62,63] This aspect has not been examined in PUVA-exposed mice.

The role of Langerhans cells in these radiation effects on DHS must be considered. Langerhans cells have been implicated as antigen-processing cells for antigens presented through the skin. UV radiation (both UV-A and UV-B) in sufficient doses can effectively abolish surface markers of Langerhans cells in human and mouse skin without destroying these cells.[64] However, UV radiation-induced reduction in ATPase-positive cells does not result in the consistent inhibition of DHS reactivity. Both UV-B and PUVA inhibit the primary epidermal cell–lymphocyte proliferation reaction, which appears to be mediated through epidermal Langerhans cells. It appears likely that the effects of UV radiation and photochemotherapy on DHS is dependent at least in part on the influences of the Langerhans cells.

6.8.3. Lymphocytes and Nonionizing Radiation[6:pp.299–303]

Lymphocytes play an essential role in both humoral and cell-mediated immune responses. UV radiation and PUVA have been shown to profoundly influence these cells *in vitro* and *in vivo*.

In vitro studies using trypan blue dye exclusion and responses to mitogens indicated that UV-C is more toxic than UV-B, which is more toxic than UV-A radiation. PUVA has also been shown to be toxic to these cells *in vitro*. T lymphocytes appear to be more sensitive to UV-C radiation than B lymphocytes. However, the sensitivity of T and B cells to UV-B radiation is approximately equal.[67]

It is obvious that UV radiation and PUVA can alter lymphocyte structure and function both *in vivo* and *in vitro*. It seems likely that effects on these cells are related to alterations in immune function that have been recorded after such irradiation. However, our information is as yet too preliminary to make any judgments concerning mechanisms of effects on human immune reactions and disease.

6.9. INDIRECT EFFECTS OF LIGHT[4:pp.231–258]

A number of miscellaneous indirect effects of light energy on animal functions have been described. These include possible control of biological rhythms (Chapter 7), gonadal function (Chapter 9), cortisol production, and the like. One of the best characterized indirect effects, other than vision, concerns the inhibition of melatonin formation in the pineal organs by light energy that impinges on the retina. What general influence this response engenders remains to be determined, though recent studies suggest that the pineal gland may play a most important physiological role in human hormonal and central nervous system function. These indirect effects appear to be mediated through a retinal receptor, as yet unidentified in mammals. Other effects include reduction in systolic and diastolic blood pressure, increase in exercise tolerance, reduction in blood cholesterol, and decrease in serum tyrosine.

6.10. ULTRAVIOLET RADIATION EFFECTS ON THE EYE[7:pp.231–262]

The ocular structures, the cornea, aqueous, lens, and vitreous have no chromophores to absorb all but the shortest of visible rays. This transparency allows the visible energy to reach the retina to allow us to see. In contrast, these tissues, especially the lens, can absorb varying amounts of UV radiation. UV radiation has been implicated in a number of adverse effects. Acute photokeratitis due to reflection of UV radiation from snow results in "snow blindness." This type of reaction may occur industrially in welders, glass workers, etc. Other conjunctival and corneal changes apparently due to UV radiation include pingueculae, pterygiums, exposure keratosis, nodular band-shaped keratopathy, dysplasia, and intraepithelial carcinoma.[7:p.232] Since the human cornea and the aqueous humor transmit almost all wavelengths longer than 300 nm, these untoward effects should be due to wavelengths of 300 nm and less.[7:pp.232,253,254] Therefore at least in humans the lens is exposed to UV radiation between 300 and 400 nm throughout the individual's lifetime. This chronic UV radiation exposure over the years leads to the generation of fluorescent chromophores, which are in part responsible for the yellow discoloration of the lens nucleus as it ages. In about one-tenth of the aged population this leads to nuclear cataracts. UV radiation induced cataracts have recently been reported supporting this concept.[68]

The vitreous and the retina are protected from shorter UV-B rays by the cornea and from the UV rays longer than 300 nm by the lens. However, in very young eyes the lens may not absorb the above noted UV radiation. In addition, when lens damage has occurred, i.e., with cataract formation, the lens is removed. Under these circumstances the vitreous and the retina are exposed to UV radiation between 300 and 400 nm. The effect of this longer UV radiation on the vitreous requires clarification, though there is some evidence suggesting that it may produce some damage to the protein and polysaccharide structures of this gel. The shorter UV rays below 295 nm, which are filtered out by the cornea, have been shown to cause shrinkage of the vitreous gel and collagen network.[69] They also can damage the polysaccharide solution due to a breakdown of the glucosidic linkages, increasing the reducing end groups and the formation of smaller fragments.[70]

The retina is also exposed to these longer UV rays as well as more of the short blue visible rays when the lens is removed. Recent data implicate the UV and short visible radiation as being able to cause retinal damage.[7:p.235]

6.11. CONCLUSION

Light energy has profound effects on mammalian cell structure and function. These influences play a key role in the relationship between humans and their environment. The primary effects, both detrimental and beneficial, occur at the site of impact, i.e., the skin. The skin is not only the most common site of photoresponse but also the most available human tissue for examination and investigation. Thus, medical photobiologists have unique opportunities. They can examine cancerous growths as they are being formed, evaluate radiation and chemical interactions as they relate to disease states, study the influence of light energy on immunological processes, and so on. Though the eye is primarily involved in vision responses, phototoxic damage also occurs to its structures. Again the eye is more available for study than other, more internal organs. Though the sophistication of methodology for the study of human responses is somewhat limited compared to other disciplines in photobiology, the significance of such studies cannot be overemphasized. The field of medical photobiology is wide open. At present we have just scratched the surface of understanding.

6.12. REFERENCES

1. B. E. Johnson, F. Daniels, Jr., and I. A. Magnus, Response of human skin to ultraviolet light, in: *Photophysiology,* Vol. 4 (A. C. Giese, ed.), pp. 139–202, Academic Press, New York (1968).
2. F. Urbach (ed.), *The Biologic Effects of Ultraviolet Radiation* Pergamon Press, New York (1969).
3. J. H. Epstein, Adverse cutaneous reactions to the sun, in: *Yearbook of Dermatology* (F. D. Malkinson and R. W. Pearson, eds.), pp. 5–43, Yearbook, Chicago (1971).
4. T. B. Fitzpatrick (consult. ed.), M. A. Pathak, L. C. Harber, M. Seiji, and A. Kukita (eds.), *Sunlight and Man,* University of Tokyo Press, Tokyo (1974).
5. J. A. Parrish, R. R. Anderson, F. Urbach, and D. Pitts (eds.), *UV-A: Biological Effects of Ultraviolet Radiation with Emphasis on Human Responses to Longwave Ultraviolet,* Plenum Press, New York (1978).
6. J. D. Regan and J. A. Parrish (eds.), *The Science of Photomedicine,* Plenum Press, New York (1982).
7. F. Urbach and R. W. Gange (ed.), *The Biological Effects of UVA Radiation,* Praeger, New York (1986).
8. I. Willis, A. Kligman, and J. H. Epstein, Effects of long ultraviolet rays on human skin: Photoprotective or photoaugmentative? *J. Invest. Dermatol.* **59,** 416–420 (1972).
9. D. S. Snyder and W. H. Eaglstein, Intradermal antiprostaglandin agents and sunburn, *J. Invest. Dermatol.* **62,** 47–50 (1974).
10. G. F. Wilgram, R. L. Kidd, W. S. Krawczyk, and P. L. Cole, Sunburn effect on keratinosomes, *Arch. Dermatol.* **101,** 505–519 (1970).
11. K. Wier, K. Fukuyama, and W. L. Epstein, Nuclear changes during light induced depression of ribonucleic acid and protein synthesis in human epidermis, *Lab. Invest.* **25,** 451–456 (1971).
12. K. Fukuyama, W. L. Epstein, and J. H. Epstein, The effect of ultraviolet light on RNA and protein synthesis in differentiated epidermal cells, *Nature* **216,** 1031–1032 (1967).
13. J. H. Epstein, K. Fukuyama, and K. Fye, Effects of ultraviolet radiation on the mitotic cycle and DNA, RNA and protein synthesis in mammalian epidermis *in vivo, Photochem. Photobiol.* **12,** 57–65 (1970).
14. G. A. Soffen and H. F. Blum, Quantitative measurements in changes in mouse skin following a single dose of ultraviolet light, *J. Cell. Comp. Physiol.* **58,** 81–96 (1961).
15. W. S. Bullough and E. B. Laurence, Tissue homeostasis in adult mammals, in: *Advances in Biology of Skin,* Vol. 7 (W. Montagna and R. L. Dobson, eds.), pp. 1–36, Pergamon Press, Oxford (1966).
16. J. J. Voorhees, E. A. Duell, and W. H. Kelsey, Dibutyryl cyclic AMP inhibition of epidermal cell division, *Arch. Dermatol.* **105,** 384–386 (1972).
17. J. E. Cleaver, Repair of damaged DNA in human and other eukaryotic cells, in: *Nucleic Acid-Protein Interactions—Nucleic Acid Synthesis in Viral Infection* (D. W. Ribbons, J. F. Woessner, and J. Schultz, eds.), pp. 87–112, North-Holland, Amsterdam (1971).
18. J. E. Cleaver, Repair processes for photochemical damage in mammalian cells, *Adv. Radiat. Biol.* **4,** 1–75 (1974).
19. P. D. Forbes, F. Urbach, and R. E. Davies, Photocarcinogenesis in hairless mice: Dose-response and influence of dose-delivery, *Photochem. Photobiol.* **34,** 361–365 (1981).
20. A. R. Lehman, S. Kirk-Bell, C. F. Arlett, M. C. Paterson, P. H. M. Lohman, E. A. de Weerd-Kastelein, and D. Bootsma, Xeroderma pigmentosum cells with normal levels of excision repair have a defect in DNA synthesis after UV-irradiation, *Proc. Natl. Acad. Sci. USA* **2,** 219–223 (1975).
21. R. L. Olson, J. Nordquist, and M. A. Everett, The role of epidermal lysosomes in melanin pigmentation, *Br. J. Dermatol.* **83,** 189–199 (1970).
22. J. H. Epstein, K. Fukuyama, and R. L. Dobson, Ultraviolet light carcinogenesis, in: *The Biologic Effects of Ultraviolet Radiation* (F. Urbach, ed.), pp. 551–568, Pergamon Press, New York (1969).

23. W. L. Epstein, K. Fukuyama, and J. H. Epstein, Early effects of ultraviolet light on DNA synthesis in human skin *in vivo*, *Arch. Dermatol.* **100,** 84–89 (1969).
24. L. H. Kligman, F. J. Akin, and A. M. Kligman, The contributions of UV-A and UV-B to connective tissue damage in hairless mice, *J. Invest. Dermatol.* **84,** 272–276 (1985).
25. H. F. Blum, *Carcinogenesis by Ultraviolet Light,* Princeton University Press, Princeton (1959).
26. J. H. Epstein, Ultraviolet carcinogenesis, in: *Photophysiology,* Vol. 5 (A. C. Giese, ed.), pp. 235–273, Academic Press, New York (1970).
27. J. H. Epstein, Photocarcinogenesis, skin cancer and aging, *J. Am. Acad. Dermatol.* **9,** 487–506 (1983).
28. A. W. Kopf, M. L. Kripke, and R. S. Stern, Sun and malignant melanoma, *J. Am. Acad. Dermatol.* **11,** 674–684 (1984).
29. J. H. Epstein, How important is the role of radiation from the sun in the rising incidence of melanomas?, in: *Controversies in Dermatology* (E. Epstein, ed.), pp. 10–12, Saunders, Philadelphia (1984).
30. M. L. Kripke, Immunology of UV-induced skin cancer: Yearly review, *Photochem. Photobiol.* **32,** 837–839 (1980).
31. R. A. Daynes, E. J. Bernhard, M. F. Gurish, and D. H. Lynch, Experimental photoimmunology: Immunologic ramifications of UV induced carcinogenesis, *J. Invest. Dermatol.* **77,** 77–85 (1981).
32. R. D. Granstein, J. A. McAuliffe, D. J. Waltenbaugh, and M. I. Greene, Immunologic inhibition of ultraviolet radiation-induced tumor suppressor cell activity, *Science* **224,** 615–617 (1984).
33. J. E. Cleaver and D. Bootsma, Xeroderma pigmentosum: Biochemical and genetic characteristics, *Ann. Rev. Genetics* **9,** 19–38 (1975).
34. V. M. Maher, L. M. Oulette, R. D. Curren, and J. J. McCormick, Frequency of ultraviolet light-induced mutations is higher in xeroderma pigmentosum variant cells than in normal human cells, *Nature* **261,** 593–595 (1976).
35. J. H. Epstein, Photoallergy: A review, *Arch. Dermatol.* **106,** 741–748 (1972).
36. J. H. Epstein and B. U. Wintroub, Photosensitivity due to drugs, *Drugs* **30,** 42–57 (1985).
37. H. W. Lim, L. Young, M. Hagen, and I. Gigli, Delayed phase of hematoporphyrin-induced phototoxicity: Modulation by complement, leukocytes and antihistamines, *J. Invest. Dermatol.* **84,** 114–117 (1985).
38. H. W. Lim, M. Hagen, and I. Gigli, Phototoxicity in mast cell deficient, C5 deficient and leukopenic mice, *Clin. Res.* **33,** 23A (1985).
39. J. H. Epstein, Photoallergy and photoimmunology, in: *Dermatologic Immunology and Allergy* (J. Stone, ed.), pp. 629–640, Mosby, St. Louis (1985).
40. H. F. DeLuca, Vitamin D: The vitamin and the hormone, *Federation Proc.* **33,** 2211–2219 (1974).
41. W. F. Loomis, Skin-pigmentation regulation of vitamin D biosynthesis in man, *Science* **157,** 501–506 (1967).
42. K. H. Kaidbey and A. M. Kligman, Photopigmentation with trioxsalan, *Arch. Dermatol.* **109,** 674–677 (1974).
43. J. A. Parrish, T. B. Fitzpatrick, L. Tanenbaum, and M. A. Pathak, Photochemotherapy of psoriasis with oral methoxsalen and longwave ultraviolet light, *N. Engl. J. Med.* **291,** 1207–1211 (1974).
44. L. E. Bockstahler, C. D. Lytle, and K. B. Hellman, A review of photodynamic therapy for herpes simplex: Benefits and potential risks, *DHEW Publication No. (FDA) 75-8013.* Available from the Bureau of Radiological Health, Rockville, MD 20852 (1974).
45. D. R. Doiron and G. S. Keller, Porphyrin photodynamic therapy: Principles and clinical applications, in: *Therapeutic Photomedicine* (H. Hönigsmann and G. Stengl, eds.), pp. 85–93, Karger, Basel (1986).
46. A. Jesionek, Tappenier zur Behandlung des Hautcarcinomes mit fluoreszierenden Stoffen, *Munch. Med. Wochenschr.* **41,** 2042–2044 (1903).
47. T. J. Dougherty, W. R. Potter, and K. R. Weishaupt, The structure of the active component of hematoporphyrin derivative, in: *Porphyrin Localization and Treatment of Tumors* (D. R. Doiron and C. J. Gomer, eds.), No. 170, pp. 301–314, Alan R. Liss, New York (1985).
48. D. R. Doiron, L. O. Svaasand, and A. E. Profio, Light dosimetry in tissue application to photoirradiation therapy, in: *Porphyrin Photosensitization* (D. Kessel and T. J. Dougherty, eds.), p. 63, Plenum Press, New York (1983).
49. *Phototherapy in the Newborn,* Final report of the Committee Division of Medical Sciences, National Research Council, 2101 Constitution Ave., Washington, D.C. 20418 (1974).
50. M. A. Pathak, Sunscreens: Topical and systemic approaches for protection of human skin against harmful effects of solar radiation, *J. Am. Acad. Dermatol.* **7,** 285–312 (1982).

51. L. Tanenbaum and D. L. Tuffanelli, Antimalarial agents: Chloroquine, hydroxychloroquine and quinacrine, *Arch. Dermatol.* **16,** 587–591 (1980).
52. M. M. Mathews-Roth, M. A. Pathak, T. B. Fitzpatrick, L. C. Harber, and E. H. Kass, Beta-carotene therapy of erythropoietic porphyria and other photosensitive diseases, *Arch. Dermatol.* **113,** 1229–1232 (1977).
53. J. H. Epstein, The effects of beta-carotene on ultraviolet induced cancer formation in the hairless mouse skin, *Photochem. Photobiol.* **25,** 211–213 (1977).
54. M. M. Mathews-Roth and N. I. Krinsky, Carotenoid dose level and protection against UV-B induced skin tumors, *Photochem. Photobiol.* **42,** 35–38 (1985).
55. C. T. Jensen, Oral carotinoid treatment in polymorphous light eruption: A cross-over comparison with oxychloroquine and placebo, *Photodermatol.* **2,** 166–169 (1985).
56. J. D. Imbraie, F. Daniels, L. Bergeron, C. E. Hopkins, and T. B. Fitzpatrick, Increased erythema threshold 6 weeks after a single exposure to sunlight plus oral methoxsalen, *J. Invest. Dermatol.* **32,** 331–337 (1959).
57. R. S. Stern, N. Laird, J. Melski, J. A. Parrish, T. B. Fitzpatrick, and H. L. Bleich, Cutaneous squamous cell carcinoma in patients treated with PUVA, *N. Engl. J. Med.* **310,** 1156–1161 (1984).
58. M. J. Ashwood-Smith, G. A. Poulton, M. Baker, and M. Mildenberger, 5-Methoxypsoralen, an ingredient in several sunscreen preparations had lethal metagenic and clastogenic properties, *Nature* **285,** 407–409 (1980).
59. L. E. Cartwright and J. F. Walter, Psoralen containing sunscreen is tumorigenic in hairless mice, *J. Invest. Dermatol.* **80,** *306* (1983).
60. W. L. Morison, Photoimmunology, *J. Invest. Dermatol.* **77,** 71–76 (1981).
61. M. L. Kripke, W. L. Morrison, and J. A. Parrish, Systemic suppression of contact hypersensitivity in mice by psoralen plus UVA radiation (PUVA), *J. Invest. Dermatol.* **81,** 87–92 (1983).
62. M. I. Green, M. S. Sy, M. L. Kripke, and B. Benacerraf, Impairment of antigen-presenting cell function by ultraviolet radiation, *Proc. Natl. Acad. Sci. USA* **76,** 6592–6595 (1979).
63. F. P. Noonan, M. L. Kripke, G. M. Pederson, and M. I. Green, Suppression of contact hypersensitivity in mice by ultraviolet radiation is associated with defective antigen presentation, *Immunology* **43,** 527–533 (1981).
64. W. Aberer, G. Shuler, G. Stingl, H. Honigsman, and K. Wolff, Ultraviolet light depletes surface markers of Langerhans cells, *J. Invest. Dermatol.* **76,** 202–210 (1981).
65. G. A. Elmets and P. R. Bergstresser, Ultraviolet radiation effects on immune processes, *Photochem. Photobiol.* **36,** 715–719 (1982).
66. K. Danno, K. Toda, and T. Horio, The effect of 8-methoxypsoralen plus long-wave ultraviolet (PUVA) radiation on mast cells: PUVA suppresses degranulation of mouse skin mast cells induced by compound 48/80 on concanavalin A, *J. Invest. Dermatol.* **85,** 110–114 (1985).
67. D. J. Cripps, S. Horowitz, and R. Hong, Spectrum of ultraviolet radiation on human B and T lymphocyte viability, *Clin. Exp. Dermatol.* **3,** 43–50 (1978).
68. S. Lerman, Human UV radiation cataracts, *Ophthalmic Res.* **12,** 303–314 (1980).
69. E. A. Balazs, T. C. Laurent, A. F. Howe, and L. Varga, Irradiation of mucopolysaccharides with ultraviolet light and electrons, *Radiation Res.* **11,** 149–164 (1959).
70. S. Lerman, K. Gardiner, J. Megew, and R. Borkman, The prevention of direct and photosensitized UV radiation damage to the occular lens, *Ophthalmic Res.* **13,** 284–292 (1981).

7

Circadian Rhythms

7.1. INTRODUCTION[1–5]

A basic concept in biology is that organisms are adapted to their environment. However, the environment is a constantly changing and highly variable milieu that exhibits dramatic fluctuations in such factors as light and temperature. An organism apparently well adapted to the environment at one time may be poorly adapted only a short time later if it cannot modify its physiology or behavior. Some of the environmental changes are quite unpredictable, such as those associated with day-to-day changes in the weather and general atmospheric conditions. However, a number of environmental changes are highly predictable because they result from specific planetary movements that exhibit precise periodicities. For example, the daily day/night cycle is caused by the earth's rotation about its axis, tidal and lunar cycles depend on the revolution of the moon around the earth, and annual cycles reflect the revolution of the earth about the sun.

Jerry F. Feldman • Thimann Laboratories, University of California, Santa Cruz, Santa Cruz, California 95064.

Organisms that can adjust to changes in the environment are likely to exhibit a greater degree of adaptiveness than those that cannot. For the unpredictable kinds of changes, an organism needs systems that respond directly to environmental changes. However, for the periodic and predictable fluctuations, specific mechanisms have evolved that generate endogenous biological oscillations whose periodicities correspond directly with certain environmental periodicities. These internal oscillations produce rhythmicities in the organism that are generated by negative feedback systems wholly within the organism, and thus do not depend directly on the fluctuations in the environment for their rhythmicity. However, these systems use the periodic information from the environment to couple the biological oscillation to the environmental cycle.

Thus, biological rhythms with daily, tidal, lunar, and annual periodicities have been identified in nearly all eucaryotic organisms, including plants, animals, and microorganisms, that affect a remarkable diversity of functions, including sleep/activity patterns, hormone levels, metabolic rate and body temperature in animals, photosynthesis, leaf position and sap exudation in plants, and cell division, phototaxis, photosynthesis, and bioluminescence in microorganisms. Among the rhythms that have been studied, those analyzed in most detail are daily rhythms known as circadian rhythms. Circadian rhythms are those daily rhythms that are the result of an endogenous timer, or circadian clock, that oscillates with a periodicity of about one day and that can be synchronized, or entrained, to the daily environmental cycle, using primarily information from the daily light–dark cycle (or sometimes the daily temperature cycle). In this chapter we will examine the general nature and properties of circadian rhythms and provide examples in different organisms. We will examine what is known about the physiological and biochemical mechanisms of the rhythmicity and the kinds of experimental approaches that are being used to study this question. Finally, we will discuss some special and unique questions that arise in the consideration of circadian rhythms in humans.

7.2. GENERAL FEATURES OF CIRCADIAN RHYTHMS[1–5]

As stated above, circadian rhythms have been the most intensively studied class of biological rhythms and exhibit a number of unique properties that distinguish them from other kinds of daily rhythms.

7.2.1. Entrainment to Artificial Light and Temperature Cycles

Circadian rhythms can be synchronized (entrained) to artificial light–dark cycles or temperature cycles that are either exactly 24 h in length or close to 24 h. However, there are rather severe limits to which the biological rhythm can be forced away from its "natural" 24-h periodicity. In some organisms, such as invertebrates, entrainment to cycles as short as 19 h or as long as 29 h can be achieved, but in mammals this range is much narrower and is rarely wider than about 22–26 h.

In most cases the environmental factor that serves as the primary entraining agent is the light–dark cycle (Fig. 7-1). It is because of the importance of the light–dark cycles and the sensitivity of circadian clocks to both visible and ultraviolet light that photobiologists have become particularly interested in circadian clock mechanisms.

Considerable progress has been made in understanding both the conceptual strategy

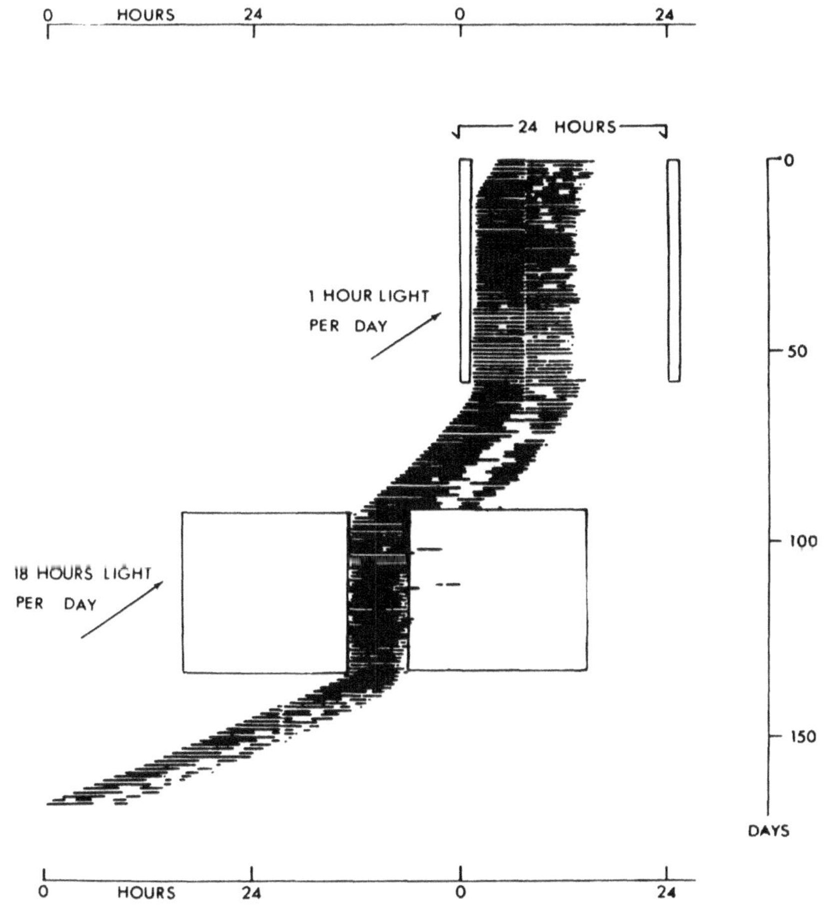

Fig. 7-1. The circadian rhythm of locomotry activity (measured on a running wheel) of the field mouse *Permoyscus maniculatus* under free-running and entrained conditions. A 1:23 light–dark cycle was imposed from day 0 to day 59. The rhythm was entrained by day 6. From day 60 to day 92, the rhythm free-ran in constant darkness with a period length slightly less than 24. From day 93 to day 132, an 18:6 light–dark cycle was imposed, and the rhythm again entrained to the 24-h cycle. After day 132, the rhythm freeran in constant darkness, again with a period length somewhat shorter than 24 h. (From Ref. 6.)

by which light–dark cycles entrain circadian clocks and the identification and nature of the photoreceptor molecules involved in the photoreception process. Indeed, because so little is known about the molecular mechanisms of the oscillator itself (Section 7.4), an understanding of the molecular mechanisms of clock photoreception offers one of the most important approaches to studying the biochemical mechanisms of circadian timing.

7.2.2. "Free-Running" (Nonentrained) Rhythms

When organisms are maintained in constant conditions in the laboratory (usually constant temperature and constant darkness, although constant light is used with some organisms) isolated from the external environmental cycle, rhythmicity persists with a period length that is close to, but not exactly equal to, the 24-h period of the earth's rotation (Fig. 7-1). This small but significant deviation from 24 h of the "free-running" (i.e., not entrained) period length results in the biological rhythm drifting out of synchrony (out of phase) with the environmental cycle and is the strongest single piece of evidence that the rhythm is caused by an endogenous oscillation rather than simply being a response to periodic changes in the environment.

It is worth mentioning at this point the origin, or rationale, for two of the terms used here. First, "circadian", coined by Halberg in 1959, comes from the Latin words *circa*, meaning about, and *diem*, meaning day, and emphasizes the importance of the free-running period length not being exactly 24 h. Second, the term "clock" is used to describe the actual timing or oscillator mechanism because it serves a specific time-keeping function for the organism, i.e., the entraining mechanism not only matches the period length of the biological oscillator to the environmental cycle but also "phases" the biological rhythmicities so that the times of maxima and minima occur at the most "appropriate" time of day for each process. For example, organisms that show a circadian rhythm of photosynthesis have their maximum about midday and their minimum about the middle of the night. On the other hand, for rhythms of bioluminescence, the phase of the rhythm is almost completely reversed, with the maximum occurring in the middle of the night and the minimum in the middle of the day.

7.2.3. Temperature Compensation

The period length of a circadian rhythm in constant conditions (i.e., the free-running period length) is the same over a wide range of physiological temperatures. This property is known as temperature compensation and is expressed methematically by a parameter called Q_{10}, which is usually given as the rate of a process at temperature t divided by the rate at temperature $t-10$. For circadian rhythms the period length of the clock, rather than its rate, is used in this calculation, and since period length is inversely related to rate, the Q_{10} is given as the period length at temperature t divided by the period length at temperature $t + 10$. The value of the Q_{10} for circadian rhythms is usually about 0.9–1.1, indicating that the period length is essentially the same at different temperatures. This Q_{10} value contrasts sharply with the temperature dependence of most chemical, biochemical, and physiological processes, whose Q_{10} is often 2 or greater, i.e., their rate approximately doubles for each 10°C increase in temperature.

The apparent temperature independence of the circadian clock theoretically could occur if the rate-limiting step in clock timing were a physical process, such as diffusion, whose rate is proportional to the absolute temperature. Within physiological temperature ranges (approximately 288–308°K), the rate would therefore not vary significantly. On the other hand, Q_{10} values close to 1 could also be obtained by a temperature compensation mechanism, in which individual rate-limiting components of the system are temperature-dependent reactions with relatively high Q_{10} values but which interact in such a way that the rate of the final output of the system does not vary much with temperature.

Several types of temperature compensation models have been proposed for circadian clocks. In one type,[7] a primary reaction A → B interacts with a secondary reaction C → D. D serves as an inhibitor of the A → B reaction. When the temperature is raised, both reactions speed up; the increased rate of the secondary reaction results in the accumulation of a higher concentration of D. As a result, the increased inhibition by D of the primary reaction compensates for the increased rate of A → B produced by the temperature increase.

A second type of temperature compensation model[8] is based on the observation that many organisms maintain a constant fluidity of their membranes at different temperatures. This is accomplished by increasing the ratio of saturated to unsaturated fatty acids as the temperature is increased. If the rate-limiting processes of the clock were dependent on the fluidity of the membrane, the clock would operate at a similar rate at different temperatures.

Several lines of evidence tend to favor temperature compensation rather than temperature independence:

1. Q_{10} values of less than 1 have been reported in a number of circadian systems.[9] This result is difficult to explain by a temperature independence model. However, if a temperature compensation system was not absolutely precise (the compensating event did not exactly match the primary event), it would be just as plausible to have a system slightly "overcompensated" ($Q_{10} < 1$) as one slightly undercompensated ($Q_{10} > 1$).

2. Circadian rhythms can be reset, or phase-shifted, by temperature steps or temperature pulses.[9] This shows that a circadian clock can detect differences in temperature and can adapt to the new conditions rather than "ignore" them. In fact one mathematical model for temperature compensation[10] specifically predicts much of the data on phase shifting by temperature steps. The primary notion is that there is a time lag associated with the adaptation process. For example, in a temperature step up, the A → B reaction (see above) initially would run faster until the new, higher steady-state level of D is achieved. This would result in a resetting of the clock "ahead," i.e., a phase advance. On the other hand, a temperature step down would result in the primary reaction running slower until the new, lower steady-state level of D was achieved and a phase delay would result. In most cases where such experiments have been done, this is exactly the result that is seen; temperature steps up cause phase advances, and temperature steps down cause phase delays.

3. At the high and low ends of physiological temperatures for many organisms, the temperature compensation is lost and the Q_{10} rises significantly above 1. This indicates that the system can operate efficiently only within certain limits. In the models above, for example, there could be upper and lower limits for the concentration of substance D or limits to the binding of D to the enzyme that controls the primary reaction A → B, or limits to the ratio of saturated to unsaturated fatty acids that control membrane fluidity.

4. Mutants have been isolated in which the temperature compensation mechanism is either altered or entirely lost, although the clock appears to run normally or nearly normally in most other respects.[11] It is difficult to imagine how this might occur in a temperature-independent clock system, but it is quite plausible for a temperature-compensated system.

7.3. ENTRAINMENT TO ENVIRONMENTAL CYCLES[12]

The synchronization, or entrainment, of the circadian clock to the environmental cycle is a complex phenomenon whose mechanisms have not yet been elucidated. There have been two major approaches to the analysis of the mechanisms of entrainment of circadian rhythms. One approach considers the clock as a "black box" and studies the general conceptual strategies by which the light information is utilized by the circadian oscillator. In such studies, pioneered primarily by Colin Pittendrigh and his co-workers, the specific cellular or molecular details of the circadian system are not at issue; rather, what have been called the "formal properties" of the system have been examined. For example, what kinds of mathematical equations or models can be constructed to explain the behavior of the system in response to different types of light cycles and light interruption, or perturbation, experiments? Does the light cycle entrain the biological oscillator by changing the rate of the biological oscillation or by periodically resetting it once or twice each cycle?

The second type of approach is complementary to the first in that it is specifically attempting to identify the cellular and molecular components involved in the photoreception mechanism and the transducing of information from the photoreceptor to the clock oscillator. A variety of studies of this type have been carried out, some of which will be described here; others are discussed in Chapter 8, "Extraretinal Photoreception."

7.3.1. Formal Aspects of Entrainment by Light[13,14]

One of the most important conclusions of these studies has been the recognition that in a normal light–dark cycle, dawn (lights on) and dusk (lights off) provide the key signals for entrainment. These transitions result in a rapid (less than 1 h) resetting of the clock mechanism, the amount and direction being determined by the time of day at which the transition occurs (dawn or dusk) and the free-running period of the circadian oscillator. Usually, but not always, dawn produces an advance (+) phase shift while dusk produces a delay (−) phase shift. Thus, if an organism has a clock whose free-running period length is 23 h, in order to entrain to a 24-h cycle, the net amount of resetting of its clock must be −1 h each day. For an organism whose clock has a free-running period of 25 h, the signals must produce a net advance of +1 h each day.

Evidence for resetting as the primary mechanism of entrainment has come from several types of experiments. One of the first involved what are called "skeleton" photoperiods. Skeleton photoperiods mimic "normal" light–dark cycles in the following way: If a normal cycle was 10 h light, 14 h dark (LD 10 : 14), a skeleton LD 10 : 14 would be one in which a brief pulse of light was given at the beginning and end of the presumptive 10-h "day" phase (to simulate the dawn and dusk signals), but between these two brief pulses, darkness would prevail. In *Drosophila* an extensive series of experiments were carried out in which the phase of emergence time of the adult fly (eclosion), which is controlled by the circadian clock, was determined for different normal and skeleton photoperiods. Remarkably, the times of eclosion under the two light regimes, which differ dramatically in the amount of light seen by the organism, were quite similar for a large number of different photoperiodic cycles, ranging from LD 1 : 23 to LD 13 : 11. For photoperiods with longer light times, the organism always interpreted the shorter interval as the day phase, but with this adjustment, agreement between the two types of cycles extended to much longer photoperiods up to LD 23 : 1.

These results indicate that the light transition provides a signal, either a strong increase in intensity (dawn) or a strong decrease in intensity (dusk), that resets the

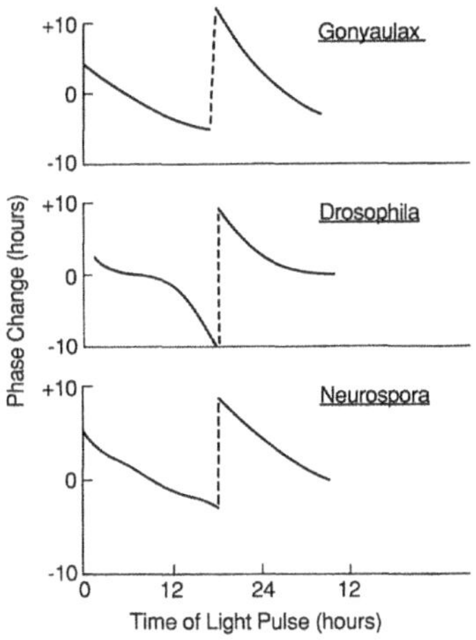

Fig. 7-2. Phase response curves from three organisms showing the phase-shifting effects of single light pulses. (abscissa) Time of day at which the light pulse was administered; (ordinate) phase shift produced by the light pulse (+ = advance; − = delay). (Redrawn from Ref. 15.)

biological clock twice a day in a manner that synchronizes it to the external environment cycle. Since the skeleton photoperiod produced entrainment in a manner similar to the normal (full) photoperiod, this hypothesis also suggests that the continued presence of light during the day phase plays only a very minimal role in the entrainment of the clock. A curious but still unexplained aspect of this model is that the direction of the light transition (on or off) is not important in conveying phase information to the clock.

Further evidence for the "resetting" role of light in entrainment has been obtained by determining the effects of single pulses of light when such pulses are given to an organism otherwise kept in constant darkness. In a typical experiment, a set of organisms (or cultures of organisms) would be subjected to a single light pulse, each culture receiving the pulse at a different phase of the cycle. The effect of the light pulse at different phases of the cycle would be determined by comparing the phase of each pulsed culture with that of a control culture that had not received any light signals during the experiment. In most cases light pulses given during the organism's late night or early day phase (organism time is called "subjective" time) produce an advance phase shift while light pulses given during the late subjective day or early subjective night produce a delay phase shift. These results can be plotted as a "phase response curve" in which the amount of the phase shift is plotted against the time in the cycle when the pulse was administered (Fig. 7-2). The data from the single pulse experiment (i.e., the phase response curve) successfully predict the behavior of the culture (i.e., the phase of the rhythm) in the skeleton photoperiods and therefore the phase of the rhythm in normal light dark cycles.

On the other hand, there are a number of experiments that indicate that the continuous action of light does have an effect, if only minor, on the behavior of the circadian clock. The best documented cases involve comparison of the period length of circadian rhythms in organisms maintained in constant darkness or at different intensities of constant light. Such experiments show that the intensity of light affects the period length, i.e., the rate at which the clock runs, with differences of several hours observed in the most extreme cases. The direction of change of period length depends on whether the animal is nocturnal or diurnal, and these relationships have been summarized in what are called

Aschoff's rules, after their discoverer, Professor Jurgen Aschoff. Aschoff's rules are discussed in more detail in Chapter 8.

7.3.2. Nature of the Photoreceptor

In Chapter 8 there is a discussion of photoreceptor pigments in several animal and plant systems, and the evidence for the role of rhodopsin, or rhodopsinlike pigments, for circadian clocks in a number of animals is presented. In these cases, identification of the clock photoreceptor has rested only on action spectrum data, in which the efficacy of different wavelengths of light in resetting the clock has been measured. Because of the inherent difficulties of measuring phase shifting with great accuracy in many organisms, such experiments at best provide a general indication of the type of molecule involved and frequently are unable to distinguish among several classes of molecules.

However, in the filamentous fungus *Neurospora crassa,* clock photoreceptor studies have been carried out in much more detail with the use of mutant strains deficient in various types of pigment molecules. *Neurospora* has a circadian rhythm of asexual spore formation (conidiation), which can be easily assayed in cultures growing across an agar medium (Fig. 7-3). Because the culture leaves a permanent record of the position, and hence the time, of formation of the conidia, no automated equipment is needed to monitor the culture continuously, and as a result a large number of cultures can be handled rapidly and inexpensively.[17]

Action spectra for the *Neurospora* clock have indicated that the primary photoreceptor belongs to a class of photoreceptors widespread in higher plants and microorganisms known as the "blue light" photoreceptor. In general, action spectra alone for blue light photoreceptors have not been able to identify a unique class of molecules as the photoreceptor, and in particular to distinguish whether the molecule is a carotenoid or a flavin. Experiments using *Neurospora* mutants point strongly to the involvement of a flavin in clock photoreception.

1. An action spectrum for the suppression of the conidiation rhythm by constant light indicated effective wavelengths of 350–520 nm, with a maximum effect at 465 nm, and other peaks at 485, 415, and 375 nm (Fig. 7-4). An albino strain carrying mutations at two different steps in carotenoid biosynthesis and containing no detectable carotenoids showed no alteration in this light response.[19]

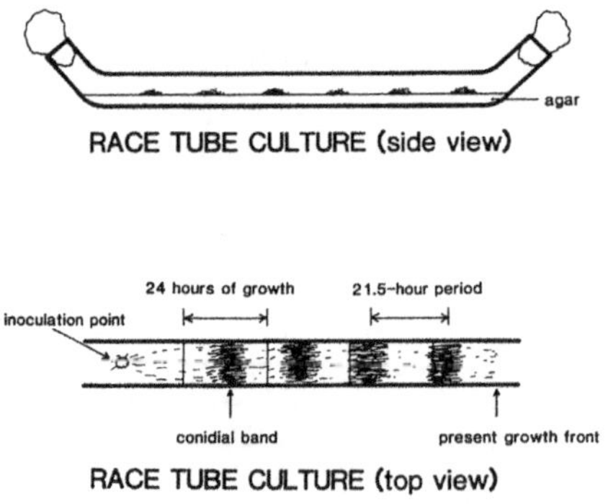

Fig. 7-3. The circadian rhythm of conidiation in the filamentous fungus *Neurospora crassa*. The culture grew from left to right in the diagrams. The "decision" to form conidia is made just behind the growth front at times of day determined by the circadian clock. Thus, the position of the conidial bands reflects the position of the growth front at the "decision time" and can therefore be used directly to determine the time of day at which the conidia were formed. (From Ref. 16.)

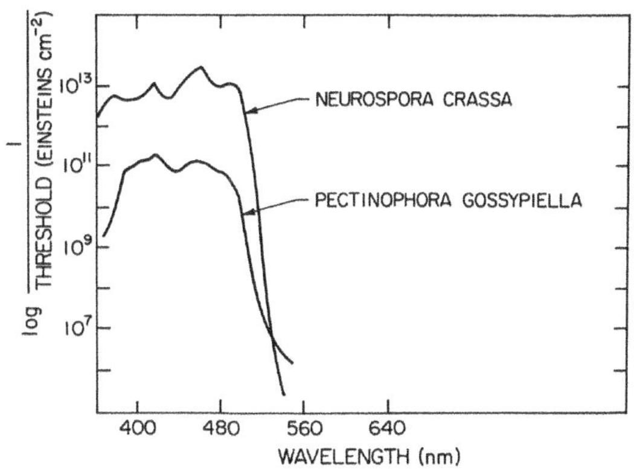

Fig. 7-4. Action spectra for setting the circadian rhythm of conidiation in the filamentous fungus *Neurospora crassa* and the egg-hatching rhythm of the moth *Pectinophora gossypiella*. (From Ref. 15, data for *Pectinophora* from Ref. 18, for *Neurospora* from Ref. 19.)

2. A plasma membrane fraction of *Neurospora* has been isolated that exhibits a flavin-mediated blue light-induced absorbance change indicative of the photoreduction of a b-type cytochrome. Similar blue light-induced absorbance changes have recently been found in other membrane fractions of *Neurospora*, including the endoplasmic reticulum and the mitochondria. This *in vitro* system has been correlated with *in vivo* clock photoreception in several ways. First, in the respiratory mutant *poky*, the threshold intensity for inhibition of banding in continuous light was about 50-fold higher than in the wild-type strain. Correlated with this reduction in light sensitivity was a reduction in the amount of nonmitochondrial cytochrome to 16% of that in the wild-type strain. The absence of a direct quantitative correlation between light sensitivity and cytochrome levels in these experiments could have been due to the use of different growth conditions under which the two measurements were made, since *poky* undergoes significant biochemical changes during an adaptation process.[20]

Second, addition of the photodynamically active dye methylene blue to the *in vitro* plasma membrane preparation resulted in the sensitization of cytochrome reduction to red light. In a similar way, addition of methylene blue to the medium of a culture growing in race tubes resulted in the sensitization of its clock to red light for both the suppression by constant light and phase shifting by light pulses.[20]

Evidence for the involvement of the flavin component of the flavin/b-cytochrome complex in the clock photoreceptor has also come from experiments with mutants. Two riboflavin auxotrophs, *rib-1* and *rib-2*, were grown under conditions where riboflavin was limiting for growth. In both mutants a reduction in the levels of FAD and FMN in the mycelia correlated with a reduction in the sensitivity of the clock to light for both the phase-shifting and damping responses. Furthermore, addition of exogenous riboflavin analogs to the medium restored light sensitivity to these strains.[21]

7.4. CELLULAR MECHANISMS OF CIRCADIAN RHYTHMS[22]

In general, two types of approaches have been used to try to identify the biochemical components of the clock oscillator system. In one approach circadian rhythms in biochemical parameters have been identified and their role in the clock analyzed. In the second approach, specific drugs or inhibitors have been tested to determine which ones can perturb circadian timing systems and thus implicate the target of the inhibitor in the clock mechanism. These approaches have some shortcomings in experimental design and interpretation that have limited their contributions to our understanding of clock mechanisms, but some important and interesting results have nevertheless emerged from them and are summarized below.

7.4.1. Biochemical Rhythms[2,17]

Biochemical rhythms have been identified in a wide variety of parameters in many different organisms. For example, in microorganisms there are rhythms in DNA, RNA, and protein synthesis, in both the substrate and enzyme associated with bioluminescence, in photosynthesis, in pathways of intermediary metabolism including glycolysis, the hexose monophosphate shunt, and the Krebs cycle, in the levels of ATP and cofactors such as NAD and NADP, and in fatty acids. A similar diversity of rhythms in higher organisms also exists. In higher plants, for example, there are rhythms in dark CO_2 fixation and protein synthesis, while in mammals there are rhythms in the levels of coenzyme A, glycogen, and succinic dehydrogenase, in tyrosine aminotransferase activity in the liver, and in tyrosine levels in the blood. In fact, this is only a small sample of the various parameters that show circadian oscillations, and there are now documented literally hundreds of enzymes, subtrates, cofactors, hormones, and other substances that show circadian oscillations in various organisms.

However, the existence of a rhythm in a particular biochemical parameter does not in itself implicate that parameter as part of the clock mechanism. An important experiment involving the phtotsynthesis rhythm in the marine dinoflagellate *Gonyaulax* illustrates this point.[23] In this experiment, photosynthesis was blocked by the addition of dichlorophenylidimethylurea (DCMU), but when the inhibitor was washed out, photosynthesis resumed in a rhythmic manner, and the phase of the rhythm was identical with that of a control culture that had not be exposed to DCMU. This result was obtained independent of the time of addition of DCMU, time of its removal, or duration of the pulse, and indicates that although the process exhibiting the overt biochemical rhythm was inhibited, the timing mechanism underlying the rhythmicity persisted throughout the experiment. Therefore, the biochemical events of photosynthesis itself are not part of the timing mechanism. As a result of this experiment and others like it, a distinction has been made between the "hands of the clock" (e.g., photosynthesis) and the clock mechanism. Similar results have been obtained with nearly all other biochemical rhythms, and as a result it is now generally accepted that the existence of a biochemical rhythm alone tells little about the mechanism of circadian timing.

On the other hand, several interesting studies have been carried out using biochemical rhythms as starting points for analyzing clock mechanisms. The rationale of such studies has been to identify the immediate cause of the biochemical fluctuation, and then to determine the cause of that rhythm and so forth, ultimately "working back" from the external hand of the clock to the actual mechanism. Although this has not yet been fully accomplished in any organism, the initial steps have been taken in two systems, photosynthesis and bioluminescence. In both *Gonyaulax* and *Euglena* circadian rhythms in photosynthesis have been traced to changes in the light reactions.[24,25] More specific identification of the specific components of the light reactions involved in photosynthetic rhythms awaits further study.

In the case of bioluminescence, rhythmicity has been found in both the level of substrate (luciferin) and the activity of the enzyme involved (luciferase). Recent studies using an antibody against purified luciferase have shown that the rhythm in luciferase activity is the result in changes in the amount of enzyme present in the cell at different times of the day, with low levels during the day phase when luminescence is at its minimum and high levels during the night when it is at its maximum. Thus, the luciferase rhythm is the result of rhythms in the synthesis and/or degradation of the enzyme.[26]

7.4.2. Studies Using Inhibitors[22]

An alternative biochemical approach has been to apply a variety of drugs or inhibitors to organisms expressing a circadian rhythm in constant conditions to determine whether the drug can perturb the circadian timing system. In such studies one asks whether the continued presence of the drug can change the free-running period length or whether a pulse of the drug can reset (phase-shift) the clock. In such experiments one hopes to be able to identify the known cellular target of the drugs as part of, or closely coupled to, the clock mechanism.

A wide variety of inhibitors has now been shown to perturb the clock in different organisms, although results vary widely from one system to another. The strongest generalization is that inhibitors of cytoplasmic (nonorganelle) protein synthesis alter circadian timing. Drugs such as cycloheximide and anisomycin induce a change in either the period length (continuous application) or the phase (pulse) of the clock in essentially all organisms tested, including *Euglena, Neurospora, Gonyaulax,* the marine mollusc *Aplysia,* and *Acetabularia.* Inhibitors of organelle protein synthesis, such as chloramphenicol and streptomycin, do not produce these effects. This result may seem to suggest that periodic synthesis of specific proteins is part of the basic oscillatory cycle. However, recent data with clock mutants of *Neurospora* (Section 7.5) favor the notion that alteration of the clock by inhibition of protein synthesis occurs because of the existence of proteins with high turnover rates that must be replenished at least once per cycle before they are needed. None of these experiments, however, suggest the identity or even the general nature of the kinds of proteins involved.

A second class of inhibitors that induces phase shifts in a number of organisms are those that affect respiratory metabolism. Drugs such as azide, cyanide, dinitrophenol, and DCCD perturb the clock of certain organisms (e.g., *Neurospora*) but not others (e.g., *Gonyaulax*). Again, in *Neurospora* one might have interpreted this result as indicating a direct involvement of some mitochondrial component or the process of oxidative phosphorylation in the generation of circadian oscillations. However, the level and duration of a cyanide pulse needed to produce a significant phase shift is about 10-fold higher than that needed to completely inhibit oxygen consumption or to significantly deplete the ATP pools. This indicates that components of oxidative phosphorylation themselves are probably not components of the clock but rather that some critical step in the clock is dependent on this process. Unfortunately, since *Neurospora* is an obligate aerobe, nearly every metabolic process in the cell shows such a dependency, and few clues as to the nature of the clock components are obtained from this result.

Finally, drugs affecting membrane structure and/or function, including various ionophores that alter ion transport, have been shown to perturb circadian timing. Anesthetics that disrupt membrane structure induce phase shifts in *Gonyaulax,* as does the K^+ ionophore valinomycin, while in *Neurospora* the Ca^{2+} ionophore A23187 also produces phase shifts. The *Neurospora* result is especially interesting since the direction of the phase shift (advance or delay) at certain times in the cycle can be correlated with whether Ca^{2+} is entering the cell or leaving it. This is the only case of this type presently known and is the kind of result expected if the oscillations in the concentration of Ca^{2+} were part of the clock mechanism. Additional data suggesting a rôle for Ca^{2+} come from phase shifts produced by antagonists of the calcium activator protein calmodulin, such as trifluoropyrizine, in both *Euglena* and *Neurospora.*[27,28]

Finally, there have been a number of studies in which inhibitors of cyclic AMP metabolism alter circadian timing. Once again, however, there is no evidence indicating that this molecule is directly involved in circadian timing.

In many of the studies reported here, investigators have proposed biochemical models for the clock based on the specific kinds of drug inhibition under study. For example, effects of protein synthesis inhibitors have led to models involving periodic synthesis and degradation of (unknown) clock proteins,[29,30] results with respiratory inhibitors have suggested models involving periodic coupling and uncoupling of oxidative phosphorylation,[31] and phase shifting by ionophores has been the basis of models involving periodic accumulation and depletion of ions as a result of rhythmic changes in membrane structure or function.[8,32] Further studies are necessary to distinguish among these hypotheses.

7.5. GENETIC APPROACHES TO CIRCADIAN RHYTHMS[33]

Because of some of the inherent difficulties associated with the biochemical approaches described above, studies have been initiated in several organisms to use genetic analysis for studying the clock mechanisms. Genetic analysis and the use of mutants have been important tools for dissecting complex biological phenomena, including biochemical pathways, gene regulation, bacteriophage assembly, cell cycles, and behavioral responses. Genetic analysis also opens the door to the powerful techniques of recombinant DNA research.

Genetic approaches to clocks have involved the isolation of clock mutants in several organisms, including *Drosophila, Neurospora,* and *Chlamydomonas.* Most of these mutants have alterations in the free-running period length of their circadian clock. In other studies, mutants with specific biochemical lesions have been isolated and their clock assayed to determine the effect, if any, of the lesion on clock function. Examples of this type of approach have been presented above in Section 7.3.2 in our discussion of the nature of the clock photoreceptor in *Neurospora.*

7.5.1. Genetic Analysis of Clock Mutants

In *Drosophila melanogaster,* approximately 15 mutants have been isolated that alter the free-running period length of its circadian clock.[34,35] In these studies both the rhythm of eclosion (adult emergence) and the rhythm of adult locomotory behavior have been studied, and in all cases both rhythms are altered in similar, although not identical, ways. The most interesting of these includes a set of five mutants that map to a single genetic locus on the X chromosome called *per*. Mutants at the *per* locus exhibit one of three phenotypes: short period, called per^{s} (19-h period), long period, called per^{l} (29-h period), or arrhythmic, called per^{0} (no rhythm).

The cells or tissues responsible for the behavior controlled by the *per* gene have been identified by a technique known as fate mapping. This procedure involves the construction of a fly that is a genetic mosaic, in which some of its cells are mutant while others are wild type. One can identify the cells carrying the mutant *per* gene with the use of visible markers, such as those affecting body color, or histological markers such as those staining for a particular enzyme, that are linked to the *per* gene. Those cells in the fly that are mutant for the visible marker will also be mutant for the clock gene, while those that are wild type for the visible marker will also be wild type for the clock gene. If a given fly exhibits a mutant clock phenotype, one can say that among those cells carrying the mutant allele are the ones that control the clock function. By examining a sufficient number of flies, it has been possible to show that clock periodicity controlled by the *per* gene is determined by brain tissue in the head of the fly. In fact, when one side of the head is mutant and the other side wild type, the fly behaves as if it had two clocks running simultaneously. Similar results indicating a duplication of clock organization on two sides of the brain have been obtained from ablation experiments in other invertebrates, such as the cockroach and cricket (Section 7.6).[36]

Additional evidence that the brain exerts control over clock periodicity in *Drosophila* also has come from experiments using mutants. Transplantation of a brain from a short period per^{s} mutant into the abdomen of an arrhythmic per^{0} host resulted in the expression of a short-period rhythm in the host.[37] This experiment also provided evidence that clock information is communicated from the brain to the rest of the animal through the secretion of a diffusible hormone.

A surprising result has been the finding that mutations at the *per* locus also altered the periodicity of a high-frequency rhythm (period length about 1 min) in the interval between pulses of the *Drosophila* courtship song. Furthermore, the *per* mutations altered this high frequency rhythm in the same manner (i.e, shorter, longer, or arrhythmic) as they did the circadian rhythm. Apparently, the *per* gene product plays a significant role in both oscillations. However, fate mapping of the effect of the *per* mutations on the courtship song rhythm has shown that the focus of the mutation is in cells of thoracic ganglia rather than the brain. Thus, different neural cell types are apparently differentiated in ways to produce different types of oscillations, but at least one gene product is common to both.[38]

In *Neurospora,* approximately 15 clock mutants have been isolated that alter the free-running period of the *Neurospora* clock. As in *Drosophila,* half of the *Neurospora* clock mutants map to a single genetic locus called *frq*. Most mutants at the *frq* locus have either short period or long period, but one mutant has lost its temperature compensation (i.e., it has a temperature-dependent circadian clock). Again, several different rhythms in *Neurospora* (DNA synthesis, RNA synthesis, CO_2 fixation, and conidiation) are all affected in the same manner by each of the mutations.[33]

An important genetic result from the *Neurospora* studies has come from the analysis of heterokaryons, strains that contain a mixture of different nuclei, some of which carry a mutant *frq* allele and some of which carry the wild-type *frq*$^+$ allele. Heterokaryons with different ratios of the two types of nuclei showed that there was a "gene dosage" effect of the *frq* gene, i.e., the change in period length was proportional to the fraction of mutant nuclei in the heterokaryon. This result suggests that the rate at which the *Neurospora* clock runs is proportional to the amount of *frq* gene product present in the cells and argues that the *frq* gene product plays an important role in the organization of the *Neurospora* clock.[39]

7.5.2. Molecular Analysis of Clock Mutants[40–44]

An important feature of genetic analysis is that it points the direction for the use of recombinant DNA techniques to clone clock genes. The cloning of such genes offers an important starting point for the eventual identification of the biochemical nature and function of the gene products of such clock genes. This approach should inevitably lead to the understanding of the role of specific biochemical events in the function of the circadian clock.

The most significant progress along these lines has been made with the *per* locus of *Drosophila*. This gene has been cloned [40,41] and its effect as a clock gene has been confirmed by transforming cloned wild-type DNA into an arrhythmic *per*^{0} recipient. [42]. Such transformed flies show restored rhythmicity, often with the normal 24-h period length. DNA sequencing of the *per* gene, as well as isolation of the *per* protein with antibodies, suggests that it is a proteoglycan. [43,44]

7.6. CIRCADIAN ORGANIZATION IN MULTICELLULAR ORGANISMS

7.6.1. Localization of Circadian Clocks in Invertebrates[45–52]

Important questions in multicellular organisms are whether each individual cell has a circadian clock, and whether there exists a master clock or clocks that control and integrate circadian rhythmicity for the entire organism. The first of these questions cannot be answered at the present time, although efforts to examine circadian rhythmicity in cell culture are proceeding along several lines.[45] For example, in the marine molluscs *Aplysia* and *Bulla,* circadian rhythms in the rate of spontaneous firing of the optic nerve have been studied in considerable detail.[46–48] When the eye is dissected along with the nerve and maintained in organ culture, the rhythm of spontaneous firing persists and can be entrained to 24-h light–dark cycles *in vitro* (Fig. 7-5). This shows that the eye and its attached optic nerve have all of the organization and machinery necessary to generate a circadian oscillator. By removing successively larger parts of the distal portion of the eye from the culture preparation, it has been shown that as few as six cells are necessary to maintain circadian output, and it seems likely that circadian clock organization exists within the individual cells of the molluscan eye.

Similar results have been obtained with cultured cells from avian pineal glands, in which rhythms in cyclic nucleotide metabolism have been observed and entrained by external light–dark cycles.[50]

On the other hand, it also seems likely that one or more master oscillators exist that control the circadian behavior of the organism and serve to couple individual rhythms in

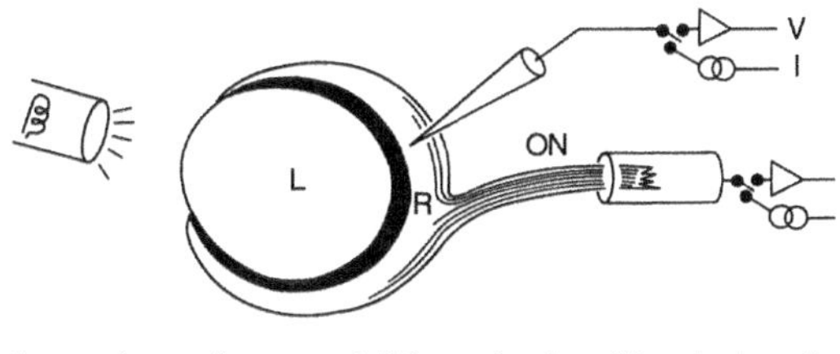

Fig. 7-5. Diagram of the *Aplysia* eye and the arrangement for recording and stimulating individual cells or the optic nerve (ON). A tubing electrode on the optic nerve can be used either for recording or for stimulation. Intracellular electrodes can be used to record (V) or stimulate (I) a single cell. The parts of the eye are the lens (L) and a pigmented retina (R). (Redrawn from Ref. 49.)

the organism to each other as well as to couple the rhythmicity of the organism to the external environment. As mentioned in Section 7.5.1, the technique of fate mapping in *Drosophila* using the clock mutants at the *per* gene has shown that neural tissue on both sides of the brain of the fly are the site at which the *per* gene product exerts its control over circadian periodicity, and that transplanting the brain from a rhythmic fly into an arrhythmic one restores rhythmicity in the arrhythmic mutant. These experiments indicate that the brain of the fly possesses a master oscillator that communicates its rhythmic information to the rest of the animal by the secretion of a neurohormone or other diffusible product.

Surgical ablation experiments have been carried out in great detail with the brain of the cockroach and have shown clearly that the optic lobes of the brain of this animal are essential for the expression of the circadian rhythm of locomotor activity, and are therefore the site of a master oscillator. Once again, the lobes on both sides of the brain seem to possess oscillator characteristics, but in contrast to the behavior of the *Drosophila* brain, the oscillators on the two sides seem much more tightly coupled than in the fly. Transplantation of the optic lobes of a rhythmic animal to the brain of an arrhythmic animal that had its optic lobes ablated restores rhythmicity.(51,52)

7.6.2. Localization of Circadian Clocks in Vertebrates(53–57)

Clock localization experiments in vertebrates have depended mostly on a variety of surgical procedures, since few clock mutants exist among any of these organisms. Early studies involving removal of a wide variety of endocrine and exocrine glands, including the adrenal, pituitary, pancreas, and thyroid, failed to eliminate rhythmicity in a number of different mammals. In recent years, however, these studies have focused on two structures, the pineal gland and the suprachiasmatic nucleus region of the hypothalamus, since their removal has abolished circadian rhythmicity in a number of different organisms.

In house sparrows, the circadian rhythms of both perching activity and body temperature are abolished by the removal of the pineal gland. The pineal exhibits a circadian rhythm of melatonin secretion, and the continuous exposure to moderate to high levels of exogenous melatonin causes a significant disruption of the perch hopping rhythm. A number of additional experiments support the general notion that the pineal possesses an endogenous oscillator that controls many of the circadian rhythms of the bird by periodic melatonin secretion: (1) Cutting the sympathetic innervation to the pineal from the superior cervical ganglion fails to abolish rhythmicity. This indicates that pineal rhythmicity is not dependent on periodic neural input from other parts of the central nervous system. (2)

If a bird is first made arrhythmic by pinealectomy, rhythmicity can be restored by transplantation of a pineal from a rhythmic bird to the anterior chamber of the eye in the pinealectomized bird. Furthermore, the phase of the newly restored rhythm in the host is similar, if not identical, to the phase of the donor pineal. This result not only supports the notion of an autonomous oscillator in the pineal but also that it exerts its control of rhythmicity by secretion of a diffusible substance, since rhythmicity returns to the host animal long before any neural connections could have been regenerated (see Chapter 8).

Further analysis of the pineal oscillator has been pursued with *in vitro* techniques, by which the pineal has been maintained in organ culture for several weeks. Under these conditions the pineal continues to exhibit a circadian rhythm of melatonin output under constant light and temperature conditions, and can be entrained to external light–dark cycles. The rhythm in melatonin secretion has been traced to a rhythm in melantonin synthesis, which itself is controlled by a rhythm in the activity of the enzyme *N*-acetyltransferase, the rate-limiting step in melatonin biosynthesis. The rhythm in *N*-acetyltransferase activity has also been detected in dispersed pineal cell cultures. It seems likely that this culture system will provide an important tool for future studies on the mechanisms of pineal circadian rhythmicity.(49)

In mammals the role of the pineal appears to be somewhat different than that in birds.(53) Although the gland exhibits a circadian rhythm in *N*-acetyltransferase activity, the rhythm does not persist when innervation to the pineal is destroyed. On the other hand, ablation of the suprachiasmatic nucleus (SCN) of the hypothalamus eliminates free-running circadian rhythmicity in the rat, including the rhythm of melatonin secretion from the pineal (Fig. 7–6).(54,55) Furthermore, the SCN itself shows a circadian rhythm of electrical activity which persists when the region is physically isolated from the rest of the brain by surgical techniques.(57)

An important result further implicating the SCN as a controlling site for circadian rhythmicity was the finding of the retinohypothalamic nerve tract (RHT), a direct innervation from the retina of the eye to the hypothalamus. It has been suggested that the transmission of light information from the environment, received by the eye, to the clock in the SCN occurs along this pathway. Perhaps most impressive of all has been the finding that transplantation of a donor SCN to animals that were made arrhythmic by ablation of their own SCN can restore rhythmicity to the host animals.(54)

7.7. Circadian Rhythms in Humans(4,58–62)

Anyone who has taken a commercial airline flight across several time zones has personally experienced the effect of circadian rhythmicity on human physiology and

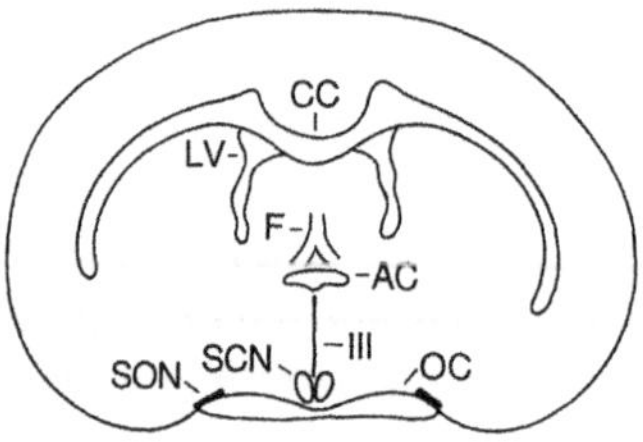

Fig. 7-6. Diagram of a coronal section through the brain of a golden hamster. SCN = suprachiasmatic nucleus; SON = supraoptic nucleus; OC = optic chiasm; III = third ventricle; AC = anterior commissure; F = fornix; LV = lateral ventricle; CC = corpus callosum. (Redrawn from Ref. 56.)

behavior. The general malaise or discomfort that is encountered is the result of a disruption in one's widespread patterns of circadian rhythmicity. In humans, as in other organisms, the circadian clock controls a vast array of biochemical, cellular, and physiological parameters that oscillate in precise phase relationships to both the external environment and to each other. Circadian rhythmicity exists in such functions as sleep/wake cycles, body temperature, and metabolic rate. Urine excretion of water, Na^+, K^+, Ca^{2+}, Mg^{2+}, and Cl^- show circadian fluctuations, while circadian rhythms are found in the levels of a many hormones, including cortisol, growth hormone, aldosterone, prolactin, testosterone, thyrotropin, luteinizing hormone (LH), and follicle-stimulating hormone (FSH).

Each of these parameters oscillates with a well-defined peak and trough, and the relationship of the oscillation among the different rhythms is critical to the normal functioning of the individual. For example, among the urinary excretion rhythms, the water, K^+, Ca^{2+}, and Mg^{2+} rhythms have maximum around midday, while the Na^+ and Cl^- rhythms have a peak at dusk. When individuals travel to a new time zone, their clocks must reset to the new environmental time, and this resetting can take 2 days to 2 weeks depending on the amount and direction (advance or delay) of the phase shift. For reasons that are completely unknown, advance phase shifts take longer to complete than delays. During the period of adjustment, the rhythms do not reset in a totally coordinated and synchronized manner. Indeed, the kinetics of resetting are quite different for different rhythms. so that certain parameters (e.g., sleep/wake cycles) may reset fairly quickly, while others, such as K^+ excretion, may reset very slowly. As a consequence, during this period the different rhythms do not maintain their normal phase relationship, and a phenomenon called *internal desynchronization* results. It is believed that this internal desynchronization is what is responsible for the feelings of ill health and poor performance exhibited by nearly everyone who must reset their clock. This phenomenon has become known as "jet lag." Although there have been a number of suggestions regarding how to avoid or minimize jet lag, at the moment the only way individuals can have significant impact on this condition when arriving at their destination is to "preadapt" by gradually resetting their clock several days prior to departure.[(58)]

Internal desynchronization has implications for other aspects of human physiology. Shift workers are a particular target of the ill effects of resetting human circadian clocks, since each change from one 8-h shift to a different one, if accompanied by a change in sleep and eating habits, is the equivalent of a flight across eight time zones and leads to a general physiological disruption for as much as 10 days to 2 weeks. It is clear that if one changes shifts frequently, one spends a considerable portion of one's life under conditions of internal desynchronization. There are also suggestions, although no real proof as yet, that animals whose cycles are continually or frequently shifted exhibit a greater tendency to develop a wide range of systemic diseases such as cancer and heart disease, and show faster rates of aging. Internal desynchronization is also a problem for individuals who have unusually long periods of work, followed by long periods of rest. These include flight crews, especially those on long international flights, who may be required to perform their functions for 20 or more hours at a time. The fact that they are then rested for a similar long period does not compensate for the disruption in their circadian physiology, and the consequent implications for their health and performance abilities.

At the behavioral level there is considerable evidence that both mental and physical

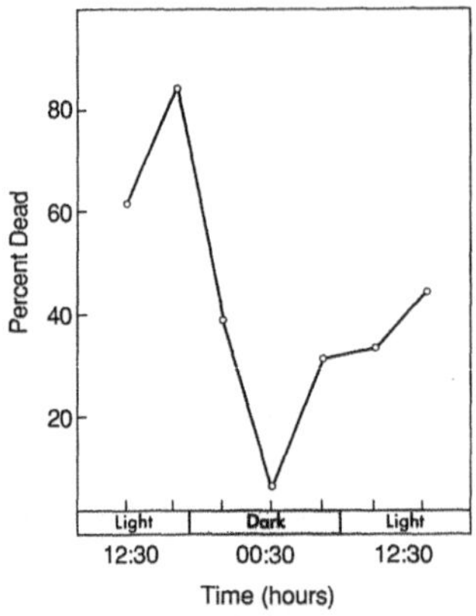

Fig. 7-7. Circadian rhythm in the sensitivity of mice to endotoxin from the bacterium *E. coli*. The same dose of endotoxin was administered at each time of day. (Redrawn from Ref. 60.)

performance shows significant circadian fluctuations, and that an individual's ability to carry out a variety of simple and complex tasks varies dramatically during the day. A number of serious disasters that have been traced to human error, including the nuclear crisis at Three Mile Island, have occurred in the early morning hours when individuals are at their lowest levels of performance ability.[58,59]

Finally, there is much documentation of circadian rhythms in the sensitivity and responsiveness to a variety of drugs or toxic agents, such as anesthetics, sedatives, antibiotics, and pain killers (Fig. 7-7). In some cases, as with anesthetics, both the desired positive effects of the drug and unwanted side-effects show rhythms, and often these two rhythms are out of phase with each other. If these rhythms are not taken into account in terms of the timing of the administration of the drug, one might be given an anesthetic at a time when a high dose is needed to sedate the individual but at the same time that maximum side-effects are seen. On the other hand, if the time of administration is altered to take into account these circadian rhythmicities, a much lower dose of drug can be applied simultaneously with the phase of least side-effects.[61]

Rhythms in drug sensitivity have been important in a new area of medicine that has developed in recent years called chronopharmacology. Most notable have been studies involving cancer chemotherapy, in which anticancer drugs such as cytosine arabinoside are administered at different levels at different times of the day. Since most of these drugs operate by inhibiting DNA synthesis in growing malignant tissue, attempts are made to take advantage of phase differences in the mitotic cycle between malignant and normal tissues. Thus, administration of an anticancer drug at relatively higher levels during the S phase of the malignant tissue and lower levels during the S phase of normal tissues allows a maximum effectiveness of inhibition of the cancer cells with a minimum inhibitory effect on the normal cells. Significant increases in survival rate using this type of approach have been reported for a number of different types of cancer.[62] In addition to rhythms in DNA synthesis, the physiological or biochemical basis for other rhythms in drug sensitivity has been traced to rhythms in liver detoxifying enzymes, in urinary excretion of the drugs, in uptake and utilization of the drug, and in blood volume that affects plasma concentration.[62]

In general the significance and impact of circadian rhythmicity in human physiology and pathology has not received widespread attention in the medical profession. Obviously, such factors significantly complicate both diagnosis and treatment of a large number of disorders as well as influencing normal patterns of behavior. However, the existing data are clearly sufficient to warrant more serious concern in all of these areas, and it can only be hoped that additional studies with more convincing and unequivocal data will gain the broader attention and concern they deserve.

7.8. CONCLUSIONS

Nearly all eucaryotic organisms exhibit daily rhythms in a wide variety of biochemical, physiological, and behavioral characteristics. These rhythms are controlled by an endogenous timing mechanism called a circadian clock that serves to couple the biological rhythms with the environmental day/night cycle. The circadian clock uses the light–dark cycle as the primary source of information concerning environmental time, and the clock photoreceptor molecules and structures have been identified in several organisms.

In order to keep proper time, the circadian clock has evolved a mechanism of temperature compensation whereby the rate of the clock remains the same over a wide range of physiological temperatures. The mechanism of temperature compensation and the mechanism of the circadian oscillator itself are not known. Several approaches to this question are being used in animals, plants, and microorganisms. Biochemical rhythms have been identified and their basis analyzed; drugs that perturb circadian timing have been used to focus attention on specific molecular events in the cell; and clock mutants have been isolated that alter circadian timing, and at least one clock gene has been cloned and the structure of the gene and the protein that it encodes have been analyzed.

In both invertebrates and vertebrates, specific parts of the central nervous system have been identified as the site of a master control oscillator that regulates circadian rhythmicity in the animal. The optic lobes of invertebrate brains, the eye of certain molluscs, the pineal gland of birds, and the suprachiasmatic nucleus region of the hypothalamus in birds and mammals have all been implicated in this regard.

Human circadian rhythms have been identified in many physiological and biochemical functions, and disruption of normal human circadian patterns due to jet travel or change in work shift produce ill effects of various sorts. Circadian rhythm of sensitivity to drugs is widespread and is being used to improve the efficacy of cancer chemotherapy and treatment of other disorders.

7.9. REFERENCES

1. E. Bunning, *The Physiological Clock,* 3rd ed., Springer-Verlag, New York (1973).
2. A. Chovnick (ed.), Biological clocks, *Cold Spring Harbor Symp. Quant. Biol.* **25** (1960).
3. B. M. Sweeney, *Rhythmic Phenomena in Plants,* Academic Press, London (1969).
4. M. C. Moore-Ede, F. M. Sulzman, and C. A. Fuller, *The Clocks That Time Us,* Harvard University Press (1982).
5. D. S. Saunders, *An Introduction to Biological Rhythms,* Blackie, London (1977).
6. C. S. Pittendrigh, Biological Clocks, in: *Science in the Sixties.* Air Force Office of Scientific Research, Washington, D.C., pp. 96–111 (1966).

7. J. W. Hastings and B. M. Sweeney, On the mechanism of temperature independence in a biological clock, *Proc. Natl. Acad. Sci. USA* **43:** 804–811 (1957).
8. D. Njus, F. M. Sulzman, and J. W. Hastings, Membrane model for the circadian clock, *Nature* **248,** 116–120 (1974).
9. B. M. Sweeney and J. W. Hastings, Effects of temperature upon diurnal rhythms, in: *Cold Spring Harbor Symp. Quant. Biol.* **25,** 87–104 (1960).
10. T. Pavlidis, W. F. Zimmerman, and J. Osborn, A mathematical model for the temperature effects on circadian rhythms, *J. Theor. Biol.* **18:** 210–221 (1968).
11. G. F. Gardner and J. F. Feldman, Temperature compensation of circadian period length in clock mutants of *Neurospora crassa, Plant Physiol.* **68,** 1244–1248 (1981).
12. V. G. Bruce, Environmental entrainment of circadian rhythms, in: *Cold Spring Harbor Symp. Quant. Biol.* **25,** 29–48 (1960).
13. C. S. Pittendrigh, The circadian oscillation in *Drosophila pseudoobscura* pupae: A model for the photoperiodic clock, *Z. Pflanzenphysiol.* **54,** 275–307 (1966).
14. C. S. Pittendrigh, Circadian organization and the photoperiod phenomena, in: *Biological Clocks in Reproductive Cycles* (B.K. Follett, ed.), Wright, Bristol, England (1981).
15. B. M. Sweeney, Chronobiology, in: *Science of Photobiology* (K.C. Smith, ed.), pp. 209–226 Plenum Press, New York (1977).
16. J. F. Feldman, Genetics of circadian clocks, *BioScience* **33:** 426–431 (1983).
17. J. F. Feldman and J. C. Dunlap, *Neurospora crassa:* A unique system for studying circadian rhythms, *Photochem. Photobiol. Revs.* **7,** 319–368 (1983).
18. V. G. Bruce and D. H. Minis, Circadian clock action spectrum in a photoperiodic moth, *Science* **163,** 583–585 (1969).
19. M. L. Sargent and W. R. Briggs, The effects of light on a circadian rhythm of conidiation in *Neurospora, Plant Physiol.* **42,** 1504–1510 (1967).
20. R. Brain, D. O. Woodward, and W. R. Briggs, Correlative studies of light sensitivity and cytochrome content in *Neurospora crassa,* in: *Carnegie Inst. Washington Yearbook 1976,* pp. 295–299 (1977).
21. J. Paietta and M. L. Sargent, Photoreception in *Neurospora crassa:* Correlation of reduced light sensitivity with flavin deficiency, *Proc. Natl. Acad. Sci. USA* **78,** 5573–5577 (1981).
22. J. W. Hastings and H.-G. Schweiger (eds.), *The Molecular Basis of Circadian Rhythms,* Dahlem Konferenzen, Berlin (1976).
23. J. W. Hastings, Biochemical aspects of rhythms: Phase shifting by chemicals, in: *Cold Spring Harbor Symp. Quant. Biol.* **25,** 131–144 (1960).
24. T. Lonergan and M. L. Sargent, Regulation of the photosynthesis rhythm in *Euglena gracilis.* I. Carbonic anhydrase and glyceraldehyde-e-phosphate dehydrogenase do not regulate the photosynthesis rhythm, *Plant Physiol.* **61,** 150–153 (1979).
25. B. Prezelin and B. M. Sweeney, Characterization of photosynthetic rhythms in marine dinoflagellates. II. Photosynthesis-irradiance curves in in vivo chlorophyll a fluorescence, *Plant Physiol.* **60,** 388–392 (1977).
26. C. H. Johnson, J. F. Roeber, and J. W. Hastings, Circadian changes in enzyme concentration account for rhythm of enzyme activity in *Gonyaulax, Science* **223,** 1428–1430 (1984).
27. K. Goto, D. L. Laval-Martin, and L. N. Edmunds, Biochemical modeling of an autonomously oscillatory circadian clock in *Euglena, Science* **228,** 1284–1288 (1985).
28. H. Nakashima, Phase shifting of the circadian conidiation rhythm in *Neurospora crassa* by calmodulin antagonists, *J. Biol. Rhythms* **1:** 163–169.
29. C. F. Ehret and E. Trucco, Molecular models for the circadian clock. I. The chronon concept, *J. Theor. Biol.* **15,** 240–262 (1967).
30. H.-G. Schweiger and M. F. Schweiger, Circadian rhythms in unicellular organisms: An endeavor to explain the molecular mechanism, *Int. Rev. Cytol.* **51,** 315–342 (1977).
31. D. P. Delmer and S. Brody, Circadian rhythms in *Neurospora crassa:* Oscillation in the level of an adenine nucleotide, *J. Bacteriol.* **121,** 548–553 (1975).
32. B. M. Sweeney, A physiological model for circadian rhythms derived from the *Acetabularia* rhythm paradoxes, *Int. J. Chronobiol.* **2,** 25–33 (1974).
33. J. F. Feldman, Genetic approaches to circadian clocks, *Ann. Rev. Plant Physiol.* **33,** 583–608 (1982).
34. R. J. Konopka, Genetic dissection of the *Drosophila* circadian system, *Fed. Proc.* **38,** 2602–2605 (1979).
35. R. J. Konopka, Genetics and development of circadian rhythms in invertebrates, in: *Handbook of Behavioral Neurobiology,* Vol. 4 (J. Aschoff, ed.), pp. 173–181, Plenum Press, New York (1981).

36. R. J. Konopka, and S. Benzer, Clock mutants of *Drosophila melanogaster, Proc. Natl. Acad. Sci. USA* **68,** 2112–2116 (1971).
37. A. M. Handler and R. J. Konopka, Transplantation of a circadian pacemaker in *Drosophila, Nature* **279,** 236–238 (1979).
38. C. P. Kyriacou and J. C. Hall, Circadian rhythm mutations in *Drosophila melanogaster* affect short term fluctuations in the male's courtship song, *Proc. Natl. Acad. Sci* **77,** 6729–6733 (1980).
39. G. F. Gardner and J. F. Feldman, The *frq* locus in *Neurospora crassa:* A key element in circadian clock organization, *Genetics* **96,** 877–886 (1980).
40. T. A. Bargiello and M. W. Young, Molecular genetics of a biological clock in *Drosophila, Proc. Natl. Acad Sci* **81,** 2142–2146 (1984).
41. P. Reddy, W. A. Zehring, D. A. Wheeler, V. Pirotta, C. Hadfield, J. C. Hall, and M. Rosbash, Molecular analysis of the *period* locus in *Drosophila melanogaster* and identification of a transcript involved in biological rhythms, *Cell* **38,** 701–710 (1984).
42. W. A. Zehring, D. A. Wheeler, P. Reddy, R. J. Konopka, C. P. Kyriacou, M. Rosbash, and J. C. Hall, P-element transformation with *period* locus DNA restores rhythmicity to mutant, arrhythmic *Drosophila melanogaster, Cell* **39,** 369–376 (1984).
43. M. Rosbash and J. C. Hall, Biological clocks in *Drosophila:* Finding the molecules that make them tick, *Cell* **43,** 3–4 (1985).
44. F. Rob Jackson, T. A. Bargiello, S. -H. Yun, and M. W. Young, Product of *per* locus of *Drosophila* shares homology with proteoglycans, *Nature* **320,** 185–188 (1986).
45. J. S. Takahashi and M. Menaker, Circadian rhythmicity. Regulation in the time domain, In: *Biological Regulation and Development* (R. F. Goldberger and K. R. Yamamoto, eds.), pp. 285–303, Plenum Press New York (1984). *Ann. Rev. Neurosci* **1,** 19–34 (1978).
46. J. W. Jacklet, Neurobiology of circadian rhythm generators, *Trends Neurosci.* **8,** 69–73 (1985).
47. G. D. Block and S. F. Wallace, Localization of a circadian pacemaker in the eye of a mollusc, *Bulla, Science* **217,** 155–157 (1982).
48. G. D. Block and D. G. McMahon, Cellular analysis of the *Bulla* ocular circadian pacemaker system. III. Localization of the circadian pacemaker, *J. Comp. Physiol.* **155A,** 387–395 (1984).
49. J. W. Jacklet, L. Schuster, and C. Rolerson, Electrical activity and structure of retunal cells of the *Aplysia* eye: I. Secondary neurones, *J. Exp. Biol.* **99,** 369–380 (1982).
50. L. M. Robertson and J. S. Takahashi, The circadian clock in cell culture, *J. Neurosci.* **8,** 12–21.
51. T. L. Page, Transplantation of the cockroach circadian pacemaker, *Science* **216,** 73–75 (1982).
52. T. L. Page, Clocks and circadian rhythms in insects, in: *Comprehensive Insect Biochemistry, Physiology, and Pharmacology. VI. Sensory Physiology* (G. A. Kerkut and L. I. Gilbert, eds.), pp. 577–652 Pergamon Press, Oxford (1984).
53. M. Menaker and S. Binkley, Neural and endocrine control of circadian rhythms in the vertebrates, in: *Handbook of Behavioral Neurobiology,* Vol. 4 (J. Aschoff, ed.), pp. 243–255, Plenum, New York (1981).
54. R. Y. Moore, Organization and function of a central nervous system circadian oscillator: the suprachiasmatic hypothalamic nucleus, *Fed. Proc.* **42,** 2783–2789 (1983).
55. M. C. Moore-Ede, The circadian timing system in mammals: Two pacemakers may preside over many secondary oscillators, *Fed. Proc.* **42,** 2802–2808 (1983).
56. F. W. Turek, Neurobiology of circadian rhythms in mammals, *BioScience* **33,** 439–444 (1983).
57. T. Deguchi, A circadian oscillator in cultured cells of chicken pineal gland, *Nature* **282,** 94–96 (1979).
58. J. Aschoff, Circadian rhythms in man, in: *Biological Timekeeping* (J. Brady, ed.), pp. 143–157, Cambridge University Press, Cambridge (1982).
59. J. Aschoff, Features of circadian rhythms relevant for the design of shift schedules, *Ergonomics* **39,** 739–754 (1979).
60. F. Halberg, E. A. Johnson, B. W. Brown, and J. J. Bittner, Susceptibility rhythm to *E. coli* endotoxin and bioassay, *Proc. Soc. Exp. Biol. Med.* **103,** 142–144 (1960).
61. M. C. Moore-Ede and F. M. Sulzman, Internal temporal order, in: *Handbook of Behavioral Neurobiology,* Vol. 4 (J. Aschoff, ed.), pp. 215–241, Plenum Press, New York, (1981).
62. F. Halberg and E. Halberg, Chronopharmacology and further steps toward chronotherapy, in: *Pharmacokinetic Basis for Drug Treatment* (L. Z. Benet, N. Massoud, and J. G. Gambertoglio, eds.), pp. 221–248, Raven Press, New York (1984).

8

Extraretinal Photoreception

8.1. INTRODUCTION

Light plays many roles in the lives of organisms other than that which we normally associate with the visual perception of patterns in the environment. Therefore, it is useful to distinguish between those situations in which (1) light is important to organisms primarily because of its energy content and (2) light functions as a signal, stimulus, or "trigger" and, from a biological viewpoint, energetic considerations are secondary.

The clearest example of the former is surely photosynthesis (Chapter 12), in which it is specifically the energy of light that the photosynthetic organism is "after"; indeed, the photosynthetic machinery is exquisitely evolved to capture and utilize this energy with maximum efficiency. On *a priori* grounds, one might suppose that the chlorophyll molecules utilized by plants to capture the energy of sunlight are also employed to perceive sunrise and sunset. As we shall see, this assumption, although plausible, is incorrect.

The most familiar signaling function of light is pattern vision. In this activity the amount of light energy involved is of relatively minor importance. Most eyes have not evolved to maximize the amount of light absorbed; in fact, in most visual situations, structures in the eye operate to control and often reduce the amount of energy reaching the pigment molecules. Spatial and temporal patterning, small variations in intensity, and

Michael Menaker • Department of Biology, University of Virginia, Charlottesville, Virginia 22901.

differences in wavelength are of much greater importance to the visual process than are large-scale changes in the energy contained in the stimulating light. It is not quite so obvious that these same considerations apply to a variety of other processes in which light plays a signaling, stimulating, or triggering role. These processes include photomorphogenesis (Chapter 10), photomovement (Chapter 11), circadian rhythms (Chapter 7), and reproductive cycles (below), and can reasonably be said to involve photoreception, as distinct from the simple absorption of light quanta by photopigments.

It is characteristic of photoreception that the biological activities initiated by it require the processing of energy in amounts completely out of proportion to the energy in the light stimulus itself. Thus, the number of joules required to escape from the tiger that you perceive to be stalking you is infinitely greater than that involved in conveying his image to your retina, and bears no necessary relationship to whether the event is illuminated by the moon or by the noonday sun. It is less evident, but equally true, that a plant bending toward the light or flowering in response to long days is processing energy in amounts that are unrelated to the energy contained in the triggering light stimuli. On the other hand, the amount of carbon fixed in photosynthesis and the number of cells killed by ultraviolet radiation depend much more directly on the number of quanta in the light that produces these effects. All these processes are mediated by the effects of light on biological molecules; however, photoreception is mediated by photopigments usually contained within photoreceptors; relatively complex organelles or organs composing a larger system that functions to amplify the biological response to light.

Having drawn the above distinction, it becomes clear that a single organism might well be expected to possess many photopigments, some of which function within photoreceptors, while others are involved primarily in energy transfer. Hence, there is no reason to assume that green plants must perceive the sunrise with their cholorphyll molecules.

The possession of multiple photopigments, and also of multiple photoreceptors that mediate different responses to light, is the rule among organisms. This should not be surprising in view of the above discussion, but our own subjective photoreceptive experience depends so heavily on our eyes that we are usually surprised to learn that most highly evolved organisms with complex image-forming eyes have, in addition, other much less obvious photoreceptors that are used to monitor crucially important aspects of the photic environment. The basic observation leading to this conclusion is that even though the eyes of an animal have been surgically removed or its optic tracts severed, the animal still responds in a variety of ways to light in the visible portion of the spectrum. In order to counteract our own pattern vision bias, let us consider the kinds of information about the environment that are available to an organism that has only simple photoreceptors incapable of forming images. The discrimination of day from night enables the entrainment of the circadian system along with the consequent adaptive regulation of the timing of many physiological and behavioral activities. The seasons of the year are most unambiguously signaled by the regular annual change in day length, which controls annual reproductive cycles, some aspects of migration, diapause, and other important physiological responses in a wide variety of organisms. (This phenomenon is called photoperiodism.) The intensity of the light to which an organism is exposed will provide cues concerning the cloud cover, whether the organism is in full light or in shadow, the time of day, and even, if moonlight can be perceived, the phase of the lunar cycle. Simple photoreceptors are also capable of detecting the spectral composition of light that impinges on them and, crudely, the direction from which it comes. These latter capacities improve rapidly with slight increases in complexity of the receptor structure.

Although plants and various simple animals extract many kinds of information from the photic environment, utilizing very simple photoreceptive structures, the discovery that vertebrates and complex invertebrates with image-forming eyes also have extraretinal photoreceptors was surprising, even to most photobiologists. After all, granted that it does not require an eye to make such responses, it would seem that once eyes had evolved as a result of selection pressure for the perception of images, they should also serve perfectly well for these other, simpler tasks.

8.2. EXTRARETINAL PHOTORECEPTION IN INVERTEBRATES

It has been experimentally demonstrated that many invertebrate species make use of extraretinal photoreception. A few examples will suffice to illustrate the diversity of activities controlled in this way.

Because of its large size and the relative simplicity of performing surgery on it, the giant silk moth has long been a favorite experimental subject of insect endocrinologists. Considerable progress has been made in understanding the hormonal basis of developmental events such as the induction and termination of diapause (a state of greatly reduced metabolic activity that allows the insect to overwinter), which are regulated by environmental light cycles. Several of the hormones involved in the control of diapause and related events are known to be produced and/or stored in the brain and other central nervous system structures. The photoreceptors that mediate these responses are also located in the brain.

In an ingenious experiment, Williams and Adkisson[1] first established that the head end of the diapausing silk moth pupa contained the photoreceptors used by the animals to perceive the long days that trigger the onset of adult development (the termination of diapause). They then showed that simply by transplanting the brain (which does not contain the developing eyes) from its normal site in the head to a new location in the tip of the abdomen, photosensitivity was transferred from the anterior to the posterior half of the animal. In a similar experiment, Truman and Riddiford[2,3] demonstrated that the brain of the silk moth contains photoreceptors that synchronize the circadian rhythm of eclosion (emergence of the adult moth from the pupal case) with the light–dark cycle, and also with the clock that times the event.

There is extensive literature on extraretinal photoreception in the control of insect photoperiodism and circadian rhythms. Indeed there are so many examples of these phenomena in diverse groups of insects that the few well-documented cases in which photoreception involves the exclusive use of the compound eyes and ocelli (e.g., cockroaches and crickets) are of particular interest.[3]

Some time ago, investigators studying the ventral nerve chord of crayfish with electrophysiological techniques observed that changes in the electrical activity of fibers in the chord could be produced by illuminating the 6th abdominal ganglion, but not by illuminating any of the other ganglia. Further work showed that the 6th abdominal ganglion contains a single pair of photoreceptive neurons. The action spectrum for these photoreceptors indicates that the photopigment involved is a rhodopsin, which is very similar to that which mediates photoreception by the crayfish eye.[4] Cave crayfish, which no longer have functional eyes, retain these so-called caudal photoreceptors. Their function remains unknown. Apparently, they are not involved in the entrainment of circadian rhythms, although there is reason to believe that they may exert some influence on locomotor behavior. On the other hand, crayfish have other extraretinal photoreceptors, not yet precisely localized (but probably in the supraesophageal ganglion), which are involved in entrainment of the circadian rhythm of locomotion. Page and Larimer[5] also studied a circadian rhythm in the electroretinogram (ERG) amplitude of the crayfish eye. They have found that even this rhythm, which is after all a rhythm of sensitivity to light of the retinal elements themselves, can be synchronized via extraretinal photoreceptors probably located in the brain.

Perhaps the most dramatic example of extraretinal photoreception among the invertebrates has been described by Block and McMahon[6] in the cloudy bubble snail, *Bulla gouldiana*. The eyes of this mollusc, like those of its better known relative *Aplysia,* can be cultured *in vitro* for a week or more, and under these conditions display a circadian rhythm in the frequency of spontaneously produced compound action potentials. This rhythm can be entrained by light cycles and phase shifted by single light pulses while the isolated eye is maintained in culture. Peculiarities in the anatomy of this preparation make it possible to remove unambiguously the entire complement of organized (retinal) photoreceptors, leaving behind a small number of electrically coupled cells at the base of the eye with no obvious specialized photoreceptive structures. Preparations surgically reduced in this manner still express circadian rhythms and, furthermore, still respond to light cycles and pulses. Thus, even within the eye itself, circadian photoreception is accomplished by a distinct set of extraretinal photoreceptors. It is interesting to compare this situation with that in the mammalian eye described in Section 8.6.

8.3. EXTRARETINAL PHOTORECEPTION IN VERTEBRATES

Extraretinal photoreception by some birds and fish has been documented for over 50 years; however, it is only recently that the extent, complexity, and importance of the phenomenon have become apparent. At least some species in all five vertebrate classes are known to regulate important aspects of their physiology and behavior with reference to extraretinally perceived light. Least is known about this phenomenon in fish (because there has been very little work) and in mammals (because it appears that extraretinal photoreception may be limited to newborn animals,[7] but see Section 8.6).

The pineals and, when present, associated structures such as the parietal eye and frontal organ of all nonmammalian vertebrates are probably photoreceptive. Solid electrophysiological evidence that this is so exists for some fish, amphibians, and reptiles. The pineal glands of some birds and lizards have been maintained *in vitro* for many days by continuously superfusing them with oxygenated culture medium.[8,9] If the culture medium is collected at frequent intervals after it has passed over the pineal, it can be assayed by radioimmunoassay for melatonin, a hormone synthesized by the gland. Melatonin synthesis by isolated, cultured pineal glands follows a circadian rhythm, and this rhythm can be entrained in the culture by a light–dark cycle. Furthermore, the entrained rhythm bears the same phase relationship to the light cycle (melatonin is high

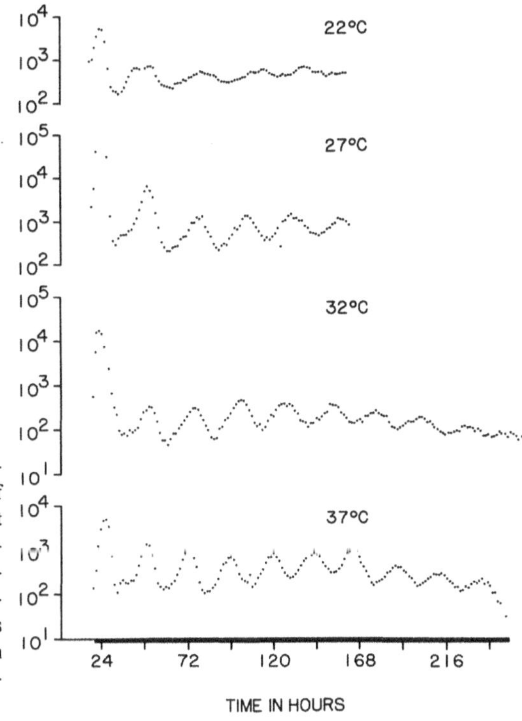

Fig. 8-1. Circadian rhythms of melatonin produced by four individual isolated pineal glands of *Anolis* held in constant darkness at four different temperatures. Each point was determined by radioimmunoassay of a superfusate sample. Superfusion was continuous and all medium was collected in 90-min aliquots. On the ordinate is plotted the log of the melatonin concentration in pg/ml. Note that periods of the rhythms are shorter at the higher temperatures. (From Ref. 9.)

during the dark portion of the cycle) as that in the intact animal (Figs. 8-1 and 8-2). These pineal glands, like the brains of silk moths and the eyes of the cloudy bubble snail, contain both circadian oscillators and photoreceptors. Pineal glands of nonmammalian vertebrates contain a variety of photoreceptive structures ranging from scattered simple outer-segment-like organelles to highly ordered arrays of photoreceptors that look much like a simple retina and, like the retina of the lateral eye, are served by a lens and are connected with the brain by a substantial nerve tract.[10–12] Whether these organized photoreceptors are in fact responsible for circadian photoreception in the pineal is not known. We are certainly not entitled to assume that they are, as the example of the *Bulla* eye makes clear.

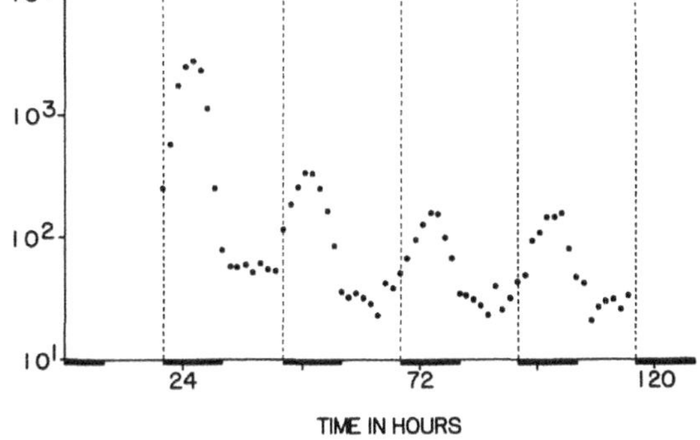

Fig. 8-2. Melatonin rhythm entrained *in vitro* by a light cycle (diagrammed on the abscissa). Temperature was a constant 27°C ± 0.5°C. Samples were obtained and plotted as described in the legend to Fig. 8-1. Note that the period of the rhythm is indistinguishable from 24 h and that its phase relationship to the light cycle is as expected, i.e., melatonin is high during the dark. (From Ref. 9.)

There have been several studies of action spectra of extraretinally mediated responses to light. Deguchi[13] found that brief light exposure inhibits melatonin synthesis in cultured chicken pineal glands. He obtained an action spectrum for this effect that strongly suggests that the photopigment involved is rhodopsinlike. Foser and Follett[14] illuminated the brains, but not the eyes, of Japanese quail, measured the effect of light on the gonads, and derived a similar action spectrum. On the assumption that the photoreceptor is hypothalamic, they calculated that its sensitivity is within an order of magnitude of that of the dark-adapted human eye and that its photopigment is rhodopsinlike. The tentative general conclusion that extraretinal photoreception is based on rhodopsinlike photopigments is supported by the effects of vitamin A deprivation on the ultrastructure of lizard parietal eye photoreceptors and by the finding of opsin immunoreactivity in the pineals of several reptiles and birds. The preliminary nature of this conclusion should be emphasized; no one has yet extracted any photopigment from an extraretinal photoreceptor nor successfully subjected one to microspectrophotometric analysis.

We know as much about extraretinal photoreception in the house sparrow (*Passer domesticus*) as in any other vertebrate species. In order to convey some idea of the complexity of this sensory capacity, a detailed description of the phenomenon in that species follows.

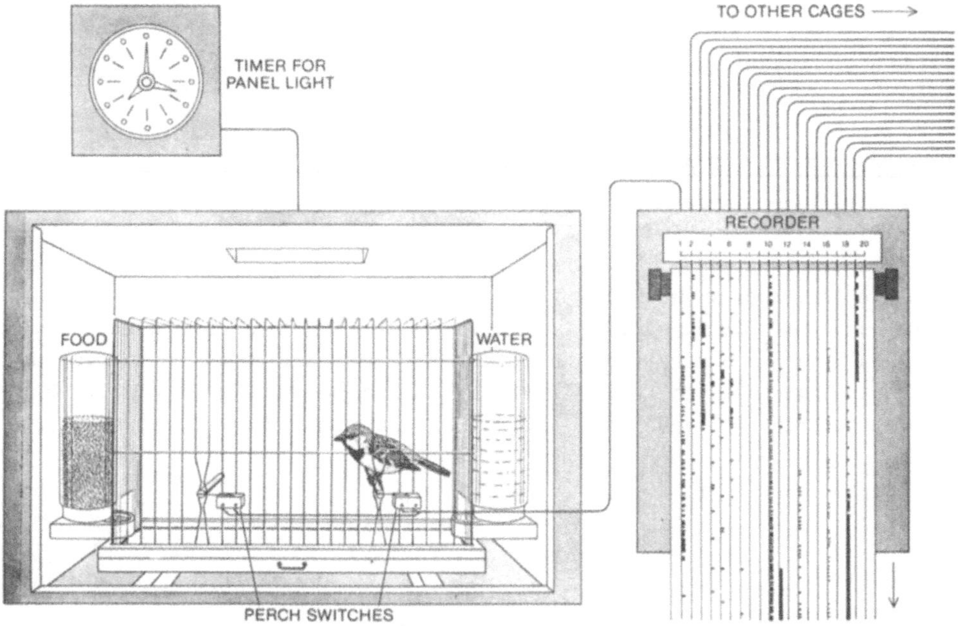

Fig. 8-3. Arrangement for the continuous recording of perching behavior in sparrows. The bird's weight depresses the switch connected to the perch and electrically activates a single pen (of a 20-pen recorder), which writes on a strip of paper moving at 18 in/day. Each day's record is separated from those of other birds, and each bird's daily records are displayed to yield cumulative data of the kind shown in subsequent figures. While such data are only semiquantitative with respect to amount of activity, they give a very accurate picture of the temporal distribution of activity over long periods of time. (From Ref. 15, Copyright 1972 by Scientific American, Inc. All rights reserved.)

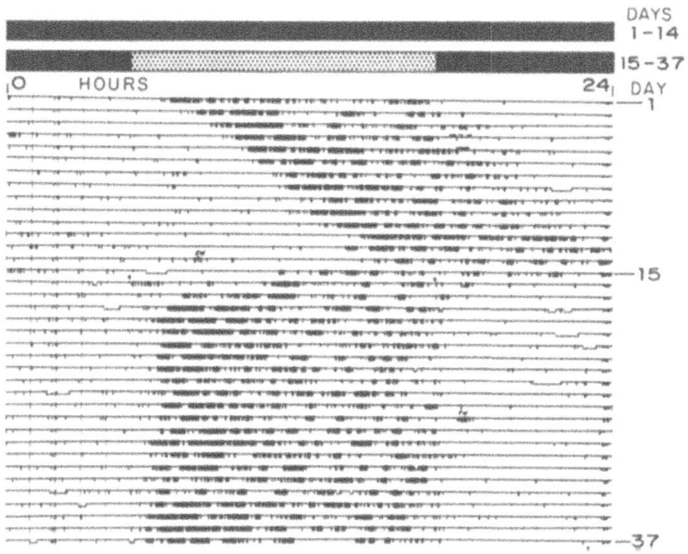

Fig. 8-4. The circadian locomotor rhythm of the house sparrow in its free-running and entrained states. During the first 14 days of the record, the bird was maintained in continuous darkness, and its rhythm free-ran with a period of about 25 h. Beginning on day 15, it was presented with a daily light cycle (diagrammed at the top of the figure; LD 12 : 12, intensity during the light portion of the cycle = 0.1 lx) to which its rhythm synchronized (entrained). Although this bird was blinded by the removal of both eyes prior to the start of the experiment, its record is indistinguishable from that of a normal sparrow. (From Ref. 16.)

8.4. EXTRARETINAL PHOTORECEPTION IN THE HOUSE SPARROW

8.4.1. Effects of Light on the Circadian Clock: Entrainment

Circadian rhythms of locomotor activity can be most easily assayed in perching birds, such as *Passer,* by continuously recording the switch closures produced when the bird hops on a perch. In this way, one can automatically record the temporal distribution of activity for weeks or months (Fig. 8-3). In the presence of a light cycle, the birds are normally active primarily when the light is on, beginning their activity at very nearly the same time each day (if the light–dark cycle has a period of exactly 24 h). They are said to be entrained (or synchronized) to the light cycle (see Chapter 7). When placed in constant darkness, the bird's locomotor activity remains conspicuously rhythmic, but since it no longer has access to an environmental cue, it "free-runs" with a period close to, but different from, 24 h. In practice, most sparrows have free-running periods longer than 24 h and begin their activity progressively later each day. The difference between entrained and free-running locomotor rhythmicity is dramatically clear in the raw perching data (Fig. 8-4). Since the bird can only entrain to light cycles it can perceive, entrainment provides an unambiguous and easily recorded assay of photoreception.

Sparrows continue to entrain to light cycles even after their eyes and pineal organ have been surgically removed, and many carefully controlled experiments have established that it is visible light rather than some other periodic feature of the regime, such as

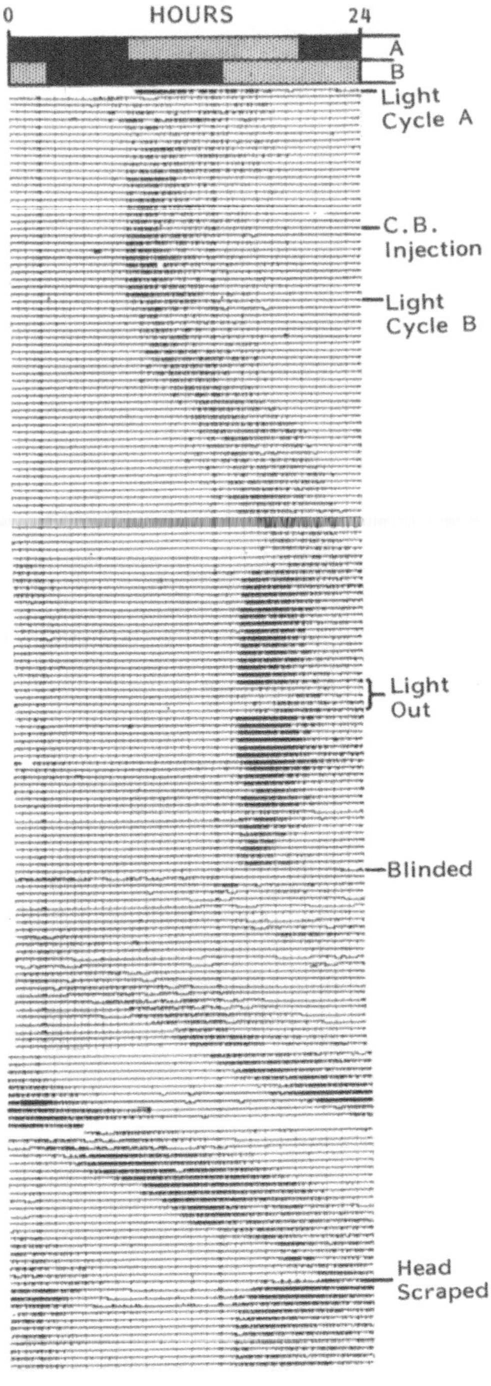

Fig. 8-5. Demonstration of the contribution of both the eyes and brain photoreceptors to entrainment of the circadian locomotor rhythm in the house sparrow. This figure is the activity record of a sparrow exposed to the light regimes (LD 12 : 12; 0.03 lx) diagrammed at the top. Entrainment to light cycle A continues after carbon black (C.B.) is injected under the skin of the head. Reestablishment of the entrained steady state following the 6-h phase delay (light cycle B) requires 36 days. After blinding, the activity rhythm free-runs through the light cycle twice indicating a contribution of retinal photoreception. When the carbon black is removed, the blind bird reentrains to the light cycle using extraretinal photoreceptors. (From Ref. 17.)

heat, noise, or electrical fields, to which they are responding. The threshold intensity to which they will entrain is surprisingly low: about 50% of blinded sparrows will entrain to a cycle consisting of 12 h of darkness alternating with 12 h of light at an intensity of 0.1 lx (approximately the intensity of bright moonlight). This level of illumination can thus be thought of as a crude threshold of extraretinally mediated entrainment.(16) Sparrows with intact eyes will entrain to cycles in which the light portion is considerably less intense, demonstrating that both the eyes and the extraretinal receptors have inputs to the entrainment mechanism, and that these inputs are, to a first approximation, additive (Fig. 8-5). Experiments in which the intensity of light reaching the brain has been either increased by plucking the head feathers or decreased by injecting opaque substances (e.g., india ink) under the skin of the head have shown that the extraretinal photoreceptors are located in the brain.(17)

8.4.2. Effects of Light on the Circadian Clock: Aschoff's Rule

Several different effects of light on circadian rhythms are discussed in Chapter 7. In addition to producing entrainment when it is presented in cycles (light–dark, LD), light affects free-running periods of circadian rhythms when it is presented continuously (light–light, LL). As the intensity of constant light is increased, the free-running periods of diurnal organisms exposed to it decrease whereas those of nocturnal organisms increase. This empirical generalization and several others related to it are known collectively as Aschoff's rule (after Dr. Jurgen Aschoff who has been responsible for much of the work in this area). The rule is well supported by data from a variety of very different organisms, although there are exceptions to it.

Sparrows obey Aschoff's rule and continue to do so after blinding. As in extraretinally mediated entrainment, the receptors responsible have been shown to reside in the brain and to act in concert with the image-forming eyes in the intact bird. The free-running period of a bird with both its eyes and its brain photoreceptors normally exposed shows greater changes with varying intensities of LL than does that of a bird in which the eyes have been removed or the brain shielded from light.(18) Figure 8-6 illustrates a brain

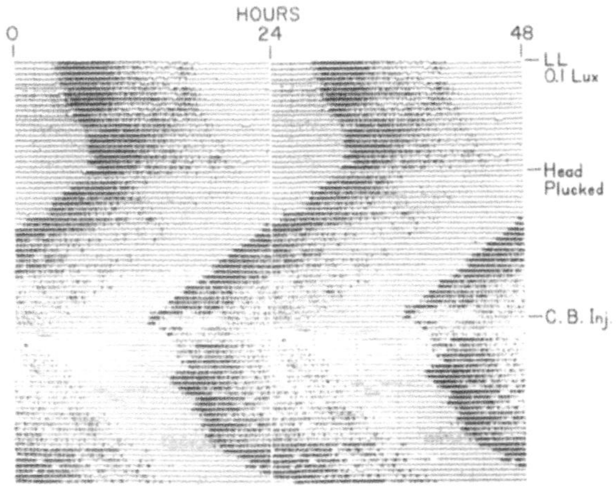

Fig. 8-6. The effects of changing the intensity of light reaching the brain on the free-running period of the locomotor rhythm of a sparrow with intact eyes held in constant light (0.1 lx). The period is initially slightly longer than 24 h. Plucking feathers from the top of the head shortens the period, whereas subsequent injection of carbon black (C.B.) under the skin of the head reverses this effect. (From Ref. 18.)

photoreceptor-mediated effect of light on free-running period. In sparrows, then, extraretinal brain photoreceptors are sufficient for both entrainment and Aschoff's rule, but in nature they probably work together with the eyes (presumably the retinal photoreceptors, although there has been no direct demonstration of this) to mediate the response to light. There is, however, another general response of circadian rhythms to LL for which, at least in sparrows, the brain photoreceptors are not sufficient.

8.4.3. Effects of Light on the Circadian Clock: Stopping the Clock with Bright Constant Light

In many organisms, the clock is stopped by constant light when its intensity is raised above some threshold, which varies considerably from one species to another. Binkley[19] showed this to be true for sparrows. One might expect that this effect represents simply an extension of the effects that LL of lower intensities has on the free-running period, but this does not appear to be so in sparrows, at least in terms of the photoreceptive mechanism. Although the intensity of constant light required to produce aperiodic locomotor activity in sparrows (the indication that the clock is stopped) lies between 10 and 100 lx, blind sparrows do not become aperiodic in constant light even when its effective intensity is raised to 20,000 lx. From these experiments, it appears that brain photoreceptors are not sufficient to mediate this effect of light on the sparrow's clock. This conclusion is especially strong in view of the fact that we know from entrainment experiments that the brain photoreceptors are sensitive to light at intensities of about 0.1 lx, so that in exposing them to 20,000 lx we are above threshold by about 2×10^5. Therefore, the eyes are apparently necessary for this particular effect of light on the circadian clock.

To further complicate the issue, it can be shown that, in intact sparrows, the brain photoreceptors are involved in the response to bright, constant light, although they are not by themselves sufficient to mediate it. If intact sparrows are brought to aperiodicity by constant light, the intensity of which is just above the required threshold ($\sim$10 lx), rhythmicity can be made to recur without changing the light intensity by simply shielding their brain photoreceptors.[20]

The biological meaning of the different patterns of interaction between the eyes and the brain photoreceptors in mediating the effects of light on the circadian system is still obscure, but the experimental results, which have been briefly described above, emphasize the complexity at the photoreceptor level of one system in one species to the photic environment. Although it is beyond the scope of this chapter to explore the comparative aspects of this phenomenon, it is worth mentioning that a similar analysis of the effects of light on the circadian clock has been performed in several species of lizard. While the results of experiments with lizards are generally similar to those with birds, several of the important details, especially concerning interaction of eyes and brain photoreceptors, are quite different not only when birds and lizards are compared, but even when one lizard species is compared with another.[21]

8.4.4. Effects of Light on the Reproductive System: Photoperiodism

Many animals and plants and most temperate zone birds use the information inherent in the cyclic annual change in day length to synchronize their reproductive cycles with the

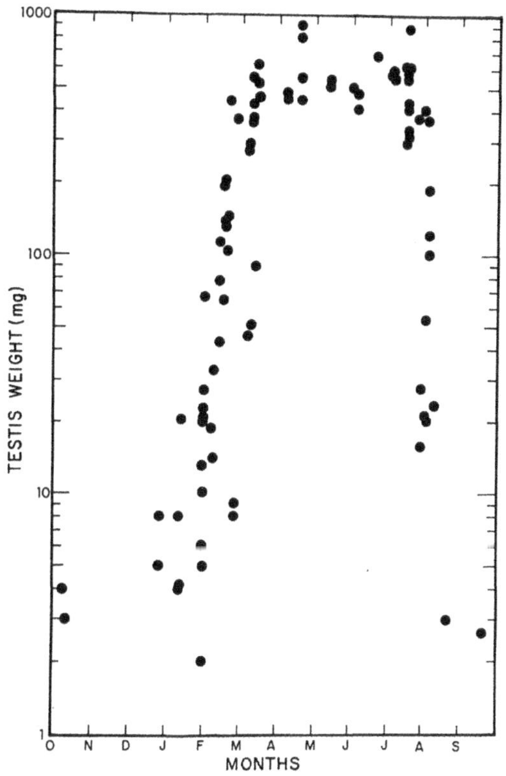

Fig. 8-7. Annual cycle of testis weight in a wild population of *Passer domesticus* in Austin, Texas. Each point represents the combined weight of both testes from a single freshly captured sparrow. Note that the testis weight is plotted on a logarithmic scale. (Data obtained by H. Underwood, from Ref. 15.)

appropriate season (Fig. 8-7). Such responses involve the ability both to perceive and "measure" day length, and are collectively termed photoperiodic. Perhaps the simplest question that can be framed concerning photoperiodic responses is, with what photoreceptors does the organism perceive the photoperiod? When Professor Jacques Benoit posed this question in 1935 with specific reference to the perception of day length, as indicated by the testicular responses of ducks, the answer that he obtained came as a great surprise to the scientific community; so great in fact that 20 or 30 years elapsed before it was fully accepted. By hindsight, and in the context of this chapter, it should not surprise you that this answer was, at least partially, that ducks use extraretinal photoreceptors in their brains to monitor the environmental light cycles that regulate their reproductive physiology.

Since Benoit's pioneering work, photoperiodic photoreception by brain photoreceptors has been directly demonstrated in several species of birds not closely related to each other, and it probably occurs in most, if not all, photoperiodic members of the class *Aves*.

Photoperiodic photoreception has been studied extensively in the house sparrow with particular attention to the possible role of the eyes. It is easy to demonstrate that sparrows employ brain photoreceptors to perceive day length. If male birds are brought into the laboratory during the winter, when their testes are small and nonfunctional, and are exposed for several weeks to artificial light cycles that simulate the long days of spring (e.g., 14 h or more of light per 24-h day), their testes grow dramatically. Not only do the gonads increase in weight by as much as 500-fold, but the testis tubules become packed with spermatids where before there were only spermatogonia. In contrast, the testes of control birds held on short days for the duration of the experiment remain unchanged. If this experiment is repeated using blind birds in both the long-day and short-day groups, the results are identical. In fact, it has not been possible to find any differences between the photoperiodic responses of blind and intact birds on the basis of gonad weight, rates of gonadal growth, or testis histology.[22,23]

These findings led to the speculation that, in this particular aspect of photoreception, the eyes might have no role at all. Some weight was lent to this suggestion by the discovery that the threshold intensity required to elicit photoperiodic testis growth in sparrows is of the order of 10 lx, or about 100 times the threshold intensity for extraretinally mediated entrainment, and thousands of times the threshold intensity for vision. A critical experiment was designed in which sparrows with their eyes intact were exposed to long days at a light intensity only slightly above this threshold. Half the birds had feathers plucked from the tops of their heads, thus increasing the amount of light that penetrated to their brains, while the other half were injected with opaque material beneath the skin of their heads, thus decreasing light to their brains. The testes of the plucked birds grew dramatically, whereas those of the injected birds did not grow at all, in spite of the fact that their eyes were exposed to long days at light intensities above threshold[24,25] (Fig. 8-8). In sparrows at least, photoperiodic photoreception is apparently performed exclusively by brain photoreceptors, perhaps by brain photoreceptors that are different from those that mediate the effects of light on the circadian clock. Note that in the course of investigating four different responses of the house sparrow to the photic environment—three responses of the clock and one of the reproductive system—three different patterns of interaction between the eyes and the brain have been uncovered. Two responses were observed for which brain photoreceptors are sufficient but to which the eyes contribute

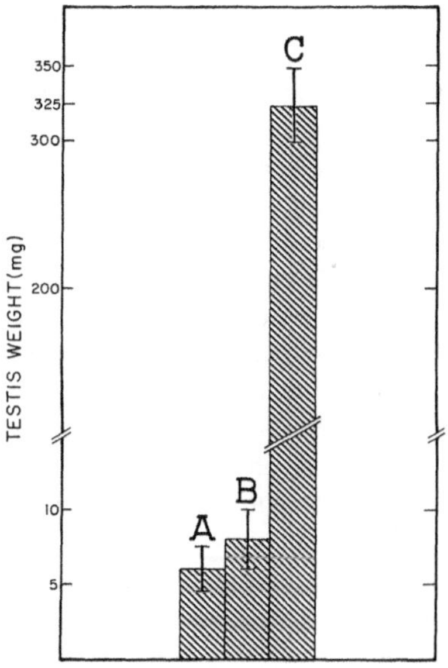

Fig. 8-8. Effects of manipulating the intensity of light reaching the brain on the testis weight of sparrows with intact eyes. A, control group sacrificed at the start of the experiment; B, india ink-injected group; C, group with head feathers plucked. See Section 8.4.4 for additional details. (Figure modified from Ref. 24.)

when present (entrainment and Aschoff's rule); one for which the eyes are necessary (arrhythmicity in bright constant light); and one in which only the brain is involved (photoperiodic reproductive response). Furthermore, there is reason to believe that there are as yet undiscovered responses to light in this as well as other species.

8.5. POSSIBLE SITES OF EXTRARETINAL PHOTORECEPTION

None of the extraretinal responses of sparrows to light that have so far been discussed depend on the presence of the pineal gland; they all proceed unaltered in its absence. On the other hand, there is direct evidence that avian pineal glands as well as those of other nonmammalian vertebrates are photoreceptive. Direct photosensitivity of the iris muscle has been reported in some vertebrates including mammals. One report in the literature describes a behavioral response of young, but not of adult, pigeons that is thought to be mediated by skin photoreceptors.[(26)] Skin photoreception is fairly common among amphibians and reptiles, and other birds may well display it when juvenile. Our list of photoreceptors in the vertebrates is thus already quite diverse even before we include the eyes and an undetermined number of brain photoreceptors.

In looking for photoreceptors that control circadian rhythms and reproduction, it is possible that we have only scratched the surface of the phenomenon of extraretinal photoreception. These responses to light are among the easiest to measure. But what of those that are a bit more difficult to assess? Perhaps important aspects of the effects of light on growth and development depend on extraretinal photoreception. Possibly light perceived in this way affects enzymatic activity in the brain and elsewhere in the body as well. The answers to these and a host of other questions await the results of future research.

8.6. DO MAMMALS HAVE EXTRARETINAL PHOTORECEPTION?

Although light penetrates the fur, skin, and skulls of mammals and reaches the brain in significant amounts, all attempts to demonstrate extraretinal photoreception in adult mammals have produced negative results.[(27)] Perhaps because they passed through a long nocturnal bottleneck in their evolutionary history, mammals appear to have consolidated all their photoreceptors in their eyes, where they are most likely to detect the dim light at dawn and dusk. If that supposition were true, then one would expect that the photoreceptors mediating the effects of light on the circadian and reproductive systems of mammals might retain some differences from those supporting vision even though now located in the same organ. This speculation receives support from the fact that retinal fibers carrying visual information arise from different kinds of retinal ganglion cells and terminate in different brain areas from those carrying circadian and reproductive information.

It was recently shown that some of the properties of the photoreceptors that mediate the effects of light on the circadian system of the golden hamster are unusual and perhaps quite different from those involved in vision.[(28)] Although their action spectrum is that of a rhodopsinlike photopigment, they are much less sensitive to light than the rhodopsin-containing rods of the hamster retina, and, unlike rods, they do not appear to light-adapt.

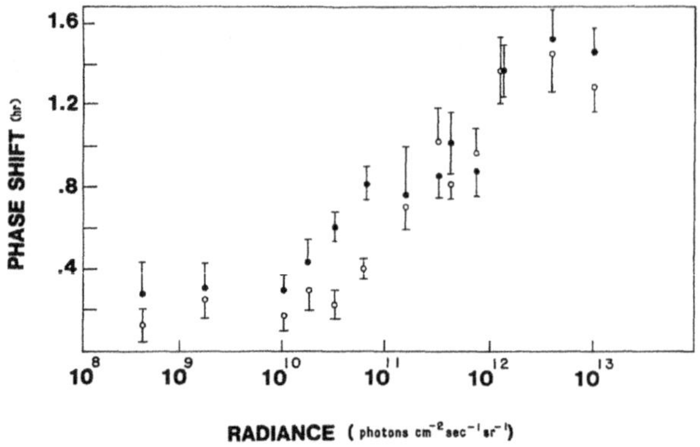

Fig. 8-9. Comparison of the effects of 15-min, monochromatic (515 nm) light pulses of increasing radiance, in phase-shifting of the circadian locomotor rhythms of mice with normal retinas (filled circles) and those with genetically degenerate retinas (open circles). The mutant animals have lost virtually all of their rods, although they retain some cones. The two data sets are remarkably similar, indicating that the rods contribute very little to circadian photoreception. The differences at radiances of 3×10^{10} and 6×10^{10} are significant, and may indicate either a minor contribution of rods to the normal process or a loss of sensitivity in the mutant animals resulting from other retinal damage. (Hudson, Pullela, and Menaker, unpublished data.)

These findings suggested that the visual rod photoreceptors, which account for 98% of photoreceptors in the retinas of nocturnal rodents, are not mediating the effects of light on the circadian system. This suggestion was confirmed by experiments that made use of a strain of mice carrying a genetic defect that results in a progressive loss of retinal rods. Mice that have lost their rods as a result of this defect are just as sensitive to the effects of light on the phase of their locomotor rhythms as mice with normal retinas (Fig. 8-9), although they are visually blind. Either the sparse, scattered cones or some retinal cell type not previously recognized as photoreceptive must be the circadian photoreceptor in these animals.

Thus, it appears that mammals, in common with all other vertebrates, may have photoreceptors that are specialized for nonvisual photoreception. Such photoreceptors, which are extraretinal in nonmammalian vertebrates, are apparently located in the retinas of mammals. As we learn more about them we may find that such receptors have common physiological properties wherever they may be localized, and share little beyond photopigments with visual photoreceptors.

8.7. CONCLUSIONS

It should be emphasized that as far as extraretinal photoreception is concerned, we do not as yet have any complete cases. Either we have identified and studied the physiology of such receptors but do not completely understand their function (e.g., the pineals of nonmammalian vertebrates), or we have identified extraretinally mediated functions, such as entrainment, without precise localization of physiological information about the receptors themselves. Not until we know both the function and the identity and physiological characteristics of several such receptor systems will we be able to proceed much further with our analysis.

It is clear from our discussion that many organisms possess and use multiple, discrete photoreceptors to monitor the photic environment. The puzzling question remains: What are the selective advantages of partitioning out photoreceptivity in this way? Although we cannot yet answer this question, we can note that the wide distribution of specialized photoreceptors and the functional complexity of some of the better known retinal–extraretinal systems, such as that of the sparrow, argues strongly for major adaptive significance. Indeed it may be most useful to think of extraretinal photoreception as a separate sensory modality independently evolved and distinct from vision.

8.8. REFERENCES

1. C. M. Williams and P. L. Adkisson, Physiology of insect diapause. XIV. An endocrine mechanism for the photoperiodic control of pupal diapause in the oak silkworm *Antherea pernyi*, *Biol. Bull.* **127,** 511–525 (1964).
2. J. W. Truman and L. M. Riddiford, Neuroendocrine control of ecdysis in silkmoths, *Science* **67,** 1624–1626 (1970).
3. J. W. Truman, Extraretinal photoreception in insects, *Photobiology* **23,** 215–225 (1976).
4. J. L. Larimer, D. L. Trevino, and E. A. Ashby, A comparison of the spectral sensitivities of caudal photoreceptors of epigeal and cavernicolous crayfish, *Comp. Biochem. Physiol.* **19,** 409–415 (1966).
5. T. L. Page and J. L. Larimer, Extraretinal photoreception in entrainment of crustacean circadian rhythms, *Photochem. Photobiol.* **23,** 245–251 (1976).
6. G. D. Block and D. G. McMahon, Cellular analysis of the *Bulla* ocular circadian pacemaker system III. Localization of the circadian pacemaker, *J. Comp. Physiol.* **155A,** 387–395 (1984).
7. M. Zweig, S. H. Snyder, and J. Axelrod, Evidence for a nonretinal pathway of light to the pineal gland of newborn rats, *Proc. Natl. Acad. Sci. USA* **56,** 515–520 (1966).
8. J. S. Takahashi, H. Hamm, and M. Menaker, Circadian rhythms of melatonin release from individual superfused chicken pineal glands in vitro. *Proc. Natl. Acad. Sci. USA* **77,** 2319–2322 (1980).
9. M. Menaker and S. Wisner, Temperature-compensated circadian clock in the pineal of *Anolis*, *Proc. Natl. Acad. Sci. USA* **80,** 6119–6121 (1983).
10. R. M. Eakin, *The Third Eye,* University of California Press, Berkeley (1973).
11. H. G. Hartwig and H. Oksche, Neurobiological aspects of extraretinal photoreceptive systems: Structure and function, *Experientia* **38,** 991–996 (1982).
12. H. Meissl and S. R. George, Electrophysiological studies on neuronal transmission in the frog's photosensory pineal organ, *Vision Res.* **24,** 1727–1734 (1984).
13. T. Deguchi, Rhodopsin-like photosensitivity of isolated chicken pineal gland, *Nature* **290,** 702–704 (1981).
14. R. G. Foster and B. K. Follett, The involvement of a rhodopsin-like photopigment in the photoperiodic response of the Japanese quail, *J. Comp. Physiol. A* **157,** 519–528 (1985).
15. M. Menaker, Nonvisual light reception, *Sci. Am.* **226,** 22–29 (1972).
16. M. Menaker, Extraretinal light perception in the sparrow. I. Entrainment of the biological clock, *Proc. Natl. Acad. Sci. USA* **59,** 414–421 (1968).
17. J. P. McMillan, H. C. Keatts, and M. Menaker, On the role of eyes and brain photoreceptors in the sparrow: Entrainment to light cycles, *J. Comp. Physiol.* **102,** 251–256 (1975).
18. J. P. McMillan, J. A. Elliott, and M. Menaker, On the role of eyes and brain photoreceptors in the sparrow: Aschoff's rule, *J. Comp. Physiol.* **102,** 257–262 (1975).
19. S. Binkley, Constant light: Effects on the circadian locomotion rhythm of the house sparrow, *Physiol. Zool.* **3,** 170–181 (1977).
20. J. P. McMillan, J. A. Elliott, and M. Menaker, On the role of eyes and brain photoreceptors in the sparrow: Arrhythmicity in constant light, *J. Comp. Physiol.* **102,** 263–268 (1975).
21. H. Underwood and M. Menaker, Extraretinal photoreception in lizards, *Photochem. Photobiol.* **23,** 227–243 (1976).
22. M. Menaker and H. C. Keatts, Extraretinal light perception in the sparrow. II. Photoperiodic stimulation of testis growth, *Proc. Natl. Acad. Sci. USA* **60,** 146–151 (1968).
23. H. Underwood and M. Menaker, Photoperiodically significant photoreception in sparrows: Is the retina involved? *Science* **167,** 298–301 (1970).

24. M. Menaker, R. Roberts, J. A. Elliott, and H. Underwood, Extraretinal light perception in the sparrow. III. The eyes do not participate in photoperiodic photoreception, *Proc. Natl. Acad. Sci. USA* **67,** 320–325 (1970).
25. J. P. McMillan, H. Underwood, J. A. Elliott, M. H. Stetson, and M. Menaker, Extraretinal light perception in the sparrow. IV. Further evidence that the eyes do not participate in photoperiodic photoreception, *J. Comp. Physiol.* **97,** 205–213 (1975).
26. M. B. Heaton and M. S. Harth, Non-visual light responsiveness in the pigeon: Developmental and comparative considerations, *J. Exp. Zool.* **188,** 251–264 (1974).
27. R. J. Nelson and T. Zucker, Absence of extraocular photoreception in diurnal and nocturnal rodents exposed to direct sunlight, *Comp. Biochem. Physio.* **69A,** 145–148 (1981).
28. J. S. Takahashi, P. J. DeCoursey, L. Bauman, and M. Menaker, Spectral sensitivity of a novel photoreceptive system mediating entrainment of mammalian circadian rhythms, *Nature* **308,** 186–188 (1984).

9

Vision

Edward A. Dratz • Chemistry Department, Montana State University, Bozeman, Montana 59717.

9.1. INTRODUCTION

9.1.1. Overview of the Vision Process

The process of vision starts with absorption of light by a chromophore, 11-*cis*-retinal, located in a pocket deep in the protein rhodopsin. The 11-*cis*-retinal has an energetically unfavorable conformation that acts like a cocked spring, which is released by light. The released spring forces the rhodopsin molecule to change conformation. The rhodopsin molecules are embedded in membranes in the photoreceptor cell. Light-excited rhodopsin stimulates a large number of guanosine triphosphate (GTP)-binding proteins on the membrane surface to bind GTP. The GTP-filled GTP-binding proteins then trigger enzymes that degrade cyclic guanosine monophosphate (cGMP). Each excited rhodopsin molecule ultimately leads to the consumption of a large number of cGMP molecules, which in turn control the conductance of a large number of ion channels in the cell surface membrane. The changing membrane conductance alters the electrical voltage across the cell membrane, and this voltage changes the rate of neurotransmitter release by the synaptic end of the photoreceptor cells. Complex control mechanisms, which are just beginning to be understood, shape and shut off the electrical response of the cell after the initial excitation. Signals from the photoreceptor cells are processed by other cells in the retina, and neurons pass signals on to the brain where visual perception occurs.

9.1.2. The Importance of Vision

A fascination with how vision works is quite natural. Light provides animals with very detailed information about their surroundings through the visual sense. Humans are especially visually oriented. There is great practical importance in understanding the inner workings of the eye, so that dysfunctions of this key sensory organ can be avoided or treated more effectively. Visual photoreceptors are also of more broad-reaching importance because they appear to function by mechanisms that are very similar to the receptors for many of the hormones and neurotransmitters. The visual receptors are packaged in specialized organelles, and are easier to isolate and study than most other receptors.

Vertebrate eyes contain two types of photoreceptor cells: the rods and the cones. The rods, responsible for dim-light vision, transmit signals to the brain that we perceive as colorless objects in shades of gray. The cones function in brighter light and enable us to perceive the colors of the spectrum. An eye is said to be "dark-adapted" when it has become well adjusted to the dark. If only 5–14 photons are absorbed in a cluster of 500 dark-adapted rods, we perceive the light. At this low photon flux, none of the rods absorbs more than one photon. The dark-adapted rod is an efficient single-photon detector! How such a sensitive detection is accomplished is a puzzle that appears to be approaching solution (Section 9.5). Coincidence between signals from several excited rods is used by the brain to discriminate against false signals due to the occasional thermal excitation of a rod in the dark. When rods and cones are exposed to a steady background light, they become light-adapted and are much less sensitive, but after returning to the dark, the maximum sensitivity soon returns. The mechanisms of this adaptation are still poorly understood, however.

The wavelength range of visible light is defined by the wavelengths that can be detected by the human eye. The limits of the visible part of the spectrum are taken to be from about 380 to 700 nm. The wavelength dependence of the sensitivity of the eye follows the absorption spectra of the visual pigments (Section 9.3.2). The maximum sensitivity of a dark-adapted vertebrate eye (called scotopic or dim-light vision) is near 500 nm, at the maximum absorption of the retinal rod cells. Under bright-light conditions (called photopic vision) the maximum sensitivity occurs at about 550 nm, which is near the absorption maximum of the most prevalent cone photoreceptors. The wavelength sensitivity of human photopic vision is shown in Fig. 1-1. The sensitivities of the visual systems of all animals fall fairly close to this curve except for insects, which can see much deeper into the near-ultraviolet region with high sensitivity.

The light is absorbed by photoreceptor cells that are located in the retina, a thin layer of tissue at the back of the vertebrate eye. The retina is considered to be an outpost of the brain, since there is a considerable amount of signal processing in the retina after light detection. Several aspects of signal processing in the nervous system have been studied profitably by the use of the accessible and easily stimulated retina.

9.1.3. Phylogenetic Organization of Visual Systems

All vertebrates have similar visual photoreceptor cells. Photoreceptors of animals as diverse as frogs, rats, and cows appear to be comparable, even identical in many aspects. Furthermore, much of the information derived from studies of animal photoreceptors appears to be directly applicable to the understanding of human photoreceptors.

Invertebrates, on the other hand, have evolved a variety of morphologically distinct visual systems that are very different from those of the vertebrates.[1,2] In the most primitive photoreceptors, light acts directly on pigmented cells without passing through a focusing apparatus. Rudimentary refractive mechanisms appear in many invertebrates. Cephalopods have a complete cornea–lens system similar to that of the vertebrates (Fig. 9-1). Invertebrate eyes develop from epithelial (skin) tissue, and are arranged in single cells or collections of cells in flat, cup-shaped, and compound or faceted structures in different species. The study of invertebrate vision is an active field. Some invertebrates have extremely large visual cells that are easier to study in detail with microelectrode and microinjection methods than the physically smaller visual cells of the vertebrates. Extensive knowledge of the genetics of some invertebrates, such as the fruit fly, provides a potentially powerful research tool. Mutations have been produced that alter the development and functional mechanisms of the invertebrate visual system. The availability of such mutants suggests numerous new approaches in research on vision. Invertebrate photoreceptors are not within the scope of this chapter, but leading references to their comparative structural organization and some of the studies of their functional mechanisms are provided.[1–5]

9.1.4. Overview of the Structural Organization and Functional Mechanisms in the Vertebrate Eye

The optical function of the vertebrate eye can be compared to a modern camera. Figure 9-1 is a diagram of a cross-section of the human eye. Light enters through a compound lens (composed of the cornea and the focusable lens), passes through a variable

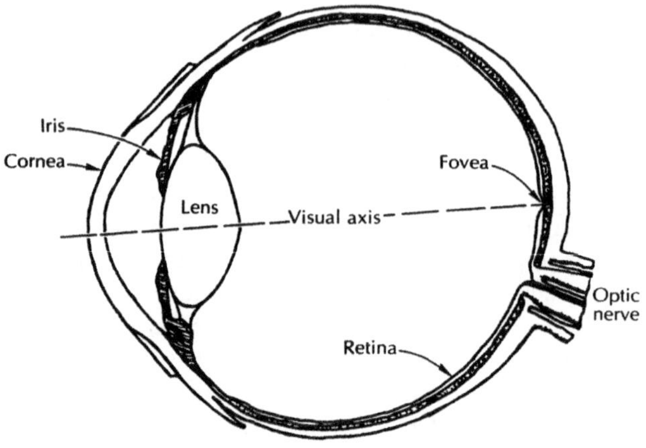

(a)

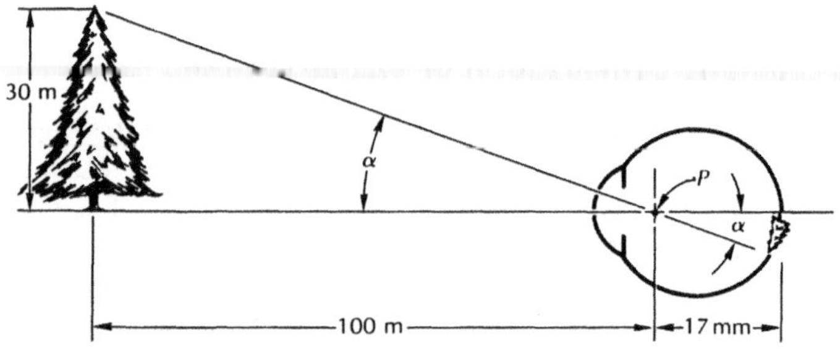

(b)

Fig. 9-1. (a) Cross-section of the human eye. The parts of the eye are discussed in the text. The space inside the eye is filled with a clear, jellylike substance called the vitreous humor. The point at which the optic nerve leaves the retina is devoid of photoreceptors and is responsible for the eye's "blind spot." (b) A simplified optical diagram of the eye looking at a tree. There is a point in the center of any lens system where light rays are not deflected from their incident paths. In the human eye this point, labeled P in the diagram, is about 17 mm in front of the retina. The size of the image on the retina can be determined using simple trigonometry: in this example the image size = 17 mm × 30 m/100 m = 5.1 mm, and the visual angle $\alpha = \arctan(30/100) \cong 16°$. (From Ref. 89.)

aperture (the iris), and is focused on the thin, light-sensitive layer (the retina). Higher (simian) primates and many birds have a tiny specialized portion of the retina, the fovea, which has the most closely spaced photoreceptors in the retina and gives the highest resolution image. The lower part of Fig. 9-1 is a simplified optical representation of an eye looking at a tree, showing an inverted image on the retina, which is analogous to the image of an object on film in a camera.

The eye is, however, functionally much more complex than a camera. It rapidly delivers an image in "real time" to the brain and can control its sensitivity to yield useful signals over a 10^{10} range of light intensities. The "mechanical" iris aperture of the eye can account for, at most, 10^2 of the intensity range of sensitivity adaptation of the eye. When completely dark-adapted, the eye is much more sensitive than camera film. Man has not constructed an apparatus to equal the eye in overall performance, but the closest type of device would be an ultrasensitive black-and-white TV camera combined with a "hi-fi" color TV camera that produces useful signals over an enormously wide range of light intensities.

The electrical response of the photoreceptor is processed by a series of interconnected cells in the retina and is coded into a train of electrical pulses that are passed on to the brain. Figure 9-2 shows a simplified drawing of the various interconnected cells that receive, process, and transmit signals from the retina. A curious feature of the vertebrate retina is that it is oriented "backward" with respect to the incident light. Light is normally incident from the bottom of the page in this drawing, and passes through several layers of other cellular material before reaching the receptors. The light receptor cells are at the top of the drawing and are located on the "backside" of the retina. The backward orientation of the retina seems to have little consequence for the efficient light detection by the

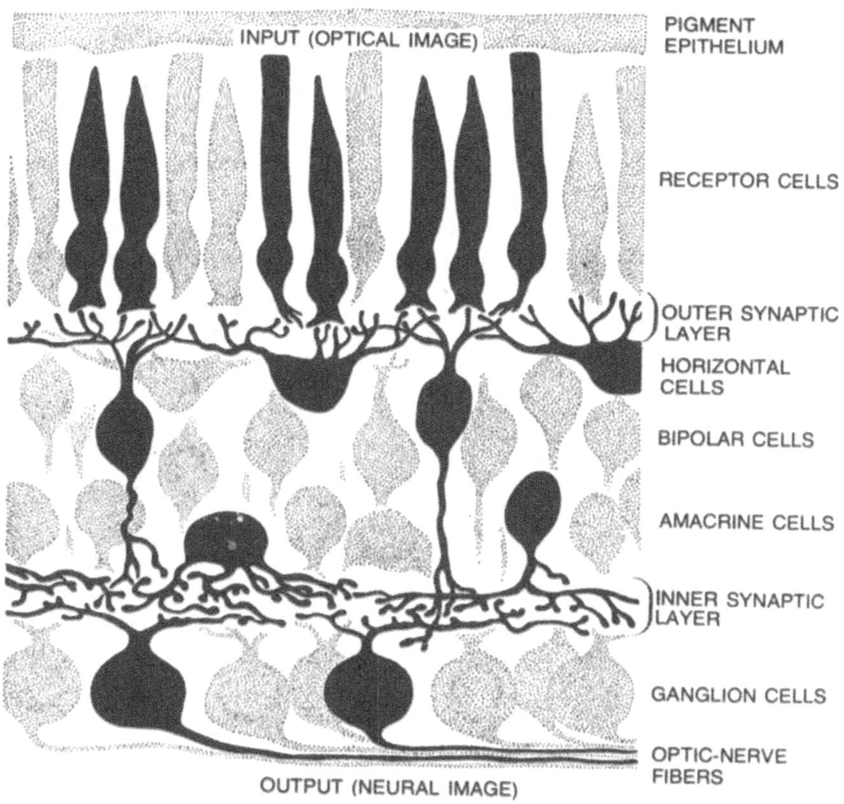

Fig. 9-2. Simplified drawing of the vertebrate retina based on a micrograph of a retina prepared by the Golgi method, which selectively stains a few cell bodies and their processes. The various cell types and the major layers of the retina are labeled. Note that light is normally incident from the bottom of the picture, as discussed in the text, and is absorbed by the rod and cone receptor cells at the top. Receptor signals are communicated across the retina by the horizontal cells, while the bipolar cells carry information to the ganglion cell layer. The amacrine cells also communicate across the retina. The ultimate output to the brain is transmitted through the axons of the ganglion cells. [From F. A. Werblin, *Sci. Am.* **228,** 70–79, (1973).]

photoreceptor cells. The layers of the retina closest to the front of the eye, which are traversed first by light, have little or no visible light absorption. The pigment epithelium cells are situated above the photoreceptors. The pigment epithelium cells supply nutrients to the photoreceptors. The pigment epithelium cells absorb much of the light that is not captured by the photoreceptor cells, and therefore cut down on the degradation of the image resolution that would be caused by light scattering between photoreceptors.

The layer below the receptors contain the bipolar and horizontal cells, each of which is connected to several photoreceptors and interconnected to one another. This layer together with the lowest layer comprises the signal-processing ''brain'' function of the retina. The processed signals are communicated to the lower region of the retina containing the amacrine cells. The ganglion cells send pulsed signals to the brain through the optic nerve fibers. After the initial detection of light, an extraordinarily varied and complex chain of neurobiological events is activated.

9.1.5. Objectives of This Chapter

In this chapter we will first consider the structure of the receptor cells; second, the current knowledge of the visual pigment rhodopsin; third, the various measurable light responses; and finally, a current view of the mechanism of action of the visual system. Useful contributions to the understanding of visual photoreceptors have come from a variety of fields that range from quantum mechanics, photochemistry, and physical chemistry, through biochemistry and electrophysiology. Much has been learned about photoreceptors in recent years, and this information has also revealed the vast territory that yet must be explored.

9.2. STRUCTURE OF THE VERTEBRATE RETINA

9.2.1. Retina Structure as Observed with the Light Microscope

The sizes and shapes of cells and their placement in a tissue may be studied with the light microscope. To prepare cells for microscopic observation, the structural integrity of the tissue is first stabilized by cross-linking it with a variety of chemicals (e.g., formaldehyde and/or glutyraldehyde). The tissue is then embedded in a wax or plastic supporting medium, and thin sections are cut with a sharp knife on a microtome. A number of different dyes or stains may be used to enhance the visibility of various components of the tissues.

Figure 9-3 shows a light micrograph of a thin section of monkey retina tissue. The retina has a layered appearance, and different types of cells or parts of cells occur in each layer. The pigment epithelium, which nourishes the photoreceptors, is the thin layer of cells (one cell thick) that extends from the edge of the retina, from 100 to about 105% on the retina depth scale in Fig. 9-3. The receptors near the top of this micrograph extend from about 45 to 100% of the width of the retina. Each cell has a single nucleus, and the relatively large number of nuclei (dark spots between 60 and 80% thickness) show the photoreceptors to be the most numerous cell type in the retina.

There is a lower density of cells in the layer below the receptors in the region of 20–45% of the width of the retina, as can be seen from the number of cell nuclei in the inner

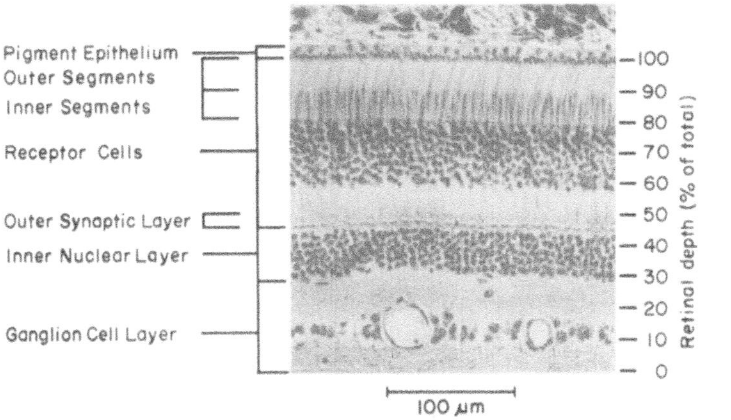

Fig. 9-3. Light micrograph of a thin section of a rhesus monkey peripheral retina stained with hematoxylin and eosin, which makes the cell nuclei appear as dark blobs. The pigment epithelium at the back of the eye is at the top of this picture, and light is normally incident from the bottom as in Fig. 9-2. The identification of the cell layers is discussed in the text. [Adapted from K. T. Brown, K. Watanabe, and M. Murikami, *Cold Spring Harbor Symp. Quant. Biol.* **30,** 457–482 (1965).]

nuclear layer (located at 30–40% of the retina depth). The cells in the layer below the photoreceptors include the bipolar and horizontal cells, each of which receives input from several receptors. The amacrine and the ganglion cells are found in the lowest layer (at 0–20% of the width of the retina), and they are relatively sparse as can be seen by the small number of nuclei around the blood vessels at around 10%. The fractional thickness of each of these layers varies from the central to the peripheral retina. The signals detected by the receptors are processed in the lower layers of the retina, and the outputs of the photoreceptors are coded into the firing rate of a smaller number of output channels, the neurons. In experimental animals it has been found that the signal processing is such that some output neurons respond to specific small moving spots, while some respond to light/dark edges of various angular orientations.

Photoreceptor cells have several morphological domains. Outer segments contain the visual pigments and are embedded in fingerlike structures that protrude from the pigment epithelium cell layer at the top of Fig. 9-3 (back of the eye). The outer segments are found in the space between 90 and 100% of the width of the retina. The so-called inner segments occupy the space between 80 and 90% of the width of the retina; the outer nuclear layer, between 60 and 80%; a fiber layer, between 50 and 60%; and a synaptic region near 45%. Under the light microscope, at least two morphological types of photoreceptors are seen in most portions of the retina in nearly every vertebrate species, although they cannot be distinguished in Fig. 9-3. Cells called rods usually have larger outer segments and thinner inner segments, and cells called cones usually have shorter outer segments and much thicker inner segments.

The distribution of rods and cones, and their morphological appearance, is not constant across the retina. Figure 9-4 shows the rod/cone distribution across the human retina. The rods predominate in the outer portions of the retina, where only a small number of cones are present. The cones in the central fovea are very closely packed and

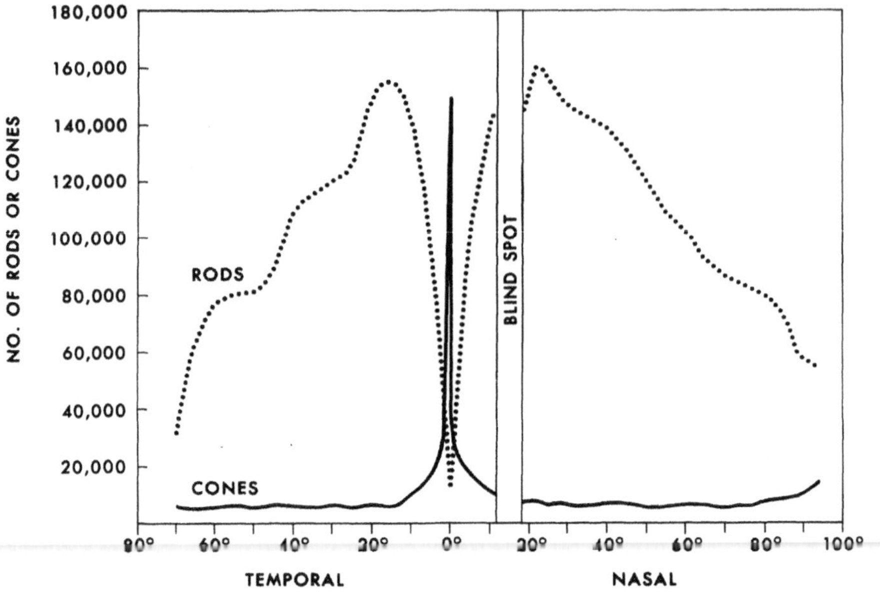

Fig. 9-4. The distribution of rods and cones across the human retina. The central point at 0° is the fovea, which is densely packed with cones. There are few cones in the peripheral retina, which is rich in rods. The blind spot is devoid of photoreceptors, since this is where the optic nerve leaves the retina, and the blood vessels pass through. [After G. Osterberg, *Acta Ophthal. Suppl.* **6,** 1–102 (1935).]

have long, thin outer and inner segments, so that they look similar to the rods in the periphery. The closely packed cells provide high spatial resolution, which produces high visual acuity. Many species show a specialized central region of the retina that is less developed than the fovea. Every retina has a blind spot, as shown in Fig. 9-4. Here there are no receptors because this is where the optic nerve leaves the retina and blood vessels pass through. The brain somehow "fills in" the image in the blind spot, so that we are normally unaware of it.

There are no blood vessels in the receptor layer. The receptors must get their nutrients from the dense bed of blood vessels that run behind the pigment epithelium located just above the point marked 100% retinal thickness in Fig. 9-3. Most mammals have retinal blood vessels that run through the deep layers of the retina far from the receptors. Two blood vessels can be seen in the section of monkey retina in Fig. 9-3. They appear as large openings in the ganglion cell layer near the level marked 10% of the thickness of the retina. Many species have no blood vessels in the retina, and all retinal nutrients are obtained from behind the pigment epithelium.

9.2.2. Photoreceptor Ultrastructure as Observed with the Electron Microscope

The electron microscope provides much more highly magnified views of cellular structure than the light microscope. Figure 9-5 shows a thin section of a portion of the photoreceptor layer of a retina. The light-sensitive outer segments are made up of dense stacks of membranes (the three large objects in the figure covered with dark lines). The adjoining portion of each photoreceptor cell above the outer segment in Fig. 9-5 is called the inner segment, and is extremely rich in cellular "power houses," the mitochondria. Numerous mitochondria are seen near the top of Fig. 9-5, and one mitochondrion is labeled M.

Figure 9-6 shows an even higher magnification view of a thin section of a rod outer segment that clearly reveals the stack of membranes surrounded by a plasma membrane that is marked with an arrow in the figure. The internal membranes occur in flattened structures called disks. That each flattened disk is hollow can be seen by the presence of open loops at the ends, where the disk membrane turns the corner, and by a thin light space down the center of some of the disks. X-ray diffraction measurements on photoreceptors in living retinas clearly show an aqueous space within the disk,[6] but this space has been partially or totally collapsed during the dehydration of the specimen in preparation for observation in the electron microscope. The space between the inner surfaces of the disk membrane is about 2.5 nm under physiological salt conditions and has been shown to respond to osmotic variations.[6] The diagram in Fig. 9-7 shows the membrane spacing. Each disk is separated from its neighbor by about 15 nm, and is about 15 nm thick, making the disk-to-disk repeat distance approximately 30 nm.

Fig. 9-5. Electron micrograph of an oblique section through a glutyraldehyde-osmium-fixed guinea pig retina showing portions of three outer segments and their attachment to the inner segments. The inner segments contain a pair of centrioles (C), one of which gives rise to a modified cilium composed of a connecting piece (CP), which gives rise to the outer segment during embryological development. The outer segments are made of dense stacks of disk membranes (D), and the inner segments are densely packed with mitochondria (M). [From A. W. Clark and D. Branton, *Z. Zellforsch. Mikrosk. Anat.* **91**, 586–603 (1968).]

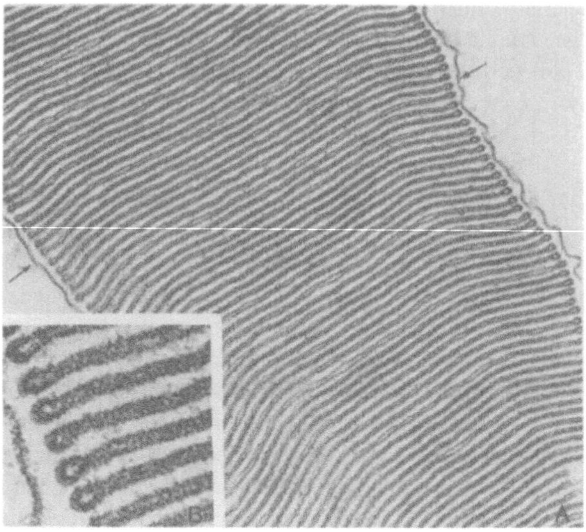

Fig. 9-6. (a) Electron micrograph of a cross-section of a portion of a rod outer segment showing the stacked arrangements of the disk membranes. The outer enveloping plasma membrane is indicated by the arrows (×24,000). (b) Magnified view of the edges of the disk membrane and the plasma membrane. Note the loop in the cross-section where the disk membranes fold back on themselves to form closed bags (×24,000). [After M. J. Hogan, J. A. Alvarado, and J. E. Weddell, *Histology of the Human Eye*, Saunders, Philadelphia (1971).]

The disk membranes are formed by infolding the plasma membrane. In the rods, most of the disks are pinched off and sealed to form closed membrane structures that are not contiguous with the plasma membrane. The disk membranes appear to be anchored in position at the edges by thin filaments that presumably are made of cytoskeletal proteins (perhaps similar to actin, tubulin, etc.). The cone disks do not pinch off, but remain contiguous with the plasma membrane. The general outlines of the synthesis and turnover of the photoreceptor membranes has been worked out using radioactive precursors and autoradiography.(7) The rod disks are replaced every 10 days. New disk components are made in the inner segment, and new disks are added to the rods at their base just above the junction with the inner segment. There is a circadian rhythm process that controls the steady-state length of the photoreceptors. Small portions of the tips of the rod outer segments are shed from the cell each morning, and are then phagocytized (internalized) and digested by the pigment epithelium. There is evidence that disk components that are still in an undamaged condition are recycled and reused for new disk synthesis. The pigment epithelium cells also phagocytize the tips of the cone outer segments, but this appears to occur in the evening.(8)

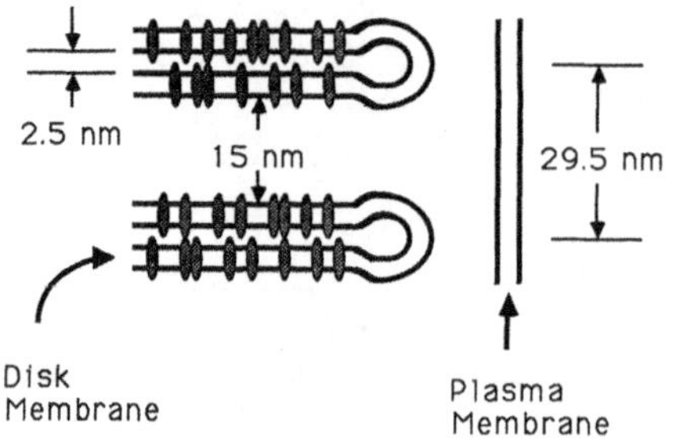

Fig. 9-7. Membrane separations in retinal rod outer segments. The disks show a 29.5-nm repeat, are separated by about 15 nm, and have about a 2.5-nm space inside the disk. The crosshatched oblongs represent the approximate disposition of rhodopsin in the membrane. The plasma membrane is diagrammed on the right, and its rhodopsin content is not shown.

The rod outer segments are larger than the cone outer segments, and in retinas of most species the rods are much more numerous; consequently, there is much more rod pigment than cone pigment in most retinas. Furthermore, the rod outer segments are more easily isolated and purified than those of the cones; consequently nearly all the work on the properties of the photoreceptor outside the intact retina has been done on rod material. Studies of the composition, structure, and mechanisms of the cone color receptors remains largely for the future.

The outer segment portion of the photoreceptor cell is highly membranous, with its membranes arranged in regularly spaced stacks. The edges of two disk membranes are shown in Fig. 9-7, which diagrams the spacings between membranes and the relation of the disk membrane to the plasma membrane. This organization of regularly stacked membranes is superficially similar to the lamellae of the chloroplasts of green plants, and both types of membranes contain light-absorbing pigments. Each membrane is very thin (4–5 nm), and in both the outer segment and the chloroplast lamellae, the probability of light absorption has been increased by the evolution of stacks of membranes. The functions of the two types of membranes are, of course, very different. The chloroplast converts the energy of absorbed photons to chemical energy (Chapter 12), while the outer segment acts as a detector and, as will be seen in Section 9.5, uses absorbed photons as signals to control the flow of metabolic energy.

Recently, a bacterial membrane protein system was characterized[9] that has several properties in common with the visual pigment in the outer segment membrane[10] and whose function is very similar to that of the chloroplast. The membranes are not stacked but occur as sheets on the surface membrane of the bacterium *Halobacterium halobium*. This organism and other related species grow in very high salt, and are called halophiles. The light-absorbing molecule, called bacteriorhadopsin, appears to have structural similarities to the visual pigment rhodopsin.[10,11] However, the primary function of bacteriorhodopsin is to pump protons across the membrane in the light in the process of photosynthesis rather than to act as a light sensor. Related membrane proteins in *H. halobium* pump chloride ions across the membrane in the light, and others function as sensory light receptors by which the motile organisms avoid blue light and are attracted to red light.[9] Further discussion of bacteriorhodopsin and related proteins is beyond the scope of this chapter, and additional information can be found in the references cited.[9–11]

9.2.3. Molecular Structure as Observed by Chemical Studies and Low-Resolution Diffraction Measurements

The electron microscope falls short of providing structural information at the molecular level. Fortunately, detailed information is obtainable by the use of other tools. The intact retina is isolated by gently peeling off the eye cup. A crude suspension of rod outer segments may be prepared by gently shaking the retina in physiological salt, sucrose, or other solution. In this sort of preparation, the outer segments often break off the retina with a portion of the inner segments attached. These preparations contain considerable mitochondrial contamination, since the inner segments are rich in mitochondria. The purest preparations of outer segment material have been obtained by a more vigorous homogenization procedure, which breaks the rod outer segments (ROS) into large fragments. The fragments are purified on sucrose density gradients, since the ROS disk membranes appear to have a unique mass density.(12) Purified disks can be separated by floatation on Ficoll (a sucrose polymer) at very low osmotic pressure.(13)

The ROS are bright pink and "bleach" to a pale yellow if exposed to light from most of the visible spectrum. The pink color is maintained under infrared or dim red light. Most of the preparations and manipulations are carried out under dim red light for structural studies. However, if very detailed studies are being carried out on physiological mechanisms, manipulations must be carried out in complete darkness using infrared image converters.

The pigment portion of the ROS cannot be extracted into aqueous buffers but remains tightly bound to the membranes. The pigment can, however, be solubilized in a number of mild detergents without changing its characteristic absorption spectrum. The pigment color is found to be due to a protein called rhodopsin. If a very powerful detergent, such as sodium dodecyl sulfate (SDS), is used to extract the pigment, the native protein structure largely unfolds, and the color of the rhodopsin solution bleaches in the dark. In SDS-treated suspensions, the components of the ROS can be separated by molecular weight using polyacrylamide gel electrophoresis. Many ROS proteins are free in the cytoplasm or are weakly bound to the membrane surface and can be washed off in low-salt solutions free of divalent ions. Figure 9-8 shows a densitometer scan of a separation of the proteins of purified ROS membranes, which have been washed free of most of the components that are not tightly membrane-bound. At least 90% of the protein of the ROS appears in a single dense band on the gel, which has been shown to be opsin. Opsin is the protein portion of the visual pigment that has been stripped of its visible light absorbing chromophore by the SDS extraction. The identities of the smaller peaks have not been fully established. There are several enzyme activities in the ROS, as will be discussed in Section 9.5.5. Some of the small peaks correspond to ROS enzymes. Cattle rhodopsin moves on the gel with an apparent molecular weight of 35,000; however, the amino acid

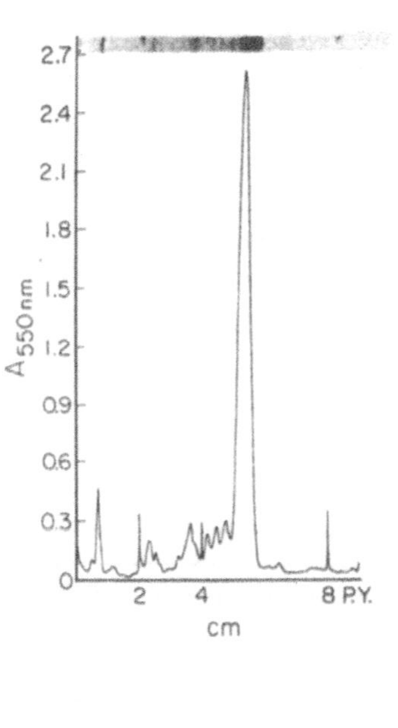

Fig. 9-8. Polyacrylamide gel electrophoresis pattern of purified bovine rod outer segment membranes dissolved in sodium dodecyl sulfate (SDS). The gels were stained with coomassie blue after electrophoresis to visualize the protein components. A photograph of the stained gel is shown at the top, and a densitometer scan of the gel (at the stain absorption maximum) is plotted below. The major peak is opsin, the protein portion of of rhodopsin. The very sharp peaks at 2, 4, and 8 cm are needle marks made for length calibration. The tiny peak at 9 cm is the Pyronine Y tracking dye (P.Y.), which indicates the distance traveled by this rapidly moving marker. The other small peaks are protein components. SDS polyacrylamide gels separate by molecular weight, the smallest components having the highest mobility. Bovine opsin moves with an apparent molecular weight of 35,000, and nearly all of the other minor components have larger apparent molecular weights. [After D. S. Papermaster and W. J. Dreyer, *Biochemistry* **13,** 2438–2444 (1974).]

sequence shows the molecular weight to be about 41,000 daltons including 2000 daltons of covalently bound carbohydrate. Numerous protein components that may be washed off the membrane are present on the left side of the gel (Fig. 9-8). The functions of some of the minor components (a GTP-binding protein, rhodopsin kinase, and a phosphodiesterase) have been identified and will be discussed in 9.5.

The disk membrane contains about 43% protein, 53% lipid, and 4% carbohydrate by dry weight. The lipid portion of the membrane can be extracted with organic solvents and separated by thin-layer chromatography. This treatment also releases the chromophore, a derivative of vitamin A (Section 9.3.1). Most of the lipids are phosphatidylcholine, phosphatidylethanolamine, phosphatidylserine, and phosphatidylinositol, which occur in the ratio 40 : 45 : 15 : 1. These lipids each have different, hydrophilic, phosphodiester "head" groups esterified to a glycerol molecule that is also esterified to two long-chain hydrophobic fatty acids. There is about 5% by weight of cholesterol present, and a trace of sphingomyelin. Small amounts of glycolipids may be present, but they have not been characterized.

The fatty acids in the phospholipids of the membrane are strikingly polyunsaturated; in cattle, rat, and frog ROS, about 45% of the fatty acids are docosahexaenoic acid (22 carbons and 6 double bonds, designated a 22 : 6 fatty acid).(14) Most of the other fatty acids found are the saturated palmitic (16 : 0) and stearic (18 : 0) acids. A typical phospholipid has a saturated fatty acid esterified to one of the hydroxyl groups of glycerol and a 22 : 6 fatty acid esterified to the other.(15)

When polyunsaturated fatty acids are removed from a rat's diet, the ROS tenaciously retain their polyunsaturated fatty acids,(17) while most other cells and organs in the animals lose their polyunsaturated fatty acids under these conditions. The exceptionally high level of 22 : 6 fatty acids and the tenacity with which they are retained by the ROS and by brain synaptic endings suggests that these fatty acids may be important for the functioning of both of these membrane systems.(17)

The regular array of stacked membranes in the ROS diffracts X-rays and neutrons. Analyses of X-ray or neutron diffraction patterns of the ROS in the intact retina have been carried out by several groups to a resolution of about 2.5 nm. These studies show the membrane spacings present in live unfixed tissue (Fig. 9-7) and indicate that much of the lipid in the disk membrane appears to be arranged in a lipid bilayer. A bilayer consists of two sheets of lipids arranged with their polar head groups in contact with water, the nonpolar hydrocarbons of the fatty acids forming an oily interior. Both X-ray and neutron diffraction analyses on rod membranes agree that about 50% of the mass of rhodopsin is embedded in the hydrocarbon region of the membrane and that a substantial mass protrudes from the membrane.(18,19) Models of the diffraction pattern are most consistent with the rhodopsin molecule fully spanning the disk membrane, as shown schematically in Fig. 9-7.

The localization of rhodopsin in the disk membrane has been studied by several other methods. Membranes have a region of structural weakness where the hydrocarbon chains meet in the center of the two halves of membrane bilayers. Freeze-fracture electron microscopy reveals the textures of the cleavage planes that pass through this region of weakness. Proteins that protrude into the hydrocarbon region are observed as bumps on the fracture surfaces. Rhodopsin remains attached to the hydrocarbon layer on the outer half of the disk membrane when samples are frozen and fractured. These experiments show that rhodopsin is asymmetrically associated with the two halves of the disk membrane.

Several laboratories have shown that treatment of the disk membrane with a number of proteases, which presumably do not penetrate the membrane, cleaves the peptide chain of rhodopsin.[18,20] Rhodopsin has two short oligosaccharide chains, consisting of mannose and *N*-acetylglucosamine, which protrude into the aqueous region.[20] Chemical labeling experiments imply that the lipids are very asymmetrically disposed across the membrane, with most of the phosphatidylethanolamine and much of the phosphatidylserine distributed toward the outside surface of the membrane.[21] The bulk of the phosphatidylcholine is presumably on the inside surface.

The amino acid sequence of rhodopsin reveals seven nonpolar stretches that are adequate to span the hydrocarbon portion of the membrane. This and other evidence has led to the model for the structure of rhodopsin (Fig. 9-9), which consists of seven slightly bent transmembrane helices with a retinal molecule oriented perpendicular to the helix axis in a "basket" of bent helices.[10]

The proposed structure has now been found to be consistent with sequences of seven different visual pigments from fruit flies, sheep, horses, pig, and human rod and three cone pigments.[11] In addition, similar structures have been proposed for the β-adrenergic and muscarinic membrane receptors.[11] The bovine sequence, fitted into the model shown in Fig. 9-10, suggests a strongly asymmetric charge distribution across the mem-

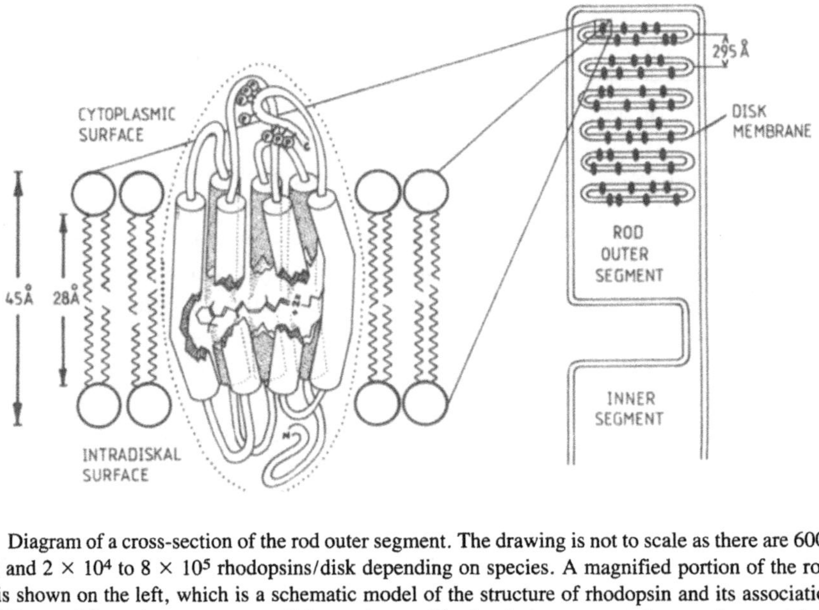

Fig. 9-9. Diagram of a cross-section of the rod outer segment. The drawing is not to scale as there are 600–2000 disks/rod and 2×10^4 to 8×10^5 rhodopsins/disk depending on species. A magnified portion of the rod outer segment is shown on the left, which is a schematic model of the structure of rhodopsin and its association with the lipid bilayer of the rod outer segment disk membrane. Rhodopsin is represented as an elongated bundle of somewhat irregular helices (bent where prolines occur in the helices) embedded in the lipid bilayer. The 11-*cis*-retinal-binding site, oriented nearly parallel to the membrane plane, is schematically shown. The seven known phosphorylation sites are marked with the letter "P." [After E. A. Dratz and P. A. Hargrave, *Trends Biochem. Sci.* **8**, 128–131 (1983).]

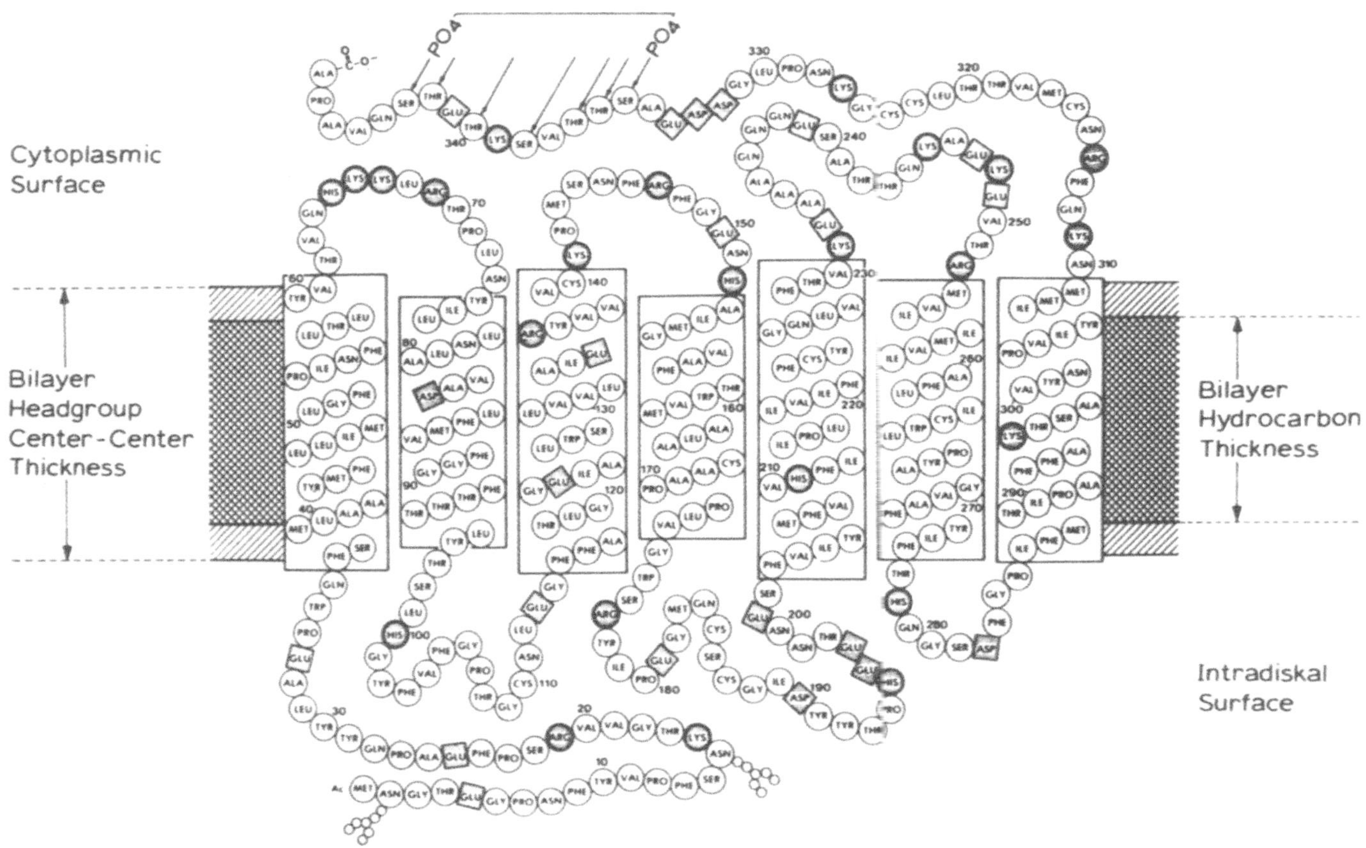

Fig. 9-10. A model for the organization of the polypeptide chain of bovine rhodopsin in the disk membrane bilayer. The carboxyl terminus is exposed to the rod cell cytoplasm. Amino acids with potentially positively charged side chains are shown as outlined shaded circles, and amino acids with potentially negatively charged side chains are shown as shaded squares. Carbohydrates are indicated by the strings of smaller circles near the amino terminal end. The crosshatched areas at either side indicate the lower polarity part of the bilayer. The central crosshatched part represents the width of the hydrocarbon region of the bilayer. [After E. A. Dratz and P. A. Hargrave, *Trends Biochem. Sci.* **8,** 128–131 (1983).]

brane with a positive charge on the cytoplasmic surface from the protein[10] and a negative charge from the lipid.[21]

The two-dimensional arrangement of rhodopsin molecules in the plane of the disk membrane appears to be a dynamic one. Spectroscopic measurements on the rhodopsin chromophore have provided information on the rotation and translation of rhodopsin in the membrane (Section 9.3.2).

9.3. PIGMENT SPECTRA AND THE CHROMOPHORE

9.3.1. Chromophore Structure and Spectra

Wald showed many years ago that the chromophore in rhodopsin is the 11-*cis* isomer of the polyene vitamin A aldehyde. This molecule is commonly called 11-*cis*-retinal, and its structure (I) together with the structures of some related molecules are shown in Fig. 9-11. When rhodopsin is exposed to light before the chromophore is isolated, retinal is found to be isomerized to the all-*trans* form (Structure III). In Wald's laboratory, Bownds found that retinal could be fixed to the ε-amino group of a lysine in rhodopsin by sodium borohydride reduction after light exposure, and therefore is bound by a Schiff base linkage

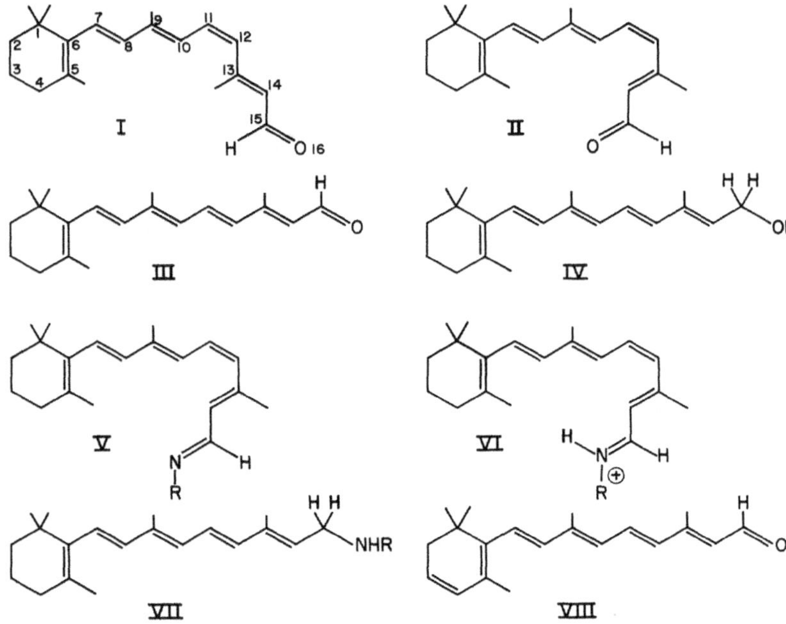

Fig. 9-11. Structures of different vitamin A derivatives important in vision. I, 11-*cis*-retinal. II, 11-*cis*-12-s-*cis*-retinal. III, all-*trans*-retinal. IV, all-*trans*-retinol, the alcohol form of this compound. V, 11-*cis*-12-s-*cis*-retinal Schiff base of 11-*cis*-N-retinylidene-alkylamine. VI, protonated 11-*cis*-12-s-*cis*-retinal Schiff base or protonated 11-*cis*-12-s-*cis*-retinylidene-alkylamine. VII, reduced all-*trans*-retinal Schiff base or all-*trans*-*N*-retinylalkyl amine. VIII, all-*trans*-3-dehydroretinal or vitamin A_2 aldehyde.

to the protein before reduction. A Schiff base linkage, shown in Structure V of Fig. 9-11, is equivalent to Structure II combined with an R-NH_2 group. The reduced Schiff base linkage of all-*trans*-retinal is shown in Structure VII of Fig. 9-11. Borohydride is water soluble and cannot reach the chromophore linkage (buried deep in the protein) in the dark, but when rhodopsin is exposed to light, the structure opens up and the borohydride can reach the linkage and reacts rapidly.

Several different geometrical isomers of retinal can be prepared, including 7-*cis*, 9-*cis*, 11-*cis*, 13-*cis*, 9,11-di-*cis*, 9,13-di-*cis*, 11,13-di-*cis*, and 9,11,13-tri-*cis*. The naturally occurring 11-*cis* isomer is energetically the least favorable mono-*cis* isomer because of considerable steric hindrance between the 13-methyl and the 10-hydrogen (Fig. 9-11, Structure I). If retinal in solution is thermally isomerized to an equilibrium mixture by iodine catalysis or is photoisomerized, much less 11-*cis* isomer is formed than any other mono-*cis* isomer, and 11-*cis* is exceeded in amount by some of the di-*cis* forms.

Theoretical calculations led to the proposal that the steric strain in 11-*cis* retinal is relieved by twisting about the 12–13 single bond. A twisted nonplanar conformation close to the 12-s-*cis* conformation (Fig. 9-11, Structure II) (the s-*cis* refers to the *cis* conformation about a single bond) is predicted to be of the lowest energy, and a nonplanar conformation close to the 12-s-*trans* (Fig. 9-11, Structure I) has a slightly higher energy. Both forms were also predicted to be highly twisted about the 6–7 bond and to be closest to a 6-s-*cis* conformation. The crystal structure of 11-*cis*-retinal was deduced soon after, and these theoretical predictions were borne out quantitatively in the crystal state. A nonplanar 12-s-*cis* conformation was found with a 40° twist out of plane, and the 6–7 bond was twisted 135° from the s-*trans* orientation, close to a 6-s-*cis* conformation. Nonplanar conformations about these bonds also persist in solution, as shown by nuclear magnetic resonance (NMR) measurements. Bacteriorhodopsin contains all-*trans*-retinal as its chromophore. Recent evidence from solid state NMR indicates that the all-*trans*-retinal in bacteriorhodopsin is in the unexpected 6-s-*trans* conformation,[(22)] and this should be examined in rhodopsin.

One other closely related polyene chromophore is found in some visual pigments. This chromophore is called 3-dehydroretinalaldehyde or vitamin A_2 aldehyde (Fig. 9-11, Structure VIII) and contains an additional double bond in the six-membered ring. A useful shorthand for vitamin A_2 aldehyde is A_2 and, correspondingly, A_1 for vitamin A aldehyde. Visual pigments from hundreds of species have now been analyzed. They all contain either the A_1 or A_2 chromophore. The retinas of some species contain both A_1- and A_2-based visual pigments, and the relative portion depends on age, season, and habitat.[(23)]

Schiff base derivatives of A_1 aldehyde chromophores may be easily prepared with aliphatic amines in organic solvents, and are found to have absorption maxima in the near-UV region. The solid and dashed lines in Fig. 9-12 show spectra for Schiff bases of A_1 in the 11-*cis* and all-*trans* conformations, respectively. Analogous compounds of the A_2 chromophore absorb maximally at somewhat longer wavelengths than those that contain the A_1 chromophore. The 11-*cis* chromophores exhibit two peaks on the short-wavelength side of the strongest band; these are called *cis* peaks and are found in every *cis* polyene. The all-*trans* chromophore shows a much stronger main band, and weaker short-wavelength bands than the *cis* forms. It retinal Schiff bases are protonated by the addition of acid, the spectral maxima are shifted to longer wavelengths. The dotted line in Fig. 9-12

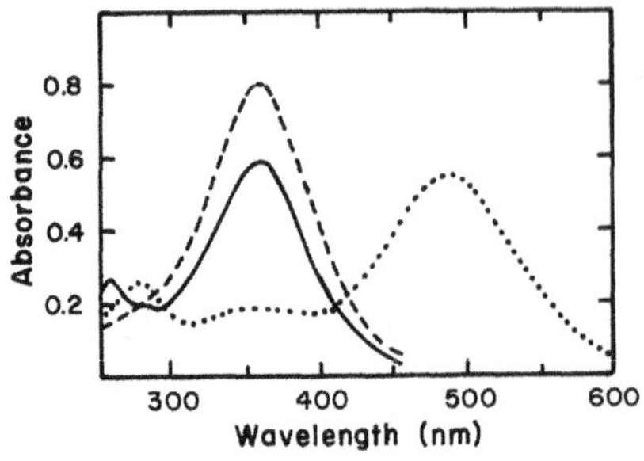

Fig. 9-12. Absorption spectra of retinal Schiff bases made with 1-amino-2-propanol. Dashed curve, all-*trans-N*-retinylidene-1-amino-2-propanol. Solid curve, 11-*cis-N*-retinylidene-1-amino-2-propanol. Dotted curve, protonated Schiff base made with equimolar $HClO_4$ added to 11-*cis-N*-retinylidene-1-amino-2-propanol. All spectra are in 1,2-dichloroethane. [Redrawn from data of J. O. Erickson and P. E. Blatz, *Vision Res.* **8**, 1367–1375, (1968).]

shows a spectrum of the protonated Schiff base of 11-*cis* A_1 whose structure is shown in Fig. 9-11, Structure VI. The chromophore in most visual pigments absorbs considerably farther to the red than do the protonated Schiff bases in solution; the additional red or blue shift due to the protein environment will be discussed in Section 9.3.3.

9.3.2. Visual Pigment Spectra

The absorption spectra of visual pigments (VP) may be measured in four ways. First, VP may be extracted from isolated retinas with a mild detergent solution. Second, they may be measured *in situ* in single-photoreceptor outer segments by direct spectrophotometry using microbeams. Third, they can be estimated in intact animals (including man) with a technique where the amount of light that is reflected from the back of the eye is measured before and after an exposure to bright light, which bleaches a large fraction of the pigment.(24) Finally, absorption spectra can be measured via the electrical responses of individual cells as a function of different colors of light.(25) All VPs show similarly shaped absorption spectra with large extinction coefficients close to 40,000 liters mol^{-1} cm^{-1}, and all of the native VPs appear to be bleached with high quantum efficiency (near 0.67).

A striking feature of VP spectra is the great range of wavelengths of maximum absorption that are observed in photoreceptors of different types in various species. The absorption maxima range from about 435 nm in the blue to about 625 nm in the red. Mixtures of different VPs are present in intact retinas of most species and in the detergent extracts of their retinas. The composition of the mixtures and the absorption maxima of the components can be estimated by a technique called "partial bleaching," if the maxima of the different pigments are not too close to one another. This method measures the difference in absorbance between visual pigment samples that have been exposed to different sets of light conditions. "Bleaching" is a term used for the loss of visible absorption that occurs after exposure to light. In the typical approach, the visual pigments in a mixture that absorb farthest to the red are first bleached by successive exposures to bright, far-red light. Difference spectra are recorded after each bleach until no further change is observed. Then the bleaching light is decreased in wavelength, and the process is repeated. The peaks in the difference spectra are close to the VP absorption maxima, if the wavelengths of the bleaching lights are varied sufficiently finely to provide reasonable assurance that predominantly one form of VP is being bleached at each stage.

Absorption spectra of purified bovine rhodopsin and of the isolated 11-*cis*-retinal are shown in Fig. 9-13. The large peak at 498 nm is called the α peak, the small peak near 340 nm is called the β peak, and the strong peak at 278 nm is called the γ peak. The γ peak is largely due to the aromatic amino acids, tyrosine and tryptophan, in the protein, although there is probably some contribution by the chromophore at these wavelengths. The wavelength sensitivity of rod vision matches the absorption spectrum of rhodopsin, with the exception that humans do not see UV light (although some insects do) because it is strongly cut off through absorption by the cornea, lens, and vitreous humor (see Fig. 9-1). Humans see rather well in UV light if the lens is removed from the eye and the focus is corrected with UV light-transmitting glasses.

In recent years microspectrophotometry on single isolated photoreceptors has been particularly useful for visual pigment spectral studies. Single cones have distinct spectra and cones with red, green, and blue absorption maxima are present in many species.[(27)] These observations support the trichromatic theory of color vision: three types of cones exist, each with a visual pigment of a different color. Some species appear to have only one or two different colors of cone visual pigments. Some species, such as birds and amphibians, also have highly colored oil droplets in the inner segment that the light must pass through to reach the cone outer segments. Different cones in the same species may have different colored oil droplets, which apparently act as color filters. Therefore, two types of spectral sensitivity controls are found in cones: wavelength regulation of the visual pigment absorption spectra and, in some cases, an oil drop color filter.

In simian primates the central region of the retina is called the macula. This region includes the cone-rich fovea and a surrounding rod-rich area. The macula contains a mixture of yellow carotenoic pigments called lutein and zeaxanthin.[26] The absorption spectra of the macular pigment shows a series of relatively sharp peaks in the blue and green, and the spectral sensitivity of the human eye shows corresponding dips that are due

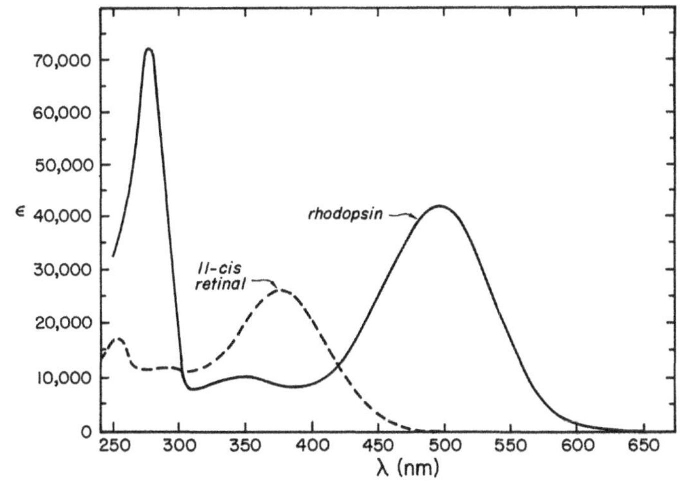

Fig. 9-13. The absorption spectrum of purified bovine rhodopsin solubilized in the detergent (solid line) and 11-*cis*-retinal (dashed line) in Ammoxy LO detergent. [From T. G. Ebrey and B. Honig, *Q. Rev. Biophys.* **8,** 129–184 (1975).]

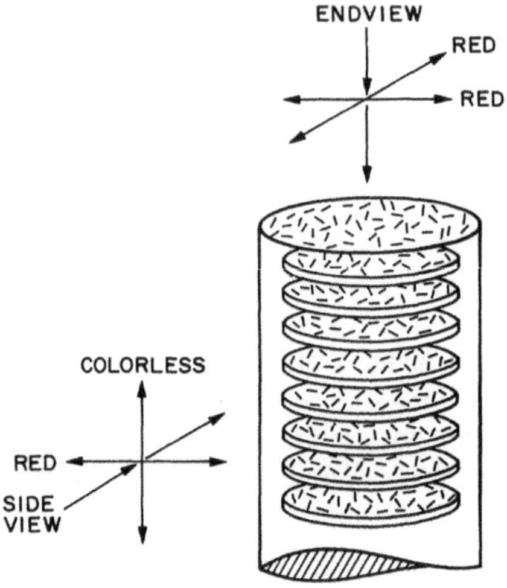

Fig. 9-14. Schematic diagram illustrating the dichroism of outer segments, when viewed from the side with polarized light; and the lack of dichroism, when viewed from the end (see text for more details). The rhodopsin chromophores are highly restricted and must lie in the plane of the disk membrane, but are randomly oriented within that plane.

to the screening of the photoreceptor by absorption by this pigment. The function of the macular pigment is not yet understood, but may protect against energetic blue light or quench toxic oxygen species.(26)

Microspectrophotometry, using polarized light, has shown that the rhodopsin chromophore is highly oriented in the disk membrane.(27) Figure 9-14 shows the geometry of viewing the rod either "end on," as light is incident on the intact eye, or "side on," as can be done experimentally in isolated photoreceptors. If the photoreceptor is viewed side on with polarized light, rhodopsin absorbs light much more strongly if the electric vector of the light is polarized in the plane of the membrane than if it is polarized perpendicular to the membrane. Viewed visually side on in the light microscope, the rods appear nearly colorless when the light is polarized parallel to the long axis of the rod. However, the rods are a distinct red when the light is polarized parallel to the plane of the disk membrane (Fig. 9-14). Viewed end on, as they normally function, rods are equally red with any direction of polarization. The red color results because the visual pigment preferentially absorbs the green and blue (see Fig. 9-13), and allows the red to pass through. The ratio of the absorbances for the parallel and perpendicular orientation of the polarized light (the dichroic ratio) is about 5. This large dichroism requires that the chromophore be strongly oriented in the plane of the membrane.

Light incident on an intact eye always hits the photoreceptor end on and travels parallel to the long axis of the photoreceptor. Rhodopsin is optimally oriented to absorb this light, for the electric vector of the end-on incident light lies in the plane of the disk membrane and has no component perpendicular to it. The rod appears uniformly red, independent of the polarization (i.e., there is no dichroism), when viewed end on, as shown in Fig. 9-14. Consequently, vertebrate eyes are not sensitive to the plane of polarization of the light, as are some insect eyes.

If a brief, bright, polarized flash is incident on the rod end on, a large transient dichroism is induced that quickly dies away.(28) The initial dichroism is due to photoselection of rhodopsins that happen to be oriented optimally for absorption of the polarized flash and are efficiently bleached. However, the rhodopsin molecules are rapidly rotating in the plane of the membrane, and therefore the dichroism quickly disappears. The half-time for the rotational disorientation is about 20 μs at 5°C. An increase in temperature of 10°C reduces the half-time for disorientation by about a factor of 2.5 (the Q_{10}). The lateral motion of rhodopsin in the plane of the membrane has also been measured spectrophotometrically.[29,30] The lateral diffusion constant is 4×10^{-9} cm^2/s at 25°C with a Q_{10} of 2. It is clear that the lipid matrix of the disk membrane presents a relatively fluid medium for rhodopsin movement. The diffusion constants of rhodopsin have been used to estimate an effective viscosity of about 2 poise in the plane of the membrane, which is 200 times that of water and is of the order of the bulk viscosity of olive oil.[28–30] Glutyraldehyde fixation stops the rotational and the translational motion of rhodopsin. The electrical response of the retina is abolished by glutyraldehyde fixation, although rhodopsin can still be bleached, and appears to undergo its normal conformational changes after fixation (see the discussion of the early receptor potential in Section 9.4.5). It is thought that movement of rhodopsin molecules in the plane of the membrane is required for photoexcitation.[34]

9.3.3. The Question of Wavelength Regulation of Visual Pigment Spectra

The wide range of absorption maxima found in visual pigments correspond to a chromophore excitation energy difference, which is "tuned" over nearly 0.75 eV in the excitation energy of the chromophore. Part of this range is spanned by the use of two slightly different chromophores, A_1 and A_2. However, the A_1 pigments themselves show a range of nearly 0.5 eV (430–575 nm), which is a very large range for a single chromophore. Different protein environments around the chromophore-binding site in different visual pigments must be able to cause a change of the absorption maximum of the chromophore from blue to red. This influence of the protein on the chromophore is called wavelength regulation.

When a VP protein is coupled with the A_2 chromophore, the VP always absorbs at longer wavelengths than if the same protein is coupled to the A_1 chromophore. If these A_2-A_1 shifts are compared for a number of VPs, the difference is largest for the pigments that absorb furthest to the red, and gets smaller for pigments that absorb to the blue. In fact, a plot of the A_2-A_1 shift vs. the absorption maxima of either the A_1 or the A_2 VP is smooth and nearly linear, and the A_2-A_1 difference converges to near zero at the short-wavelength end. This phenomenon is called "convergence."

A large red shift relative to the isolated chromophore (Fig. 9-12) can be obtained by protonation of the Schiff base linkage to rhodopsin (see Fig. 9-11, Structure VI). Resonance Raman spectroscopy experiments have shown that the Schiff base linkage in rhodopsin is protonated. The nature of the counterion to the protonated Schiff base can have a large effect on the spectral maximum. Carboxyl groups are probably the only available anions in the protein environment, and variations in the properties of the anion are probably not sufficient to explain the range of absorption maxima, although the separation distance of the counterion from the protonated Schiff base may have an important effect. Additional negative ions (or dipoles) in the oily region near the retinal have been proposed to complete the wavelength tuning.[31]

The amino acid sequences of the human red, green, and blue cone pigments and the

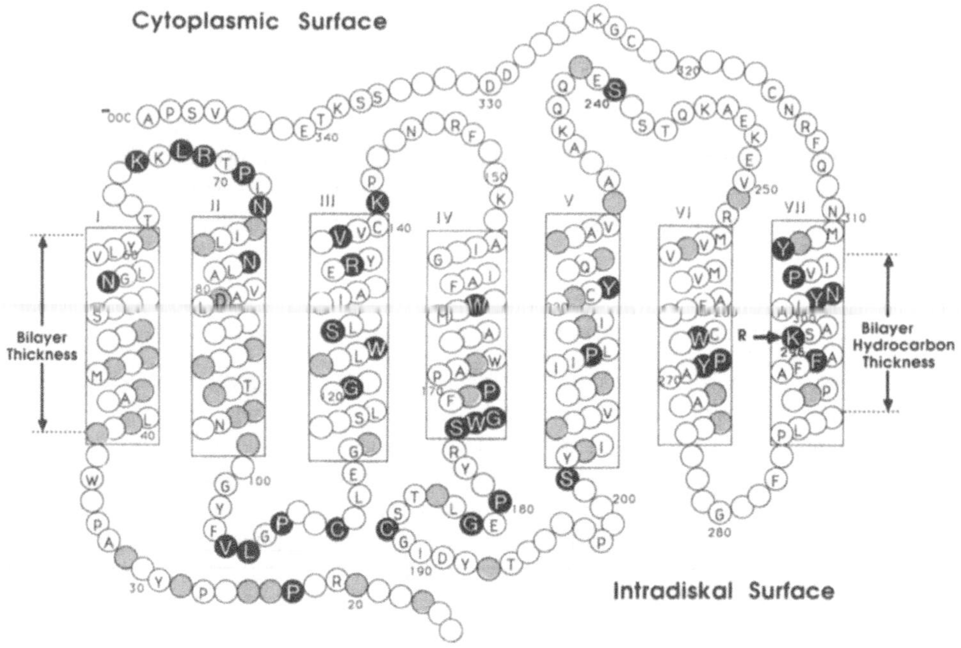

Fig. 9-15. Model for the transmembrane arrangement of amino acids in vertebrate pigments in which each amino acid is represented by a circle.[34] The structure is composed of seven transmembrane helices (I–VII) with connecting aqueous loops and N- and C-terminal "tails." The degree of shading of the different residues represents the extent of conservation of the amino acids in seven sequences, including two fly pigments, bovine rhodopsin, human rhodopsin, and three human cone pigments. Black residues are identical in all seven sequences, open residues labeled with a letter are identical in at least four sequences, and shaded residues are very similar in chemical properties in at least four sequences. The one-letter code is: A = alanine, C = cystine, D = aspartic acid, E = glutamic acid, F = phenylalanine, G = glycine, H = histidine, I = isoleucine, K = lysine, L = leucine, M = methionine, N = asparagine, P = proline, Q = glutamine, R = arginine, S = serine, T = threonine, V = valine, W = tryptophan, Y = tyrosine. **R** at Lys296 denotes retinal. [After P. A. Liebman, K. R. Parker, and E. A. Dratz, *Ann. Rev. Physiol.* **49,** 765–791 (1987).]

rod pigment have recently been obtained.[32,33] The absorption maxima of these pigments range from 440 to 540 nm. The sequences are highly homologous and are perfectly consistent with the transmembrane model originally proposed for bovine rhodopsin,[10] which has seven highly hydrophobic helices spanning the membrane with hydrophilic loops connecting the helices on both the inside and outside aqueous surfaces (Fig. 9-10). This model also accommodates several other visual pigment sequences that have now been determined.[11,31–34] Figure 9-15 shows the consistent features of the amino acid sequences in comparing seven visual pigment sequences. Most of the charged groups in the sequences are assigned to loops between helices located in the aqueous phase. Very few charged groups are present in the regions assigned to the intramembrane helices, but a few charged groups are thought to be present.[11,34]

The Glu-Arg or a Asp-Arg at 134–135 are found in the region assigned to helix III in every pigment. The opposite paired charges are stabilized in the oily membrane region. The negative end of the dipole from the paired charges may be oriented toward the retinal, but since the dipolar pair of charges is present in visual pigments absorbing over the full range of colors, it cannot have a differential effect on the colors of the different pigments.

The strongest support for the role of hydrophobic charges in contributing to the color of the visual pigments comes from analyzing the human rod, blue-cone, green-cone, and red-cone sequences.[32,34] Figure 9-16 shows a diagram of the charge location superimposed on the transmembrane model of the sequence. The blue-cone visual pigment has only the dipolar pair of charges on helix III and lysine 296. The human rod pigment adds two negative charges on helices II and III, and possibly a positive charge on helix V (a histidine that may or may not be protonated under physiological conditions). A doubly negative charge site on helix II is added to both the green and red pigments. Both pigments lose a negative site on helix III and lose the potentially positive histidine charge site on helix V.

The general picture that charges in the hydrophobic regions around retinal control visual pigment color is attractive, but some other mechanism must also contribute, e.g., to differentiate between the green and red pigments. If aromatic amino acid side chains are placed in the protein-binding pocket near retinal, they can also contribute to significant shifts in the absorption maximum.[35,36] Therefore, aromatic side chains might be used for "fine-tuning" the colors of the visual pigments.

9.3.4. Chromophore Analogs

Comparisons of the properties of the VPs formed from isomers of the naturally occurring A_1 and A_2 retinals provide some testable criteria for theories of VP spectra. The 9-*cis* and 9,13-di-*cis* isomers of these chromophores form pigments with the protein, in addition to the native 11-*cis* isomer. For some time chemists have been making synthetic vitamin A aldehyde analogs that differ in chain length and steric and conformational properties to test the restrictions in the retinal-binding site. It has been found that the rhodopsin-binding site is quite restrictive as to what sorts of alterations of the chromophore it will accept. Longer or shorter polyene molecules do not form VPs, and the six-membered ring must be present. Anywhere from zero to three double bonds in the ring is accepted. None of the methyl groups is required for the formation of VPs; however, some of the analogs that are missing methyl groups are not bound as tightly as the native

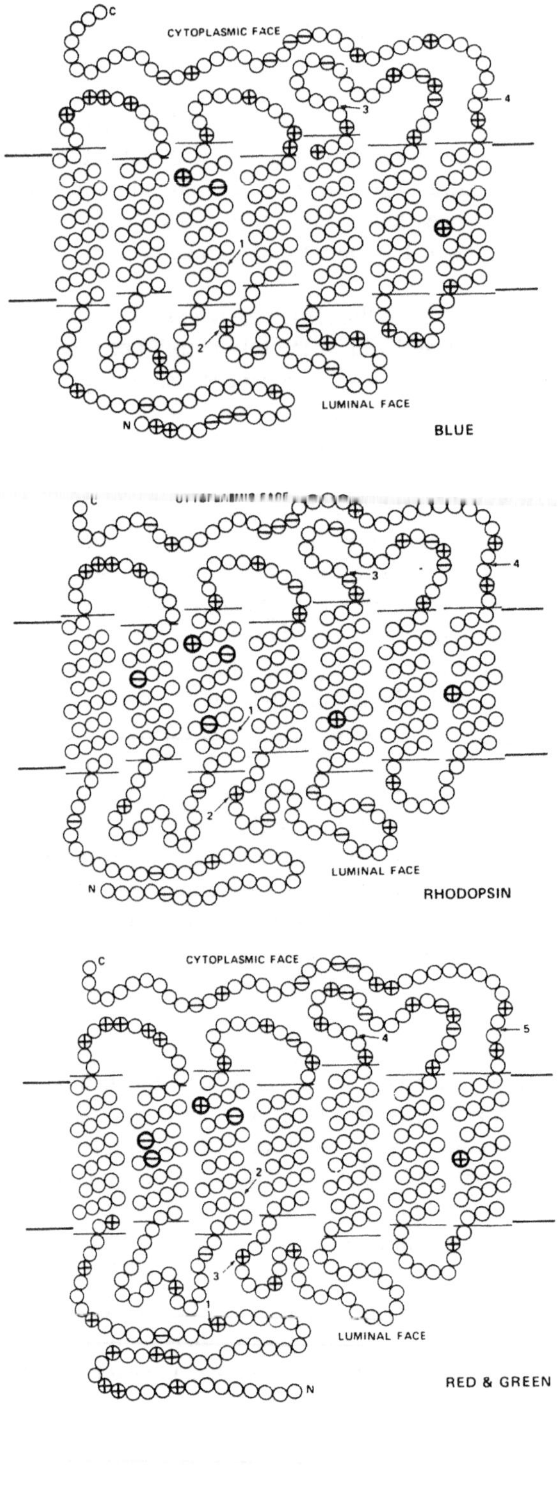

Fig. 9-16. The charged amino acids in the human blue, green, and red cone pigments (after Ref. 32) and rod pigment (after Ref. 33) are shown by the ⊕ and ⊖ symbols in the sequence. The sequence is arrayed according to the transmembrane folding model proposed for bovine rhodopsin (Fig. 9-10), and fit by all known visual pigment sequences (e.g., Fig. 9-15). Most of the charged groups are in contact with the aqueous media outside the membrane. The few charges in the hydrophobic region of the membranes are proposed to contribute to "tuning" the color of the visual pigment. The small numbers 1–5 denote the intron/extron boundaries in the genes.

chromophores. The pigment formed from the 9-desmethyl (lacking the 9-methyl in Fig. 9-11, Structure I) has a significantly different absorption maximum and circular dichroism intensity from that of the native chromophore. Most of the structural variations of the chromophores produce less stable pigments with opsin and also show a reduced photosensitivity compared with the native pigment. It is beyond the scope of this chapter to discuss these results in detail, but much of the research has been extensively reviewed in other works.[37–40] Retinal analogs may be very useful in future research on the interactions between the protein and the chromophore.

9.3.5. Photochemistry

The quantum efficiency of stimulation of rhodopsin is very high, near 0.67. It was pointed out in Section 9.3.1 that the bleaching of rhodopsin causes the isomerization of the chromophore from the 11-*cis* isomer (Fig. 9-11, Structure I) to the all-*trans* conformation (Fig. 9-11, Structure III). Conformational changes initiated by photon absorption trigger a series of changes that are discussed in the next section. After a photon is absorbed, experiments show that all the other reactions in vertebrate photoreceptors occur in the dark. The photochemistry in the system then centers around the isomerization mechanism, and whether the isomerization occurs by a singlet or a triplet pathway. The first detectable spectral change in rhodopsin appears in less than 6 ps (6×10^{-12} s). This observation does not explain the mechanism, but it excludes the involvement of large simultaneous changes in the protein-binding site. While 11-*cis*-retinal isomerizes predominantly by a triplet mechanism in solution, the protonated Schiff base isomerizes much more rapidly through a singlet mechanism.[40] These observations indicate that retinal may be a poor model system of rhodopsin photochemistry, and that the protonated Schiff base has important, different properties.

9.4. MEASURABLE RESPONSES TO LIGHT STIMULI

9.4.1. Spectra and Properties of "Bleaching" Intermediates

When rhodopsin is exposed to light, it undergoes a series of color changes, from its original reddish-pink to yellow in a series of steps. As has been mentioned, this entire process is called "bleaching." Figure 9-17 shows the series of intermediate steps in the bleaching process that have been identified, their absorption maxima, and the approximate half-times of interconversion in the retina or in isolated outer segments. The first change from rhodopsin to prelumirhodopsin is light-induced, all other changes being dark reactions. All the intermediates may photobackreact to reform rhodopsin; however, it has been claimed that a form of metarhodopsin I does not photobackreact.[41] A single bright flash, no matter how bright the flash, does not bleach all the rhodopsin because of the photobackreactions. The amount of bleaching in a saturating flash depends on the relative absorbance of rhodopsin and the intermediates integrated over the spectral distribution of the flash, the photobackreaction quantum efficiency, and the lifetimes of the intermediates relative to the flash length. The lifetimes of the intermediates, and presumably the photobackreaction quantum efficiency, depend on the environment of rhodopsin. For exam-

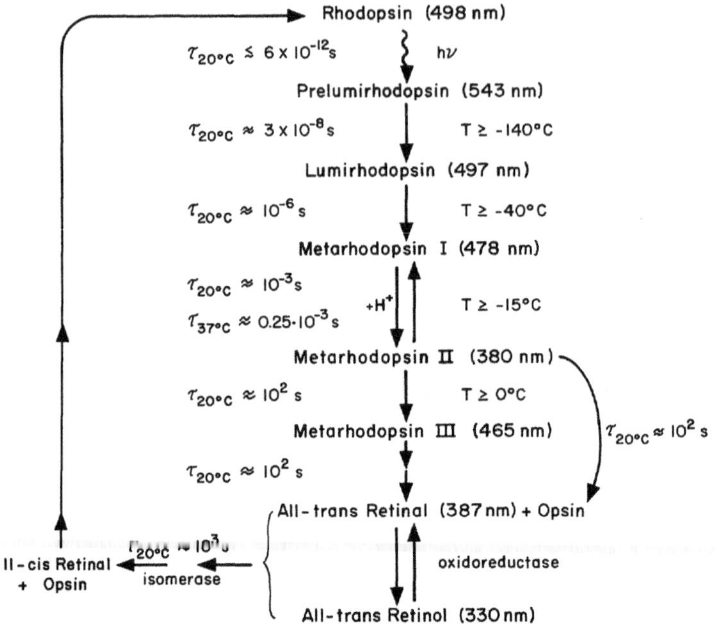

Fig. 9-17. Stages in the bleaching and regeneration of rhodopsin. The initial photoevent is shown as a wavy line, and the subsequent thermal, dark reactions are shown as solid lines. The absorption maximum of each species is shown in parentheses after the name of the species, and is the value observed for bovine rhodopsin. The approximate half-times for interconversion of each species at 20 and 37°C are denoted by 20°C and 37°C, respectively. The approximate temperatures above which the interconversions occur are indicated by "T." More than one step may be involved in the interconversions of metarhodopsin I to metarhodopsin II, metarhodopsin III to all-*trans*-retinal + opsin, and the interconversion of all-*trans*-retinal to 11-*cis*-retinal.

ple, in many detergents the meta I → meta II rate is accelerated about 1000-fold.[(42)] In these detergents, nearly complete bleaching can be obtained in a single 1-ms flash. In contrast, in the lipid environment of the intact membrane it is difficult to obtain more than a 50% bleach in a single bright flash.

The kinetics of the spectral changes are rather strongly temperature-dependent. At 37°C, meta II appears with a half-time of 0.25 ms in the rat retina, whereas below 0°C meta II appears very slowly, and meta I may be thermally trapped below −15°C. Bovine rhodopsin, which has been studied in most detail, exhibits the same general pattern of temperature-dependent kinetics, with some quantitative differences. For example, lumirhodopsin can be trapped below about −40°C, and prelumirhodopsin below −140°C. Prelumirhodopsin can also be formed at liquid nitrogen temperature (−196°C), and can be photobackreacted to rhodopsin at this temperature; however, some 9-*cis* pigment (called isorhodopsin, λ_{max} = 486 nm in cattle rhodopsin) is also formed in the backreaction under these conditions. The prelumirhodopsin formed at −196°C completes the bleaching sequence when it is warmed to room temperature in the dark, resulting in isomerization of the retinal to the all-*trans* form. The chicken cone pigment, called iodopsin, behaves differently; prelumi-iodopsin formed at −196°C backreacts to iodopsin upon warming, and does not therefore get distorted enough in prelumi-iodopsin to attain isomerization.

A proton is taken up by rhodopsin during the meta I → meta II transition. This is surprising because the absorption spectrum of meta II rhodopsin is similar to that of an unprotonated Schiff base and that of meta I rhodopsin is similar to that of a protonated Schiff base. In fact, Raman and Fourier transform infrared vibrational spectroscopy shows that the chromophore loses a proton upon meta II formation. The protein must take up this proton, as well as one additional proton during the meta I → meta II transition.

A dependence of the ratio of meta I/meta II on pH can be observed at 5°C, where bovine meta II has a lifetime of several hours. Meta II predominates at acid pH, and the ratio shows a simple Henderson–Hasselbach equilibrium relationship with a pK near 6.5. Meta III and later products form at higher temperatures. The color of the bleaching product, formed after meta III, is also pH-dependent in a manner that appears to reflect variable protonation of a retinal Schiff base linkage.

The meta I/meta II ratio is pressure-dependent at 5°C, when rhodopsin is located in the lipid environment of the disk membrane.[43] Application of pressure increases the proportion of meta I at the expense of meta II. Complete reversal of the meta I ⇌ meta II equilibrium to meta I is obtained at about 3×10^6 kg/m^2. These observations show that the formation of meta II results in a volume expansion of the membrane. This volume expansion depends on the lipid–protein interaction in the membrane because the pressure effect is absent if rhodopsin is extracted into detergent micelles. In detergent extracts, rhodopsin also exposes some additional sulfhydryl groups to reaction with sulfhydryl reagents upon formation of meta II. In the disk membrane, however, rhodopsin does not appear to expose additional sulfhydryl groups upon the formation of meta II.[44] These observations are some of several indicating that detergent extracts are not adequate models for studying the behavior of the formation of meta II, and that the membrane environment is very important. The steps up to and including the meta I to meta II interconversion are particularly significant because spectral changes subsequent to meta II formation are much too slow to account for the brief time delay of visual excitation after light absorption.

It was mentioned in Section 9.2.3 that the retinal photoreceptor membranes contain a large concentration of highly unsaturated fatty acids with 22 carbons and 6 double bonds (22:6), which are also rich in synaptic endings.[17] It has recently been found that 22:6 is required in membrane phospholipids to reconstitute the pressure dependence of the meta I ⇌ meta II reaction.[45] The double bonds are thought to coil up like a spring, and are able to accommodate conformational changes of the protein at meta II that expand the area of the protein in the membrane plane.

9.4.2. The Visual Cycle

The bleaching of rhodopsin leads to the slow liberation of all-*trans*-retinal. In the eye, bleached rhodopsin is regenerated by the enzymatic conversion of all-*trans*-retinal into 11-*cis*-retinal. The 11-*cis*-retinal spontaneously combines with bleached rhodopsin (opsin) to regenerate rhodopsin. There are many intermediate steps in the formation of 11-*cis*-retinal that are diagrammed in Fig. 9-15. The entire process from the bleaching of rhodopsin to its regeneration is called the visual cycle. In isolated retinas, purified ROS, or detergent extracts, there is little or no spontaneous regeneration. If 11-*cis*-retinal is added to the bleached rhodopsin in a membrane environment or in one of a very few detergents (e.g., digitonin, Tween-80, octylglucoside, nonylglucoside, or dodecyl maltoside), regeneration of the visual pigment color may be accomplished. Retinal and other aldehydes make membranes leaky to ions if the membranes contain the lipid phosphatidylethanolamine, but the corresponding alcohols do not.[46] The ROS membranes are rich in phosphatidylethanolamine. Retinal does not build up in the retina upon bleaching because an oxidoreductase activity (Fig. 9-17) reduces it rapidly to retinol (Fig. 9-11, Structure IV) with the consumption of reduced nicotinamide adenine dinucleotide phosphate (NADPH). Some of the retinal-reducing activity survives in isolated ROS, but the NADPH is largely depleted.

The events in the visual cycle, after retinal formation, have not been fully established in detail. If a large amount of pigment is bleached, the retinal is reduced to retinol; the retinol is esterified to a fatty acid and transported to the pigment epithelium. The visual cycle is completed by the isomerization to 11-*cis*, return to the ROS, ester hydrolysis, and oxidation of retinol to retinal; but the localization and sequences of the events are not entirely clear. There is some evidence for a "short visual cycle" where retinal is isomerized without leaving the ROS, if bleaching of the visual pigments occurs at fluences closer to physiological light levels.[47]

9.4.3. Psychophysics and Action Spectra

One powerful way to study the sensory system is to ask an intact animal to report in some way what it detects, or if it detects a carefully controlled stimulus. This method is called psychophysics. Another powerful method, widely used in photobiology, is to measure the quantum efficiency of a response as a function of wavelength of excitation. The resulting curve is called the action spectrum of the response (Section 1.5.5).

Using psychophysical methods, it has been shown in numerous experiments that the action spectrum of vision matches precisely the absorption spectrum of the visual pigments present. These measurements are straightforward in pure rod or pure cone retinas, but are more complex in mixed retinas that have both rods and cones. The sensitivity of a mixed retina matches the absorption of rod visual pigment under dim light, and matches the absorption spectra of cone visual pigment under brighter light stimuli. After stimulation with bright light, the photoreceptors show a marked decrease in sensitivity; this phenomenon is called light adaptation. It is a familiar experience to be momentarily blinded by going from a dim to a brightly lighted environment, then to adapt rapidly to seeing well in the bright environment. Similarly, in going from bright to dim light the photoreceptors regain sensitivity. This phenomenon is called dark adaptation and is also familiar from everyday experience.

Dark adaptation shows complex kinetics. The cones recover sensitivity rapidly; however, while the cones are active, they seem to repress the recovery of the rods.[24] The rods recover sensitivity much more slowly than the cones, with a half-time of about 5–8 min. Complete dark adaptation of the rods takes about 30 min. Understanding the mechanisms of light and dark adaptation will require more knowledge of the internal mechanisms of the photoreceptors (Section 9.5).

9.4.4. The Electroretinogram (ERG)

The electrical responses of the photoreceptor cells to light stimuli may be measured in a number of ways. A simple and surprisingly informative method is to measure the potential generated across the whole eye in intact animals, excised eyes, or across isolated retinas, in response to light stimuli. Plots of these potentials are called the electroretinogram (ERG) and may be measured with gross electrodes (cotton wicks soaked in saline) or with extracellular microelectrodes. These responses have been separated into components that come from the photoreceptor cells (the a wave) or P III, from deeper layers of the retina (b wave), and from the pigment epithelium (c wave). Therefore, the a wave of the ERG allows the functioning of the photoreceptors to be studied in intact retina.

The a wave can be isolated in retina preparations by the addition of the amino acid, sodium aspartate, which serves to deactivate the horizontal cells below the receptors.[48] The a wave is highly sensitive to anoxia and to the removal of glucose, and therefore requires oxidative energy sources. Furthermore, the a wave is suppressed by the removal of Na^+ or by the addition of Ca^{2+}. The appearance of the a wave shows a several-milliseconds delay after a stimulus flash.

9.4.5. Early Receptor Potential (ERP)

Very bright stimulus flashes produce a potential that appears with no detectable delay after the stimulus.[49] This response has been called the early receptor potential (ERP). In contrast with the a wave of the ERG, the ERP is highly resistant to anoxia and is not dependent on the identity of the salts present. The ERP depends linearly on the amount of rhodopsin bleached, whereas the other potentials generated by the retina are highly nonlinear in the amount of rhodopsin bleached.[50] Utilizing the photobackreactions of the bleaching intermediates, it has been shown that the ERP appears to monitor the interconversion of rhodopsin intermediates.[51]

The ERP (which Hagins calls the fast photovoltage, or FPV) appears to correspond to charge displacements across the plasma membranes of the rod.[52] Presumably rhodopsin changes its conformation during interconversion between bleaching intermediates, and these conformational changes lead to a change in the distribution of charges across the membrane. The rhodopsin in the pinched-off disk membranes cannot contribute any net voltage to the ERP, so the ERP is due only to changes of the rhodopsin in the enveloping plasma membrane (including, of course, the disks that have not yet pinched off at the base of the outer segment). The spectral changes after bleaching result primarily from the rhodopsin that is in the disk membrane, since 97–99% of the ROS membrane area is due to the disk. The close similarity of the kinetics of the spectral changes and the ERP changes, in both the forward and backward directions, indicate that rhodopsin in the disk and plasma membranes behave very similarly.

The ERP tells us little about the mechanism of visual excitation but does indicate that rhodopsin undergoes a conformational change associated with charge movement after bleaching and that a conformational change is particularly evident upon formation of meta II rhodopsin. Evidence for conformational changes upon meta II formation have been detected in a number of experiments,[34] including the chemical reactivity experiments described in Section 9.4.1. All the intermediates up to, but not including, meta II form in dehydrated gelatin films prepared from digitonin micelles. If the light-exposed films are hydrated in the dark, meta II is formed and is therefore the first intermediate that requires an aqueous environment. Meta II formation is also blocked in lipid-stripped rhodopsin.[42] All these considerations suggest that the formation of meta II requires a substantial conformational rearrangement and the formation of meta II is a leading candidate for the step that triggers visual excitation.[34]

9.4.6. Intracellular Recording of Photoreceptor Responses

Intracellular recording involves piercing a cell with a glass micropipet electrode of very small diameter and directly measuring the electrical potential across the cell membrane. Electrical contact is provided by filling the micropipet with a concentrated salt solution. Intracellular recording has been very useful in understanding the electrical properties and ion flows in a variety of cells, the most prominent examples being nerve axons and muscle cells. The relatively small size of photoreceptor cells has made intracellular recording difficult. However, evolution of the technique has made it possible to obtain intracellular recordings from several different species of amphibian retinas, which happen to have particularly large photoreceptors. Numerous studies have also been made of the intracellular responses of other cells deeper in the retina that are stimulated by the photoreceptors.[53]

The photoreceptors do not give spike responses like axons but give smoothly graded responses to graded light stimuli.[48,54] This behavior is not unusual for short neurons. The photoreceptors have a negative potential inside, as do all cells; however, they are unusual in that they hyperpolarize (become more negative inside) rather than depolarize in response to stimuli. The graded hyperpolarizing property upon stimulation is also shown by the horizontal and bipolar cells deeper in the retina. The ganglion cells behave like "standard" neurons; they depolarize in response to stimuli and have an output that consists of a train of spikes.

Intracellulear recordings have been made from both cones and rods, as shown by spectral sensitivities, characteristic saturation behavior, and intracellular dye marking after the recordings. The rods show much slower electrical kinetics than cones, particularly in the recovery phase after a brief stimulus. Rods have been shown to be electrically connected to neighboring rods in many species and to receive additive signals from them. This additive behavior contributes to the high sensitivity of rod vision, although it must cause reduced acuity (spatial resolution) in rod vision. With broad-field stimulation of neighboring rods, individual rods give detectable signals at much less than 1 quantum absorbed per rod. Rods give responses that increase with intensity up to about 300 quanta absorbed per rod. Above this intensity, the rod response saturates in amplitude and shows increasingly slow recovery after brighter supersaturating stimuli. The voltage response amplitude of photoreceptors follows the relation $V/V_{max} = I^n/(I^n + K^n)$, where V is the voltage response, I is the stimulus intensity, n is a constant that ranges from 0.8 to 2.0 depending on the photoreceptor; K is the stimulus intensity that produces half-saturation, and has a value of about 30 quanta for rods.

Dark-adapted cones give detectable responses from about 100 photons absorbed per cone, and increase with intensity up to about 10^4–10^5 photons per cone.[54] The half-saturating intensity is about 1200–1300 photons per cone. These numbers are only approximate, but it is clear that the cones are much less sensitive than the rods. Upon exposure to conditioning background lights, cones are desensitized and give detectable responses to flashes above the background, up to about 10^8 quanta absorbed per cone. The two types of receptors, rods and cones, together cover a useful dynamic range of intensities of about 10^{10}.

9.4.7. Extracellular Microelectrode Recording

The spatial localization of currents may be measured in slices of retina.(55) This method measures voltage gradients between two relatively low-resistance extracellular microelectrodes. The extracellular medium has a small but significant resistance, and currents between two regions of the tissue produce a small voltage between two electrodes placed in these regions. The spatial distribution of voltage gradients along the photoreceptors has been used to locate sources and sinks of current, and to measure the magnitude of these currents.

In the invertebrate retina there is an increase in Na^+ current into the photoreceptor outer segment in the light,(56) but the opposite result is found in vertebrate photoreceptors.(5) For example, in the rat retina there is a large Na^+ current ($\approx 10^7$ Na^+ ions rod $^{-1}$ s^{-1}) from the inner segment to the outer segment in the dark, and this "dark current" is decreased by light. One photon absorbed per rod decreases the current by about 1%. The decrease is linear up to about half the maximum response, approximately 30 photons absorbed per rod. The full response curve to flash illumination follows the relation $V/V_{max} = I/(I + 30)$,(59) which is similar to that found by intracellular recording on other rods.

A high concentration of Ca^{2+} (about 10 mM) reversibly mimicked the effect of light in repressing the photoreceptor Na^+ current.(57) It was proposed that Ca^{2+} was the intracellular signal that controls Na^+ permeability but that extracellularly applied Ca^{2+} did not get into the cell very effectively. The Na^+ dark current runs down in the presence of ouabain (a Na^+-K^+ ATPase inhibitor) and the rate of rundown is slowed by light exposure or by extracellular Ca^{2+}. The "calcium-coupling" hypothesis proposes that Ca^{2+} is the intracellular signal for visual excitation. More will be said about this in Section 9.5.3.

9.4.8. Rod Outer Segment Ion Currents Measured by Osmotic Responses

Fresh isolated frog or rat ROS shrink rapidly after hyperosmotic shock.(58) The rods remain shrunken in most salt solutions. However, in NaCl solution in the dark, the rods rapidly recover to their native length. If shrunken rods in NaCl are illuminated or exposed to high Ca^{2+} concentrations (10 mM), they do not recover their length. These and other experiments show that light and extracellular Ca^{2+} can control the Na^+ flux into outer segments. The recovery after a hyperosmotic NaCl shock is easily lost if the preparation is not kept highly concentrated in ROS or if the ROS age more than about 15 min after the retina is dissected.

The rods act as if they have a "spring" in the 15-nm space between the disks; this space collapses in high salt but returns to 15 nm in NaCl solutions in the dark, as shown by freeze-fracture measurements of the disk membrane spacing. The ROS plasma membrane appears to be permeable to NaCl in the dark, which allows recovery from the osmotically induced shrinkage. From the rate of recovery, it has been calculated that 10^9 Na^+ ions/s pass through the plasma membrane of each frog rod in the dark and that the absorption of 1 quanta/rod decreases the Na^+ permeability by 1–3%.(58)

9.4.9. Direct Measurements of the Photocurrent of Single Rods

Studies of the photocurrent of retinal rods has been facilitated by a technique whereby the outer segment is sucked into a snug-fitting pipet tip.[59,60] Most of the plasma membrane current of the rod passes through the suction pipet. This large pipet has a much lower resistance than the tiny intracellular electrodes, and therefore the suction pipet has a far superior signal-to-noise ratio. The rod response to single photons is very reproducible,[60] and this indicates that a sufficiently large number of internal transmitter molecules (greater than many hundreds) must be produced per photon to keep the statistical fluctuation in the response so low. A single absorbed photon decreases the light sensitive membrane current by 3–5%,[55,60] and fewer than 100 photons saturate the response of the dark-adapted cell. This same method has been adapted to cones and has been used to obtain excellent data on the spectral sensitivity of cones.[61]

9.4.10. Patch-Clamp Experiments Are Powerful Tools

Recently, ''patch-clamp'' experiments have been utilized to great advantage in the study of retinal rods. Large pipets are tightly sealed with a very high electrical resistance (so called ''gigaohm seals'') to the rod plasma membrane.

The plasma membrane patch may be pulled off to study its electrical properties.[62] Alternatively, the plasma membrane patch may be blown out of the pipet while maintaining the seal to allow diffusion of substances from the large patch pipet in the so-called ''whole-cell'' configuration.[63,64] Results of this method will be discussed more fully in the next section on the mechanism of visual excitation.

9.5. THE MECHANISM OF VISUAL EXCITATION

9.5.1. The Internal Transmitter Hypothesis

The topological organization of the rod cell requires that an intracellular ''transmitter'' carry a signal from the site of light absorption in the disk membrane to the site of Na^+ conductance change on the outer plasma membrane. Only a few newly synthesized disks at the base of the rod are connected with the plasma membrane, and most of the disks have pinched off and are not connected. The plasma membrane accounts for only about 1–3% of the total outer segment membrane area depending on the species. The remaining 97–99% of the membrane area is made up of the disks. Therefore, the disks must contain most of the rhodopsin, since rhodopsin could not be much more closely packed in the plasma membrane than it is in the disk. Since the rod is so sensitive, the disks not connected to the plasma membrane must be used to cause the cell to respond. This argument leads to the requirement for a signal or transmitter between the disks and the plasma membrane. It is highly unlikely that this signal is a brute force electrical coupling, which would require the disks to take up or release sufficient amounts of an ion to make a large contribution to the membrane potential. A more detailed discussion of the implausibility of electrical coupling can be consulted.[5] The high sensitivity of the rod is consistent with the intracellular transmitter being a chemical with specific receptors on the plasma membrane. A specific chemical transmitter could effect the plasma membrane

Na^+ permeability by binding to specific sites, even if it were released or taken up in very small amounts.

9.5.2. Requirements for a Transmitter

A transmitter would have to be produced or depleted sufficiently rapidly in the light to change the concentration at the plasma membrane. The time involved is about the time for a peak response to a dim flash (100–200 ms).

After a single photon is absorbed, a transmitter would have to be produced or depleted in sufficient quantity to detect the change in concentration above the background leakage of the rod and above the thermal noise in the conductivity of the plasma membrane.

A transmitter would have to bind to the plasma membrane and the binding must affect the Na^+ conductance. Experimental manipulation of the transmitter concentration in the dark should mimic the effect of light.

In the dark, the transmitter concentration must return to its initial level rapidly enough to explain the restoration rate of plasma membrane voltage. Depending on the species, the restoration rate has a half-time of 0.15–2.0 s with dim stimuli and gets much longer with bright stimuli.

Excited rhodopsin in the disk membrane must control the level of the transmitter in the light. Since the photoreceptor cell has such a high sensitivity, it is likely that the dark-adapted level of the intracellular transmitter is very low. In this way a relatively small change in the absolute amount of transmitter could cause a relatively large fractional change in the transmitter. From studies of photovoltages in turtle cones, Baylor and Fuortes[(65)] proposed that light produces a "substance which decreases the permeability of membrane channels acting as a shunt of the membrane in darkness." Yoshikami and Hagins[(66)] proposed that this substance was free calcium ions in the cytoplasm.

9.5.3. Evidence for the Role of Calcium

Calcium clearly plays a role in photoexcitation (see Ref. 67 for a general review of the roles of calcium in cellular regulation and Ref. 68 for a review of the role of calcium in photoreceptors). For a time it seemed likely that free cytoplasmic calcium was the internal transmitter because of the following observations:

1. Elevated external calcium mimics light effects on the dark current and calcium ionophores potentiate this effect.[(66)]
2. Electrophoretic injection of calcium hyperpolarizes the rod membrane just as does light.[(69)]
3. In very low external calcium solutions the rod depolarizes greatly to about 10 mV, the potential at which the light-sensitive current reverses as would be expected if the light-sensitive conductance became larger and dominant. Light then drives the membrane potential to the same hyperpolarized level as measured at normal calcium levels, indicating that only the light-sensitive conductance has been altered by calcium (see Ref. 68).
4. Injection of calcium chelators increases the outer segment membrane current,[(64)]

which is expected if the lowering of internal free calcium opens the light-sensitive channels.

5. Light causes an increase in extracellular calcium,[70] which is expected if light causes a transient rise in internal calcium. Thus, the calcium hypothesis seemed rather well established, but several results since 1984 have cast very serious doubt on the calcium hypothesis (Section 9.5.4).

9.5.4. Evidence against the Calcium Hypothesis for Visual Excitation

The light-sensitive channels were found to be quite permeable to calcium, and large Ca^{2+} influxes do not immediately mimic the light-induced dark current suppression.[71] There is an enormous flood of Ca^{2+} that enters through the light-sensitive channels in the dark current that would dwarf the light-induced Ca^{2+} release required by the Ca^{2+} hypothesis.[68]

Infusion of large amounts of calcium into the rod through whole-cell patch pipets (Section 9.4.10) does *not* decrease the light sensitivity.[64] Infusion of calcium chelators *increased* rather than decreased light sensitivity,[64] a finding in total conflict with the calcium hypothesis.

Free calcium has been found to *decrease* during the light response,[72] a result consistent with considerable other evidence.[68] Taken together, the evidence against Ca^{2+} as the internal transmitter is very strong.[68] However, Ca^{2+} has profound and rapid effects on rod transduction and this must be accounted for in any complete theory of transduction.

9.5.5. Cyclic GMP as Internal Transmitter

The rod outer segment contains a set of enzymes that regulate cGMP. Light-excited rhodopsin activates a highly amplified series of enzymatic reactions, the "cGMP cascade" that leads to the destruction of cGMP and the hyperpolarization of the cell.[34]

Relationships between the components of the cGMP cascade hypothesis for visual excitation are given in Fig. 9-18, where part a shows the activation reactions and part b shows the deactivation phase.

Light-excited rhodopsin (R*) interacts transiently with a GTP-binding protein[73] on the cytoplasmic face of the membrane in vertebrate photoreceptors. The GTP-binding protein in the rod photoreceptor is denoted G_V in Fig. 9-18. The GTP-binding protein associated with R* exchanges GDP for GTP[74,75] which excites the GTP-binding protein (G_V-GTP). Large numbers of GTP-binding proteins are excited per R*,[75] and many of the excited GTP-binding proteins turn on cGMP phosphodiesterase.[76] The excited cGMP phosphodiesterase in turn hydrolyzes cyclic GMP.

The presence of cGMP opens a cation channel in the outer plasma membrane[62,77] and in the disk membrane.[34,75,79] This channel is not controlled by Ca^{2+} nor is phosphorylation required to control it. The plasma membrane channel opened by cGMP is thought to be closed in the light due to transient depletion of cGMP by the light-stimulated phosphodiesterase, resulting in hyperpolarization of the photoreceptor cell membrane.

Figure 9-18b shows steps in the shutoff of R*. R* is heavily phosphorylated (7–9 phosphates/R*).[80] The phosphorylation of R* is essential to rapid reduction in R*

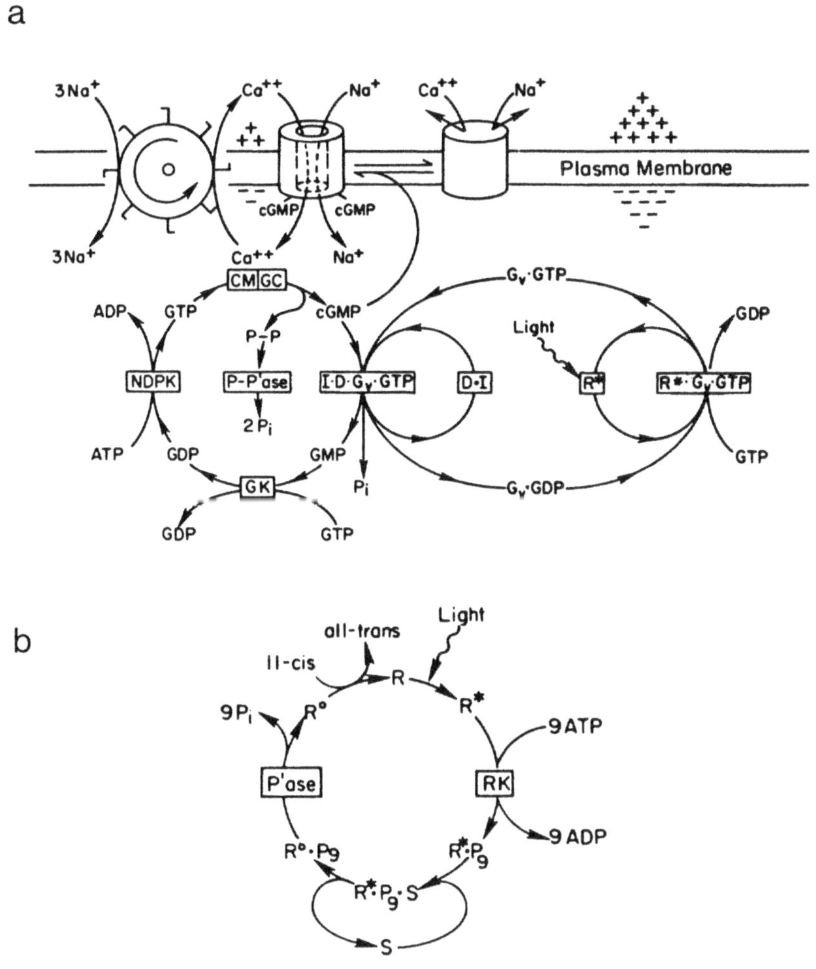

Fig. 9-18. (a) The cGMP cascade model for the activation of the electrical response of the vertebrate rod. Light converts rhodopsin to R*, which forms a transient complex with the GTP-binding protein (G_V) and catalyzes the exchange of GTP for GDP. The G_V-GTP complexes the inhibited phosphodiesterase ($D•_I$) to activate the hydrolysis of cGMP. The reduced level of cGMP closes the Na^+-Ca^2 /,xX^1 channels on the plasma membrane, which hyperpolarizes the cell and stops the entry of Ca^2 /,xX^1. A Na^+-Ca^{2+} exchanger in the plasma membrane reduces the Ca^{2+} concentration in the cytoplasm, which accelerates the production of cGMP, presumably by activating the guanylate cyclase–calmodulin complex (CM-GC). Other activities required to maintain the cGMP and GTP levels are guanylate kinase (GK), nucleotide diphosphokinase (NDPK), and pyrophosphatase (P-Pase). G_V-GTP is spontaneously inactivated by an endogenous GTPase activity to form G_V-GDP plus P_i. (b) The inactivation of the excited receptor (R*). R* is rapidly phosphorylated by ATP and rhodopsin kinase (RK) with up to nine phosphates/R* (R*-Pg). R*-Pg is fully inactivated by binding the S antigen (48°K or arrestin). The detailed pathway is poorly known after this point, but probably involves first the spontaneous inactivation of R*-P_9 to R°-P_9 (R° denotes the inactive conformation of R), loss of S antigen, loss of phosphate by phosphatase (Pase), and replacement of all-*trans*-retinal by 11-*cis*-retinal. [Taken from P. A. Liebman, K. R. Parker, and E. A. Dratz, *Ann. Rev. Physiol.* **49**, 765–791 (1987).]

activity and to ultimate blockage by binding a 48-kDA protein (arrestin).[81] A hypothetical complete restoration cycle is shown but the later steps are poorly understood.

Evidence for a large increase in metabolic flux of cGMP in whole retina in the light and for an increase in free cGMP in the light has been presented.[82] This recent work implies that the simple form of the model presented in Fig. 9-18a that requires a decrease in cGMP in the light might not be adequate.[82] Changes in the flux of cGMP rather than the change in level of cGMP may be important in photoreceptor physiology, although the detailed mechanism is obscure.[82]

The cGMP cascade hypothesis for visual excitation (Fig. 9-18) is now very widely accepted. The detailed mechanism of operation of the cGMP cascade and its control is a rich field of investigation.[34,68] There has been considerable speculation as to whether the powerful effects of calcium and calcium chelators on the photoreceptor (Section 9.5.3) might be explained by interaction with the cGMP cascade. The leading hypothesis for this mechanism is that raising calcium may inhibit the guanylate cyclase (CM/GC in Fig. 9-18a), which would drop the concentration of cGMP and thus mimic light. Lowered Ca^{2+} would then stimulate guanylate cyclase, raising cGMP and opening plasma membrane channels.[68] A potential problem with this mechanism is that the isolated cyclase is only weakly inhibited by calcium.[83]

The kinetics of many of the enzymatic reactions in the retinal rod are strongly concentration-dependent, as if weakly bound factors that can affect rates are released when the cells are disrupted and diluted.[84] In particular, the GTP-binding protein, which shuts itself off by hydrolysis of GTP, hydrolyzes GTP very slowly (1–3/min) when studied in dilute form, but speeds up about 100-fold near physiological concentration.[84] All the excitation reactions (Fig. 9-18a) are concentration-independent, thus all the interacting components appear to be membrane-bound. In contrast, all the shutoff reactions (Fig. 9-18b) appear to be concentration-dependent, and evidently require weakly bound factors that may be diluted out.[84]

It is likely that Ca^{2+} acts as a negative feedback signal derived from the dark current. Ca^{2+} enters the light-regulated channel and is extruded by the Na/Ca exchanger (Fig. 9-18a). Cytoplasmic Ca^{2+} drops in the light because the influx is stopped and the exchanger keeps pumping it out. Calcium chelates alter the time constant of the feedback signal and can give an electrical overshoot beyond the normal steady-state voltage.[64] The calcium feedback may cut off the excitation and account for the faster electrical properties of cones, and also provide a basis for explaining light and dark adaptation.

There may be other mechanisms for Ca^{2+} regulation in photoreceptors. The membrane lipid, phosphatidylinositol diphosphate (PIP_2), is reported to decrease in the light in both vertebrate[85] and invertebrate[86] photoreceptors. The PIP_2 hydrolysis product, inositol trisphosphate (IP_3), increases in the light in invertebrate photoreceptors.[86] There are reports of protein kinase C activation in the light[87] and inositol trisphosphate modulation of calcium metabolism in retinal rods.[88]

9.6. LIGHT DAMAGE TO THE RETINA

As basic knowledge of the visual system biochemistry is collected, we may hope to gain insight into medical problems of the retina. For example, many types of retinal photoreceptor degeneration are known. Some types are quite common in older people, but some types, such as retinitis pigmentosa, all too often afflict young and old alike. One striking property of normal photoreceptors is that they can be severely damaged by exposure to excessive light.[89,90] In many different animal species the light levels for severe damage are only slightly above the levels commonly encountered by humans. Damaging effects are marked in some animals exposed to the light levels equivalent to a bright beach for 8 h or to normal ambient illumination for 2–3 days.[91–93] In experimental animals the light-induced damage is largely reversible, but studies have not yet been done to see if there is a cumulative effect of repeated light damage.

The highly unsaturated membrane fatty acids in isolated, purified rod photoreceptor membranes are extremely sensitive to damage by atmospheric oxygen in the dark.[94] The regenerability of opsin is irreversibly lost by incubation in the presence of oxygen in the dark. The photoreceptor membranes are clearly very sensitive to damage by oxygen. Furthermore, light might be expected to facilitate oxygen damage to the photoreceptors by a singlet oxygen-mediated photodynamic mechanism.[26]

High levels of vitamin E in the isolated membrane are helpful in retarding this oxygen damage. However, vitamin E is consumed in air and other potent mechanisms must be present in the intact system to oppose oxygen damage.[95] The photoreceptors are the most metabolically active tissues in the body, containing unusually large numbers of mitochondria, and must absorb large quantities of oxygen from the surrounding medium.[26] Oxygen metabolism by mitochondria can leak reactive free-radical intermediates, which pose a threat to the tissue.[26]

Even potent protective mechanisms against oxygen and light damage must have limits. This may be the reason that light levels slightly above those commonly encountered are damaging to animal photoreceptors. Potentially damaging light levels are encountered by some people, and might be expected to produce cumulative damage that is evident in advanced age. Nutritional factors (e.g., vitamin E and other antioxidants) would also be expected to have a role in maintaining the long-term health of the photoreceptors.[26] More drastic forms of photoreceptor degeneration may be due to genetic or disease-related deficiencies in the normally potent protective mechanism against lipid oxidation. The above hypotheses have potential medical importance that can be tested with the present knowledge of photoreceptors.[26]

9.7. REFERENCES

1. R. A. Eakin, Structure of invertebrate photoreceptors, *Handbook Sensory Physiol.* **7**(1), 625–684 (1972).
2. T. H. Goldsmith, The natural history of invertebrate visual pigments, *Handbook Sensory Physiol.* **7**(1), 685–719 (1972).
3. M. F. Moody, Photoreceptor organelles in animals, *Biol. Rev.* **39,** 43–86 (1964).
4. M. G. F. Fuortes (ed.), Physiology of photoreceptor organs, *Handbook Sensory Physiol.* **7**(2) (1972).
5. W. A. Hagins, The visual process: Excitatory mechanisms in the primary receptor cells, *Ann. Rev. Biophys. Bioeng.* **1,** 131–158 (1972).
6. A. E. Blaurock and M. H. Wilkins, Structure of frog photoreceptor membranes, *Nature* **223,** 906–909 (1969).
7. R. W. Young, Biogenesis and renewal of visual cell outer segment membranes, *Exp. Eye Res.* **18,** 215–223 (1974).

8. M. J. Hogan, I. S. Wood, and R. H. Steinberg, Phagocytosis by pigment epithelium of human retinal cones, *Nature* **252,** 305–307 (1974).
9. W. Stoeckenius and R. A. Bogomolni, Bacteriorhodopsin and related pigments of halobacteria, *Ann. Rev. Biochem.* **52,** 587–616 (1982).
10. E. A. Dratz and P. A. Hargrave, The structure of rhodopsin and the rod outer segment disk membrane, *Trends Biochem. Sci.* **8,** 128–131 (1983).
11. M. Applebury and P. A. Hargrave, Molecular biology of the visual pigments, *Vision Res.* **26,** 1881–1895 (1986).
12. D. S. Papermaster and W. J. Dreyer, Rhodopsin content in the outer segment membrance of bovine and frog retinal rods, *Biochemistry* **13,** 2438–2444 (1974).
13. P. P. Nemes and E. A. Dratz, The preparation of isolated osmotically intact bovine rod outer segment disk membranes, *Methods. Enzymol.* **81**(H), 116–123 (1982).
14. W. L. Stone, C. C. Farnsworth, and E. A. Dratz, A reinvestigation of the fatty acid content of bovine, rat and frog rod outer segments, *Exp. Eye Res.* **28,** 387–397 (1979).
15. G. P. Miljanich, L. A. Sklar, D. L. White, and E. A. Dratz, Disaturated and dipolyunsaturated phospholipids of the bovine retinal rod outer segment disk membranes, *Biochim. Biophys. Acta.* **552,** 294–306 (1979).
16. R. M. Benolken, R. E. Anderson, and T. G. Wheeler, Membrane fatty acids associated with the electrical response in visual excitation, *Science* **182,** 1253–1254 (1973).
17. E. A. Dratz and A. J. Deese, The role of docosahexaenoic acid (22:6w3) in biological membranes: Examples from photoreceptors and model membrane bilayers, in: *Health Effects of Polyunsaturated Fatty Acids in Seafoods* (A. Simopolous and R. Kifer, eds.), pp. 319–351, Academic Press, New York (1986).
18. E. A. Dratz, G. P. Miljanich, P. P. Nemes, J. E. Gaw, and S. Schwartz, The structure of rhodopsin and its disposition in the rod outer segment disk membrane, *Photochem. Photobiol.* **29,** 661–671 (1979).
19. M. Chabre, X-ray diffraction studies of retinal rods. I. Structure of the disk membrane, effect of illumination, *Biochim. Biophys. Acta.* **382,** 322–335 (1975).
20. P. M. Hargrave, Rhodopsin chemistry, structure and topography, *Prog. Ret. Res.* **1,** 1–51 (1982).
21. G. P. Miljanich, M. F. Brown, S. Mabrey-Gaud, E. A. Dratz, and J. M. Sturtevant, Thermotropic behavior of retinal rod membranes and dispersions of extracted phospholipids, *J. Membrane Biol.* **85,** 79–86 (1985).
22. G. S. Harbison, S. O. Smith, J. A. Pardoen, C. Winkel, J. Lugtenburg, J. Herzfeld, R. Mathies, and R. G. Griffin, Dark-adapted bacteriorhodopsin contains 13-*cis,* 15-*syn* and all-*trans,* 15-anti retinal Schiff bases, *Proc. Natl. Acad. Sci. USA* **81,** 1706–1709 (1984).
23. C. D. B. Bridges, The rhodopsin-porphyropsin visual system, *Handbook Sensory Physiol.* **7**(1), 417–480 (1972).
24. W. A. H. Rushton, The Ferrier lecture: Visual adaptation, *Proc. Roy. Soc. B* **162,** 20–46 (1965).
25. A. L. Zimmerman and D. A. Baylor, Single channel currents from the cyclic GMP sensitive conductance of retinal rod outer segments, *Nature* **321,** 70–72 (1986).
26. G. J. Handelman, E. A. Dratz, C. C. Reay, and F. J. G. M. van Kuijk, Carotenoids in the human macala and whole retina, *Invest. Ophthal. & Vis. Sci.* **29,** 850–855 (1988).
27. P. A. Liebman, Microspectrophotometry of photoreceptors, *Handbook Sensory Physiol.* **7,** 481–528 (1972).
28. R. A. Cone, Rotational diffusion of rhodopsin in the visual receptor membrane, *Nature [New Biol.]* **236,** 39–43 (1972).
29. M. Poo and R. A. Cone, Lateral diffusion of rhodopsin in the photoreceptor membrane, *Nature* **247,** 438–441 (1974).
30. P. A. Liebman and G. Entine, Lateral diffusion of visual pigment in photoreceptor disk membranes, *Science* **185,** 457–459 (1974).
31. B. Honig, U. Dinu, K. Nakanishi, V. Balogh-Nair, M. Gawinowicz, M. Arnaboldi, and M. Motto, The external point charge model for wavelength regulation in visual pigments, *J. Am. Chem. Soc.* **101,** 7084–7086 (1979).
32. J. Nathans, D. Thomas, and D. S. Hogness, Molecular genetics of human color vision: The genes of inherited variation in human color vision, *Science* **232,** 193–202 (1986).
33. J. Nathans and D. S. Hogness, Isolation and nucleotide sequence of the gene encoding human rhodopsin, *Proc. Natl. Acad. Sci. USA* **8,** 4851–4855 (1984).

34. P. A. Liebman, K. R. Parker, and E. A. Dratz, The molecular mechanism of visual excitation and its relation to the structure and composition of the rod outer segment, *Ann. Rev. Physiol.* **49,** 765–791 (1987).
35. D. S. Kliger, S. J. Milder, and E. A. Dratz, Solvent effects on the spectra of retinal Schiff bases. I. Models for the bathochromic shift of the chromophore spectrum in visual pigments, *Photochem. Photobiol.* **25,** 277–286 (1977).
36. H. Kakitani, T. Kakitani, H. Rodman, and B. Honig, On the mechanism of wavelength regulation in visual pigments, *Photochem. Photobiol.* **41,** 471–479 (1985).
37. B. Honig and T. G. Ebrey, The structure and spectra of the chromophore of the visual pigments, *Ann. Rev. Biophys. Bioeng.* **3,** 151–177 (1974).
38. A. Kropf, B. P. Whittenberger, S. P. Goff, and A. S. Waggoner, The spectral properties of some visual pigment analogs, *Exp. Eye Res.* **17,** 591–606 (1973).
39. R. S. H. Liu, A. E. Asato, M. Denny, and D. Mead, The nature of restrictions in the binding site of rhodopsin: A model study, *J. Am. Chem. Soc.* **106,** 8298–8300 (1984).
40. E. L. Menger and D. S. Kliger, Photoisomerization kinetics of 11-*cis* retinal, its Schiff base, and its protonated Schiff base, *J. Am. Chem. Soc.* **98,** 3975–3979 (1976).
41. T. P. Williams, An isochromic change in the bleaching of rhodopsin, *Vision Res.* **10,** 525–533 (1970).
42. M. L. Applebury, D. M. Zuckerman, A. A. Lamola, and T. M. Jovin, Rhodopsin purification and recombination with phospholipids assayed by the metarhodopsin I $\rightleftharpoons$ metarhodopsin II transition, *Biochemistry* **13,** 3448–3458 (1974).
43. A. A. Lamola, T. Yamane, and A. Zipp, Effects of detergents and high pressures upon the metarhodopsin I $\rightleftharpoons$ metarhodopsin II equilibrium, *Biochemistry* **13,** 738–745 (1974).
44. W. J. DeGrip, G. L. M. Van der Laar, F. J. M. Daemen, and S. L. Bonting, Biochemical aspects of the visual process. XXIII. Sulfhydryl groups and rhodopsin photolysis, *Biochim. Biophys. Acta* **325,** 315–322 (1973).
45. E. A. Dratz, N. Ryba, A. Watts, and A. Deese, The role of highly unsaturated fatty acids in membranes, submitted for publication.
46. S. L. Bonting and A. D. Bangham, On the biochemical mechanisms of the visual process, *Exp. Eye Res.* **6,** 400–413 (1967).
47. W. Zimmerman, M. Yost, and F. J. M. Daemen, Dynamics and function of vitamin A compounds in the rat retina after a small bleach of rhodopsin, *Nature* **250,** 66–67 (1974).
48. T. Tomita, Electrical activity of vertebrate photoreceptors, *Q. Rev. Biophys.* **3,** 179–222 (1970).
49. K. T. Brown and M. Murakami, Biphasic form of the early receptor potential of the monkey retina, *Nature* **204,** 739–740 (1964).
50. R. A. Cone and W. L. Pak, The early receptor potential, *Handbook Sensory Physiol.* **1,** 345–365 (1971).
51. R. A. Cone, Early receptor potential: Photoreversible charge displacement in rhodopsin, *Science* **155,** 1128–1131 (1967).
52. W. A. Hagins and H. Ruppel, Fast photoelectric effects and the properties of vertebrate photoreceptors as electric cables, *Fed. Proc.* **30,** 64–78 (1971).
53. F. S. Werblin and J. E. Dowling, Organization of retina of the mudpuppy, *Necturus maculosus.* II. Intracellular recording, *J. Neurophysiol.* **32,** 339–355 (1969).
54. D. Baylor, A. Hodgkin, and T. Lamb, The electrical response of turtle cones to flashes and steps of light, *J. Physiol. (Lond.)* **242,** 685–727 (1974).
55. W. A. Hagins, R. D. Penn, and S. Yoshikami, Dark current and photocurrent in retinal rods, *Biophys. J.* **10,** 380–412 (1970).
56. W. A. Hagins, H. V. Zona, and R. G. Adams, Local membrane current in the outer segments of squid photoreceptors, *Nature* **194,** 844–847 (1962).
57. S. Yoshikami and W. A. Hagins, Control of the dark current in vertebrate rods and cones, in: *Biochemistry and Physiology of Visual Pigments* (H. Langer, ed.), pp. 245–255, Springer-Verlag, Berlin (1973).
58. J. E. Korenbrot and R. A. Cone, Dark ionic flux and the effects of light in isolated rod outer segments, *J. Gen. Physiol.* **60,** 20–45 (1972).
59. D. A. Baylor, T. D. Lamb, and K. W. Yau, The membrane current of single rod outer segments, *J. Physiol.* **288,** 589–611 (1979).
60. D. A. Baylor, T. D. Lamb, and K.-W. Yau, Responses of retinal rods to single photons, *J. Physiol.* **288,** 613–634 (1979).

61. B. J. Nunn, J. L. Schnapf, and D. A. Baylor, Spectral sensitivity of single cones in the retina of Macaca fascicularis, *Nature* **309,** 264–266 (1985).
62. E. E. Fesenko, S. S. Kolesnikov and A. L. Lyubarsky, Induction by cyclic GMP of cationic conductance in plasma membrane of retinal rod outer segment, *Nature* **321,** 66–70 (1986).
63. W. H. Cobbs and E. N. Pugh, Jr., Cyclic GMP can increase rod outer-segment light-sensitive current 10-fold without delay of excitation, *Nature* **313,** 585–587 (1985).
64. T. D. Lamb, H. Matthews and V. Torre, Changes in the photocurrent of salamander retinal rods during incorporation of calcium buffers, *J. Physiol.* **372,** 315–349 (1986).
65. D. A. Baylor and M. G. F. Fourtes, Electrical response of single cones in the retina of the turtle, *J. Physiol.* **207,** 77–92 (1970).
66. S. Yoshikami and W. H. Hagins, Ionic bases of dark current and photocurrents of retinal rods, *Biophys. J.* **10,** 60a (1970).
67. H. Rasmussen, *Calcium and cAMP as Synarchic Messengers,* Wiley, New York (1981).
68. E. N. Pugh, Jr., The nature and identity of the internal excitation transmitter of vertebrate phototransduction, *Ann. Rev. Physiol.* **49,** 715–741 (1987).
69. J. E. Brown, J. A. Cole and L. H. Pinto, Effects of injections of Ca^{2+} and EGTA into the outer segment of the retinal rods of Bufo Marinus, *J. Physiol.* **269,** 707–722 (1977).
70. G. H. Gold and J. I. Korenbrot, Light-induced calcium release by intact retinal rods, *Proc. Natl. Acad. Sci. U.S.A.* **77,** 5557–5561 (1980).
71. K.-W. Yau and K. Nakatani, Electrogenic Na-Ca exchange in retinal rod outer segment, *Nature* **311,** 661–663 (1984).
72. P. A. McNaughton, B. J. Nunn and L. Cerretto, Measurement of intracellular free Ca^{2+} concentration in salamander rod outer segments, *Nature* **322,** 261–263 (1986).
73. H. Kuhn, Light- and GTP-regulated interaction of GTPase and other proteins with bovine photoreceptor membranes, *Nature* **283,** 587–589 (1980).
74. B. B.-K. Fung, and L. Stryer, Photolyzed rhodopsin catalyzes the exchange of GTP for bound GDP in retinal rod outer segments, *Proc. Natl. Acad. Sci. U.S.A.* **77,** 2500–2504 (1980).
75. P. A. Liebman, and E. N. Pugh, Jr., Gain, speed and sensitivity of GTP binding vs. PDE activation in visual excitation, *Vision Res.* **22,** 1475–1480 (1984).
76. P. A. Liebman, and E. N. Pugh, Jr., The control of phosphodiesterase in rod disk membranes: Kinetics, possible mechanisms and significance for vision, *Vision Res.* **19,** 375–380 (1979).
77. J. H. Stern, U. B. Kaupp, and P. R. Macleish, Control of the light-regulated current in rod photoreceptors by cyclic GMP, calcium and 1-*cis*-diltiazem, *Proc. Natl. Acad. Sci. U.S.A.* **83,** 1163–1167 (1986).
78. A Cavaggioni and R. T. Sorbi, Cyclic-GMP releases calcium from disk membranes of vertebrate photoreceptors, *Proc. Natl. Acad. Sci. USA* **78,** 3964–3968 (1981).
79. K.-W. Koch and U. B. Kaupp, Cyclic GMP directly regulates a cation conductance in membranes of bovine rods by a cooperative mechanism, *J. Biol. Chem.* **260,** 6788–6800 (1985).
80. U. Wilden and H. Kuhn, Light-dependent phosphorylation of rhodopsin: Number of phosphorylation sites, *Biochemistry* **21,** 3014–3022 (1982).
81. U. Wilden, S. W. Hall, and H. Kuhn, Phosphodiesterase activation by photoexcited rhodopsin is quenched when rhodopsin is phosphorylated and binds the intrinsic 48-kDa protein of rod outer segments, *Proc. Natl. Acad. Sci. USA* **83,** 1174–1178 (1986).
82. N. D. Goldberg, A. Ames, III, J. E. Gander, and T. F. Walseth, Magnitude of increase in retinal cGMP metabolic flux determined by ^{18}O incorporation into nucleotide α-phosphoryls corresponds with intensity of photic stimulation, *J. Biol. Chem.* **258,** 9213–9219 (1983).
83. D. Fleischman and M. Denisevich, Guanylate cyclase of isolated bovine retinal rod axonemes, *Biochemistry* **18,** 5060–5066 (1979).
84. E. A. Dratz, J. W. Lewis, L. E. Schaechter, K. R. Parker, and D. S. Kliger, Retinal rod GTPase rate increases with concentration: A key to the control of visual excitation?, *Biochem. Biophys. Res. Commun.* **146,** 379–386 (1987).
85. A. Ghalayini and R. E. Anderson, Phosphatidylinositol 4,5-bisphosphate: Light-mediated breakdown in the vertebrate retina, *Biochem. Biophys. Res. Commun.* **124,** 503–506 (1984).
86. J. E. Brown, L. J. Rubin, A. J. Ghalayini, A. P. Tarvar, and R. F. Irvine, Myo-inositol polyphosphate may be a messenger for visual excitation in *Limulus* photoreceptors, *Nature* **311,** 160–163 (1984).
87. C. L. Kapoor and G. J. Chader, Endogenous phosphorylation of retinal photoreceptor outer segment

proteins by calcium-phospholipid-dependent protein kinase, *Biochem. Biophys. Res. Commun.* **122,** 1397–1403 (1984).
88. K. R. Parker, J. A. Briggs and E. A. Dratz, Inositol triphosphate stimulates Ca^{2+} release from toad retinal rod outer segment preparations, *Biophysics* **49,** 31a (1986).
89. W. K. Noell, U. S. Walker, B. S. Kang and S. Berman, Retinal damage by light in rats, *Invest. Ophthalmol.* **5,** 450–472 (1966).
90. T. Kuwabara and R. A. Gorn, Retinal damage by visible light, *Arch. Ophthalmol.* **79,** 69–78 (1968).
91. K. V. Anderson, F. P. Coyle and W. K. O'Steen, Retinal degeneration produced by low intensity colored light, *Exp. Neurol.* **35,** 233–238 (1972).
92. T. Lawwill, Effects of prolonged exposure of rabbit retina to low-intensity light, *Invest. Ophthalmol.* **12,** 45–51 (1973).
93. J. Marshall, J. Mellerio, and D. A. Palmer, Damage to pigeon retina by moderate illumination from fluorescent lamps, *Exp. Eye Res.* **14,** 164–169 (1972).
94. C. C. Farnsworth and E. A. Dratz, Oxidative damage of the rod outer segment (ROS) disk membrane and the role of vitamin E, *Biochim. Biophys. Acta* **443,** 556–570 (1976).
95. G. J. Handelman and E. A. Dratz, The role of antioxidants in the retina and retina pigment epithelium and the nature of prooxidant-induced damage, *Adv. Free Rad. Biol. Med.* **2,** 1–89 (1986).

10

Photomorphogenesis

10.1. INTRODUCTION

10.1.1. Light as a Source of Information

An organism can utilize the energy of light in either of two ways (Fig. 10-1). Light can be used as a source of free energy to facilitate a process, such as photosynthesis (see Chapter 12), that otherwise would be thermodynamically unfeasible. Alternatively, light can be used as a source of information. It is this second aspect of light with which we will be concerned.

Lee H. Pratt • Department of Botany, University of Georgia, Athens, Georgia 30602. **Marie-Michèle Cordonnier** • Biotechnology Research, CIBA-GEIGY Corporation, Research Triangle Park, North Carolina 27709-2257.

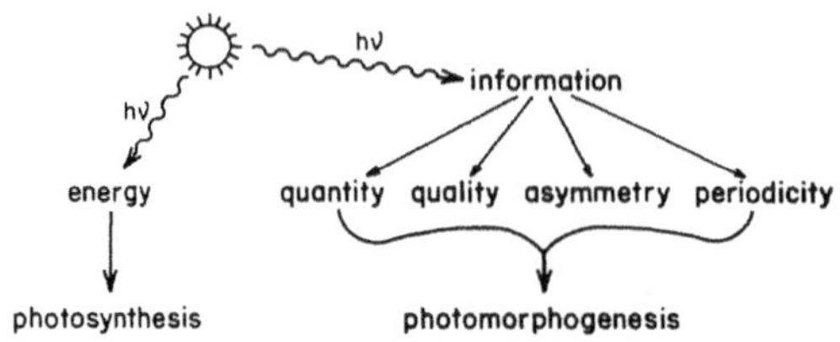

Fig. 10-1. Light as a source of both energy and information.

It would seem reasonable that an organism can be best adapted to its environment only if it can sense that environment and respond appropriately to this sensory input. For example, we can sense potentially harmful ultraviolet radiation and respond to it by accumulating more melanin in our epidermis, thus protecting ourselves against a subsequent exposure to this radiation (Chapter 6). As a motile organism, however, we can also respond by moving to another, potentially more suitable environment. In contrast, a nonmotile organism such as a plant or fungus does not have this second alternative available to it. Instead, it must adapt to its existing environment rather than move to a better one. The ability to sense incident light energy and to respond to it in appropriate ways is therefore especially important to such organisms. This ability is most important to a green plant, since via photosynthesis it also utilizes light as its sole source of free energy.

Photomorphogenesis encompasses phenomena whereby an organism senses and responds appropriately to incident light energy. The precise meaning of this term derives from its three Greek roots. Photomorphogenesis is therefore the development (*genesis*) of form (*morph*) of an organism as influenced by light (*photo*). Consequently, photomorphogenesis deals with events that occur at both cellular and multicellular levels. The underlying phenomena that lead to changes in morphology, however, occur at subcellular and molecular levels. Hence, this chapter will consider responses to light that occur at all levels, including the molecular. Furthermore, because of the unique importance of photomorphogenesis to nonmotile organisms, coverage here will be restricted to photomorphogenesis among fungi and plants, including algae.

10.1.2. Diversity of Photomorphogenesis

There are at least four independent aspects of light that can serve as a source of information. They are its (1) quantity, (2) quality, (3) spatial asymmetry (i.e., direction from which the light comes), and (4) periodicity (Fig. 10-1). In a few instances, organisms also respond to the plane of polarization of incident light. A notable example of such a response is polarotropism of the chloronema of *Dryopteris felix-mas*. The direction of the growth of the chloronema, which is a filamentous growth stage of the haploid generation of this fern, can under certain growth conditions be strictly determined by the plane of polarization of incident light. When this is the case, the chloronema grows in such a direction that its long axis is normal to the plane of vibration of the electrical vector of incident light. Because such responses to polarization of light are limited to rather special cases, and because their significance in the natural environment is generally uncertain, they will not be included here.

A photomorphogenic response to the *quantity,* or fluence rate, of incident light can be, in the simplest case, a response to its presence or absence. Most notably, a plant exhibits vastly different morphology depending on whether it is grown in the presence or absence of light (Section 10.2.2.1). The soybean seedling serves as a representative example of this difference among dicotyledonous angiosperms, which are those flowering plants that possess two cotyledons, or seed leaves. The cotyledons are typically used for storage of food reserves that sustain the seedling until it develops the ability to perform photosynthesis and become self-sufficient. When grown in total darkness (Fig. 10-2c), such a seedling has relatively little pigmentation; a long, spindly hypocotyl, which is that part of the shoot that is below the cotyledons; and small leaves still enclosed within the cotyledons, together with the apical meristem that will eventually give rise to the mature shoot. When grown in the light, however, the same seedling will have a shorter hypocotyl, expanded cotyledons and leaves, and considerably more pigmentation (Fig. 10-2d). The situation in the monocotyledonous angiosperms, which are flowering plants with a single cotyledon, is somewhat different. A typical monocotyledonous grass seedling when grown in darkness (Fig. 10-2b) also contains relatively little pigmentation. Moreover, its coleoptile, which is a hollow sheath that covers the growing tip and the primary leaves of the seedling, remains intact for a considerable period of time and its mesocotyl, which is that part of its stem below the node from which the coleoptile and leaves of the seedling arise, is quite long. In the light, however, the mesocotyl is vastly reduced in size and the bulk of the plant body consists of highly pigmented, expanded leaves (Fig. 10-2a). More subtly, an organism can also respond to differences in the fluence rate of incident radiation, as in the case of mustard seedlings that accumulate anthocyanins in amounts that are proportional to the fluence rate of light that is incident upon them (Section 10.2.2.2).

The capacity to respond to the *quality,* or wavelength distribution, of incident light energy is of special importance to photosynthetically active organisms (Sections 10.2.2.3 and 10.5). By doing so, these organisms can improve the efficiency with which they trap incident radiant energy and make it available for photosynthesis. Some cyanobacteria, for example, have the ability to produce photosynthetically active pigments that absorb light maximally in different spectral regions, a process known as photochromic adaptation (Section 10.5). Presumably, it would be to their advantage to synthesize pigments that absorb most efficiently the wavelengths of light that are available to them, which is precisely what they do.

Organisms can also sense the direction from which incident light comes. An example of a response to this *asymmetry* in incident light is a leaf that changes its orientation throughout the day so that its surface is always at a right angle to the direction from which incident light comes.[1] This "tracking" of the sun, which is observed with plants such as cotton, soybean, and alfalfa, presumably maximizes the absorbance of photosynthetically active light by the pigments in the leaf. In addition to movements of this type, an organ can also move as a consequence of differential rates of cell elongation on the sides nearest to and furthest away from a source of light. These irreversible growth responses, which are generally toward (in the case of shoots) or away from (in the case of roots) the source of light, are known as phototropisms. Because photomovement is dealt with as a separate topic elsewhere in this volume (Chapter 11), coverage here is restricted.

The final aspect of light to be considered is its *periodicity.* The rotation of the earth leads to an ubiquitous diurnal variation in the light environment. Moreover, the length of a

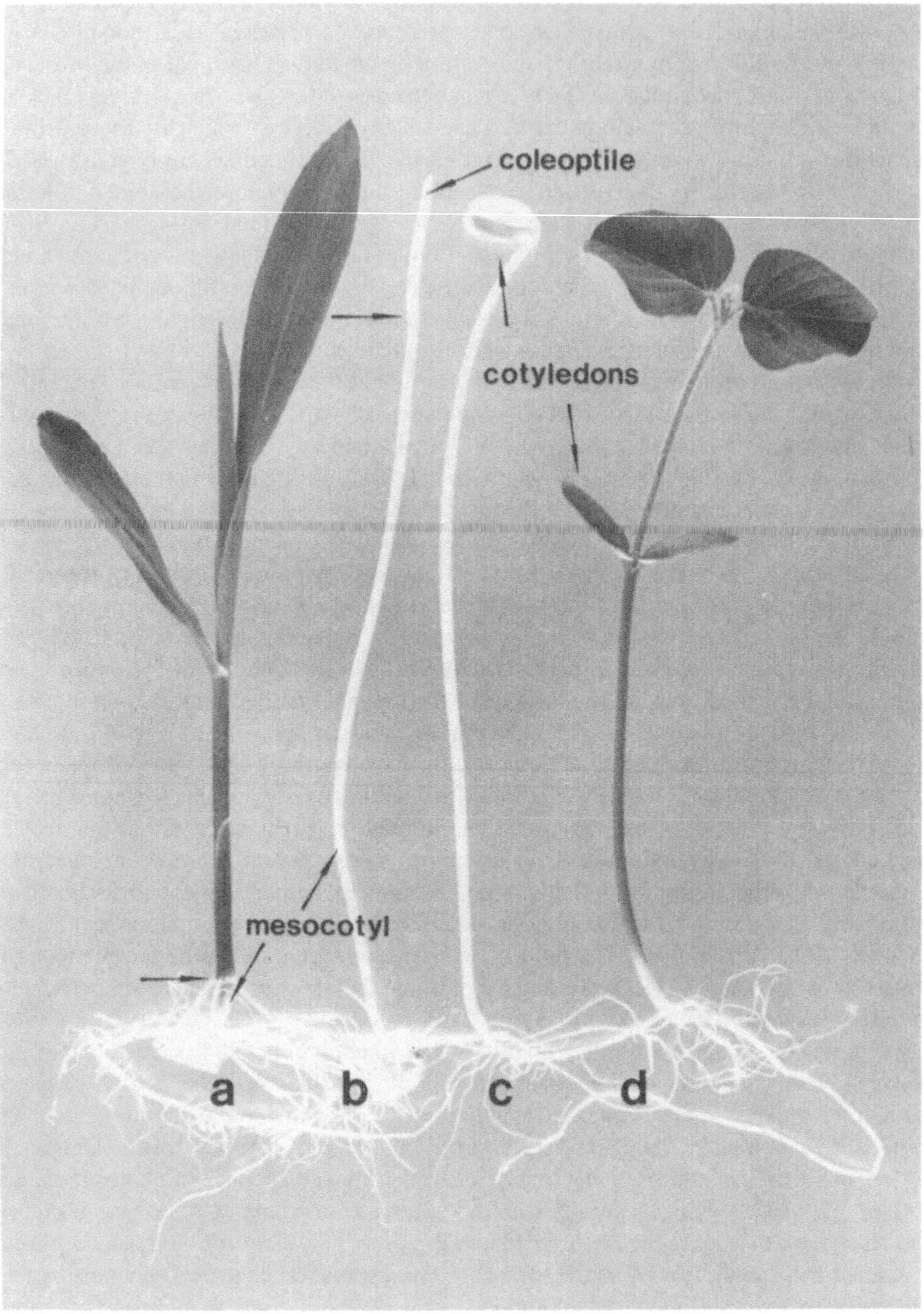

Fig. 10-2. Representative examples of seedlings grown for 10 days at 25°C in the presence or absence of light. Light-grown (a) and dark-grown (b) maize (*Zea mays* L.) seedlings. The position of the node from which the coleoptile and primary leaves arise is indicated with an unlabeled arrow. Dark-grown (c) and light-grown (d) soybean (*Glycine max* L.) seedlings.

daily photoperiod varies as a function of both latitude and time of year. Since these variations are highly reproducible, an organism can use this information as a dependable environmental cue. For example, plants often respond to defined changes in photoperiod by flowering, a photomorphogenic response that involves a change from vegetative to reproductive morphology.(2) Because this aspect of light is dealt with elsewhere (Chapters 7 and 8), discussion of photoperiodism here is minimal.

The ultimate goal in studying photomorphogenesis is to explain how the absorbance of light leads to the expression of morphogenic responses. To achieve this goal, one must (1) identify the photoreceptor, (2) determine the initial biochemical event that follows photoreceptor excitation, and (3) elucidate the subsequent chain of events that leads to the response. Since the energy content of the light to which an organism responds is insufficient by itself to bring about a response, the explanation must include a mechanism for signal amplification. Furthermore, since a single exposure to light typically leads to a vast array of sometimes seemingly unrelated responses, the explanation must also include one or more mechanisms whereby the transduction chain branches. Since photoreception itself and the events that occur most immediately after the absorption of light are of most general interest, the focus here will be on the identification and characterization of photomorphogenically active pigments, and on suggestions of how they might function at the molecular level.

The expression of photomorphogenesis is, of course, modulated by a host of parameters other than light. For example, the outcome of a potentially effective exposure to light may depend on the time at which it is given during a diurnal cycle. Similarly, whether a given light treatment will result in a given response depends on the ability of the organism to respond. As emphasized by Mohr,(3) during the development of a young seedling this potential to respond varies in a predictable fashion. Thus, even though photomorphogenesis will be treated here as an isolated topic, the reader should bear in mind that all responses to light are modified by other factors.

In contrast to the difficulty of identifying photoreceptors and the initial steps leading to photomorphogenic responses, it is relatively easy to describe the responses themselves. Even a decade ago, it was possible to reference several hundred photomorphogenic responses.(4) It is therefore important to subdivide them into appropriate categories rather than try to deal with them individually. While it would be convenient to group them by photoreceptor, phytochrome is the only one that has been identified. It is therefore necessary to categorize other photomorphogenic phenomena in some alternative fashion. Thus, responses to blue light will be dealt with together, as will be those to ultraviolet light. It should become apparent, however, that this classification is artificial. Not only are there extensive interactions among all photomorphogenic systems, but there are probably multiple photoreceptors for each of these spectral regions. Because this approach to subdividing the subject omits the interesting phenomenon of chromatic adaptation, it will be discussed separately in Section 10.5. And, while green light can apparently also be morphogenically active (e.g., 5) so little is known about its effects and how it might be sensed that they will be ignored here.

The diversity of photomorphogenesis requires that it be covered here in cursory fashion. In particular, the wide range of pigments that might serve as receptors for photomorphogenesis will not be evaluated independently. Although these pigments, which include flavins and flavoproteins, carotenoids and carotenoproteins, biliproteins, cytochromes, and chlorophylls, represent a diverse range of chromophores, it will become evident from the following discussion that their status as photomorphogenically active pigments remains, for the most part, uncertain. Phytochrome receives the greatest attention here only because it is the photomorphogenic system about which most is known. More extensive coverage of both phytochrome and other photomorphogenic systems is available elsewhere.[6]

10.2. PHYTOCHROME-MEDIATED MORPHOGENESIS

10.2.1. Discovery

The experiments that led to the discovery of phytochrome were performed at the U.S. Department of Agriculture laboratories in Beltsville, Maryland.[7–9] H. A. Borthwick, S. B. Hendricks, and their colleagues were interested in the question of how photoperiodically sensitive plants detect day length. They worked with both long-day plants, which flower in response to nights that are shorter than a critical length, and short-day plants, which flower in response to nights that exceed a critical length. In the case of a long-day plant such as barley, interruption of a long-night period by a brief exposure to light induces flowering that otherwise would not occur. Conversely, in the case of a short-day plant such as cocklebur, interruption of a long-night period prevents flowering that otherwise would occur. Action spectra (see Chapter 1 for definition and discussion) for these seemingly opposed effects of light indicated that red light was most effective (Fig. 10-3). Shortly thereafter, the same investigators obtained similar action spectra for the inhibition of barley stem elongation, initiation of pea leaf expansion, and promotion of lettuce seed germination, indicating the existence of a common photomorphogenic system that controls a wide variety of phenomena.[9]

It was the study of lettuce seed germination, however, that resulted in the critical observation that led directly to the discovery of phytochrome. Following up on observations reported in 1935 by Flint and McAlister[10] and in 1936 by Meischke,[11] Borthwick and Hendricks found that not only would far-red light suppress the germination of light-sensitive lettuce seed, but it would also reverse the promotive effect of red light (Fig. 10-3). Moreover, these antagonistic effects of red and far-red light were repeatedly reversible through as many as 100 cycles. The final outcome was a function solely of the

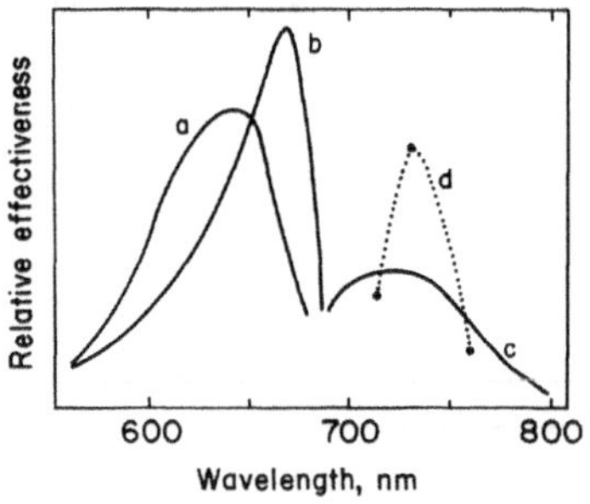

Fig. 10-3. Effectiveness of light as a function of wavelength on (a) preventing flower induction in the short-day plant cocklebur (*Xanthium saccharatum* Wallr.), (b) promoting germination of light-sensitive lettuce seed (*Lactuca sativa* L.), (c) reversing red-light-induced germination of lettuce seed, and (d) reversing red-light-induced suppression of flowering in cocklebur (since only three datum points are available, they are connected by a dotted line). Relative effectiveness values beyond 690 nm have been multiplied by 10. (Adapted from Ref. 9.)

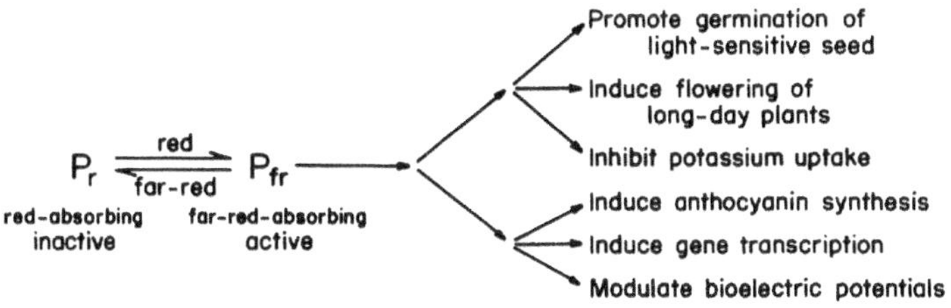

Fig. 10-4. Schematic representation of phytochrome function, showing its photoreversibility between an inactive, red-absorbing form (P_r) and an active, far-red-absorbing form (P_{fr}). A few examples are given of the diverse range of photoreversible responses that have been investigated.

color of light given last. Subsequently, it was learned that other photomorphogenic responses to red light were similarly photoreversible, reinforcing the notion that a single photoreceptor was responsible for all of them and that this photoreceptor must exist in two photointerconvertible forms (Fig. 10-4). One form (P_r) should absorb red light most strongly, while the other (P_{fr}) should absorb far-red light best.

Given the existence of this photoreversible pigment, two possibilities existed. Either the red-absorbing, P_r form was active, preventing the display of photomorphogenic responses, or the far-red-absorbing, P_{fr} form was active, promoting their display. As Hendricks[9] explains elegantly for dark-grown plant tissues, it must be P_{fr} that is active for the following reason. The photoconversion of less than 0.1% of P_r to P_{fr} can lead to a significant biological response. Since this photoconversion represents an insignificant change in the relative P_r level (100% → 99.9%), it is difficult to imagine how the change could be sensed by the plant. This same change, however, represents a relatively large increase in P_{fr} level that is from nothing to a detectable value. Thus, the plant must respond to the appearance of P_{fr}, not the disappearance of P_r (Fig. 10-4).

A summary of the characteristics of classical phytochrome-mediated responses, of which hundreds have been described,[4,6] includes the following points: (1) they are potentiated by red light; (2) the effects of red light are reversed by far-red light; (3) the red and far-red light effects are repeatedly reversible; (4) both red and far-red light are effective at low fluences; and (5) the requirements for both red and far-red light exhibit reciprocal relationships between fluence rate and time of exposure, such that the outcome reflects the product of the two parameters.

Consideration of action spectra for the red and far-red light effects (Fig. 10-3) led Warren Butler, who was then associated with Borthwick and Hendricks, to search for a blue pigment that would undergo photoreversible changes in its absorbance properties.[7] Utilizing a newly designed spectrophotometer created for this purpose, Butler and his colleagues were able to detect such a pigment, which they shortly thereafter named phytochrome.[12] They were then able to isolate phytochrome and to begin its purification and characterization.[13]

Phytochrome has since been found in representatives of all major groups within the plant kingdom, including angiosperms, gymnosperms, ferns, bryophytes, and algae. While occasional reports have indicated that it may be present in at least some fungi, these reports remain unconfirmed.

10.2.2. Events under Phytochrome Control

Phytochrome is often associated only with photoreversible responses that exhibit the characteristics summarized above. It has become evident, however, that plants also "use" phytochrome in other ways. It is therefore preferable to discuss phytochrome-mediated morphogenesis with respect to the type of information to which a plant is responding (Fig. 10-1) rather than with respect to specific morphogenic changes. From this perspective, phytochrome mediates responses to two aspects of light quantity: its onset and its fluence rate. Those that occur as a consequence of the onset of even dim light are the classical photoreversible responses that led to its discovery. Those that are proportional to fluence rate are referred to as high-irradiance responses. In addition to these two functions, phytochrome also provides a mechanism whereby a plant can sense light quality. While these different sensory functions will be treated independently, it is important to note that in some cases they can all contribute to the same photomorphogenic outcome.

10.2.2.1. Detection of the Onset of Light

Phytochrome can mediate responses to exceedingly low levels of light, as in the examples of enhancement of coleoptile elongation and suppression of mesocotyl elongation (Fig. 10-2) in grass seedlings.[14] Responses such as these, which are called very low irradiance or very low fluence responses, can be to the quantity of light provided by a few firefly flashes. Many responses to the onset of light, however, require an irradiance that is 10^4 times greater. Nevertheless, since even these responses are to low fluences, they are known as low-irradiance or low-fluence responses. All of these responses to the onset of light have obvious significance at two stages in the life cycle of a plant: (1) at the time of seed germination and (2) when deetiolation occurs as a seedling emerges from the soil into the light.

Many seeds remain dormant in the absence of light, as would be the case, for example, when buried deeply in the soil. Following hydration, only a brief exposure to light is required to initiate germination, which can then proceed in total darkness.[15] This simple observation emphasizes an important point. It is the induction of a process, rather than its expression, that requires light, at least when considering those responses that occur as a consequence of the onset of light. It is this photosensitivity of many seeds that gives rise to the gardener's frequent observation that each time the soil is turned over, a new crop of weeds seems to develop. Teleologically, one can explain this phenomenon as a survival mechanism. It presumably would prevent the germination of a seed unless it were near enough to the soil surface to allow the resultant seedling to become photosynthetically competent before its food reserves were exhausted, as might happen if the seed germinated too far below the surface.

A more general photomorphogenic response to the onset of light is the process of deetiolation. A seedling normally begins life as a heterotrophic (i.e., etiolated) organism, which means that it cannot manufacture its own organic compounds and must therefore subsist on stored food reserves in the seed. Upon emerging from the soil into the light, however, it undergoes a dramatic change to a photoautotrophic mode of existence (Fig. 10-2), which is to say that it develops the capacity to utilize light energy via photosynthesis to manufacture from inorganic sources all of the organic compounds that it requires. All of the changes that accompany this morphogenic transformation can be grouped under the heading of deetiolation. Among the more prominent of these changes are the synthesis of a host of new gene products that are required for the development of photosynthetically competent chloroplasts,[16] as well as those that are required for the visually obvious morphogenic changes that occur (Fig. 10-2). Many of the morphogenic features of an etiolated seedling can be explained in terms of their unique value to a seedling as it emerges from the soil. For example, the hook near the shoot apex of an etiolated dicotyledonous seedling (Fig. 10-2c) presumably provides protection for the sensitive apex, which is still enclosed within the cotyledons, as the shoot pushes up through the soil. Deetiolation thus involves hook opening, which produces the normal upright shoot apex that is associated with a green plant (Fig. 10-2d). Interestingly, in some instances these photomorphogenic events reflect acceleration of developmental events that ultimately would occur even in the absence of light. As a result it seems important that an etiolated seedling be able to modify its response to a given light exposure as a function of its developmental stage, such that its response at any given time is most appropriate to its presumed needs.[17] There is considerable evidence that these changes in the competence of a seedling to respond do take place.[3,17]

While the vast majority of photomorphogenic responses to the onset of light fall into the category of deetiolation, others have been equally well studied. In particular, there is the photoreversible effect of light given in the middle of a night period on the photoperiodic behavior of a plant.[2,9] This effect of light is in simplest terms interpreted as the P_{fr}-dependent division of a night into two parts. If an otherwise photoperiodically inductive night is divided by a pulse of light into two shorter dark periods, neither of which is long enough to be perceived by the plant as a "long night," then the inductive effect of that dark period is lost. In the case of a long-day plant, the outcome is the induction of flowering; in the case of a short-day plant, the inhibition of flowering.

10.2.2.2. High-Irradiance Responses

Some photomorphogenic responses, known as the high-irradiance responses (HIR),[18] are to irradiations given over long periods of time. The magnitude of a HIR is typically a function of both the fluence rate and duration of the incident light. A reciprocal relationship between fluence rate and time of exposure is normally not observed. While phytochrome-mediated responses to the onset of light generally result from considerably less than one one-hundredth of the energy in all visible wavelengths provided by sunlight during a 1-min period, the HIR require at least 100 times more energy per unit time, given over a much longer period. Even though the photoreversibility that is typically associated with phytochrome-mediated morphogenesis (Fig. 10-4) is not observed, there is compelling evidence that these responses are mediated at least in part by phytochrome.[18,19]

One of the most productive model systems for the study of the HIR has been light-induced synthesis of anthocyanins (Fig. 10-5),[20] which are red to blue pigments found throughout a plant, most obviously in many brightly colored flowers and fruits. The dependence on fluence rate is well illustrated by this system. As is generally the case for the HIR, light in both the red/far-red and blue spectral regions is effective. It is well established that phytochrome is solely responsible for mediating the response to red/far-

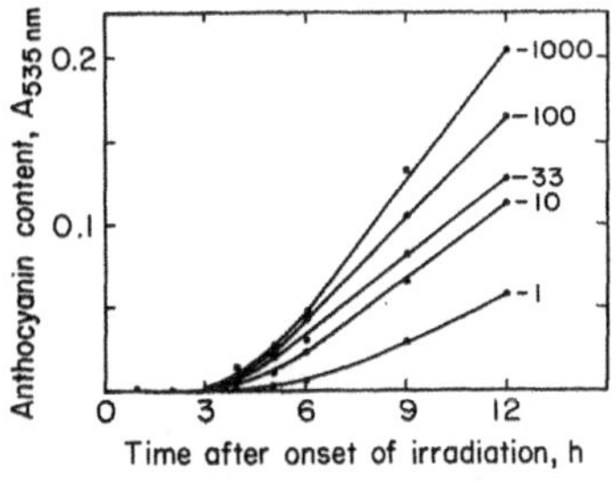

Fig. 10-5. Time course for anthocyanin accumulation in mustard (*Sinapis alba* L.) seedlings as a function of fluence rate during continuous irradiation with broad-band far-red light. Onset of irradiation was at 36 h after sowing. The relative irradiance is indicated next to each curve; a relative irradiance of 1 is 3.5 mW/m^2. (Adapted from Ref. 21.)

red light,[19] but the photoreceptor for the blue region has not yet been identified, although in at least some cases it too may be phytochrome.[20]

The most detailed action spectrum for an HIR is that for inhibition of lettuce hypocotyl elongation (Fig. 10-6). Since those wavelengths that are most effective are absorbed well by both P_r and P_{fr} (Section 10.2.3.2), and since the magnitude of a typical HIR is both fluence rate- and time-dependent, hypotheses to explain how phytochrome mediates the HIR have generally included a requirement for cycling of the pigment between its two forms. Thus, the magnitude of a response would be proportional to the extent of cycling that had occurred. While a variety of hypotheses along this general theme have been proposed,[19,20,23] none has yet been found acceptable in all respects. Moreover, action spectra for the HIR sometimes indicate a lack of involvement of red/far-red light, and, in at least some cases, the action spectrum for a given response changes markedly during the development of a seedling.[24]

10.2.2.3. Sensing of Light Quality

Probably the most pervasive function of phytochrome, even though it has been least studied, is the ability of a green plant to sense the wavelength distribution of incident light,[25,26] an observation that derives from early work on seed germination by Meischke.[11] This function is of most obvious benefit to a green plant that must compete with other green plants for available photosynthetically active radiation. When sunlight is filtered through a green-plant canopy, the pigments within the leaves forming the canopy remove most of the photosynthetically active radiation (Fig. 10-7). In contrast, relatively

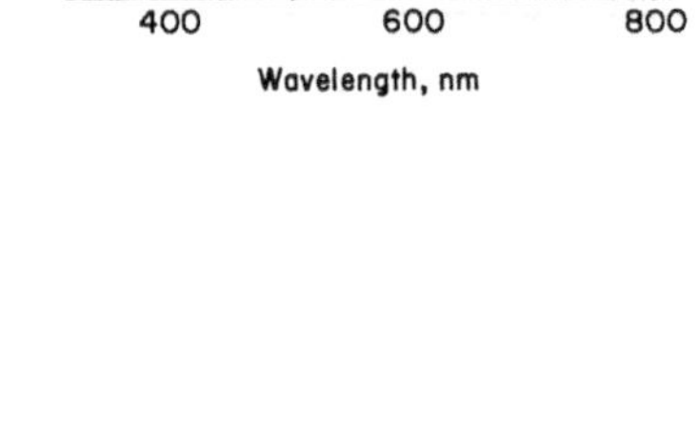

Fig. 10-6. Effect of light via the high-irradiance response (HIR) on the inhibition of lettuce (*Lactuca sativa* L.) hypocotyl elongation. (Adapted from Ref. 22.)

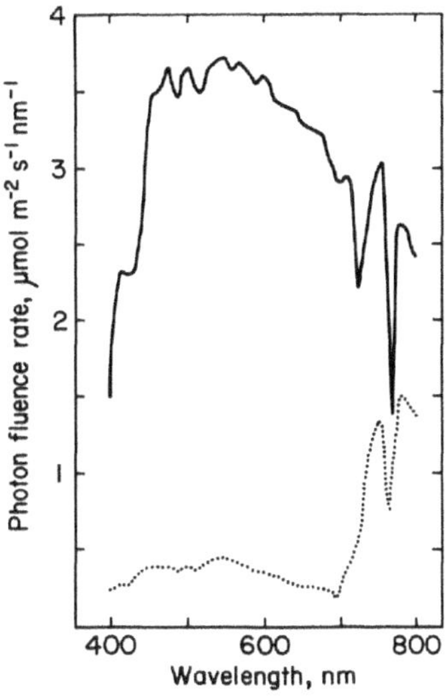

Fig. 10-7. Wavelength distribution of normal daylight (——) and of shadelight beneath a canopy of broadleaf deciduous vegetation (······). (Adapted from Ref. 26.)

little far-red light is removed, which markedly reduces the ratio of red to far-red light. The effect of such a change on phytochrome should be to decrease significantly the rate of phototransformation of P_r to P_{fr}, while decreasing only slightly the back transformation of P_{fr} to P_r (Fig. 10-4). Consequently, at photoequilibrium the steady-state concentration of P_{fr}, expressed as the ratio of P_{fr}/P_{tot} ($P_{tot} = P_r + P_{fr}$), should be reduced. The amount by which the ratio is reduced should in turn be a function of the density of the canopy. Direct measurements of P_{fr}/P_{tot} ratios indicate that this expectation is fulfilled (Table 10-1). In fact, under a dense sugar beet canopy, the steady-state level of P_{fr} approaches that produced by monochromatic far-red light.

The biological significance of these phytochrome-sensed changes in light quality has been documented largely by the work of H. Smith and his colleagues.[25,26] For example, they have shown that the appearance of *Chenopodium album* plants changes dramatically

TABLE 10-1. Proportion of Phytochrome Present as P_{fr} at Photoequilibrium[a]

Lighting	R/FR[b]	P_{fr}/P_{tot}[c]
Midday daylight	0.99–1.17	0.64–0.67
Incandescent	0.71	0.60
Fluorescent	13.5	0.83
Red	20.1	0.88
Far-red	0.002	0.04
Under a wheat canopy	0.21	0.36
Under a sugar beet canopy	0.033–0.045	0.07–0.11

[a]Data from reference 29.

[b]R/FR is the ratio of incident fluence rates at 660 and 730 nm, respectively.

[c]P_{fr}/P_{tot} is the ratio of P_{fr} to total phytochrome at photoequilibrium under the indicated lighting. The data have been recalculated to reflect a value of 0.88 under monochromatic red light.[33]

Fig. 10-8. *Chenopodium album* L. seedlings grown in white light as a function of the red to far-red fluence rate ratio, which decreases from left to right. The proportion of phytochrome that is calculated to be present as P_{fr} (ψ_c) decreases from 0.65 to 0.36, as indicated. (Adapted from Ref. 26.)

as a function of the ratio of red to far-red light that is incident upon them (Fig. 10-8). All of the changes are those that would be predicted to help the plants compete for available photosynthetically active radiation. Moreover, recent work(27,28) has shown that plants respond in similar fashion to a change in light quality that occurs as the result of nearby, neighbor plants. With phytochrome, a plant can thus determine not only whether it is in the shade of another plant, but whether it is growing at high or low plant density. The most obvious change in both situations is an increase in the rate of internode elongation, which presumably would help the plant outgrow its competitors in order to reach open sunlight. While only some species respond in this way to a changing red-to-far-red ratio, it nevertheless seems likely that this function of phytochrome is pervasive, perhaps being manifested in other ways in other species. For example, in order to germinate, many light-sensitive seeds require a level of P_{fr} that is higher than that produced under a dense green canopy. These seed would germinate only when the canopy is eliminated, as happens naturally under a deciduous plant canopy in the autumn or after a fire. Thus, seedlings would develop only when sufficient photosynthetically active radiation would be available to sustain their growth.(15,30)

10.2.3. Properties of Phytochrome

10.2.3.1. Physicochemical Properties

Characterization of phytochrome as a molecule has until recently been restricted to the chromoprotein as it is isolated from etiolated plants.[31,32] There are two reasons for this situation. First, green plants have 10- to 100-fold less phytochrome than etiolated plants of the same genotype. Second, the abundance of chlorophyll in a green plant prevents the spectral assay of phytochrome in them. Unfortunately, the characterization of phytochrome even from etiolated plants has been difficult because of its lability to proteases in crude extracts. Thus, the first report concerning purification of phytochrome to homogeneity indicated that its monomer size was 60 kDa, whereas subsequent reports indicated a size of about 118 kDa. It is now evident, however, that full-size phytochrome, at least from oats, is 124 kDa in size and that earlier reports described partially degraded molecules.[31,32]

Full-size phytochrome is now understood to be a dimer of two very similar, if not identical, monomers, each bearing a linear tetrapyrrole chromophore joined to the protein moiety via a thioether linkage (Fig. 10-9). It is the variable interaction between the P_r and

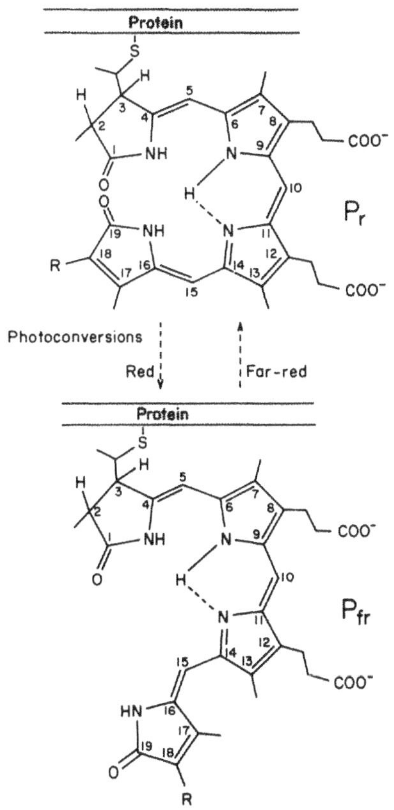

Fig. 10-9. Proposed structures for the phytochrome chromophore in the P_r and P_{fr} forms, indicating its thioether linkage to the protein moiety.

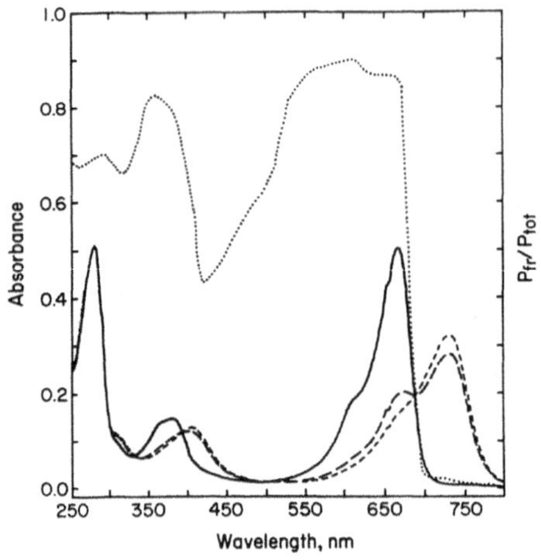

Fig. 10-10. Absorbance spectra for highly purified oat phytochrome after saturating far-red (——) and red (— —) irradiation. A calculated spectrum for a solution of P_{fr} equimolar to the P_r solution obtained by irradiation with far-red light is also shown (-----). The P_{fr}/P_{tot} ratio that is established at different wavelengths (······) is calculated from the absorbance spectra that are shown. (Adapted from Ref. 33.)

P_{fr} forms of the protein moiety with the two forms of the chromophore that give rise to their different absorption spectra (Fig. 10-10). The absorption bands in the red/far-red and the blue regions are due to the chromophore, whereas that at 280 nm arises primarily from the protein moiety. Apart from the presence of a chromophore and about one phosphate residue per monomer, there is no indication that the protein moiety undergoes posttranslational modification. The amino acid composition of phytochrome is typical of a water-soluble protein, indicating that it does not function as an integral membrane protein. This assumption is reinforced by recent sequence data, which indicate that the protein moiety does not possess a hydrophobic domain large enough to permit its insertion into the lipid core of a membrane.[34]

10.2.3.2. Photochemical Reactions

Both photoconversion pathways, $P_r \rightarrow P_{fr}$ and $P_{fr} \rightarrow P_r$, occur via several intermediates (Fig. 10-11). Following the initial excitation of either P_r or P_{fr}, the first intermediate that is produced in each case decays to the opposite form of phytochrome via different sets of thermally unstable intermediates (I_r's and I_{fr}'s). The slowest reactions are those leading to P_r, which occur on a millisecond-to-second time scale. Since the thermal decay of intermediates represents the rate-limiting step in phototransformation, appreciable levels of phytochrome can be present in intermediate form at steady state under bright illumina-

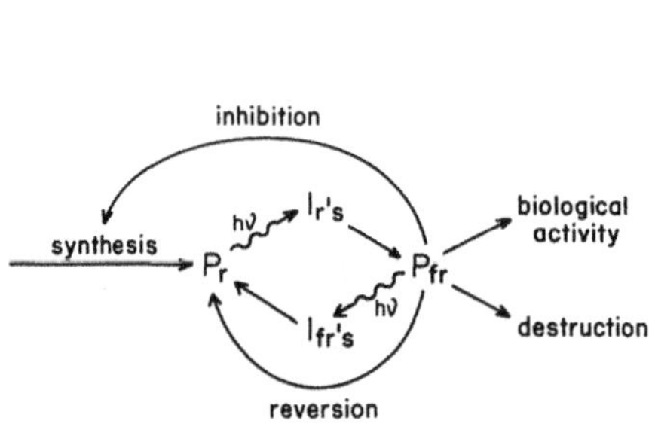

Fig. 10-11. The phytochrome system. Synthesis of phytochrome, which occurs as P_r, is inhibited by P_{fr}. The photoconversion pathways between P_r and P_{fr} involve several intermediates (I_r's and I_{fr}'s). P_{fr}, which yields biological activity by an as yet unknown mechanism, disappears from the cell via both thermal reversion to P_r and enhanced proteolytic degradation of its protein moiety, a process that is referred to as destruction.

tion.[32,35] In principle, therefore, the possibility exists that under certain circumstances, such as in the case of the HIR, a phototransformation intermediate may have biological activity.

A second feature of the photochemical reactions of phytochrome that has attracted considerable interest is the proportion of phytochrome present as P_{fr} (P_{fr}/P_{tot}) at photoequilibrium as a function of wavelength (Fig. 10-10). Except in the far-red region, where P_r absorbs negligibly, the two forms of phytochrome have overlapping spectra. Thus, while it is possible to obtain essentially pure P_r, it is impossible to obtain pure P_{fr}. Instead, under conditions of constant illumination, phytochrome will be cycling continuously between its two forms. The proportion that is present as P_{fr} is then strongly wavelength-dependent (Fig. 10-10; see Sections 10.2.2.2 and 10.2.2.3). It is these considerations that provide a sensible explanation for the ability of phytochrome to sense light quality, as discussed in Section 10.2.2.3.

10.2.3.3. Nonphotochemical Reactions

Phytochrome is involved in four important, nonphotochemical processes. These include (1) *de novo* synthesis of P_r, (2) destruction, preferentially of P_{fr}, (3) nonphotochemical reversion of P_{fr} back to P_r, and (4) the as yet unknown reaction(s) that lead to biological activity (Fig. 10-11). Because of its importance, this fourth aspect will be dealt with separately (Section 10.2.4).

Phytochrome is synthesized in darkness as P_r and accumulates until a plateau is reached. This plateau presumably corresponds to a balance between synthesis and turnover of the chromoprotein. Interestingly, one of the most rapid P_{fr}-mediated responses is an inhibition of P_r synthesis.[36] A 5-s irradiation with red light induces within 15 min a measurable decline in the level of phytochrome mRNA. Within 5 h of a pulse of red light, about 80% of the phytochrome mRNA is gone. Restriction map polymorphisms, together with partial sequence data, indicates that there are at least four different genes expressing phytochrome mRNA in etiolated oat seedlings.[34] It is not known, however, whether these genes are expressed differentially, either with respect to tissue or to age. If either were the case, the physiological implications could be considerable.

The photoconversion of P_r to P_{fr} *in situ* leads to a rapid decline in the level of spectrophotometrically detectable phytochrome; this decline is commonly referred to as destruction (Fig. 10-11).[37] It was subsequently learned that destruction also results in a loss of immunochemically detectable phytochrome. The process is apparently enzymatically mediated, exhibiting an appropriate temperature dependence and requiring an input of free energy. It presumably results from a targeted, proteolytic degradation of phytochrome, although the precise nature of this process has not been elucidated. It has recently been suggested that phytochrome is targeted for destruction by ubiquitin,[38] a ubiquitous protein that in other instances is known to serve this function. However, this potential role of ubiquitin in phytochrome destruction remains to be firmly established. Furthermore, while it has often been said that destruction is specific for P_{fr}, this is not the case. In some instances, when light conditions are such that P_r is first converted to P_{fr} and then back to P_r, this cycling of the pigment leads to a limited, but comparably enhanced, rate of destruction of P_r.[37] It is as though, in some cases, phytochrome molecules that had been targeted for destruction as P_{fr} remain marked for destruction even after being converted back to P_r, such that P_r will also undergo limited destruction. Together with the P_{fr}-induced decrease in phytochrome mRNA level, it is this P_{fr}-initiated destruction of phytochrome that results in such low levels of this pigment in light-grown plants.

Thermal reversion of P_{fr} to P_r, which generally occurs over a span of hours and which was once widely implicated as a timing function to explain the role of phytochrome in photoperiodism,[9] is a puzzling event. It has been described for phytochrome *in vivo* in etiolated dicotyledonous seedlings, but then involves only a limited pool of the pigment. It has never been observed *in vivo* in etiolated grass seedlings, nor, apart from one exception, has phytochrome isolated from etiolated grass seedlings been found to exhibit reversion *in vitro,* unless the protein has been partially degraded.[32] The exception derives from the observation that a monoclonal antibody, when bound to a domain on phytochrome that is apparently about 50 amino acids from its amino terminus,[39] induces thermal reversion of P_{fr} to P_r. This domain had previously been shown to be important for the maintenance of phytochrome in a spectrally native P_{fr} form.[32] This observation leads to the speculation that plant cells might regulate thermal reversion, and thereby control the level of the active P_{fr} form of phytochrome during the night, by controlling the availability of a ligand that interacts with phytochrome at this location. Apart from the observation that only a limited fraction of the phytochrome in dicotyledonous cells undergoes thermal reversion, however, there are no independent data to support this speculation. At present, it is simply impossible to assess the significance, if any, of this process. Nevertheless, it is evident that thermal reversion, as well as phytochrome destruction, provides a mechanism whereby a cell can rid itself of the active form of phytochrome during the night.

10.2.3.4. Differences between P_r and P_{fr}

Since P_{fr} is active and P_r is not (Fig. 10-4), considerable attention has been given to the question of how the two forms of phytochrome differ from one another. Such differences might provide a clue as to how P_{fr} functions.[31,32,37] Unfortunately, in spite of the effort expended on answering this question, little of substance has been learned. The bulk of the evidence indicates that there is no gross conformational difference between P_r and P_{fr}, although recent data do indicate a slightly different behavior by size exclusion chromatography, with P_{fr} exhibiting a slightly greater hydrated radius than P_r.[32] While many discrete differences between the two forms have been reported, they have for the most part not been related to the structure of the chromoprotein. The recent production in several laboratories of domain-specific monoclonal antibodies to phytochrome should facilitate identification of differences between P_r and P_{fr}. Already a limited number of monoclonal antibodies have been identified that recognize domains that change upon phototransformation.[32,40] For example, three antibodies that bind several fold more tightly to P_r than to P_{fr} have been found.[39] These antibodies bind to a domain that is apparently about 50 amino acids from the amino terminus of phytochrome, a region that previously had been shown to be important for proper interaction between the protein and chromophore moieties of this pigment.[32,40] Interestingly, it was one of these antibodies that was also found to facilitate reversion of P_{fr} to P_r, as mentioned above. Much more will have to be learned, however, before even this information will assist us in understanding how P_{fr} functions (e.g., see Section 10.6).

Two other significant differences between P_r and P_{fr} have been described. While P_r

behaves as a soluble protein in crude plant extracts, P_{fr} associates preferentially with particulate, subcellular debris.[41–43] It has been suggested that this enhanced association of P_{fr} with sedimentable subcellular material, which is termed enhanced pelletability, results from an interaction of phytochrome with membranes.[44] There is, however, little hard evidence to support this conclusion.[42,43] Independently, immunocytochemical observations have led to the conclusion that while P_r is uniformly distributed throughout the cytosol, consistent with its presence as a soluble protein, P_{fr} is associated with discrete subcellular regions.[37] This redistribution of phytochrome after photoconversion to P_{fr}, which has been called sequestering, has also led to the suggestion that it results from the binding of P_{fr} to a membrane receptor. Again, there is no hard evidence to support this suggestion. Instead, recent electron microscope observations of immunogold-labeled phytochrome indicated that both sequestering and enhanced pelletability of P_{fr} involve an association of phytochrome with structures that appear not to be bounded by a membrane.[45,46] Even though there seems to be no detectable association between P_{fr} and a membrane, both sequestering and enhanced pelletability occur in oats within 5–10 s after the appearance of P_{fr} within the cell, which is sufficiently rapid to leave open the possibility that they result from an initial step in the mode of action of phytochrome.

10.2.3.5. Phytochrome from Green Plants

The development of sensitive immunochemical assays for phytochrome which eliminate interference by chlorophyll has recently made it practical to begin the physicochemical characterization of phytochrome from green-plant tissue. Although indirect evidence had led to the suggestion that phytochrome in green plants would be different from that in etiolated plants, the recent discovery that phytochrome from green oats was both spectrally and immunochemically distinct from phytochrome from etiolated oats was nevertheless somewhat surprising.[47,48] With respect to spectral properties, P_{fr} from green oats is similar to that from etiolated oats, whereas the absorbance maximum for P_r is shifted about 10 nm to shorter wavelengths. With respect to immunochemical properties, the majority of monoclonal antibodies to phytochrome from etiolated oats fail to recognize phytochrome from green oats[48,49] and vice versa.[50] Moreover, phytochrome from green oats is not only different from that from etiolated oats, but preliminary indications are that it may itself consist of multiple populations.[40,48,49] Even though phytochrome from green oats is about the same size as that from etiolated oats, they yield different peptide maps upon proteolysis, reinforcing the conclusion that they are different proteins.[47,49] Whether these two proteins are products of different genes or are differentially modified products of the same gene is presently a matter of intensive investigation.

The significance of two kinds of phytochrome in the same plant is unknown. While it is evident that one of the two, namely, that which has previously been well characterized, is only abundant in etiolated tissue, tools with which to determine how the abundance of the other might be regulated are only now becoming available.[50] Initial evidence does indicate that the levels of the two phytochromes are differentially regulated.[51] It may even be that these two phytochromes have different biological activities. It would be interesting to determine, for example, whether the chromoprotein that is abundant in etiolated oats is responsible for detecting the onset of light (Section 10.2.2.1) whereas the chromoprotein detected in green oats is responsible for sensing light quality (Section

10.2.2.3). In addition, it will be important to determine whether the observations made with oats can be generalized. Initial work with peas indicates that they can be.[48,52]

10.2.4. Primary Mode of Action

It has long been evident that P_{fr} is the active form of phytochrome.[9] While this conclusion has occasionally been challenged, the challenges have never withstood scrutiny. It has also been proposed that there are multiple primary reactions for P_{fr}.[53] While this hypothesis has merit, there is at present no compelling reason to accept it. In some respects, it would seem more prudent to begin with the simpler assumption, at least with respect to phytochrome itself, that it has a single primary function. The multiplicity of events controlled by phytochrome would then arise by branching in the transduction chain between P_{fr} and the responses, after the initial function of P_{fr} has been expressed (Fig. 10-4). Alternatively, however, it is conceivable that there are multiple pools of phytochrome (e.g., Section 10.2.3.5) and that, while each pool exerts only a single primary function, that function is different for each pool. Irrespective of these considerations, the key question to be asked is, what is it that P_{fr} does that P_r does not or cannot do? The answer to this question must incorporate answers to other, more specific questions as well: (1) How is the signal that is detected by phytochrome amplified? (2) How does the primary event involving P_{fr} result in such a vast array of seemingly unrelated responses? (3) How is it that the formation of P_{fr} within a cell gives rise to responses that are a function of both the tissue and the morphological age of the cell in which they occur?

Three major hypotheses concerning the primary action of P_{fr} have been proposed. While each of these three hypotheses has its unique merits, the following discussion is intended to make clear that none has been proven correct.

The first hypothesis derives directly from the observation that phytochrome is a chromoprotein. Since proteins are often enzymes, it seemed sensible at the outset to propose that P_{fr} exhibits an enzyme activity that P_r lacks.[9] No such activity, however, has yet been described for P_{fr}. While interest in this hypothesis has not been very great, there is presently no good reason to discard it. Indeed, the recent evidence that highly purified phytochrome is associated with a protein kinase activity[54] is rekindling interest in this hypothesis. Nevertheless, as the authors of this work note, it remains to be determined whether the kinase activity belongs to phytochrome or an otherwise undetected contaminant.

The origin of the second hypothesis coincided with the rapid increase in knowledge concerning the differential expression of genetic information. With respect to phytochrome, for a given cell at a given developmental stage there should be four components of the genome[53]: (1) genes that are induced by P_{fr}, (2) genes that are repressed by P_{fr}, (3) genes that are never expressed, irrespective of the form of phytochrome, and (4) genes that are expressed independently of phytochrome control. It has become abundantly clear that phytochrome does influence gene expression as hypothesized,[55] but whether it does so by interacting directly with the genome or indirectly by means of a second messenger has not been established. The existence of promoter regions responsible for phytochrome-mediated gene expression has been established,[55,56] but direct control of such promoters by phytochrome has not been demonstrated. While it is conceivable that phytochrome does regulate gene expression directly, such a primary function for phytochrome cannot account for the most rapid phytochrome-mediated events that have been described (see below), nor has there been any verifiable demonstration that phytochrome exists in the nucleus.

The third hypothesis, for which experimental support is the strongest, arose at a time when interest in, and knowledge of, the organizational and functional complexity of membranes was increasing rapidly. Logically, one should focus on those events that occur most rapidly after the appearance of P_{fr} if one wants to learn about its primary function. By doing so, one is almost invariably led to the study of phytochrome-mediated responses that indicate an early effect of P_{fr} on membrane properties.(44,57) For example, transmembrane ion flux changes, notably of K^+, begin within at most a few minutes after the formation of P_{fr}. Effects of phytochrome on bioelectric potentials are even more rapid. In one of the most elegant of such studies, Newman(58) demonstrated that P_{fr} initiates an electric potential change in the etiolated oat coleoptile beginning only 4.5 s after its appearance within the cell. One is thus led to the hypothesis that phytochrome functions in a primary sense by altering one or more membrane properties or functions.(57) Unless phytochrome has different primary activities, it is evident that it cannot be directly regulating the expression of genetic information.

Numerous variants of the membrane hypothesis have been suggested. It has been proposed that phytochrome might function as an integral membrane protein, although its biochemical properties (Section 10.2.3.1) are inconsistent with this possibility. Alternatively, it has been suggested that phytochrome might interact with a membrane by binding to a receptor on its surface and that this interaction is specific for the P_{fr} form. Superficially, both red-light-enhanced phytochrome pelletability and immunocytochemically detected sequestering of P_{fr} appear to be consistent with this possibility.(44) Nevertheless, as already discussed (Section 10.2.3.4), neither of these phenomena is likely to involve phytochrome binding to a membrane receptor. A third variant, which at present would seem to be the most reasonable, is that the effect of P_{fr} on membranes may be mediated via a second messenger such as Ca^{2+}. As yet, however, there are no compelling data to indicate that such a second messenger is, in fact, obligatorily involved in phytochrome function.

Thus, while the hypothesis that phytochrome exerts its primary effect at the membrane level is presently the most acceptable of the three that have been considered, even this hypothesis does not enjoy overwhelming experimental support. It is at least equally possible that P_{fr} functions as an enzyme, as originally proposed, or in some as yet unimagined fashion.

10.3. BLUE-LIGHT-MEDIATED MORPHOGENESIS

Since blue light is absorbed by phytochrome, it should not be surprising that phytochrome can serve as a photoreceptor for blue-light-mediated morphogenesis. Nevertheless, there are blue-light-absorbing photoreceptors quite independent of phytochrome. Morphogenic phenomena that are mediated specifically by blue light differ in a few important aspects from those mediated by phytochrome: (1) They are induced most effectively by blue and, in many instances, near-UV radiation, but not by light in other spectral regions, notably the red. (2) Whereas mediation by phytochrome of blue-light effects requires relatively high fluence rates of blue light, mediation by a specific blue-light photoreceptor often, but not always, involves much lower incident energies. (3) Unlike many phytochrome-mediated events, responses to low fluence rate blue light are not photoreversible.

No chemically defined blue-light photoreceptor has yet been identified. At least three reasons can be given for this failure: (1) The discovery of phytochrome was facilitated greatly by the fact that it is a reversibly photochromic pigment, thus permitting the development of a sensitive spectrophotometric assay for its detection *in vitro*.(7,13) Conversely, there is no firm evidence that any blue-light photoreceptor is comparably photochromic, which means that development of an assay for its detection *in vitro* will be difficult at best. (2) A typical cell contains an abundance of blue/near-UV-absorbing pigments. A search for a physiologically significant blue-light photoreceptor is thus equivalent to looking for a needle in a haystack. (3) There are likely to be many blue-light photoreceptors,(59) perhaps even within the same organism and involved in the same response.(60) It is for this last reason that the term cryptochrome(61) is not used here. Inherently, this term implies that there exists a single, chemically definable pigment that is responsible for blue-light-mediated morphogenesis. Since this almost certainly is not the case, more generic terminology will be used to minimize the possibility that a reader will come to the conclusion that a single pigment is responsible for the wide range of blue-light-mediated responses that have been described.

10.3.1. Events under Blue-Light Control(62)

The number of morphogenic responses to blue light that have been characterized is considerably more limited than the number known to be mediated by phytochrome. Conversely, the taxonomic distribution of blue-light photoreceptors is more extensive. As is true for phytochrome, blue-light photoreceptors are found in angiosperms, gymnosperms, and other nonflowering plants, including green algae. Unlike phytochrome, however, they are also widespread among the fungi. In part because of the absence of a single well-defined photoreceptor, it is difficult to provide a uniform description of a typical blue-light photoresponse. In general, such a photoresponse is insensitive to wavelengths of light longer than the blue, and it typically exhibits sensitivity to near-UV radiation as well. Given this rather loose definition, the following are selected examples of blue-light-mediated morphogenesis.

One of the best studied blue-light responses is the phototropic curvature of plant shoots toward the light (Chapter 11). Positive phototropism of the coleoptile, which is an organ that is best developed in young grass seedlings grown in total darkness (Fig. 10-2b), has long served as a model system.(63)

A second growth response of shoots to blue light, which appears to be distinct from that leading to phototropism, is the blue-light-mediated inhibition of stem elongation, best described for the etiolated cucumber hypocotyl.(64) In this instance, blue light of relatively high fluence rate initiates a decrease in growth rate that begins after a lag of as little as 20–30 s. This reduced growth rate persists as long as the blue light is on (Fig. 10-12). The fluence rate required is roughly 1000-fold greater than that required to induce a phototropic curvature of the grass coleoptile. Since the response begins after such a brief lag,

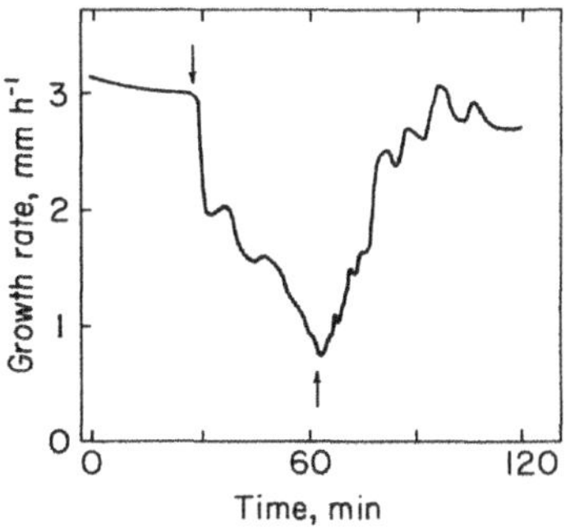

Fig. 10-12. Growth response of the cucumber (*Cucumis sativus* L.) stem to irradiation with blue light (5 W/m^2). A rapid decrease in growth rate is seen beginning within less than 1 min after the onset of light (↓). The growth rate returns to the dark level within several minutes after the blue light is turned off (↑). (Adapted from Ref. 64.)

it is evident that it is relatively closely coupled to photoreceptor excitation, although the photoreceptor and the nature of the transduction chain itself are completely unknown. Simple shading experiments indicate that the photoreceptor is in the region of the hypocotyl that exhibits the response, indicating that there may be no need for signal transmission in this instance as there is, say, in at least some cases of phototropism.[63]

The study of blue-light-induced pigment synthesis has also provided considerable information about blue-light-mediated morphogenesis.[62,65] Accumulation of both carotenoids and anthocyanins are at least in part dependent on blue light. Light-induced anthocyanin synthesis is an especially good example of the interactions that can occur among different photoreceptor systems. The extent of blue-light-mediated anthocyanin synthesis is itself regulated in a photoreversible fashion by phytochrome. If red light is given after the termination of a prolonged exposure to a relatively high fluence rate of blue light, such that the level of P_{fr} is high, then the quantity of anthocyanin that is accumulated is high. In contrast, if the blue-light exposure is terminated by far-red light such that the level of P_{fr} is low, then little anthocyanin is accumulated. If only blue light is provided, the response is then a function of the amount of P_{fr} that happens to be present when the exposure to blue light is terminated. Since red light given in the absence of a preceding exposure to blue light is by itself relatively ineffective, it is evident that a blue-light photoreceptor is in some way interacting synergistically with phytochrome to produce the final response.

Not all interactions are synergistic. For example, red light that is absorbed by phytochrome reduces by a factor of 10 the sensitivity of the maize coleoptile to phototropically active blue light, even though the red light itself is phototropically inactive.[66]

Other well-studied blue-light-mediated responses include (1) induction of a transition from one-dimensional to two-dimensional growth of the fern protonema, which is an early, filamentous growth stage of the haploid generation of this group of plants, (2) formation in certain fungi of fruiting bodies known as perithecia, (3) regulation in higher plants of stomatal apertures, which are variably sized pores in the epidermis of leaves through which gas exchange occurs, and (4) changes in membrane potentials of cells near the apex of young bean shoots.[59,61,62] It is therefore evident that blue light induces a diverse range of morphogenic responses.

10.3.2. Photoreceptor(s)

As already noted, it is evident that not all blue-light-mediated morphogenesis involves the same photoreceptor. Nevertheless, it seems reasonable that most blue-light photoreceptors might belong to the same pigment category.

Action spectra for typical blue-light-mediated responses, such as phototropism of the oat coleoptile and carotenogenesis in the fungus *Fusarium* (Fig. 10-13), indicate that the photoreceptor in each case has three absorbance bands in the blue region of the spectrum, in addition to a band in the near-UV region. What pigments are candidates for such a photoreceptor? Two possibilities have been given serious consideration: carotenoids, which have a typical three-peaked absorbance spectrum in the blue region, and flavins, which exhibit absorbance in both the blue and near-UV regions. Carotenoids typically do not have absorbance in both the blue and the near-UV regions, nor do flavins usually exhibit the fine structure in the blue that is seen in many action spectra. Examples of each group that possess all requisite absorbance properties to serve as a blue-light photoreceptor do exist, however. Given these two distinctly different possibilities, it is perhaps only natural that there has been considerable controversy over the question of whether blue-light photoreceptors are flavins or carotenoids.

A consensus of opinion now suggests, however, that blue-light photoreceptors are most likely flavins in association with protein, that is to say flavoproteins.[59,61,62] For example, flavins are well known as photosensitizers (Chapter 3). In addition, chemicals that interact with photoexcited flavins, such as phenylacetic acid, inhibit specifically one of the classic blue-light-mediated events, namely, phototropism.[69] Moreover, "carotenoid-deficient" mutants have recently been found to exhibit a phototropic response to blue light, albeit with slightly decreased photosensitivity. Thus, while a blue-light photoreceptor has not yet been identified, it appears that the search should focus on flavoproteins as likely candidates.[59]

One approach to the question of how to identify a blue-light photoreceptor is potentially fruitful. In 1970, Berns and Vaughn[70] proposed that one consequence of light absorption by a blue-light photoreceptor might be a change in absorbance of either the blue-light photoreceptor itself or another, closely associated pigment. This *l*ight-*i*nduced *a*bsorbance *c*hange (LIAC) might then serve as the basis of an assay for the isolation, purification, and characterization of a blue-light photoreceptor. In order for such a LIAC

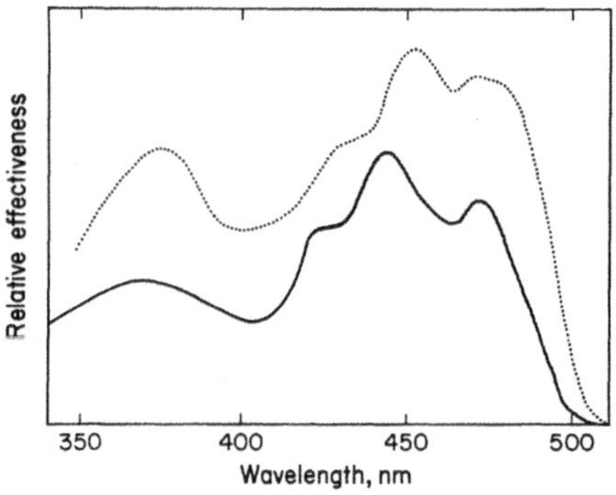

Fig. 10-13. Action spectra for first positive phototropic curvature of the oat (*Avena sativa* L.) coleoptile (——) and for carotenogenesis in the fungus *Fusarium* (······). (Adapted from Refs. 67 and 68.)

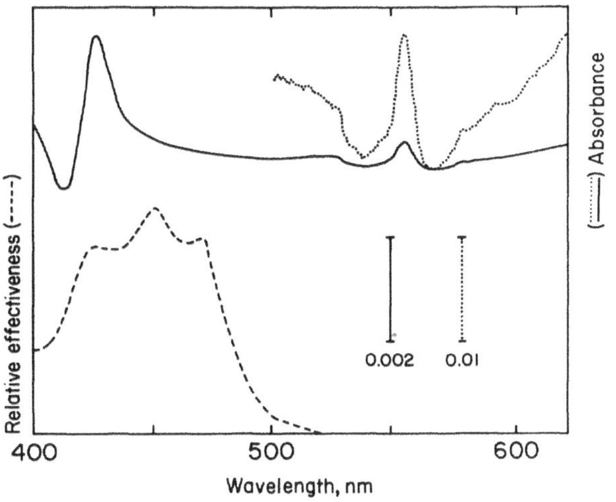

Fig. 10-14. Difference spectrum (——) showing a blue-light-induced absorbance change (LIAC) in a membrane-enriched fraction from cauliflower (*Brassica oleracea* L.). The LIAC was obtained by subtracting an absorbance spectrum measured in the absence of any irradiation from one measured after irradiation for 2 min with blue light at a fluence rate of 15 W/m^2. Both spectra were measured at 77K. Part of the LIAC is expanded to show greater detail (······). Calibrations in absorbance units for the original (——) and the expanded (······) difference spectrum are given at the lower right. The action spectrum that gives rise to the LIAC is shown for comparison (----). (Adapted from Ref. 72.)

to serve as an assay for a potential blue-light photoreceptor, it should have an action spectrum like that for blue-light-mediated photomorphogenic responses (Fig. 10-13). Since the LIAC need not result from a change in absorbance of the blue-light photoreceptor itself, however, it could be displayed at any wavelength.

LIACs with these appropriate characteristics have been identified from both fungal and higher plant sources.[71] The action spectra are similar to those for typical blue-light-mediated morphogenic responses, while the LIACs themselves are indicative of the reduction of a b-type cytochrome (Fig. 10-14). Isolation and partial purification of subcellular fractions that exhibit a blue-light-induced absorbance change indicate that all necessary components are membrane-associated and that the membrane with which they are associated is most likely the plasma membrane. Purified, putative plasma membrane fractions exhibiting the LIAC have been shown to possess flavin by both fluorescence excitation and emission assay. It thus seems apparent that the LIAC is a flavin-mediated photoreduction of a b-type cytochrome.[59,71,72]

The question that remains is whether this LIAC derives from the activity of a biologically relevant blue-light photoreceptor. Unfortunately, this question is exceedingly difficult to answer unequivocally. Given that commercially obtained horse heart cytochrome c and either flavin mononucleotide or riboflavin can exhibit a comparable LIAC under wholly artificial conditions,[73] it is possible to suggest that the naturally isolated LIACs might be biologically irrelevant. Alternatively, it is evident that even in purified membrane fractions, it is a specific b-type cytochrome that is photoreduced. The bulk of the cytochromes that are present are not photoreduced. Moreover, a *Neurospora* mutant that is deficient in cytochrome b is more than 100-fold less sensitive than the wild-type strain to blue light, which in this instance induces conidiation. Similarly, methylene blue sensitizes several biological systems to red light, such that action spectra for typical blue-light-mediated events are extended artificially to include red light that is absorbed by the methylene blue. Of interest is the correlated observation that this photosensitizer also facilitates the production *in vitro* of a LIAC in response to red light and that the LIAC is specific for the same b-type cytochrome that is photoreduced by the endogenous system. It is evident, therefore, that there is considerable, albeit indirect, evidence to support the hypothesis that the LIACs that have been described might serve as an assay for a blue-light photoreceptor, eventually allowing its complete purification and chemical identification.[71] Nevertheless, it is important to bear in mind that the evidence is far from conclusive. Whereas the indefinite photoreversibility of phytochrome, which is also expressed as a LIAC, provides a ready assay for its identification, the LIAC in question here cannot be associated so unambiguously with a biologically relevant blue-light photoreceptor.

10.4. UV RADIATION-MEDIATED MORPHOGENESIS

That part of the UV spectrum that is significant with respect to photomorphogenesis is from about 300 to 390 nm. Wavelengths between 320 and 390 nm constitute the UV-A region and, as we have already seen, can be active via a blue-light photoreceptor and phytochrome, as well as via a separate UV radiation photoreceptor. The region from 280 to 320 nm is known as the UV-B region. This region is of special interest because of its potentially harmful effects on both plants and animals (Chapters 5 and 6). Normally, UV-B radiation is for the most part selectively removed from solar radiation by ozone present in the stratosphere. Recent indications that human activities are resulting in a depletion in the amount of ozone in this layer, which would lead to increased UV-B radiation reaching the surface of our planet, have generated increased interest in the effects of UV-B radiation on biological systems.

As in the case of blue-light-mediated morphogenesis, there appear to be multiple UV-absorbing photoreceptors that control a variety of morphogenic events.[74] Since specific photomorphogenic responses to UV-A radiation are unknown in higher plants,[74] the emphasis here will be on the effects of UV-B radiation. Unfortunately, since there is no good candidate for a specific UV-B radiation photoreceptor, it is difficult to provide a focus for even this limited discussion of UV radiation-mediated morphogenesis. As in the case of blue-light-mediated morphogenesis, it will only be possible to indicate with a few examples the wide range of responses to UV radiation that have been described and to consider the possible nature of what is perhaps the best described UV radiation photoreceptor system, which has been termed mycochrome since it is present uniquely in the fungi. The extensively characterized photochemical effects of UV radiation, especially on nucleic acids, and the photorepair of this damage, both of which are perhaps best excluded from a discussion of photomorphogenesis in any case, are covered in Chapter 4.

10.4.1. Events under UV Radiation Control

Although UV radiation-mediated morphogenesis is best known among the fungi, UV-B radiation does have significant morphogenic effects on higher plants. It not only interferes with the photosynthetic activity of isolated chloroplasts,[75] but it also decreases the photosynthetic activity of intact plants.[76] Furthermore, UV-B radiation leads to a general decrease in plant growth, including decreases in leaf area and overall dry weight, as well as enhanced flavonoid synthesis.[74] Since the newly synthesized flavonoids are located in outer cell layers, and since they absorb solar UV-B radiation strongly, it has been speculated that their function is to protect underlying tissues from the harmful effects of UV-B radiation, such as impairment of the photosynthetic machinery.[74]

In contrast to UV radiation-induced changes in higher plants, the mycochrome system in fungi has been better defined, although it is still far from being understood.[77] At a particular stage of the development of spore-producing structures known as conidiophores, UV radiation and blue light have opposing effects on the development of spores, referred to in this instance as conidia. If UV radiation and blue light are given in alternating fashion, conidia develop if the last exposure is to UV radiation, whereas they do not develop if blue light is given last. This photoreversibility is reminiscent of the red/far-red photoreversibility seen with phytochrome. Nevertheless, the UV/blue effects do not result from the photointerconversion of a single pigment between two forms,[77] as is the case with phytochrome.

10.4.2. Photoreceptor(s)

There are serious practical difficulties associated with obtaining action spectra in the UV region. Both the relatively low fluence rates that are available from monochromatic sources and the massive screening that results from the absorption of UV radiation by plant tissues interfere significantly. For these and other reasons,[74] very few detailed action spectra have been obtained. Consequently, it has not even been possible in most instances to predict with any precision the absorbance properties of potential photoreceptors.

The mycochrome system has been the one that is most amenable to analysis.[77] An action spectrum for the induction of conidia formation exhibits a single, reasonably well-defined maximum at about 300 nm, with a shoulder at about 320 nm (Fig. 10-15). The reversion of this UV-B radiation effect by blue light exhibits maximum activity near 450 nm, with a shoulder or subsidiary peak on either side and with a second peak of activity at about 380 nm. If both the UV-B radiation and blue-light effects were mediated by a single photoreversible pigment analogous to phytochrome, it might have been relatively easy to use this photoreversibility as an assay to permit its isolation and purification. As already noted, however, it seems evident that separate pigments are responsible for the opposing effects of light. The chemical nature of these two pigments is completely unknown. The obvious similarity between the action spectra for the effect of blue light in the my-

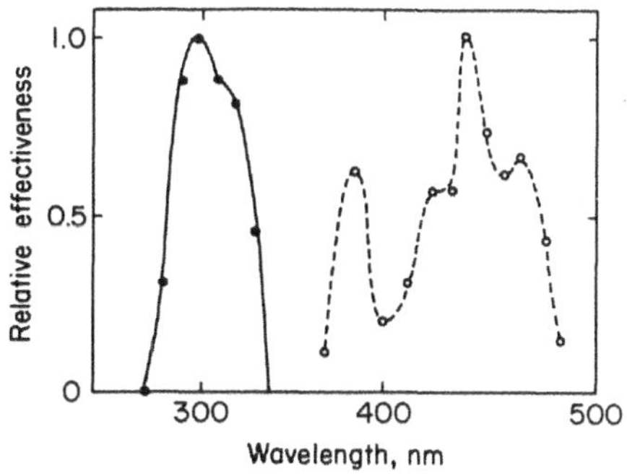

Fig. 10-15. Action spectra for the induction of conidiation (———) and its reversal by blue light (– – – – –) in *Alternaria tomato*. (Adapted from Ref. 78.)

cochrome system and in other blue-light-mediated morphogenic responses (Fig. 10-13) indicates that a flavoprotein would be a good candidate for this photoreceptor. It has been proposed that the two pigments involved in the mycochrome system interact in a coupled oxidation–reduction process,(77) but the evidence so far available in support of this hypothesis does not appear to be very strong.

10.5. PHOTOCHROMIC ADAPTATION

Cyanobacteria, which are phototrophic procaryotes, utilize phycobiliproteins as their major light-harvesting pigments in photosynthesis (Chapter 12). Both the phycocyanins and the phycoerythrins absorb light maximally at wavelengths that chlorophyll absorbs poorly. The ability of the cyanobacteria to produce selectively pigments that are complementary to the color of light incident upon them at any given time would be expected to help them utilize available photosynthetically active radiation with maximum efficiency. Indeed, as early as 1883, Engelmann(79) observed just such an adaptation among the cyanobacteria, which was detected as changes in cell color corresponding to changes in the color of ambient illumination. These color changes, now known as complementary chromatic adaptation, were later shown to result from modifications in the relative amounts of the red-colored phycoerythrins and the blue-colored phycocyanins that were synthesized.(80)

It is customary to divide the cyanobacteria into three groups, based on their ability to exhibit chromatic adaptation.(81) Group I includes those organisms that do not adapt chromatically. Organisms in group II exhibit unidirectional adaptation in that they can control the synthesis only of phycoerythrin. Within group III, however, the synthesis of both phycoerythrin and phycocyanin is regulated by the wavelength distribution of incident light. These organisms are said to exhibit bidirectional chromatic adaptation. Photocontrol of phycoerythrin synthesis in this case is strictly a function of green light. Control of phycocyanin synthesis, however, is more complicated. A total of four phycocyanin subunits are found, each of which appears to be a unique gene product. Two subunits, which are 17.8 and 20.2 kDa, are expressed constitutively and when assembled yield phycocyanin that is designated as PC-1. The other two subunits, which are 17.4 and 21.7 kDa, are found only in cells grown in red light or in darkness. Phycocyanin composed of these two subunits is known as PC-2. It appears, therefore, that phycocyanin synthesis results from the differential expression of four genes, only two of which are light-regulated.

Action spectra for chromatic adaptation of representatives from both group II and group III indicate maxima between 540 and 550 nm for potentiation of phycoerythrin synthesis by green light, and between 640 and 660 nm for potentiation of PC-2 synthesis by red light. In every case, only a few minutes of actinic light is required. When successive exposures to green and red light are given, it is always the last exposure that determines the response. This system is thus analogous to the photoreversible responses that are mediated by phytochrome.

Regulation of chromatic adaptation occurs at the level of transcription.(81) While it is evident that the photoreceptor is not involved in energy transduction, it has not yet been identified. A chromoprotein that exhibits absorbance changes consistent with a role as a photoreversible detector of chromatic adaptation was first described by Scheibe.(82) Subsequently, Björn and Björn[83] described four photoreversible pigments that they called phycochromes a, b, c, and d. Recent work, however, has led to the conclusion that these photochromic pigments, which are derived from native phycobiliproteins, are not involved in chromatic adaptation.[81,84]

In a recent synthesis of available data, Tandeau de Marsac[81] proposed a model for regulation of chromatic adaptation that assumes the existence of a photoreceptor in two interconvertible forms, one absorbing primarily in the green (PG) and the other in the red (PR). According to her model, both forms of the photoreceptor are active, with each regulating a different event, in marked contrast to the phytochrome system, in which only one form of the chromoprotein is active. An alternative hypothesis (Fig. 10-16) suggests that only the PR form is active and that it functions both as a repressor of the genes coding for PC-2 and as an inducer of phycoerythrin synthesis. This alternative model accounts for the existence of PC-2 both in complete darkness and in red light, since expression of PC-2 would occur whenever PR is absent. Conversely, PR would act as an activator or derepressor of phycoerythrin synthesis.

To learn more about the control of complementary chromatic adaptation at a molecular level, it is evident that methods for the functional incorporation of foreign genetic material into the cyanobacteria are needed. As these methods have just become available, it is likely that they will soon be used for this purpose. A few pigmentation mutants, which may reveal important information concerning chromatic adaptation, have already been isolated. Information obtained with eight "blue mutants," each of which simultaneously either lost or had impaired ability both to photoinduce phycoerythrin and to photorepress phycocyanin,[84] is consistent with the model presented here (Fig. 10-16). Ultimately, however, isolation and characterization of the photocontrolled genes, and in particular their promotor/repressor regions, will be required to determine the mode of action of the photoreceptor system.

10.6. THE FUTURE

Apart from those photomorphogenic responses that can be ascribed to phytochrome, the most compelling need is for the identification of photoreceptors. This task will be difficult both for blue-light and UV-B radiation photoreceptors for reasons that have already been mentioned. Since a direct approach to identifying photoreceptors appears unlikely to be successful, it is evident that a different strategy is needed. Given the rapid advances made in recent years in molecular genetics, a genetic approach would appear to be most promising. Since strategies are being developed for the identification of defective genes in the absence of prior knowledge of their products, it may become possible to identify "photoreceptor genes" in the absence of any information about the chemical

Fig. 10-16. Hypothesis for the control of complementary chromatic adaptation by green and red light via a hypothetical photochromic pigment that exists in both an inactive, green-absorbing form (PG) and an active, red-absorbing form (PR). PR would function by simultaneously promoting synthesis of phycoerythrin and repressing synthesis of PC-2, which is phycocyanin that is composed of the two subunits that are photoregulated.

identity of the photoreceptor. Once relevant genes have been identified, methods for subsequent determination of their products are already available. Thus, it should eventually be possible to identify gene products that are in some way related to a photoreceptor without knowing anything in advance about its chemical structure. The most serious limitation to this approach at present is that it will be necessary to work with an organism for which there are available highly efficient methods for the introduction of foreign genetic material under circumstances that will permit development of a mature organism that has stably integrated the introduced DNA into its own genome. Keeping in mind the present state of technology in this area, it is unlikely that this approach will be undertaken until that technology has improved.

With respect to phytochrome, two questions seem most pressing. First, phytochrome present in green oat shoots has been shown to be different from that which predominates in etiolated oat shoots.[47,48] Similarly, pea shoots contain at least two distinct populations of phytochrome, one that is relatively more abundant when plants are grown in the dark, the other more abundant when they are grown in the light.[52] Thus, it becomes important to determine whether the chromoproteins that are most abundant in dark- and light-grown tissues, respectively, are differentially modified products of the same gene, or products of different genes. Moreover, it is evident that all of the information gained about phytochrome so far, virtually all of which has been obtained with the chromoprotein that is abundant only in plants grown for extended periods in darkness, might be of little value when considering phytochrome-mediated morphogenesis in photosynthetically competent plants grown under natural environmental conditions. Clarification of the question of the nature of phytochrome in green plants is needed.

Second, we still know far too little about phytochrome itself, even from etiolated plant tissues. What, precisely, are the differences between P_r and P_{fr}? On which part(s) of the molecule are these differences to be found? What domains on phytochrome are likely to be important with respect to its biological function? Since these domains are likely to have been highly conserved throughout evolution, it is important to identify those regions that are common to phytochromes isolated from a wide variety of taxonomically distinct organisms. Monoclonal antibodies, which recognize well-defined, relatively small regions on an antigen, are ideally suited to both purposes. Antibodies that discriminate between P_r and P_{fr} have already been described,[39,40] as has been an antibody that recognizes an antigenic determinant on a polypeptide the size of phytochrome from both angiosperms and algae.[85] This work, however, represents only a beginning. Finally, even though the primary structures of phytochrome from etiolated oats and zucchini have recently been predicted from their cDNA sequences,[34,86] equally significant advances will also have to be made concerning knowledge of its secondary, tertiary, and quaternary structures if we are eventually to understand how it might function in a primary sense. Consequently, continued biochemical and biophysical investigations of phytochrome itself are required.

10.7. REFERENCES

1. A. Schwartz and D. Koller, Phototropic response to vectorial light in leaves of *Lavatera cretica* L., *Plant Physiol.* **61,** 924–928 (1978).
2. D. Vince-Prue, *Photoperiodism in Plants,* McGraw-Hill, London (1975).

3. H. Mohr, Pattern specification and realization in photomorphogenesis, *Encycl. Plant Physiol., New Ser.* **16A,** 336–357 (1983).
4. D. L. Correll, J. L. Edwards, and W. Shropshire, Jr., *Phytochrome: A Bibliography with Author, Biological Materials, Taxonomic and Subject Indexes of Publications Prior to 1975,* Smithsonian Institution Press, Washington, D.C. (1977).
5. Z. Lechowski and J. Białęzyk, Interaction between green and far-red light on the low fluence rate chloroplast orientation in *Mougeotia, Plant Physiol.* **85,** 581–584 (1987).
6. W. Shropshire, Jr., and H. Mohr (eds.), *Encycl. Plant Physiol., New Ser.* **16A and 16B,** Springer-Verlag, Berlin (1983).
7. W. L. Butler, Remembrances of phytochrome twenty years ago, in: *Photoreceptors and Plant Development* (J. A. DeGreef, ed.), pp. 3–7, Antwerpen University Press, Belgium (1980).
8. H. A. Borthwick, The biological significance of phytochrome, in: *Phytochrome* (K. Mitrakos and W. Shropshire, Jr., eds.), pp. 3–23, Academic Press, London (1972).
9. S. B. Hendricks, Photochemical aspects of plant photoperiodicity, *Photophysiology* **1,** 305–331 (1964).
10. L. H. Flint and E. D. McAlister, Wavelengths of radiation in the visible spectrum inhibiting the germination of light-sensitive lettuce seed, *Smithson. Misc. Collect.* **94,** 1–11 (1935).
11. D. Meischke, Über den Einfluss der Strahlung auf Licht- und Dunkelkeimer. *Jahrb. Wiss. Bot.* **83,** 359–405 (1936).
12. H. A. Borthwick and S. B. Hendricks, Photoperiodism in plants, *Science* **132,** 1223–1228 (1960).
13. W. L. Butler, K. H. Norris, H. W. Siegelman, and S. B. Hendricks, Detection, assay, and preliminary purification of the pigment controlling photoresponsive development of plants, *Proc. Natl. Acad. Sci. USA* **45,** 1703–1708 (1959).
14. D. F. Mandoli and W. R. Briggs, Phytochrome control of two low-irradiance responses in etiolated oat seedlings, *Plant Physiol.* **67,** 733–739 (1981).
15. B. Frankland and R. Taylorson, Light control of seed germination, *Encycl. Plant Physiol., New Ser.* **16A,** 428–456 (1983).
16. P. Schopfer and K. Apel, Intracellular photomorphogenesis, *Encycl. Plant Physiol., New Ser.* **16A,** 258–288 (1983).
17. V. Gaba and M. Black, The control of cell growth by light, *Encycl. Plant Physiol., New Ser.* **16A,** 358–400 (1983).
18. A. L. Mancinelli and I. Rabino, The high irradiance responses of plant photomorphogenesis, *Bot. Rev.* **44,** 129–180 (1978).
19. K. M. Hartmann, A general hypothesis to interpret high energy phenomena of photomorphogenesis on the basis of phytochrome, *Photochem. Photobiol.* **5,** 349–366 (1966).
20. A. L. Mancinelli, Light-dependent anthocyanin synthesis: A model system for the study of plant photomorphogenesis, *Bot. Rev.* **51,** 107–157 (1985).
21. H. Lange, W. Shropshire, Jr., and H. Mohr, An analysis of phytochrome-mediated anthocyanin synthesis, *Plant Physiol.* **47,** 649–655 (1971).
22. K. M. Hartmann, Ein Wirkungsspektrum der Photomorphogenese unter Hochenergie-bedingungen und seine Interpretation auf der Basis des Phytochroms (Hypokotyl-wachstumshemmung bei *Lactuca sativa* L.), *Z. Naturforsc. Ser. B* **22,** 1172–1175 (1967).
23. L. Fukshansky and E. Schäfer, Models in photomorphogenesis, *Encycl. Plant Physiol., New Ser.* **16A,** 69–95 (1983).
24. C. J. Beggs, M. G. Holmes, M. Jabben, and E. Schäfer, Action spectra for the inhibition of hypocotyl growth by continuous irradiation in light and dark-grown *Sinapis alba* L. seedlings, *Plant Physiol.* **66,** 615–618 (1980).
25. H. Smith, Light quality, photoperception, and plant strategy, *Annu. Rev. Plant Physiol.* **33,** 481–518 (1982).
26. H. Smith and D. C. Morgan, The function of phytochrome in nature, *Encycl. Plant Physiol., New Ser.* **16B,** 491–517 (1983).
27. M. J. Kasperbauer, Far-red light reflection from green leaves and effects on phytochrome-mediated assimilate partitioning under field conditions, *Plant Physiol.* **85,** 350–354 (1987).
28. C. L. Ballaré, R. A. Sánchez, A. L. Scopel, J. J. Casal, and C. M. Ghersa, Early detection of neighbour plants by phytochrome perception of spectral changes in reflected sunlight, *Plant Cell Environ.* **10,** 551–557 (1987).

29. M. G. Holmes and H. Smith, The function of phytochrome in plants growing in the natural environment, *Nature* **254,** 512–514 (1975).
30. J. D. Bewley and M. Black, *Physiology and Biochemistry of Seeds, Vol. 2, Viability, Dormancy and Environmental Control,* Springer-Verlag, Berlin (1982).
31. L. H. Pratt, Phytochrome: The protein moiety, *Annu. Rev. Plant Physiol.* **33,** 557–582 (1982).
32. J. C. Lagarias, Progress in the molecular analysis of phytochrome, *Photochem. Photobiol.* **42,** 811–820 (1985).
33. J. M. Kelly and J. C. Lagarias, Photochemistry of 124-kilodalton *Avena* phytochrome under constant illumination in vitro, *Biochemistry* **24,** 6003–6010 (1985).
34. H. P. Hershey, R. F. Barker, K. B. Idler, J. L. Lissemore, and P. H. Quail, Analysis of cloned cDNA and genomic sequences for phytochrome: Complete amino acid sequences for two gene products expressed in etiolated *Avena, Nucl. Acids Res.* **13,** 8543–8558 (1985).
35. R. E. Kendrick and C. J. P. Spruit, Phototransformation of phytochrome, *Photochem. Photobiol.* **26,** 201–214 (1977).
36. H. P. Hershey, R. F. Barker, J. T. Colbert, J. L. Lissemore, and P. H. Quail, Phytochrome: Molecular properties, feedback regulation of mRNA levels and genomic clone isolation, in: *Molecular Form and Function of the Plant Genome* (L. van Vloten-Doting, G. S. P. Groot, and T. C. Hall, eds.), pp. 101–111, Plenum Press, New York (1985).
37. L. H. Pratt, Phytochrome: Function and properties, *Photochem. Photobiol. Rev.* **4,** 59–124 (1979).
38. J. Shanklin, M. Jabben, and R. D. Vierstra, Red light-induced formation of ubiquitin-phytochrome conjugates: Identification of possible intermediates of phytochrome degradation, *Proc. Natl. Acad. Sci. USA* **84,** 359–363 (1987).
39. M.-M. Cordonnier, H. Greppin, and L. H. Pratt, Monoclonal antibodies with differing affinities to the red-absorbing and far-red-absorbing forms of phytochrome, *Biochemistry* **24,** 3246–3253 (1985).
40. M.-M. Cordonnier, Recent advances in the study of the protein moiety of phytochrome, in: *From Photobiophysics to Photobiology* (A. Favre, R. Tyrrell and J. Cadet, eds.), pp. 229–244, Elsevier, Amsterdam (1987).
41. P. H. Quail, D. Marmé, and E. Schäfer, Particle-bound phytochrome from maize and pumpkin, *Nature [New Biol.]* **245,** 189–191 (1973).
42. L. H. Pratt, Molecular properties of phytochrome, *Photochem. Photobiol.* **27,** 81–105 (1978).
43. P. H. Quail, Intracellular localization of phytochrome, in: *Trends in Photobiology* (C. Hèléne, M. Charlier, and T. L. Garestier, eds.), pp. 485–500, Plenum Press, New York (1982).
44. D. Marmé, Phytochrome: Membranes as possible sites of primary action, *Annu. Rev. Plant Physiol.* **28,** 173–198 (1977).
45. D. McCurdy and L. H. Pratt, Immunogold electron microscopy of phytochrome in *Avena:* Identification of intracellular sites responsible for phytochrome sequestering and enhanced pelletability, *J. Cell Biol.* **103,** 2541–2550 (1986).
46. V. Speth, V. Otto, and E. Schäfer, Intracellular localisation of phytochrome in oat coleoptiles by electron microscopy, *Planta* **168,** 299–304 (1986).
47. J. G. Tokuhisa, S. M. Daniels, and P. H. Quail, Phytochrome in green tissue: Spectral and immunochemical evidence for two distinct molecular species of phytochrome in light-grown *Avena sativa* L., *Planta* **164,** 321–332 (1985).
48. Y. Shimazaki and L. H. Pratt, Immunochemical detection with rabbit polyclonal and mouse monoclonal antibodies of different pools of phytochrome from etiolated and green *Avena* shoots, *Planta* **164,** 333–344 (1985).
49. M.-M. Cordonnier, H. Greppin, and L. H. Pratt, Phytochrome from green *Avena* shoots characterized with a monoclonal antibody to phytochrome from etiolated *Pisum* shoots, *Biochemistry* **25,** 7657–7666 (1986).
50. L. H. Pratt and M.-M. Cordonnier, Phytochrome from green *Avena,* in: Phytochrome and Photoregulation in Plants (M. Furuya, ed.), pp. 83–94, Academic Press, Tokyo (1987).
51. J. G. Tokuhisa and P. H. Quail, The levels of two distinct species of phytochrome are regulated differently during germination in *Avena sativa* L., *Planta* **172,** 371–377 (1987).
52. H. Abe, K. T. Yamamoto, A. Nagatani, and M. Furuya, Characterization of green tissue-specific phytochrome isolated immunochemically from pea seedlings, *Plant Cell Physiol.* **26,** 1387–1399 (1985).
53. H. Mohr, *Lectures on Photomorphogenesis,* Springer-Verlag, New York (1972).
54. Y.-S. Wong, H.-C. Cheng, D. A. Walsh, and J. C. Lagarias, Phosphorylation of *Avena* phytochrome in vitro as a probe of light-induced conformational changes, *J. Biol. Chem.* **261,** 12089–12097.

55. E. M. Tobin and J. Silverthorne, Light regulation of gene expression in higher plants, *Annu. Rev. Plant Physiol.* **36,** 569–593 (1985).
56. G. Morelli, F. Nagy, R. T. Fraley, S. G. Rogers, and N.-H. Chua, A short conserved sequence is involved in the light-inducibility of a gene encoding ribulose 1,5-bisphosphatase carboxylase small subunit of pea, *Nature* **315,** 200–204 (1985).
57. S. B. Hendricks and W. J. Van der Woude, How phytochrome acts: Perspectives on the continuing quest, *Encycl. Plant Physiol., New Ser.* **16A,** 3–23 (1983).
58. I. A. Newman, Rapid electric responses of oats to phytochrome show membrane processes unrelated to pelletability, *Plant Physiol.* **68,** 1494–1499 (1981).
59. H. Senger and W. R. Briggs, The blue light receptor(s): Primary reactions and subsequent metabolic changes, *Photochem. Photobiol. Rev.* **6,** 1–38 (1981).
60. G. Löser and E. Schäfer, Are there several photoreceptors involved in phototropism of *Phycomyces blakesleeanus*? Kinetic studies of dichromatic irradiation, *Photochem. Photobiol.* **43,** 195–204 (1986).
61. J. Gressel, Blue light photoreception, *Photochem. Photobiol.* **30,** 749–754 (1979).
62. H. Senger (ed.), *Blue Light Effects in Biological Systems,* Springer-Verlag, Berlin (1984).
63. W. R. Briggs, The phototropic responses of higher plants, *Annu. Rev. Plant Physiol.* **14,** 311–352 (1963).
64. D. J. Cosgrove, Rapid suppression of growth by blue light: Occurrence, time course, and general characteristics, *Plant Physiol.* **67,** 584–590 (1981).
65. A. L. Mancinelli, The photoregulation of anthocyanin synthesis. *Encycl. Plant Physiol., New Ser.* **16B,** 640–661 (1983).
66. H. P. Chon and W. R. Briggs, Effect of red light on the phototropic sensitivity of corn coleoptiles, *Plant Physiol.* **41,** 1715–1724 (1966).
67. K. V. Thimann and G. M. Curry, Phototropism and phototaxis, in: *Comparative Biochemistry,* Vol. 1 (M. Florkin and H. S. Mason, eds.), pp. 243–306, Academic Press, New York (1960).
68. W. Rau, Untersuchungen über die Lichtabhängige Carotinoidsynthese. I. Das Wirkungsspektrum von *Fusarium aquaeductuum, Planta* **72,** 14–28 (1967).
69. R. D. Vierstra and K. Poff, Mechanism of specific inhibition of phototropism in corn seedlings by phenylacetic acid, *Plant Physiol.* **67,** 1011–1015 (1981).
70. D. S. Berns and J. R. Vaughn, Studies on the photopigment system in Phycomyces, *Biochem. Biophys. Res. Commun.* **39,** 1094–1103 (1970).
71. W. R. Briggs, A blue light photoreceptor system in higher plants and fungi, in: *Photoreceptors and Plant Development* (J. A. DeGreef, ed.), pp. 17–28, Antwerpen University Press, Belgium (1980).
72. S. Widell, R. J. Caubergs, and C. Larsson, Spectral characterization of light-reducible cytochrome in a plasma membrane-enriched fraction and in other membranes from cauliflower inflorescences, *Photochem. Photobiol.* **38,** 95–98 (1983).
73. W. Schmidt and W. L. Butler, Flavin-mediated photoreactions in artificial systems: A possible model for the blue-light photoreceptor pigment in living systems, *Photochem. Photobiol.* **24,** 71–75 (1976).
74. E. Wellmann, UV radiation in photomorphogenesis, *Encycl. Plant Physiol., New Ser.* **16B,** 745–756 (1983).
75. W. Iwanzik, M. Tevini, G. Dohnt, M. Voss, W. Weiss, P. Graber, and G. Renger, Action of UV-B radiation on photosynthetic primary reactions in spinach chloroplasts, *Physiol. Plant.* **58,** 401–407 (1983).
76. W. B. Sisson and M. M. Caldwell, Photosynthesis, dark respiration, and growth of *Rumex patientia* L. exposed to ultraviolet irradiance (288 to 315 nanometers) simulating a reduced atmospheric ozone column, *Plant Physiol.* **58,** 563–568 (1976).
77. T. Kumagai, Mycochrome system in the induction of fungal conidiation, in: *Blue Light Effects in Biological Systems* (H. Senger, ed.), pp. 29–38, Springer Verlag, Berlin (1984).
78. T. Kumagai, Action spectra for the blue and near ultraviolet reversible photoreaction in the induction of fungal conidiation, *Physiol. Plant.* **57,** 468–471 (1983).
79. T. W. Engelmann, Farbe und Assimilation, *Botanische Zeitung* **41,** 1–13 (1883).
80. K. Boresch, Die komplementäre chromatische Adaptation, *Arch. Protistenk.* **44,** 1–70 (1922).
81. N. Tandeau de Marsac, Phycobilisomes and complementary chromatic adaptation in cyanobacteria, *Bull. Inst. Pasteur* **81,** 201–254 (1983).
82. J. Scheibe, Photoreversible pigment: Occurrence in a blue-green alga, *Science* **176,** 1037–1039 (1972).
83. G. S. Björn and L. O. Björn, Photochromic pigments from blue-green algae: Phycochromes a, b, and c, *Physiol. Plant.* **36,** 297–304 (1976).

84. J. G. Cobley and R. D. Miranda, Mutations affecting chromatic adaptation in the cyanobacterium *Fremyella diplosiphon, J. Bact.* **153,** 1486–1492 (1983).
85. M.-M. Cordonnier, H. Greppin, and L. H. Pratt, Identification of a highly conserved domain on phytochrome from angiosperms to algae, *Plant Physiol.* **80,** 982–987 (1986).
86. R. A. Sharrock, J. L. Lissemore, and P. H. Quail, Nucleotide and amino acid sequence of a *Cucurbita* phytochrome cDNA clone: Identification of conserved features by comparison with *Avena* phytochrome, *Gene* **47,** 287–295 (1986).

11

Photomovement

Pill-Soon Song • Department of Chemistry and Section of Molecular Plant Biology, School of Biological Sciences, University of Nebraska, Lincoln, Nebraska 68588-0304. **Kenneth L. Poff** • MSU/DOE Plant Research Laboratory, Michigan State University, East Lansing, Michigan 48824-1312.

11.1. INTRODUCTION

Organisms are subjected to a variety of environmental stimuli to which they respond and adapt. A schematic relationship between the stimulus and the response is illustrated in Fig. 11-1. Each stimulus (e.g., light, chemical) in some way interacts (e.g., absorption) with a discrete receptor in a step referred to as perception. The perceived stimulus signal is then processed through a chain of reactions, leading to an outcome such as a change in motility. This entire process from the perception of a stimulus to the response is referred to as sensory transduction.

As might be expected from the variety of stimuli listed in Fig. 11-1, there is variation in the complexity of the transduction chains, and even more variation in the level at which we understand the particular response mechanism. For example, bacterial chemotaxis (swimming toward or away from a chemical stimulus) is perhaps the best characterized and understood of the sensory transduction systems.[1] Thermotaxis (moving toward or away from heat) of certain bacteria and amoebae reflects the surprising ability of an organism to respond to extremely small temperature changes using some presently un-

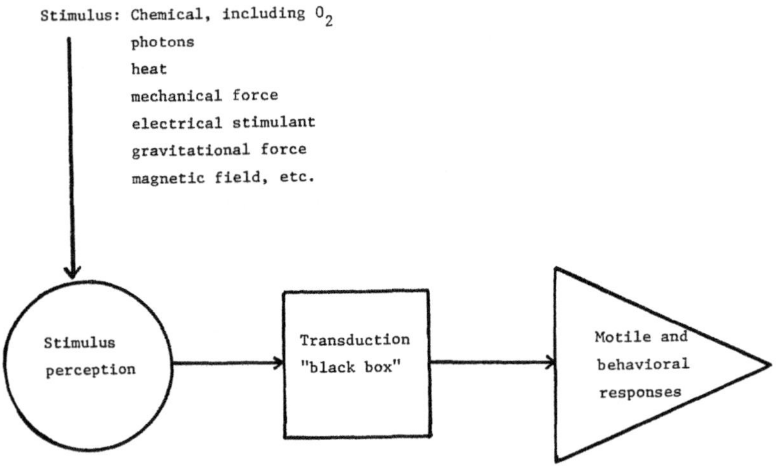

Fig. 11-1. Transduction processes of different stimuli through a "black box" that may or may not be commonly shared for different stimulus–response systems. Stimulus perception (circle) is carried out by a respective receptor (e.g., photoreceptor for photons as the stimulus).

known biothermometer.[2] Some stimuli may evoke a single, highly specific response. In contrast, in some cases, other stimuli such as a gravitational force, or mechanical and electrical stimulation may evoke several different sensory responses. [3,4] No matter what the stimulus, it appears that at least one organism can be found with a behavioral response to that stimulus. For example, some organisms are able to align themselves with respect to the magnetic field of the earth in the so-called magnetotaxis phenomenon.[5]

In this chapter, we will deal exclusively with that aspect of photosensory transduction known as photomovement. Photomovement refers to any light-induced motility or behavioral response involving the spatial displacement of all or part of an organism. Most, if not all, procaryotic and eucaryotic organisms exhibit some form of motile/behavioral reaction to light direction, intensity, and wavelength (for a review, see Ref. 6). However, the reactions to light stimuli known as photomovement are not confined to freely motile organisms. For example, many sessile plants respond to light stimuli through the movement or spatial displacement of a portion of the organism, the movement of cells, or the movement of cellular organelles. Phototropic curvature by a plant toward or away from the light stimulus (cf. Section 11.3.3) is included as a well-known example of the displacement of a portion of an organism (for reviews, see Refs. 7–16). In eucaryotic organisms, cells and cellular organelles move as a result of conformational changes of certain critical proteins of the contractile elements and membranes involved. For example, cytoplasmic streaming in plant cells is well known. Many of these cell and organelle movements are sensitive to light. One of the best characterized responses of cell organelles to light is an intracellular chloroplast rotation, which is discussed in Section 11.3.2 for the green alga *Mougeotia*.

In addition to the diversity in photomovement responses between organisms, there may also be diverse responses in the same organism. Many organisms ranging from procaryotes and unicellular eucaryotes to higher animals, such as birds and fish, exhibit several responses under different wavelengths of light or different light intensities.

The reader should keep in mind throughout this chapter that the various examples of photomovement discussed have an adaptive advantage for the organism. For example, an exposure to light can be either beneficial or deleterious to an organism. In such a case, one can understand the advantage to the organism of sensing the light and coupling this perception with a movement response such that the organism is optimally positioned in light or darkness. Thus, an organism may use light through photosynthesis for its energy content, but to optimize the process of photosynthesis, it may use a nonphotosynthetic photomovement mechanism and position itself close to the optimum light intensity for photosynthesis.

11.2. TERMINOLOGY AND METHODS

11.2.1. Terminology

Although there are many problems with definitions, it is necessary to briefly cover the commonly used terms to make our discussion of photomovement easier. One specific problem of which the student should be wary is the tendency to use a term to describe a phenomenon or mechanism instead of seeking an understanding of that phenomenon or mechanism. However, the student of photomovement frequently encounters a major problem with terminology, since the most widely accepted terminology is based on mechanism. Thus, if the precise mechanism of a photomovement response under observation is not clearly known, it is difficult to define such a photomovement response with a generally acceptable term. Moreover, there is frequently no general agreement among the workers in this field for an acceptable set of definitions to describe a particular photomovement phenomenon.

In 1976, a group of photobiologists proposed a set of terms describing the various movement responses in photosensitive organisms.[17] In this chapter, we use these fundamental photomovement terms, which are based on the concept of a photoresponse.

A *photoresponse* may be defined as *any light-induced change in the organism's motor activity, resulting in an alteration of the movement or orientation of the organism.* The three major classes of photoresponses of motile organisms are termed *photokinesis, photophobic responses,* and *phototaxis.* Each of these types of photoresponse may lead to exactly the same photomovement, since the term refers not to the final response but rather to the mechanism for achieving that response. A photomovement that is restricted to a cellular organelle is termed a *photodinesis,* while the orientation to light of a nonmotile organism is termed *phototropism.* These terms are defined below, and the photoresponses are represented diagrammatically in Fig. 11-2.

11.2.1.1. Photokinesis

Imagine a population of organisms swimming faster in dim light than in bright light, and in an environment that is half shaded. Each individual would swim quickly through the shaded portion but would decrease its speed upon entering the unshaded portion. The organism would swim through the unshaded portion and back into the shaded portion, and so on, but, as a result of the change in swimming speed, would spend a greater percentage of time in light than in shade. Thus, if this population were observed after a time, more organisms would be in bright light than in dim light. This is an example of photokinesis.

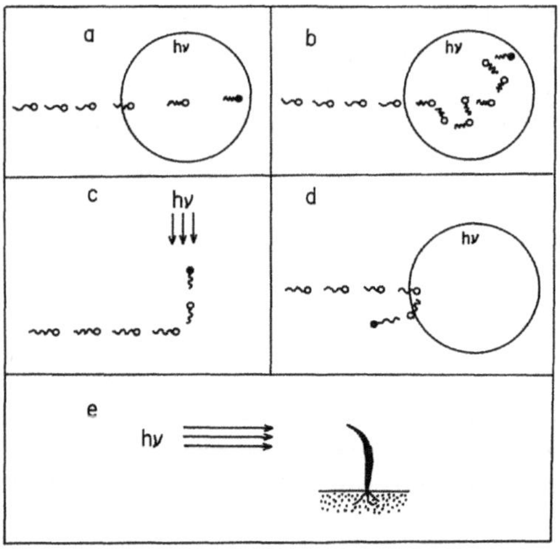

Fig. 11-2. Types of photoresponses: (a) Positive photokinesis. The time interval between the successive positions of a cell is constant; thus, the cell is moving faster in the light (inside the circle) than in darkness. (b) Photoklinokinesis. The organism changes direction more frequently in light (inside the circle) than in darkness. (c) Positive phototaxis. The organism turns toward the light source. (d) Step-up photophobic response. The cell stops, and changes direction upon entering a light field (inside the circle). (e) Phototropism. A plant seedling grows toward light.

Photokinesis is a response in which the steady-state rate of activity of the organism is affected, without adaptation, by the steady-state light stimulus. If the activity parameter measured is swimming velocity, it may be called "photoorthokinesis," but this is commonly referred to simply as photokinesis. If the activity parameter measured represents the frequency of directional change, then it may be referred to as "photoklinokinesis" (e.g., a bacterium may tumble both in light and in darkness, but tumble more frequently in light).

A positive photokinesis refers to an activity rate, e.g., swimming rate, that is higher in the presence of the stimulus light than in its absence (dark control). A negative photokinesis refers to an activity rate, e.g., swimming rate, which is lower in the presence of the stimulus light than in its absence. Thus, in the example given above in which the organism swam slower in bright light than in dim light, the photoresponse would be called a negative photokinesis.

11.2.1.2. Photophobic Responses

Imagine another population of organisms in an environment that is half shaded. Imagine also that each organism swims smoothly when going from dim to bright light. However, upon going in the opposite direction, from bright light to dim light, the swimming pattern is altered; the organism stops, tumbles briefly, and then continues swimming. The tumble results in a change in swimming direction, which may be into the dimly lighted area but may also be into the brightly lighted area. This will result in a higher probability for the organism entering the brightly lighted area than for the organism leaving that area. Thus, if this population were observed after a time, more organisms would be in bright light than in dim light. This is an example of a photophobic response. Notice that the end result is the same for this example as for that given above for photokinesis, but that the mechanism is different.

Photophobic response refers to a transient alteration in the activity, e.g., directional change or swimming velocity, of the organism elicited by a change in the stimulus light intensity. The photophobic response most commonly observed is, as in the example, a stop response, typically followed by a change in movement direction. As a consequence, an organism either moves into the area of higher light intensity or away from the area, depending on the given organism, the light stimulus, and other variables.

A photophobic response may be due either to a step-up stimulus (a sudden increase in light intensity) or a step-down stimulus (a sudden decrease in light intensity) stimulus. Thus, in the example given above, the organism exhibited a photophobic response upon encountering a decrease in light intensity so the photoresponse could be referred to as a step-down photophobic response.

11.2.1.3. Phototaxis

Imagine yet another population of organisms in an environment that is half shaded. Imagine that each organism can in some way determine the light direction, and orient itself so that it swims toward the light, and thus stay within the brightly lighted area. If this population were observed after a time, more organisms would be in bright light than in dim light. This is an example of a phototactic response. The end result, an accumulation of organisms in light, is the same for this example as for those given above for photokinesis and for the photophobic response; only the mechanism is different.

Phototaxis is the oriented movement of an organism toward or away from the source of a light stimulus, referred to as positive or negative phototaxis, respectively. In the example above, the organism exhibited a positive phototaxis, moving toward the light source. The direction of the taxis often depends on the stimulus intensity, with many organisms showing positive phototaxis at low light intensity and negative phototaxis at high light intensity.[18] At an intermediate light intensity, the organism may in fact move perpendicular to the source of the light stimulus; such a response is called transverse taxis or diaphototaxis.[17]

11.2.1.4. Photoaccumulation and Photodispersal

In the above examples, the populations of organisms, through different photoresponses, accumulated in a lighted area. The terms *photoaccumulation* and *photodispersal* are used to describe the end result of photokinesis, a photophobic response, or phototaxis on *populations* of organisms. Responses that result in organisms accumulating or dispersing from a region of higher fluence rate are referred to as photoaccumulation and photodispersal, respectively. Photodispersal is a behavioral consequence of positive photokinesis, a step-up photophobic response, or negative phototaxis. Similarly, photoaccumulation is a behavioral consequence of negative photokinesis, a step-down photophobic response, or positive phototaxis. Note that the terms photoaccumulation and photodispersal describe the effect of a photoresponse on a population of organisms, and not the photoresponse itself.

11.2.1.5. Photodinesis

The term *photodinesis* refers to the phenomenon in which light triggers or modulates the movement of *intracellular organelles.* A photodinesis can be an acceleration of existing streaming (as in cytoplasmic streaming) or an induction of movement of resting organelles in response to light stimuli. An example of photodinesis will be discussed in Section 11.3.2.

11.2.1.6. Phototropism

Charles Darwin and his son Francis observed that canary grass seedlings grew toward a lighted window.[19] This *growth-mediated directional response* of a sessile organism or organ to an asymmetric external light stimulus is phototropism. The organ may bend toward or away from the direction of light stimulus in what are termed positive and negative phototropism, respectively. (The bending of an organ relative to the electric vector of a polarized light stimulus is *polarotropism,* and is usually excluded from the definition of phototropism in the strict sense.[20])

11.2.1.7. Terminology Summary

Photoaccumulation or photodispersal by a population of organisms may result from either one of three different photoresponses: photokinesis, phototaxis, or a photophobic response. If the photomovement is restricted to a cellular organelle, it is referred to as a photodinesis. If, however, a sessile organism or organ orients with respect to light through a growth response, it is called phototropism. The reader is referred to the treatise by Haupt and Feinleib[21] for a discussion of terminological problems in the field of photomovement. However, at the same time, the reader is cautioned that the objective is not to concentrate on a set of terms but rather to understand the mechanisms of photomovement. Terms used incorrectly may obscure that goal.

11.2.2. Methods

A suitable method of observation and an appropriate choice of movement parameter(s) is critical for a study of photomovement. Not surprisingly, a considerable number of methods for the measurement of photomovement have been described in the literature. These methods can be divided into two categories: those based on the response of a single cell or organ to the light stimulus, and those based on the response and/or behavior of a population of cells in response to the light stimulus. Some of the most commonly used methods for measuring photomovement are described below.

11.2.2.1. Individual Cell Methods

An individual cell and its photoresponse can be observed using a microscope and recorded manually, or recorded automatically using a videomicroscopy system, such as the one shown in Fig. 11-3. In this device, actinic light (i.e., light of a proper wavelength and fluence rate to produce a response) is provided from the side of a culture chamber, and the directional movement of the cells is then photographed or video-recorded using a nonactinic monitoring beam from under the transparent culture chamber. A typical result is presented in Fig. 11-4, showing six individual cells of the biflagellate *Chlamydomonas* moving randomly in the control exposed only to nonactinic red light (a), and moving away from the source of stimulus with actinic blue-green light, i.e., exhibiting negative phototaxis (b). Large numbers of cells can be traced for their tactic response using such a videomicroscopy recording system. In addition, automated motion analysis systems have been developed to analyze the precise movements of the video-recorded cells.

To quantitate the phototaxis of a population of cells or organisms, one can measure the direction of each organism with respect to the direction of the stimulus light. For example, one organism may be directed 45° from the stimulus light. Another organism may be directed directly toward (0°) the stimulus light. A number of such data may be combined using the orientation mathematics described by Batschelet[24,25] to arrive at the mean direction of the population and, more importantly, the directedness of the population. The mean direction of a population is between 0 and 360°, and directedness ranges from 0.0 for a completely random distribution to 1.0 for a distribution in which every individual is oriented in exactly the same direction.

The application of such an analysis to the recordings shown in Fig. 11-4 yields a value for directedness of 0 and 0.94 for the dark control and light-stimulated cells, respectively. Thus, these represent a random distribution and a highly directed distribution, respectively.[22,24,25]

Figure 11-5 shows another example of videomicroscopy applied to observe and record a step-up photophobic response by cells of the ciliate *Stentor coeruleus* using a

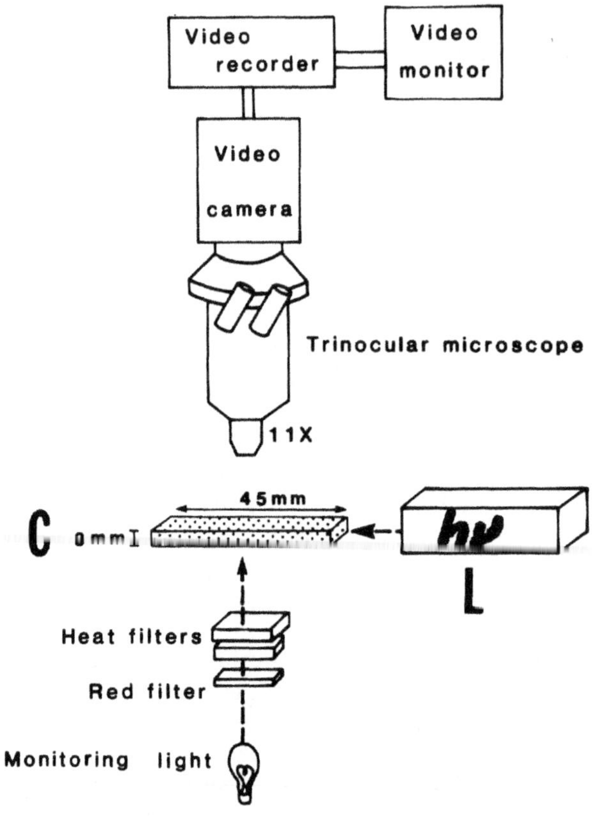

Fig. 11-3. Block diagram of a videomicroscopic recording system for observing the photomovement of individual cells. The monitoring light is usually from a dim infrared source that does not act as an actinic stimulant but is sufficient for videorecording by an IR-sensitive camera. C: sample chamber; L: actinic light source (e.g., laser or Xe arc plus monochromator or interference filter). (Taken with minor modification from Ref. 22.)

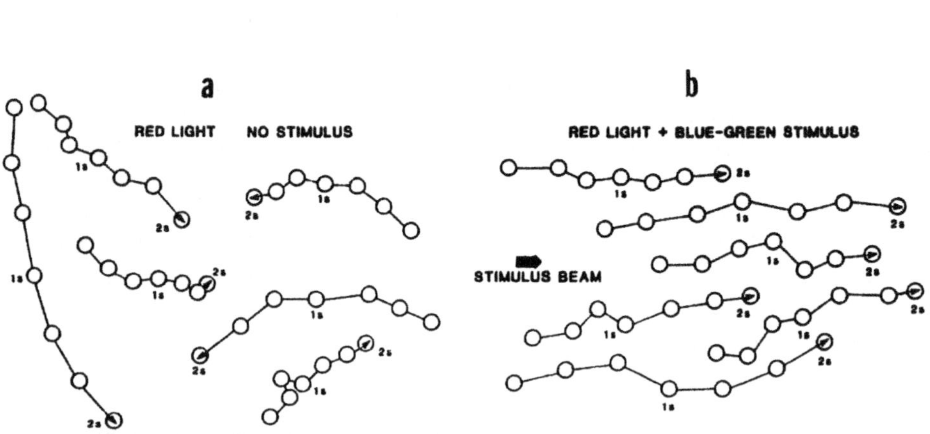

Fig. 11-4. Videomicroscopic tracings of swimming paths (for 2 s) exhibited by cells of *Chlamydomonas*. Each circle represents the cell position, at intervals of 1/3 s. (a) Control with nonactinic red background light. (b) In actinic blue-green light incident from the left. (Adapted from Ref. 23.)

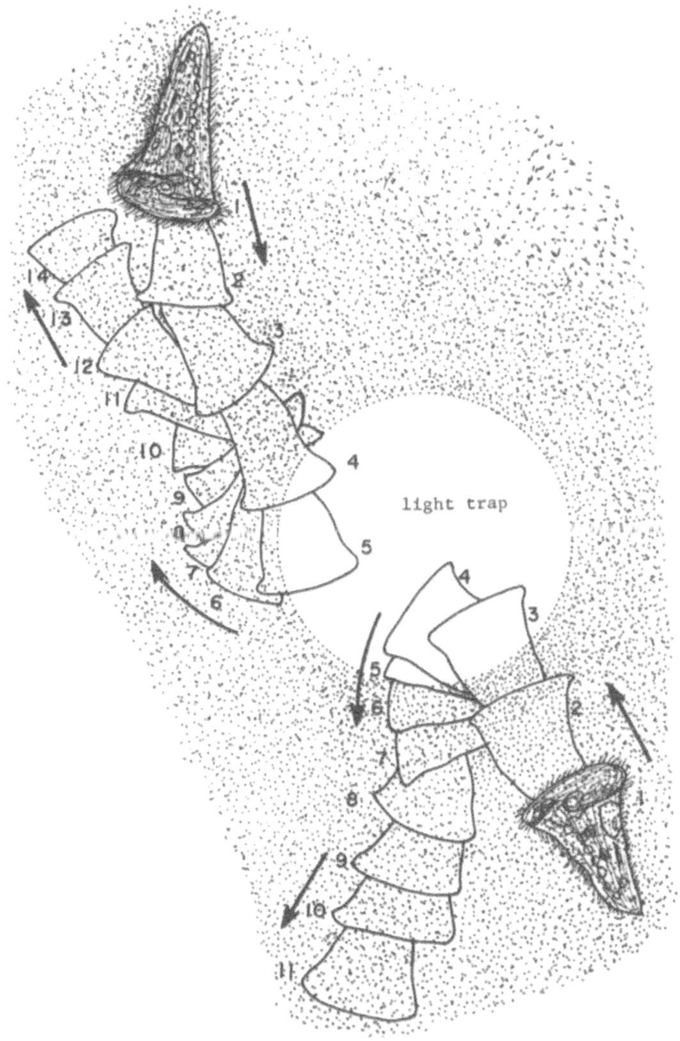

Fig. 11-5. Step-up photophobic response of *Stentor coeruleus*. The position of two cells has been traced every 0.24 s from a videomicroscopic recording. The stippled area represents darkness, while the clear circle represents an area lighted at 700 W/m^2. (Drawing by Il-Hyun Kim.)

setup similar to that shown in Fig. 11-3. Response parameters such as the fraction of responding cells, stop response at the dark–light border, rotation and swimming speeds, and reversal of the ciliary beating direction, etc., can be determined from video recordings of a sample of *Stentor coeruleus*.

Stentor coeruleus also shows negative phototaxis.(26) However, to establish the phototactic response, it is necessary to distinguish between a true phototactic orientation and a photophobic response of the organism. A *Stentor* cell might exhibit movement away from a light source, with direction maintained through a series of step-up photophobic

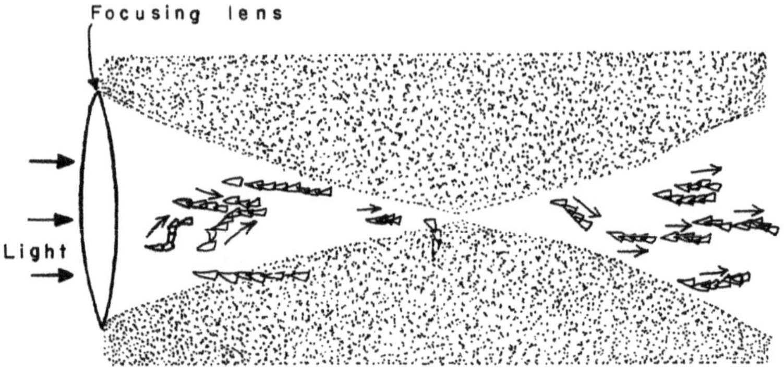

Fig. 11-6. Focused light beam used to distinguish between a phototactic response and a photophobic response. Cells of *Stentor coeruleus* move away from the light source even through and beyond the focal point of the lens. The stippled area represents darkness. (Modified from Ref. 27.)

responses. Thus, movement in the "wrong" direction would elicit a step-up photophobic response permitting a course correction. On the other hand, the cell might have the capacity for phototaxis, having a mechanism for measuring the direction of the light, and maintaining its course relative to that light direction. A focused light beam, such as that shown in Fig. 11-6, can be used to distinguish between phototaxis and a step-up photophobic response as the mechanism for movement of cells away from the light. Light is focused by a convex lens such that the focal point is within the culture chamber. The light direction is from the left in this apparatus. However, in the absence of an absorbing medium, the fluence rate is highest at the focal point of the lens. It then follows that a cell possessing only a step-up photophobic response, and lacking a negative phototactic response, would be incapable of moving through the region of highest fluence rate at the lens focal point. Thus, since cells move within the light beam through the focal point, they must in fact measure light direction itself, and therefore exhibit phototaxis in addition to the more easily demonstrated step-up photophobic response.[28]

A second technique to demonstrate a true phototactic response is illustrated in Fig. 11-7. In this case, an organism, such as *Chlamydomonas,* moving toward one light source

Fig. 11-7. Test for a phototactic response in *Chlamydomonas*. The alga abruptly changes its swimming direction from the original light direction ($h\nu_1$) to the new light direction ($h\nu_2$) turned on normal to the first source at the same time that the first source is turned off.

can be subjected to a second light source applied from a direction normal to the direction of the first at the same time that the first source is turned off. If the intensity of the two sources is the same, then the fluence rate experienced by the cell is constant, so no photophobic response should be elicited. If the cell changes its direction smoothly toward the second light source, then the response is a true positive phototactic response. Similar tests can be applied to other organisms that exhibit phototaxis.

11.2.2.2. Cell Population Methods

There is a variety of population methods for measuring the photomovement responses of motile organisms. One classical method, devised by Engelmann,[29] uses a light-trap. A small spot of light is projected onto an otherwise dark field of organisms. After a period of time, the cells may accumulate within or disperse from the lighted area, and the results can be easily seen with the unaided eye. Keep in mind that this demonstrates only the presence or absence of a photoresponse, and gives no information concerning the mechanism for that response.

One of the more common methods for measuring photomovement automatically measures the photoaccumulation of cells or organisms through the use of a device called a phototaxigraph.[30] In the phototaxigraph, an infrared monitoring beam passes through a chamber containing a population of cells and is photoelectrically detected by means of a photomultiplier tube. Accumulation or dispersal of the cells is induced by an actinic light

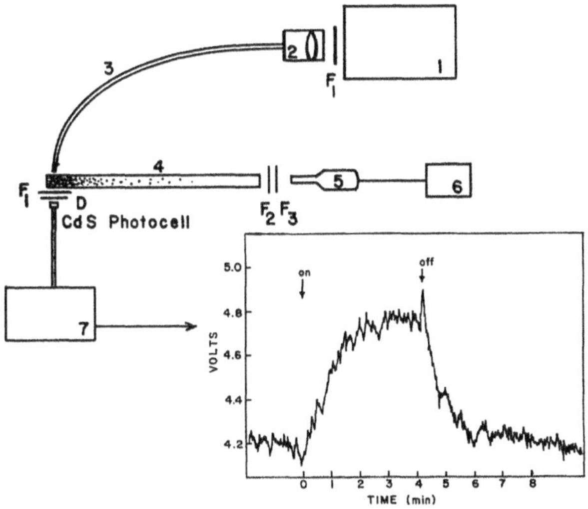

Fig. 11-8. A device for measuring phototaxis of *Stentor* utilizing an infrared monitoring beam. (1) IR light source. The light emerges through a circular hole in the side of the lamp housing, is filtered by F_1, and focused by condensing optics (2) onto an optical fiber bundle (3). The IR beam is directed through a cuvette (4) containing *Stentor* cells, passes through a diffuser plate (D), and is detected with a CdS photocell connected to an appropriate amplifier. The cells can also be exposed to an actinic light beam from light source (5) connected to a transformer (6). The wavelength of the actinic light is defined by filters F_2 and F_3. Inset shows a typical x-y recorder (7) tracing from the photocell amplifier. (Modified after Ref. 31.)

and measured as a change in the amount of radiant energy from the monitoring beam striking the photomultiplier tube.

A relatively simple device has been used to measure the photoaccumulation and photodispersal of *Stentor coeruleus* (Fig. 11-8). For the measurement of cell density at the distal portion of the chamber away from the stimulus source, a monitoring beam from an optical fiber is detected with a photocell. The inset in this figure is a typical recording, showing that the cells accumulate maximally within about 2 min at a given fluence rate of stimulus light. As expected, the microscopic counting of the cells in the accumulated population matched the photocell tracing shown in Fig. 11-8. Such a device can also be used to measure the average swimming speed of the cells.

It should be cautioned that the population methods employing photoaccumulation alter the stimulus strength and directionality due to the absorbance caused by the accumulating cells. Therefore, the net result will be a compromise that is time- and concentration-dependent. Moreover, these methods necessarily result in high cell concentrations. Since the cells are certainly metabolically active and modifying their environment, and since the response of the cell is expected to be sensitive to its environment, the photomovement responses of a population of cells should not be expected to always reflect the photomovement responses of single cells.[31,32] In addition, it must be remembered that little information concerning mechanism may be obtained using a population method. The major value of the population methods is their simplicity. Their use is important but should not replace studies of the behavior of single cells.

11.2.2.3. Methods for Measuring Phototropism

Phototropic curvature of an organ is measured by determining the angle of curvature toward or away from the light stimulus direction. Figure 11-9 shows a time lapse display (from a photographic recording) of the phototropic bending of the fungus *Phycomyces* toward a blue-light source. From such data, the angle of curvature can be measured with a protractor, while elongation-type growth can be measured quite simply with a ruler. Transducers have been described with which one can automate the measurement of curvature and growth.[33,34]

11.2.2.4. Methods Summary

Several of the more commonly used methods for measuring photomovement have been described. In general, those methods based on the direct observation of individuals

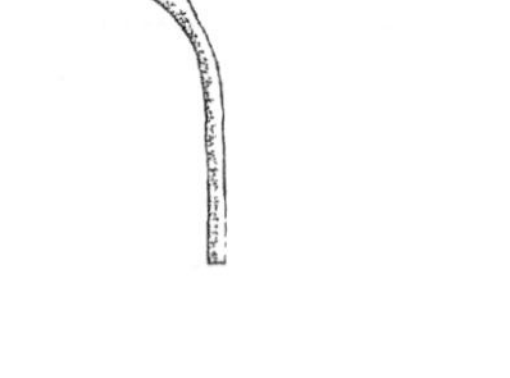

Fig. 11-9. Phototropic bending of the sporangiophore of the fungus *Phycomyces blakesleeanus* toward the source of actinic blue light. Position shifts of the sporangiophore represent 5-min intervals. (Adapted from Ref. 33 as modified by Ref. 34.)

yield the most information relevant to mechanism. Those methods based on the response by a large population are the most easily automated. Choice of the appropriate method may be critical for the study of any particular photomovement. A review of some of the methodology for measuring photomovement should be consulted for more detail.[35]

11.3. PHOTOSENSORY TRANSDUCTION MECHANISMS

A very general scheme of photosensory transduction was shown in Fig. 11-1. Let us now consider a somewhat more complex version (Fig. 11-10) of that scheme, recognizing at least some of the complexity that we know must exist. In the perception step, the light stimulus is absorbed by a photoreceptor pigment that then participates in some biochemical reaction, setting into operation a sequence of events that includes at least some amplification and processing and ends in a response. In most cases of sensory physiology, it is not solely the stimulus that results in a response. Rather, the organism is in some manner extracting some piece of information from the stimulus, and responding based on that information. For example, in phototaxis or phototropism, the organism is not responding solely to light but to light direction. Similarly, the organism may respond to fluence rate or a change in fluence rate. Thus, processing in many cases must include a mechanism for the extraction of information from the signal.

The physical and chemical basis of the absorption of light is described in Chapter 1. For the purposes of this discussion, we need only remind the reader that absorption of a quantum by a photoreceptor pigment causes an electron in that pigment to be raised to the so-called excited state. Since the energy requirements for the electron transitions are specific, only specific wavelengths of light are absorbed by a particular pigment. Because absorption must occur to initiate the photosensory transduction process, it follows that the response is inherently wavelength-dependent. The spectral features of light absorption by various photobiological receptors have been discussed elsewhere.[36,37]

The excited photoreceptor pigment must interact with some biochemical to convert the perceived stimulus into a signal that can act as the trigger to elicit the subsequent transduction chain reactions. The generation of such a signal usually begins with an ultrafast primary reaction, k_{rx}, of the excited photoreceptor pigment. The specific nature of the primary reaction depends on the photoreceptor pigment system involved. To ensure a high photosensitivity of the organism, the primary reaction must be fast enough to effectively compete with the other relaxation processes of the excited state of the pigment molecules. If this were not the case, a large percentage of the absorbed quanta would be "wasted" in one of these other relaxation processes, and not productively elicit the response. Extending this line of reasoning, one can also argue that primary reactions are more likely to originate from the singlet excited state (S_1 in Fig. 11-11), than the triplet

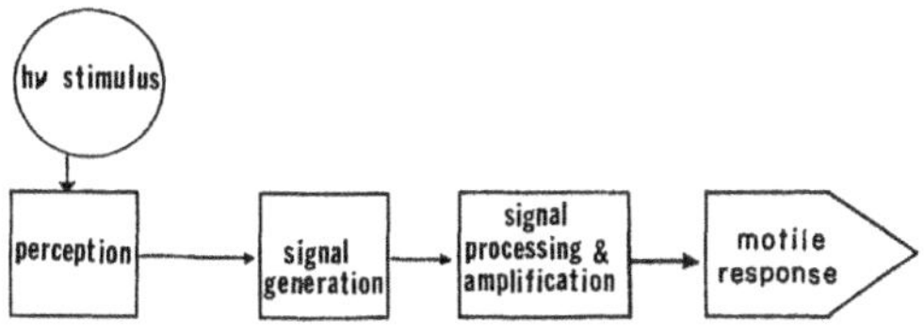

Fig. 11-10. Perception, signal generation, and signal processing in photosensory transduction. The black box in Fig. 11-1 is divided into signal generation and signal processing and amplification.

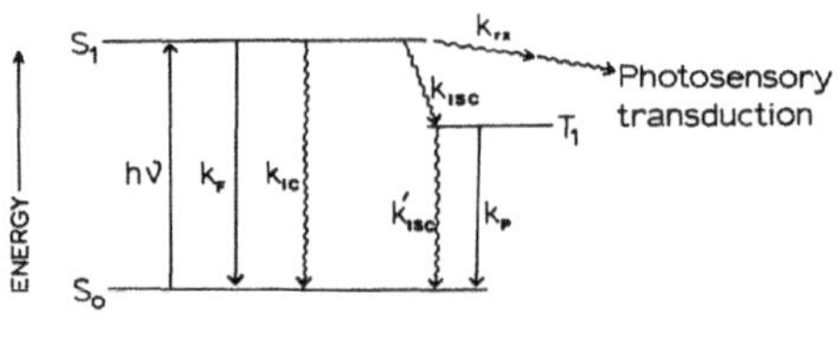

Fig. 11-11. Modified Jablonski energy level diagram for a photoreceptor pigment. k_F: rate constant for fluorescence, 10^8–10^9 s^{-1}; k_{ic}: rate constant for internal conversion, 10^7–10^{10} s^{-1}; k_{isc}, k'_{isc}: rate constants for intersystem crossing, 10^6–10^8 s^{-1}; k_P: rate constant for phosphorescence, 10^{-2}–10^2 s^{-1}; k_{rx}: rate constant for biochemical reaction; S_o: ground state; S_1: lowest singlet excited state; T_1: lowest triplet excited state.

state (T_1). This is because the efficiency of the triplet state reaction is limited by the relatively slow rate of intersystem crossing from the S_1 to the T_1 state. On the other hand, the triplet state reaction will be more important in bimolecular processes in which the reaction requires the longer lifetime of the triplet state.

Once the signal is generated, it must be processed and sufficiently amplified to trigger the cellular/organelle responses, usually mechanical in nature, such as reversal of the ciliary beat direction in protozoa and reversal in flagellar rotation in flagellates and bacteria. The processing and amplification steps, where known, also depend on the specific system involved.

In the following sections, we will discuss a few selected examples of photomovement systems rather than describing a great number of diverse organisms that display photomovement. We hope to show the student a bit of the diversity of photosensory transduction mechanisms and illustrate the use of various experimental approaches in elucidating these mechanisms.

11.3.1. Photomovement of Whole Cells

The initial act of photosensory transduction, leading to a photomovement response in an organism, is the absorption of stimulus light quanta by the photoreceptor(s) (Fig. 11-10). When one begins work with a previously undescribed photoresponse, the photoreceptor pigment and the transduction chain are unknown and, in addition, cannot be directly observed; only the response itself can be directly observed or measured. Fortunately, this may permit one to approach the identification of the photoreceptor pigment through the measurement of an action spectrum. Such a spectrum is a plot of the response, usually expressed in relative quantum efficiency or the reciprocal of fluence rate needed for a given response, e.g., 50% response, as a function of wavelength. An action spectrum is of value if it reflects the spectral absorption characteristics of the photoreceptor pigment *in vivo*. Because of various errors in the action spectral measurements and optical biases *in vivo*, such as light scattering, reflection, refraction, absorption by screening pigments, and polarization of incident light, the action spectrum usually does not quantitatively match the absorption spectrum of the photoreceptor pigment *in vitro*. Nevertheless, action spectroscopy permits one to eliminate many of the pigments in the organism, leaving a few or, optimally, one candidate for the physiological photoreceptor pigment. For this reason, action spectroscopy is the primary method used by the photobiologist for the identification of the photoreceptor pigment for any photobiological process, including photomovement responses (refer to Chapter 1 for a more detailed discussion of action spectroscopy). Action spectra for some photomovement responses are described below, and the possible photoreceptor chromophores are discussed.

The mechanisms of photosensory transduction in whole cells will be illustrated with two procaryotes, photosynthetic *Cyanobacteria* (blue-green algae) and the nonphotosynthetic flagellate *Halobacterium halobium,* and two eucaryotes, the photosynthetic flagellate *Chlamydomonas* and the nonphotosynthetic ciliate *Stentor coeruleus.* For each of the four organisms, we will describe the photoresponse, the photoreceptors, signal generation and amplification, and the mechanisms of movement.

11.3.1.1. Cyanobacteria

11.3.1.1a. Photoresponses

Cyanobacteria exhibit the three major classes of photomovement responses. They show positive and negative phototaxis and photophobic responses, and also show photokinesis in response to light of photosynthetically active wavelengths. Reviews of these photomovement responses in *Cyanobacteria* such as *Phormidium* and *Anabaena* should be consulted for a more extensive discussion.[18,38]

11.3.1.1b. Receptor Pigments for Photomovement

Figure 11-12 shows action spectra for three photoresponses in the filamentous cyanobacterium *Phormidium uncinatum.*[39–41] The action spectrum for positive photokinesis exhibits maxima in the red and blue regions of the spectrum. This spectrum is very similar to the absorption spectrum of chlorophyll a, which shows absorption bands in the red and blue. These data are consistent with the suggestion that this pigment is the photoreceptor chromophore for the photokinesis response. However, the absorption of light by chlorophyll a drops rapidly at wavelengths longer than 700 nm, in contrast with the photokinesis action spectrum, which shows a relatively high efficiency at these wavelengths. There are three possible explanations for this discrepancy: (1) The efficiency of wavelengths of light below 700 nm may be suppressed because of the screening effect of other bulk pigments, such as the photosynthetic accessory pigments. This would disproportionately exaggerate the efficiency of the far-red wavelengths where screening by other

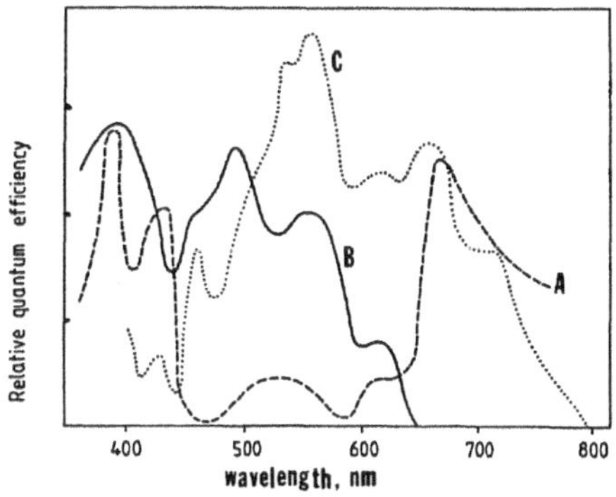

Fig. 11-12. Action spectra for (A) photokinesis, (B) phototaxis, and (C) photophobic response in *Phormidium uncinatum.* (Modified after Refs. 39–41.)

pigments is minimal. (2) The chlorophyll involved in photokinesis may represent a special type of chlorophyll a rather than the bulk chlorophyll a in the organism. It has been suggested that the chlorophyll involved here is that of photosystem I.[18] (3) The photokinesis action spectrum could be some pigment other than chlorophyll a. There is inadequate information at present to decide which of these possibilities is correct. Not surprisingly, other *Cyanobacteria* also show action spectra for photokinesis suggestive of chlorophyll as the photoreceptor chromophore.

Although the shapes of the action spectra for photokinesis and the photophobic responses are dramatically different (Fig. 11-12), chlorophyll a also has been suggested as a photoreceptor pigment for the photophobic responses in *Phormidium uncinatum*. However, for the photophobic responses, the suggestions have been that both photosystem I and photosystem II chlorophyll a function as the primary photoreceptor pigment, with other photosynthetic accessory pigments (particularly chlorophyll b and the phycobilins with phycocyanobilin and phycoerythrobilin as typical chromophores) contributing to the action spectrum in the 500- to 650-nm region (Fig. 11-12). It is not known why these accessory pigments contribute to the action spectrum of the photophobic responses but not to the action spectrum of photokinesis, if indeed photosystem I chlorophyll a is responsible for the photokinetic response in *Phormidium uncinatum*.

In contrast to the action spectra for photokinesis and for the photophobic responses, the phototaxis action spectrum shows maxima in the green, blue, and near-UV regions (Fig. 11-12). Flavins absorb over much of this spectral region. However, the phototaxis action spectrum extends well beyond 550 nm, where flavins do not absorb. Thus, it is unlikely that a flavin is the primary photoreceptor pigment for phototaxis in *Phormidium*. It is possible that phototaxis in *Phormidium uncinatum* uses a pigment complex consisting of a primary photoreceptor pigment and antenna pigments. If the concentration of the primary photoreceptor were negligibly low compared to that of antenna pigments, its absorbance would not contribute significantly to the action spectrum (Fig. 11-12) relative to the contribution of the antenna pigments. Thus, a pigment absorbing at wavelengths longer than 600 nm, such as chlorophyll a, could be the primary photoreceptor pigment without any significant indication of this in the action spectrum. The action spectrum would then represent the contribution of the light-harvesting antenna pigments, which for the green, blue, and near-UV region could be flavins, carotenoids, and/or phycobilins. It is interesting that the cyanobacterium *Anabaena* shows different action spectra for positive and negative phototaxis, with the former probably mediated by C-phycocyanin and the latter mediated by chlorophyll a as the primary photoreceptor pigment.[42,43]

11.3.1.1c. The Primary Photoreaction

In the photokinetic responses of *Cyanobacteria*, if one accepts the action spectrum (Fig. 11-12) as reflecting the absorption characteristics of chlorophyll a in the photosynthetic systems, the primary photoreaction is most likely to be photoionization and charge separation in the photosynthetic reaction center. This primary reaction leads to the photophosphorylation of ADP (Chapter 12). The ATP formed then serves as an energy source and accelerates the motile activity of *Cyanobacteria* in the photokinetic response.[40,41]

The nature of the primary reactions in the phototactic and photophobic responses of *Cyanobacteria* are more ambiguous than that for photokinesis. In addition, the action spectra for phototaxis and the photophobic responses are substantially different. Therefore, the nature of their primary photoreactions cannot be discussed with any certainty. Perplexingly, in the *Anabaena*, the positive and negative phototaxes exhibit different action spectra, suggesting that the photoreceptors for these opposing phototactic responses are not identical (see Section 11.3.1.1b).

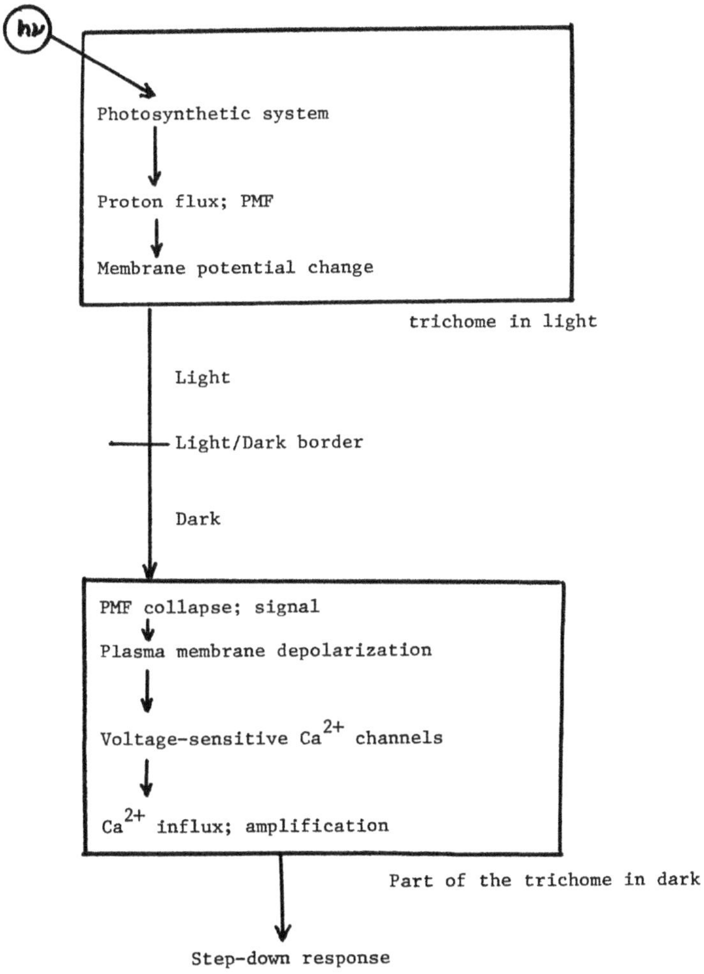

Fig. 11-13. Proposed transduction steps involved for the step-down photophobic response in *Phormidium*. As part of the trichome moves into darkness (represented by the lower rectangle), the photosynthetically generated membrane potential and PMF collapse, electrically polarizing the two ends (one in light and one in darkness) of the trichome. A membrane depolarization in the darkened part leads to an opening of a Ca^{2+} channel and amplification of the electric signal.

Based on the similarity between the photophobic action spectrum (Fig. 11-12) and that for CO_2 fixation, a model has been proposed in which both photosystems I and II reaction centers (Chapter 12) participate in the primary photoreaction in the photophobic response in *Phormidium*. The basic unit in *Phormidium* is a trichome, a linear array of cells that is motile. As a *Phormidium* trichome crosses from a lighted area into a dark area, the trichome stops and the reversal movement is elicited, i.e., a step-down photophobic response occurs. In the lighted area, the trichome's photosynthetic electron transport system is activated. Thus, a pH gradient and a membrane potential (proton motive force, or PMF) can be expected to have been generated across the thylakoid membrane. Accompanying the generation of the PMF, a change in plasma membrane potential occurs establishing a steady-state potential that is different from the resting potential in darkness. As part of the trichome moves into the dark side, the photosynthetic electron transport ceases, and the PMF collapses. It is suggested that the PMF collapse is taken as a signal by the trichome, which induces depolarization of the plasma membrane. At this point, the trichome's front end and rear (shaded) end are electrically polarized.

The membrane depolarization causes a voltage-sensitive Ca^{2+} channel to be opened and Ca^{2+} ions to flow in from the extracellular medium. The light signal is thus amplified; the change in lighting resulting in a massive Ca^{2+} ion influx. This influx triggers the contraction of microfilaments necessary for a stop response (Section 11.3.1.3). This model is schematically illustrated in Fig. 11-13.

Indirect evidence supports the transduction scheme shown in Fig. 11-13.[44–46] Four main lines of evidence can be cited: (1) The photophobic response in *Phormidium* can be stimulated by an abrupt external pH change in the absence of a light stimulus. This supports the involvement of a change in PMF, possibly as a primary signal.[47] (2) Light-induced membrane potential changes can be measured in *Phormidium*. The action spectrum for the induction of these changes is similar to the action spectrum for the photophobic response shown in Fig. 11-12.[48,49] These data are consistent with membrane depolarization being involved as in the amplification step shown in Fig. 11-13. (3) A change in the direction of movement can be induced by an external DC electrical field. This is further evidence that electrical signals are involved in the transduction sequence. (4) Light-induced calcium uptake has been shown in *Phormidium uncinatum* using radioactive Ca^{2+}.[50] This result is consistent with the involvement of Ca^{2+} ion influx as in the amplification of the sensory signal. Thus, considerable evidence supports the model in Fig. 11-13 for the photophobic response in *Phormidium*.

Whether or not a common transduction mechanism is utilized by both photophobic and phototactic responses in *Cyanobacteria* should be considered an open question.

11.3.1.1d. Motor Responses

Cyanobacterium moves using a gliding mechanism that is not well understood. Exactly how the reversal of the gliding direction is dictated by a front–rear electrical polarization (Section 11.3.1.2) (Fig. 11-13) is not known. One can conjecture that the contractile microfilaments undergo a significant conformational transformation upon binding Ca^{2+} ions. Such a hypothesis is worthwhile if it generates thoughts and experiments that take us closer to a true understanding of the mechanism.

11.3.1.2. Halobacterium halobium

11.3.1.2a. Photoresponses

Halobacterium halobium shows a unique wavelength dependence of its photomovement which has been called "color sensing."[(51)] This flagellate bacterium exhibits responses to both step-up and step-down stimuli. On a cellular level, the response is a change in the direction of flagellar rotation from clockwise to counterclockwise or vice versa.[(52)] If the stimulus is in the blue-UV region of the spectrum, a step-up stimulus results in the photodispersal ("repellent response") of the population. However, if the step-up stimulus is in the yellow-green region of the spectrum, the result is a photoaccumulation ("attractant response") of the population. In contrast, a step-down stimulus in the blue-UV region results in photoaccumulation, while a step-down stimulus in the yellow-green region results in photodispersal. Thus, the bacteria accumulate in yellow-green light and disperse from blue-UV light (Fig. 11-14). Reviews of these photomovement responses in *Halobacterium* should be consulted for a more extensive discussion.[(51–53)]

11.3.1.2b. Receptor Pigments for Photomovement

As a result of the presence of rhodopsinlike compounds on the *Halobacterium halobium* membrane, these membranes are purple. Two of these pigments probably mediate the two opposing photoresponses in this organism. Thus, the photoreceptor pigment chromophore for each of these photoresponses would be a retinal, linked to an opsin protein via a Schiff's base.

A proposed scheme that could account for the sensory responses of *Halobacterium halobium* is shown in Fig. 11-15. In addition to bacteriorhodopsin and halorhodopsin, which are involved in proton and chloride ion transport, respectively, a new rhodopsinlike molecule has been detected in isolated purple membrane preparations, based on data from flash photolysis experiments.[(54)] This new species is called slow-cycling rhodopsin (sR_{587}) because of its relatively slow photochemical cycle. It has been proposed[(51)] that sR_{587} is the photoreceptor pigment for the attractant response of the bacterium, and that the blue-UV photoreceptor pigment for the repellent response of the cell is S_{373}, a photocycling intermediate of sR_{587}. Supporting evidence for sR_{587} and S_{373} as the pho-

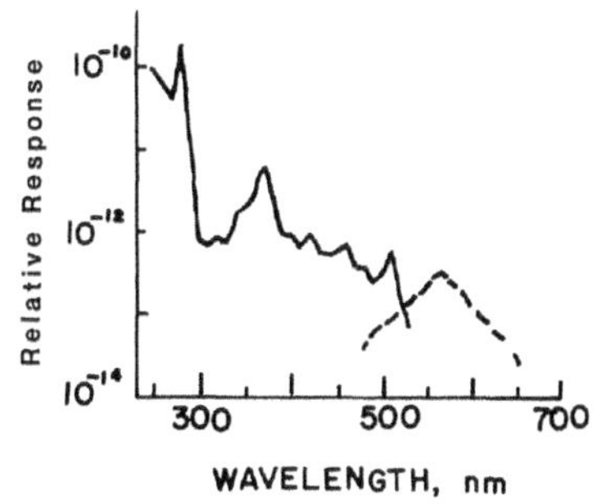

Fig. 11-14. Action spectra for the attractant (broken line) and repellent responses (solid line) in *Halobacterium halobium*. The ordinate gives the sensitivity for the photophobic reaction (i.e., reciprocal of photons/mm²/s required for a given response). (Redrawn from Ref. 53.)

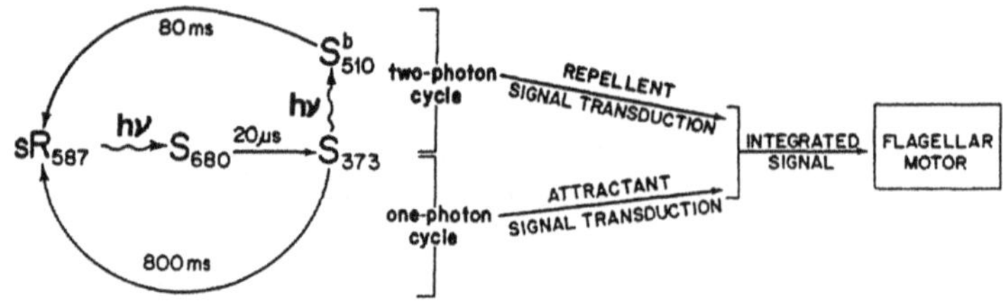

Fig. 11-15. Photochemical cycle for slow-cycling rhodopsin (sR_{587}) and its relationship to the mechanism of color sensing based on the dual role of sR as the attractant and repellent photosensory receptors. (Modified after Ref. 51.)

toreceptor pigments has been provided by the observation that sensitivity to near-UV light is reduced when the photoresponse is observed under conditions that cause less excitation of sR_{587}. Hence, in this experiment less S_{373} intermediate is formed to act as the photoreceptor for the repellent response.(51) Though this hypothesis is attractive, others do not agree that the S_{373} and sR_{587} pigments are the primary photoreceptor pigments.(52)

11.3.1.2c. The Primary Photoreaction

Assuming that sR_{587} (Fig. 11-15) is the photoreceptor pigment for the attractant response, one can conjecture that the primary reaction of sR_{587}, leading to its intermediate S_{373}, is likely to be a configurational isomerization of the retinylic chromophore in the excited state of the sR_{587}. An alternative photoisomerization path can then be invoked for the repellent response mediated by S_{373}. The signal generated from the primary reaction could then be a conformational change in the sR protein embedded in the membrane. This could then modulate an ionic flux as a processible form of the signal. It should be emphasized that the molecular details of the primary reactions are not known. However, these possibilities are consistent with what is known about rhodopsin in higher organisms.

The flagella motors in procaryotes appear to be driven by a PMF.(55) By analogy with cyanobacterial photosensory transduction, we speculate that the signal generated by the photoreceptors, e.g., sR_{587}, is processed, resulting in a change in the PMF, and is possibly amplified through a changed Ca^{2+} ion flux. This process ultimately triggers a change in the direction of rotation of the flagella, from clockwise to counterclockwise rotation or vice versa.

The photoreceptor pigment in its excited or metastable state may interact with a methyl-accepting protein (MaP), which acts as the site of signal integration. This is based on the observation that specific membrane proteins in *Halobacterium halobium* become methylated upon exposure to light, which would function as an attractant. In contrast, repellent light causes demethylation of these proteins.(52) It is suggested that the MaP then lowers the level of a hypothetical regulator, which serves as the flagellar rotor switch. It is known that Ca^{2+} ions are essential for the reversal of swimming direction, presumably through their involvement in the methylation/demethylation cycle. Whether or not Ca^{2+} ions directly affect the flagella switch regulator and its conformation, and whether or not any role is played by PMF is subject to conjecture.

11.3.1.2d. Motor Responses

Bacteria employ one or more flagella as the motor apparatus. Each flagellum is driven by a rotary motor called the M ring (Fig. 11-16). As mentioned above, PMF's are the driving force for the rotary motor. Protons flow through the interstice between the M ring and the S ring (a stator), which powers the rotary motion of the motor apparatus. In *Halobacterium halobium*, rhythmic spontaneous reversals are controlled by an oscillator,[56] and the photoresponse is either a suppression or an induction of a reversal in the flagellar rotary motor. It is not known whether Ca^{2+} alone or Ca^{2+} plus protons are directly involved in the possible conformation changes of the proteins of the rotary motor in *Halobacterium halobium*.

11.3.1.3. Chlamydomonas

11.3.1.3a. Photoresponses

Chlamydomonas reinhardtii, a green flagellate, exhibits step-up photophobic responses and a negative phototactic response. Thus, these cells disperse from a lighted area. Several comprehensive reviews[22,57] of the photoresponses of *Chlamydomonas* are available for the interested reader.

11.3.1.3b. Receptor Pigments for Photomovement

Chlamydomonas shows a maximum positive phototactic response to light of 500–510 nm and 440 nm, as shown in Fig. 11-17. For comparison, an action spectrum for photoaccumulation of *Euglena gracilis*[30] is also shown. The action spectrum for *Euglena* closely matches the absorption spectrum of flavin, but the differences between the absorption spectrum of flavin and the action spectrum for *Chlamydomonas* are considerable. Although the identity of the *Chlamydomonas* photoreceptor pigment is uncertain, two suggestions have been made.

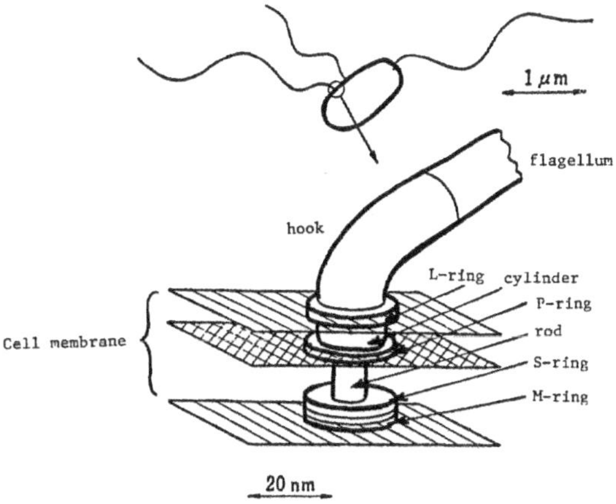

Fig. 11-16. The flagellar rotor of a bacterial flagellum. Top: Flagellum on a gram-negative bacterium such as *E. coli*. Bottom: the basal end of a bacterial flagellum representing the flagellar rotor apparatus consisting of disks (rings) that function in the flagellar rotation. The flagellum is composed of the protein flagellin assembled in a helical filament wound around the central core. The bacterium is propelled by rotation of the flagella about their long axes. The flagellar rotor is driven by energy (PMF) supplied from the cytosol. (Redrawn from Ref. 55.)

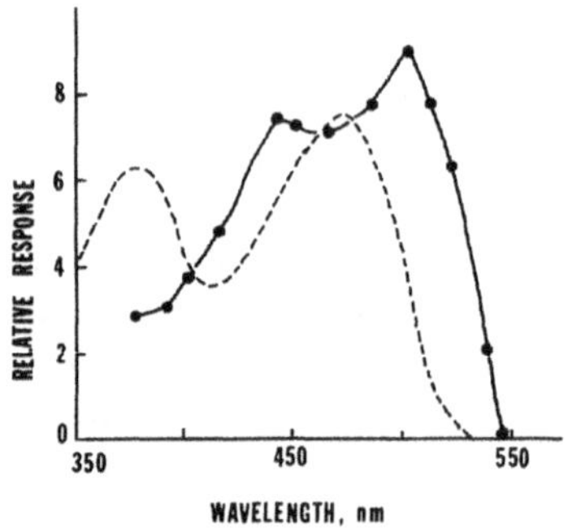

Fig. 11-17. Action spectra for phototaxis of *Chlamydomonas reinhardtii* (solid line) and photoaccumulation of *Euglena gracilis*[30] (broken line). (Redrawn from Ref. 44.)

1. The *Chlamydomonas* photoreceptor pigment may be a naturally occurring flavin other than the most common flavins such as riboflavin, FMN, and FAD. These more common flavins show a broad, single absorption band around 450 nm. Others, e.g., deazaflavins and hydroxyflavins,[58] show absorption maxima on either side of the 450-nm peak. Furthermore, optical biases such as reflection and scattering, and screening by other light-absorbing substances, may result in an action spectrum that is different from the absorption spectrum of the photoreceptor pigment. Thus, a flavin cannot be ruled out as the primary photoreceptor.

2. A quarter-wavelength interference reflector may radically alter the light absorbed by the photoreceptor pigment such that the action spectrum for the response differs dramatically from the absorption spectrum of the pigment. A quarter-wavelength interference reflector is any structure composed of layers of alternating refractive indices and of approximately the thickness of a quarter of the incident light wavelength. Light reflected from such a structure is constructive, resulting in a reinforced light intensity for a particular wavelength, and destructive, resulting in a decreased light intensity for other wavelengths. Such quarter-wavelength interference phenomena in living organisms have been described by Land.[59] To explain this suggestion, we first must describe the *Chlamydomonas* stigma.

The stigma or "eye spot" is an optically dense body at the base of the flagellum on one side of the *Chlamydomonas* cell. Historically, the stigma in *Chlamydomonas* and other flagellates has been thought to have a role in light sensing. In the classic model, the stigma was thought to function as a shading device. More recently, it has been suggested that the stigma may function as a quarter-wavelength interference reflector. Although both models ascribe a role to the stigma, that role is dramatically different for the two models.

In the shading model, the stigma functions solely as an optically dense body, periodically shading the photoreceptor pigment as the cell rotates about its long axis. The intensity transiently perceived by the photoreceptor pigment during the periodic shading by the stigma causes a shift in swimming direction. Thus, the cell effectively measures light direction through two measurements with the same photoreceptor pigment at two times in its rotation.[22]

In the quarter-wavelength model, the stigma serves as a directional antenna, which is rotated somewhat like a radar antenna as the cell rotates while swimming.[60] This model is based on the observation that the granular layers of the stigma can provide alternating layers of different refractive index media, with the thickness or the spacing between the layers being about 120 nm (Fig. 11-18). Thus, this could be a quarter-wavelength interference reflector for a wavelength of 480 nm, the maximum in the *Chlamydomonas* action spectrum (Fig. 11-17). It has been suggested that the photoreceptor pigment is located in the plasma membrane in front of the granular rows of the stigma.[61] If this is so, then the two stacked layers of stigma granules shown in Fig. 11-18 can in fact act as a reflector, and the photoreceptor apparatus in the membrane in front can be excited by the incident and the reflected light beams. Since the latter is strongly dependent on the direction of the incident light beam, the mechanism would be highly directional.

Both stigma-based models would result in action spectra different from the absorption spectrum of the photoreceptor pigment. On the one hand, if the shading model is correct, then the action spectrum must be a function not only of the absorption by the photoreceptor pigment, but also of the shading pigment. On the other hand, if the quarter-wavelength interference reflector model is correct, the action spectrum must be a function of absorption by the photoreceptor pigment, reflectance of the stigma granular layers, and absorption by those layers. There are insufficient data at present to unequivocally say whether or not the quarter-wavelength interference reflector model is valid for *Chlamydomonas*. If it can be demonstrated that this model is valid, it will be the first example that such a mechanism serves in an aneural organism as it does in the eye of animals.[62]

Returning to the original question, what is likely to be the primary photoreceptor pigment of *Chlamydomonas*? The phototaxis action spectrum (Fig. 11-17) resembles the absorption spectrum of a carotenoid more closely than that of a flavin. However, a flavin chromophore can be accommodated given substantial screening, etc., or a "nontypical" flavin.[58]

The photomovement response (a photodispersal) can be restored in a carotenoidless mutant of *Chlamydomonas* by exogenously added retinal isomers.[60] Retinals are the photoreceptor chromophores of animal vision and their photochemistry is well established, so it would not be surprising if the photoreceptor pigment in *Chlamydomonas* were a retinal. Unfortunately, these experiments do not completely rule out photodynamic and thermal effects of the exogenous retinals in the alga. For example, *Paramecium* without apparent pigmentation can respond to light when artificial dyes such as methylene blue and eosin are added to the cells.[63,64] Similarly, the nonphotosensitive, colorless species *Stentor polymorphus* becomes phototactic when pigmented with symbiotic *Chlorella*.[57]

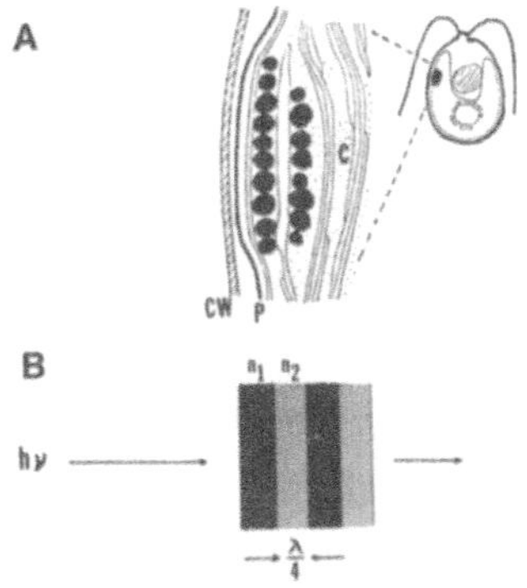

Fig. 11-18. (A) Schematic diagram showing the array of stigma globules of the eye spot or stigma in *Chlamydomonas*. C: chloroplast, CW: cell wall, P: plasmalemma. (B) Multilayer stack as a reflector. The stack consists of alternative quarter-wavelength layers of high (n_1) and low (n_2) refractive index media. The refractive index of air is assumed to be $n_0 = 1$.

These examples illustrate the point that an exogenous pigment can induce or modify the photoresponses of the organisms. In summary, the best evidence at present indicates that a retinal is the photoreceptor pigment for phototaxis of *Chlamydomonas*. Caution should always be exercised in the interpretation of action spectra, since the difficulties encountered in the identification of the *Chlamydomonas* photoreceptor pigment are similar to problems involved in other systems.

11.3.1.3c. The Primary Photoreaction

A step-up light stimulus causes a transient reversal in the flagellar strokes in *Chlamydomonas*. Although the primary reaction and the subsequent signal processing are not well understood, what is qualitatively known at present is summarized in Fig. 11-19. Two important steps in the photosensory transduction chain are depolarization of the plasma membrane following a primary reaction of the photoreceptor, and Ca^{2+} ion influx, which triggers a reversal of the flagellar rotor. A similar scheme has been proposed for the phototactic response in *Chlamydomonas*.[44]

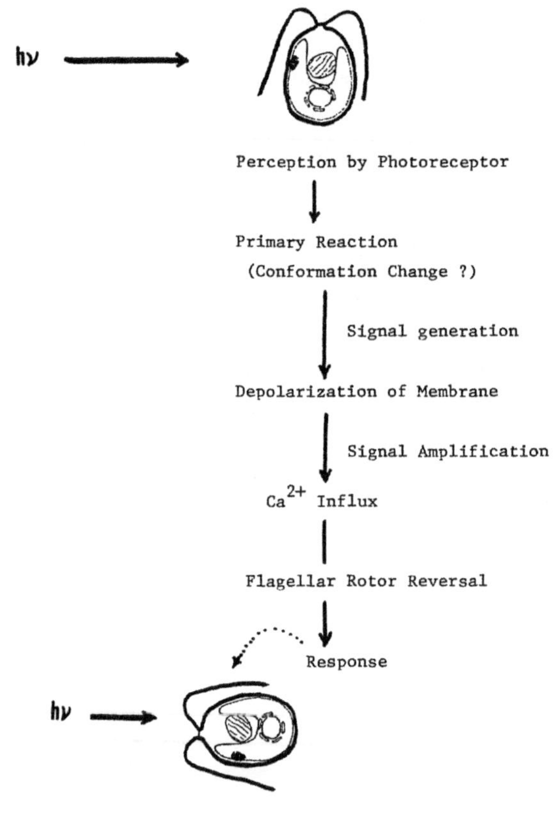

Fig. 11-19. Schematic diagram of photosensory transduction in *Chlamydomonas*. Note the key role of the Ca^{2+} ion influx in reversing the flagellar rotor in response to the light stimulus.

11.3.1.3d. Motor Responses

The "motor" of the flagellated *Chlamydomonas* is considerably more complex than that of the flagellated bacterium. In general, the flagella and cilia of eucaryotic cells possess a characteristic protein assembly composed of nine doublet microtubules surrounding a pair of singlet microtubules ("9 + 2 axoneme"). It is reasonable to suggest that a light-induced influx of Ca^{2+} ions (Fig. 11-19) triggers a conformational change in the axoneme, resulting in the reversal in flagellar stroke in *Chlamydomonas*. For further information concerning the flagellar motor of eucaryotic cells, consult the excellent review volume edited by Brokaw and Verdugo.(65)

11.3.1.4. Stentor coeruleus

11.3.1.4a. Photoresponses

Cells of the ciliate *Stentor coeruleus* exhibit a photophobic response and also a phototactic response (Section 11.2.2.1). Comprehensive reviews on the photomovement responses of this organism(22,57) should be consulted for further information.

11.3.1.4b. Receptor Pigments for Photomovement

The action spectrum for the step-up photophobic response in *Stentor coeruleus* (Fig. 11-20) is similar to the action spectrum for the negative phototactic response, suggesting that the same photoreceptor chromophore is involved in both photoresponses.(26,66,67) Moreover, these action spectra closely match the absorption spectrum of whole *Stentor coeruleus* cells, the absorption of which is mainly due to the pigment stentorin. This compound, which has a hypericinlike chromophore, is located in pigment granules, presumably in the cell membrane.(68) Although the photoreceptor chromophores of the photophobic and phototactic responses in *Stentor coeruleus* appear to be identical, it is not established whether or not the same set of photoreceptors is utilized in both responses. If the phototactic response is a consequence of a series of photophobic responses in *Stentor coeruleus,* as discussed below, it then follows that the ultimate basis for the two responses is the same.

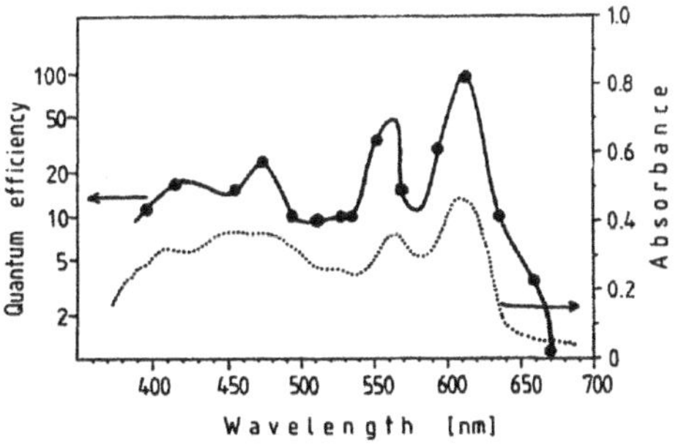

Fig. 11-20 Action spectrum (solid circles; left ordinate) for the step-up photophobic response in *Stentor coeruleus*. An absorption spectrum of the cell (right ordinate) at 77°K is also shown for comparison. (Redrawn from Ref. 66.)

11.3.1.4c. The Primary Photoreaction

Based on the available data, it has been proposed that the photophobic response in *Stentor coeruleus* is triggered by a light-induced proton release from the photoreceptor pigment in the pigment granules, which are located in the ectoplasm.(57) The intracellular and/or intraciliary pH drop causes a depolarizing effect that triggers the opening of Ca^{2+} channels, as outlined in Fig. 11-21.

It is suggested that the photoreceptor pigment ejects protons efficiently during its excited state lifetime.(57,69) If this proton release serves as an initial signal to trigger the photosensory transduction chain as proposed here, then the photoresponse of the cells should be pH-dependent. Results indeed show that the photoresponse is pH-dependent.(32) A sharp drop in response is seen at pH values lower than 6.0, with the midpoint of the drop being at 5.4. Less than 5% of the maximum photoresponse is observed at pH values below 5.0. It is not certain whether the midpoint of the pH profile represents an apparent pK_1 of the photoreceptor chromophore or its protein residue(s), but the pH effect itself is striking. Given the strong dependence of the photoresponse on pH, one would expect proton ionophores to show strong inhibitory effects on the photoresponse of *Stentor coeruleus*.

A slow cytoplasmic pH drop under steady-state photic stimulation has been observed.(57) The fact that the photoresponse occurs only at $pH_{out} > 6$ (extracellular pH) suggests that proton release from the excited photoreceptor organelles (pigment granules) must be sufficient to generate a transient $pH_{in} < 6$ to be an effective transducing signal. Thus far, no experiments have been carried out to monitor transient cytoplasmic pH changes upon flash photic stimulation of *Stentor coeruleus*. Such experiments could test the hypothesis that a change in pH acts as the initial transducing signal following the photoexcitation of the photoreceptor.

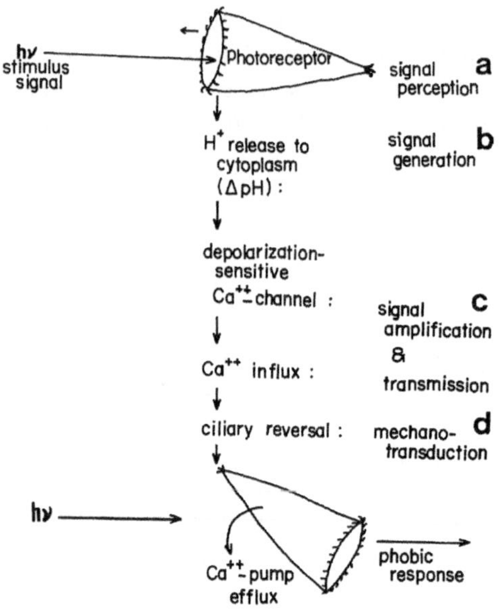

Fig. 11-21. Tentative scheme for a step-up photophobic response in *Stentor coeruleus*. (a) The initial step, signal perception, involves absorption of light by the photoreceptor chromophore located in the pigment granules. (b) The excited state chromophores release protons into the cytoplasm, thus generating a signal in the form of a pH gradient. (c) The pH produced in step b activates the depolarization-sensitive Ca^{2+} channel in the plasma/ciliary membranes, resulting in Ca^{2+} influx that further depolarizes the membranes, thus amplifying the signal. (d) Ciliary beating is reversed as a result of the depolarization, causing the organism to turn and swim in the opposite direction, thus giving a phobic response. It is assumed that a Ca^{2+} pump operates to cause the efflux of Ca^{2+} ions following the transient ciliary reversal, thus causing the resumption of clockwise ciliary beating. (Redrawn from Ref. 57.)

TABLE 11-1. Summary of the Effects of Inhibitors on the Photophobic Response of *Stentor coeruleus*[26,31,57,67]

Drug	Specificity	Inhibition	Effect on motility
Valinomycin	K^+	None	None
2,4-Diaminopyridine[a]	K^+	None	None
Ouabain[a]	Na^+	None	None
$TPMP^+$ (triphenylmethylphosphonium)	PMF[b]	Strong	Weak
FCCP (carbonylcyanide-*p*-trifluoromethoxyphenylhydrazone)	H^+	Strong	Weak
CCCP (carbonylcyanide-*m*-chlorophenylhydrazone	H^+	Strong	Weak
Nigericin	H^+	Strong	Weak
Gramicidin	H^+/K^+	Strong	Weak
Calimycin	Ca^{2+}	Strong	None
Ruthenium red	Ca^{2+}	Strong	Weak
Methoxyverapamil	Ca^{2+}	Mild	Weak
Verapamil	Ca^{2+}	Mild	Weak
Lanthanum	Ca^{2+}	Mild	Weak
Poly-L-lysine	Ca^{2+}	Mild	Weak

[a]Unpublished data (I. H. Kim and R. K. Prusti).
[b]An electric potential component of proton motive force (PMF).

The role of Ca^{2+} in the transduction sequence has been examined using Ca^{2+} inhibitors and ionophores. Ca^{2+} blockers including ruthenium red, methoxyverapamil, and the Ca^{2+} ionophore calimycin specifically inhibit both the step-up photophobic and phototactic responses in *Stentor*.[57] Results from the inhibition studies suggest that during the response, depolarization-sensitive Ca^{2+} channels are opened, permitting a sudden influx of Ca^{2+} ions into the cell. Table 11-1 summarizes the inhibitory effects of different drugs on the photophobic response in *Stentor coeruleus*.

Several photoresponse-enhancing agents also exist (Table 11-2). Calimycin may stimulate the photoresponses of *Stentor coeruleus* under certain conditions. Plausible

TABLE 11-2. Agents That Enhance the Photophobic Response of *Stentor coeruleus*[31,66,70,71]

Agent	Specificity	Enhancement	Effect on motility
Calimycin	Ca^{2+} ionophore	Mild[a]	None
D_2O (50% v/v)	Primary reaction (?)	Strong	Weak
Caffeine	Ca^{2+} permeability	Strong	None/weak
α-Phosphatidate	Ca^{2+} permeability	Strong	None

[a]Only at low concentrations ($\leq$1 nM). At higher concentrations of calimycin, a ciliary reversal occurs even in the dark without a light stimulus.

mechanisms for the observed effects of D_2O, caffeine, and α-phosphatidate have been described.[66,70,71] It is particularly noteworthy that D_2O substantially sensitizes the cell to a step-up phobic stimulus, while it inhibits negative phototaxis. It appears that D_2O influences both the primary photoprocess of stentorin (i.e., H^+ dissociation) and the rate of cell body rotation and action potentials.[70] Though less likely, the enhanced photophobic response in heavy water might arise from a higher concentration of singlet oxygen in D_2O than in H_2O (Chapter 3), as a photophobic response can be induced by photodynamic action in some protozoa.[63,64]

Both caffeine and α-phosphatidate enhance the photophobic response of *Stentor coeruleus*, apparently by increasing the permeability of Ca^{2+} across the cell membrane. This interpretation is also consistent with a calimycin-induced ciliary reversal in the dark. The effects of caffeine as a Ca^{2+} permeability promoter and α-phosphatidate as a Ca^{2+} ionophore are well known in neurobiology.[66,71]

Is the scheme shown in Fig. 11-21 applicable to the negative phototactic response in *Stentor coeruleus?* If the phototactic response results from a series of photophobic responses, the same transduction mechanism would account for phototaxis and the photophobic response. However, while heavy water enhances the step-up photophobic response in *Stentor coeruleus* (Table 11-2), it inhibits the negative phototactic responses. In addition, if the same transduction mechanism accounts for phototaxis and the photophobic response, one would expect any inhibitor of the step-up photophobic response to inhibit phototaxis also. In addition, ions of lanthanum, which is competitive with calcium, strongly inhibit the step-up photophobic response (Table 11-1), but the phototactic response is not affected. In fact, the inhibition of the phototactic response by heavy water is partially restored by lanthanum ions.[70] These results also are consistent with the existence of different transduction mechanisms for the photophobic and phototactic responses in *Stentor coeruleus*. Thus, there is no evidence at present that the photophobic response is the basis of the phototactic response, so the scheme of Fig. 11-21 for the photophobic response must not be applicable to the phototactic response.

11.3.1.4d. Motor Responses

The "motor" of the ciliated *Stentor,* like that of the flagellated *Chlamydomonas,* is considerably more complex than that of the flagellated bacterium. As stated above (Section 11.3.1.3d), the eucaryotic cilium is composed of nine doublet microtubules surrounding a pair of singlet microtubules (9 + 2 axoneme). It is suggested that a light induced influx of Ca^{2+} ions triggers a conformational change in the axoneme, resulting in the reversal in ciliary stroke.

11.3.2. Photomovement of Organelles (Photodinesis)

The cellular organelles of eucaryotes are structurally dynamic, and often exhibit varied forms of movement within the cell. Even nonmuscle cells and their intracellular organelles are capable of movement, largely controlled by filamentous proteins, e.g., through actin–myosin interactions or assembly to disassembly cycles of actin molecules. Light strongly affects many of these intracellular movements. Photodinesis, the light-induced or modulated movement of cytoplasm, and other intracellular organelles such as

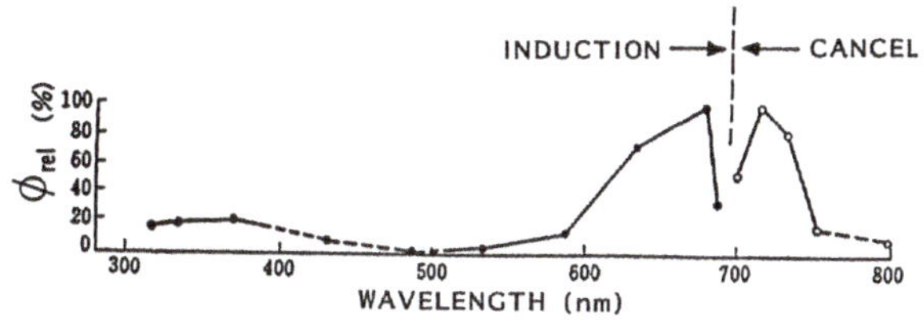

Fig. 11-22. Action spectra for the induction of chloroplast movement in *Mougeotia* at low fluence rate (red light) and the cancellation of the movement induction (far-red light). Ordinate, ϕ_{rel}, is relative quantum efficiency. (Redrawn from Ref. 79.)

the nucleus and chloroplasts will be briefly described using chloroplast movement in the photosynthetic alga *Mougeotia* as a specific example. (See Ref. 72 for a review of chloroplast and nuclear movements.)

11.3.2.1. Photoreceptor Pigments for Photodinesis

Generally, there are two regions of spectral sensitivity in the photodinesis of several algae that have been investigated. Chloroplast migration or local accumulation in *Vaucheria sessilis*,[73] *Mougeotia*,[74] *Mesotaenium*,[75] *Selaginella martensii*,[76] *Lemna trisulca*,[77] and *Vallisneria spiralis*[78] show an action spectrum maximum in the blue region at 440–480 nm. A flavin is most likely the chromophore.

Chloroplast movement in many algae also exhibits a maximum in the red region at 600–800 nm. The red sensitivity is particularly important at low fluence rates. Some algae such as *Mougeotia* exhibit a maximum sensitivity of chloroplast movement in the blue region at higher fluence rates, as shown in Fig. 11-22. However, even these algae show red sensitivity at low fluence rates.[79] The fact that chloroplast movement in response to polarized light, "polarotropism," in *Mougeotia* is red/far-red-reversible suggests that the photoreceptor pigment is phytochrome.[80]

11.3.2.2. Polarotropism of Chloroplasts

Each cell of a trichome of the filamentous alga *Mougeotia* contains one flat chloroplast. The rectangular plate-shaped chloroplast turns its flat face toward light of relatively low intensity within minutes of irradiation, so that the plate-shaped chloroplast faces approximately perpendicular to the direction of actinic red light of 660 nm ("face

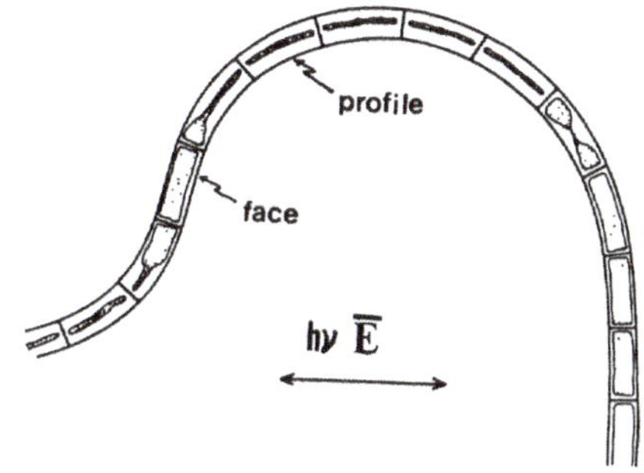

Fig. 11-23. Chloroplast rotation in *Mougeotia* induced by low fluence rate red light ($h\nu$) polarized along the axis shown by the double-headed arrow ($\bar{E}$). The chloroplasts in the cells oriented perpendicular to the electric vector $\bar{E}$ turn to the face orientation. (Modified from Ref. 80.)

orientation''; Fig. 11-23). However, at high fluence rates, the flat chloroplast rotates such that its edge faces the actinic beam (the ''profile orientation'' shown in Fig. 11-23). The low fluence rate response (rotation to the face orientation) shows a maximum spectral sensitivity in the red region (Fig. 11-22), whereas at high fluence rates the profile orientation is most effectively induced by blue light. Thus, the orientations of chloroplasts are opposite at high and low fluence rates, and different photoreceptor pigments appear to regulate the mechanism at high and low fluence rates. In darkness, *Mougeotia* chloroplasts remain in the orientation established by the previous exposure to light. Intensity-dependent chloroplast movement/rotation in *Mougeotia* and other algal cells has been reviewed elsewhere.[81]

Photoreversibility of the low fluence response has been established (Fig. 11-22), and the action spectra for induction and reversal approximate the absorption spectra of phytochrome in solution (Chapter 10). Phytochrome from higher plants has been isolated and substantially characterized (Chapter 10). Since this chromoprotein serves as the red-light receptor for numerous photomorphogenic and developmental responses in plants, it is reasonable to assume that the basic structural and functional characteristic of phytochromes are probably common to both higher plants and *Mougeotia*.

There is a distinct possibility that phytochrome in *Mougeotia* is membrane- or organelle-bound. This suggestion is based on the fact that chloroplast rotation in *Mougeotia* is polarotropically controlled (Fig. 11-23). Thus, red light polarized perpendicular to the long axis of the cell is preferentially absorbed at the front and back sides of the cells, converting the oriented and dichroic phytochrome (P_r) to its P_{fr} form (Figs. 11-22 and 11-23). As P_{fr} builds up at the front and back sides of the cell, the chloroplast edges are turned away from the higher P_{fr} gradient (Fig. 11-24). On the other hand, red light polarized parallel to the long axis of the cell is not effective for chloroplast rotation.

By means of microbeam irradiation, it has been shown that phytochrome is not freely diffusable, translationally or rotationally. The face orientation of chloroplasts achieved with red light shown in Fig. 11-24 can only be reversed with far-red light of parallel polarization ($E^{\|}$), followed by a dark period. Perpendicularly polarized light ($E_{\perp}$), is not effective in reversing the low fluence rate response of *Mougeotia*.[82]

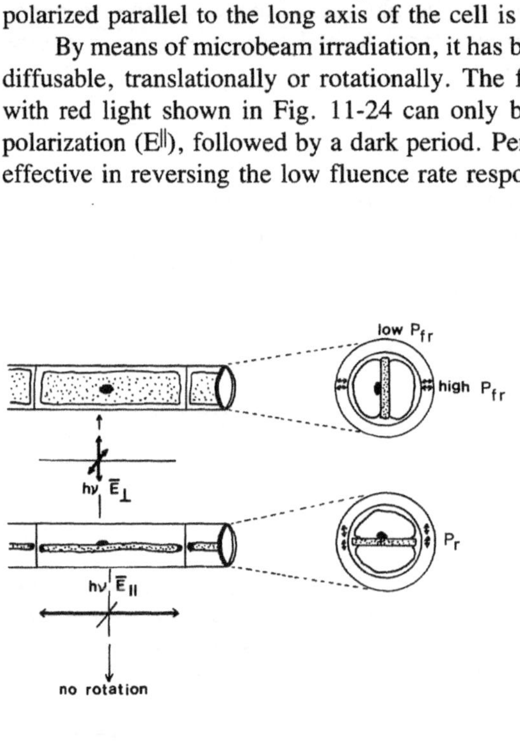

Fig. 11-24. Polarotropic chloroplast rotation. In response to red light ($h\nu$) polarized along the short axis of the *Mougeotia* cell ($D_{\perp}$), the chloroplast rotates with its edges turning toward the low P_{fr} concentration. The difference in P_{fr} concentration is established as the result of preferential excitation of the P_r transition dipoles (double arrows parallel to the cell surface) in the proximal and distal sides of the cylindrical cell. Note that the P_r transition dipoles in the upper and lower flanks of the cylindrical cell are not favorably oriented for the absorption of the short-axis polarized light, resulting in low P_{fr} concentrations at the flanks. Red light polarized along the long axis of the cell is not effective in inducing the chloroplast rotation because the P_r transition dipoles are orthogonal to the long axis. (Modified from Ref. 80.)

Based on the results with polarized light and those obtained using microbeam irradiations, the dichroism of phytochrome most likely results from the transition moments of P_r and P_{fr} that are parallel and perpendicular to the cell surface, respectively. Thus, it is likely that phytochrome is bound to the plasma membrane.[80]

11.3.2.3. Transduction Mechanisms of Polarotropism

The polarotropic chloroplast movement in *Mougeotia* cells involves at least two features: (1) inversion of the dichroism of phytochrome and (2) the primary event following the phototransformation of the dichroically oriented phytochrome molecules.

There are several mechanisms by which this inversion of dichroism could occur. Assuming that the Q_y-transition dipole (660-nm absorption band) of P_r lies parallel to the cell surface of *Mougeotia*,[80] the following mechanisms can be suggested for the Q_y-transition dipole of P_{fr} orienting perpendicular to the cell surface: (1) The direction of the Q_y-transition dipole changes by about 90° upon phototransformation from P_r to P_{fr}, with the phytochrome molecule fixed to the membrane either as a peripheral or an integral membrane component. (2) The whole protein rotates by about 90° upon phototransformation, with the chromophore Q_y-transition dipoles of P_r and P_{fr} remaining fixed with respect to the chromophore binding site/crevice. (3) The chromophore moves and reorients itself (by about 90°) relative to the chromophore binding site/crevice, with the protein moiety more or less fixed to the membrane. (4) An extensive conformational change of the protein moiety occurs, resulting in the P_{fr} chromophore oriented in space perpendicular to the original orientation of the P_r chromophore.

Let us now consider the above mechanisms in relation to the action dichroism of chloroplast movement in *Mougeotia*. The Q_y-dipole orientation (mechanism 1) can change upon phototransformation of phytochrome if the chromophore undergoes a gross conformational/geometric isomerization. Analysis of the absorption spectra of P_r and P_{fr} suggests that such a gross conformational isomerization is unlikely,[83] although a localized geometric isomerization at either end of the tetrapyrrole chromophore ring system can occur.[84] Calculations of the Q_y-transition dipole orientation in P_r and the Q_y-transition dipole orientation in several assumed structures of P_{fr} do not change significantly with respect to the axis of polarization in P_r.

Recently, it has been reported that the direction of the Q_y-transition dipole changes by a substantial angle upon phytochrome phototransformation from P_r to P_{fr}.[85] This suggests that there is a significant movement of the chromophore during the phototransformation of phytochrome. These results are qualitatively consistent with the observation that the chromophore topography also changes in the phototransformation from P_r to P_{fr}.[86]

The phototransformation of phytochrome from P_r to P_{fr} does not seem to increase the membrane fluidity of liposomes.[87] If membrane fluidity changes significantly in *Mougeotia* plasmalemma, it is unlikely that phytochrome dichroism will be preserved or that the dipole orientation will be changed significantly upon phototransformation. One can postulate that P_r as a peripheral membrane protein rotates upon its phototransformation to the P_{fr} form, which then binds to the lipid bilayer core via the newly generated hydrophobic surface. Since both electrostatic and hydrophobic forces contribute to the binding of phytochrome to membrane,[88–90] an anisotropic rotation of the protein molecule can be envisioned. Mechanism 3 is an alternative to the above mechanism, in that only the chromophore reorients; the topography of membrane binding remains fixed for P_r and P_{fr}. In view of the chromophore movement,[85,86] it is likely that both the chromophore and the protein moiety rotate upon phototransformation. Mechanism 4, involving an extensive conformational change of the protein moiety upon the phototransformation from P_r to P_{fr}, is not likely but cannot be ruled out entirely. It should be noted that the chromophore reorientation depicted in mechanism 3 implicitly entails at least local conformational changes around the chromophore and/or binding crevice.

Following the phototransformation of P_r to P_{fr} by red light, chloroplast movement is elicited in *Mougeotia* cells. The primary molecular event leading to chloroplast movement remains largely obscure, although the phytochrome phototransformation accompanies a rapid Ca^{2+} influx and accumulation within the *Mougeotia* cell. It has been suggested that phytochrome elicits chloroplast movement in *Mougeotia* by modulating the level of bound Ca^{2+} ions.[91]

Based in part on the observation that Ca^{2+} uptake is enhanced by red-light irradiation of *Mougeotia,*[92] a tentative scheme for the transduction chain involved in chloroplast movement can be formulated. Haupt and Wagner[80] proposed that the phototransformation from P_r to P_{fr} triggers changes in ion fluxes, with an enhanced Ca^{2+} influx and a release of Ca^{2+} from an internal sequestering system. Calcium ions then regulate actomyosin filament activity in eliciting chloroplast movement (Fig. 11-25). However, the regulatory mechanism of the actin–myosin interactions and their coupling to the membrane dynamics are not yet understood.

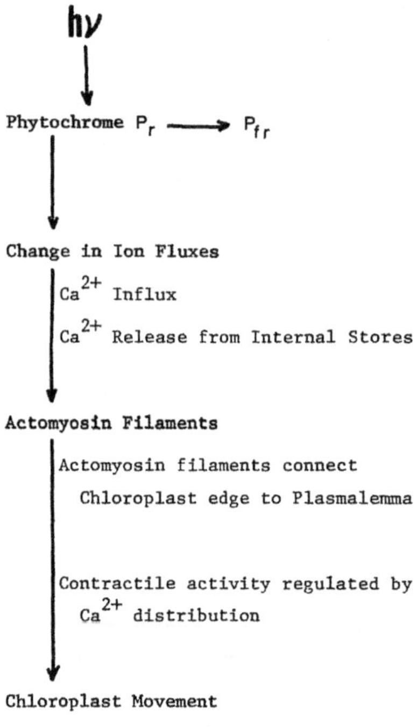

Fig. 11-25. A schematic outline of the possible transduction steps in the chloroplast movement of *Mougeotia*. (Adapted from Ref. 82.)

11.3.3. Phototropism

Over 100 years ago, Charles Darwin accurately described many aspects of plant phototropism, the process whereby plants and many fungi grow toward, or orient with respect to, a source of light.[19] This process permits a given plant to optimize its capacity to harvest light through photosynthesis, and may enable the fungus to position its spore-bearing structure to facilitate spore dispersal. The description by Darwin provided the beginning of the field of plant physiology, and also showed an appreciation for the great importance of phototropism. Although phototropic responses may be observed in flowering plants, ferns, mosses, algae, and fungi, the phenomenon has been studied mainly in the seedlings of a few grasses (corn and oats) and in the fungus *Phycomyces blakesleeanus* (Fig. 11-9).

The mechanism through which the plant directs its growth with respect to light direction can be divided into three aspects: photoreception, the transduction sequence between photoreception and the regulation of growth on the two sides of the plant, and, finally, the procedure through which the plant measures light direction itself. Although considerable information is available concerning all three areas, the last is best understood. These topics will be discussed in sequence.

11.3.3.1. Photoreception

If one exposes the shoots of young seedlings, such as corn, to light from one side, after a brief delay period the seedling shoots will begin to grow toward the source of the light. The final curvature attained is a function of the intensity, the duration of the exposure, and the wavelength of the light, and the opposing tendency of the plant to grow upward in response to the force of gravity. Given that gravity is constant in an earth-based experiment (assuming no special centrifugal forces), we can study phototropism by varying the intensity, duration, and wavelength of light, while measuring the curvature made by the plant in response to the stimulus. In this way, we can determine an action spectrum for phototropism that should be roughly equivalent to the absorption spectrum that we would measure for the photoreceptor pigment regulating the process.

The maxima from several such action spectra for phototropism are presented in Table 11-3 together with absorption maxima for β-carotene as an example of a carotenoid and riboflavin as an example of a flavin, representing the two classes of compounds that agree best with the data from action spectra. Two points are evident from these data. First, the action spectra for phototropism for many different organisms are surprisingly similar. In fact, even the action spectrum for the fungus *Phycomyces* is essentially the same as the action spectra for phototropism in flowering plants. All of the action spectra show a peak at about 370 nm, and a series of peaks between 400 and 490 nm. Based on these similarities, it is generally assumed that a single photoreceptor pigment regulates phototropism in all of these organisms. Second, it is evident that neither carotenoid nor flavin absorption spectra exactly match these action spectra.

The peak separation in the blue is mimicked either by a carotenoid or by a flavin in a hydrophobic environment, such as oil, but the 370-nm peak is best explained by a flavin. Neither carotenoids nor flavins show a 280-nm maximum, but the latter absorb in the 260-

TABLE 11-3. Comparison of the Action Maxima for Phototropism and Absorption Maxima for Riboflavin and All-*trans* β-Carotene

Material	Action/absorption maxima (nm)[a]							
Phycomyces (phototropism)	280[b]	370[c]	(385[d])	43[illegible][c]	445[c]	(455[d])	470[c]	(485[d])
Avena (phototropism)	280[e]	385[f]		425[f]	445[f]		475[f]	
β-Carotene in ethanol, room temp. (absorption)[g]				427	449		475	
β-Carotene in ethanol, 77°K, all-*trans*[k]	388[l]	411		434	468		501	
β-Carotene in ethanol, 77°K, 15,15′-*cis*[k]	351	414		432	465		493	
Riboflavin in aqueous buffer at 23°C (absorption)	266[h]	373[h]		435[i]	445[h]		485[i]	
Riboflavin in ethanol at 77°K (absorption)[j]	270	350		413	447		472	

[a]Numbers in parentheses refer to shoulders. [b]Ref. 93. [c]Ref. 94. [d]Ref. 95. [e]Ref. 96. [f]Ref. 97. [g]Ref. 98. [h]Ref. 99. [i]Ref. 100. [j]Ref. 101. [k]Ref. 102. The visible absorption band of β-carotene is resolved into sharp vibrational peaks at 77°K. [l]A weak absorption band.

nm region, and the 280-nm action maximum could result from the attachment of the pigment moiety to a protein or be due to screening by proteins and nucleic acids. Based on data such as these, it is thought that a flavin attached to a protein and constrained in the hydrophobic matrix of a membrane is the most probable candidate for the photoreceptor pigment. This conclusion, i.e., that the photoreceptor pigment is a flavoprotein, is supported by data from other experiments. For example, those compounds known to specifically inhibit phototropism all bind to or in some way interfere with the action of flavins.[103] These and other data strongly support the flavin hypothesis, but none of the data can be considered conclusive proof. Ultimately, this could be provided by a mutant with a single gene change, lacking or with a reduced amount of one specific flavoprotein, and a comparable decrease in the phototropic response.

11.3.3.2. Signal Generation and Transduction

Although very little work has been done on the transduction sequence in *Phycomyces,* enough information exists to suggest a possible sequence for converting photoreception into the directed growth response. If light were directly stimulating the activity of a rate-limiting step in growth, a higher light intensity at the distal wall of the organism would increase growth on the side away from the light, resulting in growth toward the light. The cell wall is composed primarily of chitin, the synthesis of which is stimulated by light.[104] If light were directly stimulating the activity of a rate-limiting enzyme in chitin synthesis, and if chitin synthesis were the rate-limiting step in growth, a higher light intensity at the distal wall of the organism would result in increased chitin synthesis on the side away from the light, and growth toward the light. Unfortunately, although light is focused onto the distal side (Section 11.3.3.3), there is some internal screening, so that the total light energy absorbed is higher on the proximal side than on the distal side. Hence, any asymmetry in chitin synthesis must depend on a more complex spatial and temporal pattern of light intensity in *Phycomyces.*

For many years, there has been convincing evidence that growth in plants is regulated by hormones such as indoleacetic acid (IAA). This growth regulator is produced by cells at the tip of the shoot and is transported down the shoot, regulating growth of the cells below. The Cholodny–Went theory for phototropism suggests that IAA also is transported across the shoot from the illuminated side to the shaded side. (See Ref. 14 for a more extensive discussion of this theory.) As a result of this redistribution, cell elongation is inhibited on the illuminated side and stimulated on the shaded side, and thus the shoot grows toward the lighted side or twoard the unilateral light. The best evidence supporting this theory are data showing that growth is indeed stimulated on the shaded side and inhibited on the lighted side of the shoot.[105]

11.3.3.3. Measurement of Light Direction

One of the most interesting aspects of phototropism is the mechanism used to measure light direction. This is also the aspect that is generally considered to be best understood. To measure light direction, some gradient must exist in light absorption, i.e., there must be some method that results in a greater number of quanta absorbed on one side of a unilaterally irradiated organism than on the other side. Two obvious models accom-

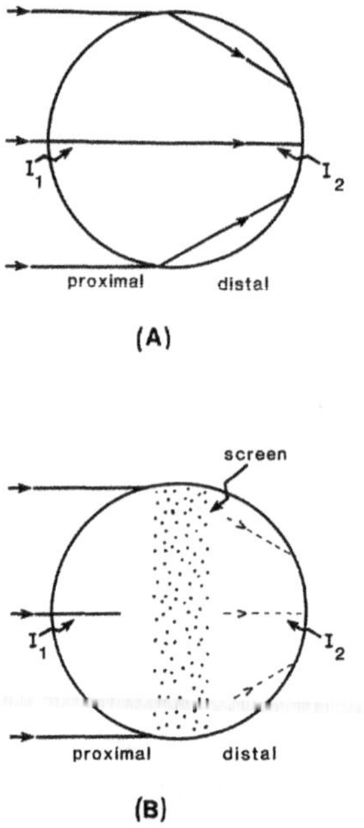

Fig. 11-26. (A) Diagram demonstrating the path of light in the cross-section of a cylindrical organism such as *Phycomyces* with low internal absorption. The straight lines with arrows represent three rays of light incident from the left. Refraction occurs for each ray not normal to the surface of the organism as that ray passes from air with a low index of refraction to cytoplasm with its relatively high index of refraction. With the low light losses within the cytoplasm, the intensity of light is higher at the distal surface (I_2) than at the proximal surface (I_1). (B) Diagram demonstrating the path of light in the cross-section of a cylindrical organism such as a corn seedling with high internal absorption. As in Part A, the straight lines with arrows represent three rays of light incident from the left. Screening is indicated by the stippled area. As a result of the screening by absorption or scattering within the organism, the intensity of light is lower at the distal surface (I_2) than at the proximal surface (I_1).

plish this, one through refraction and one through screening. Either could be employed by the organism to determine light direction.

In the refraction or "lens" mechanism, light is refracted as it passes from air with a relatively low index of refraction (1.0) to cytoplasm, which is mostly water with its relatively high index of refraction (1.34), and is thereby focused onto the distal side of the cylindrical organism (Fig. 11-26A). Given low light losses across the organism (e.g., low absorption and low scattering), the intensity of light will be higher on the distal side than on the proximal side of the cylindrical organism (Fig. 11-26A). If the organism could measure the light intensity in those two places, this "information" could be translated into light direction. For instance, wall deposition could be related to the number of quanta absorbed, and would be greater on the distal side resulting in growth toward the light.

The refraction mechanism is at least the dominant mechanism used by *Phycomyces* to measure the direction of visible light. It is evident that the difference in index of refraction between the air and the organism is the crucial element in the refraction model. It should not, then, be surprising that manipulation of the index of refraction has been used to obtain the most convincing evidence for the refraction mechanism in *Phycomyces*. The index of refraction has been changed by immersing the cylindrical sporangiophore in oil with an index of refraction roughly equal to that of the cytoplasm. Without the difference in index of refraction, focusing does not occur, and the phototropic response is lost.[106] The effect of the focusing within the organism has also been changed by placing a cylindrical glass rod between the organism and the unilateral light. The light, if focused by the additional lens such that the normal focusing by the organism does not result in an increased light intensity on its distal side but rather on the proximal side, causes the organism to grow *away* from the light.

In the screening mechanism, light is attenuated as it passes through the organism such that the intensity of light is highest on the side of the organism facing the light (Fig. 11-26B). This attenuation may be from absorption by pigments within the organism absorbing over the same wavelength range as the photoreceptor pigment, or though scattering of the light within the organism by subcellular particles such as mitochondria or surfaces such as cell walls. It should be clear from a comparison between Fig. 11-26A and B that the effect of attenuation is opposite for the two mechanisms. For the refraction mechanism, light attenuation across the organism would diminish the gradient established through focusing. For the screening mechanism, light attenuation across the organism would supplement the gradient. This is the basis for the most convincing evidence that the screening mechanism is used by plants to measure light direction. Using inhibitors, genetic mutants, and exogenous dyes, the absorption-based screening has been varied, and found to be directly related to the capacity of the plant to respond phototropically.[107] Had the plant used the refraction mechanism, the relationship between absorption and phototropism would have been inverse.

In summary, phototropism in both plants and fungi is thought to use a flavoprotein as a photoreceptor pigment. The end result of the transduction sequence in *Phycomyces* may be an increase in chitin synthesis, while in flowering plants, light may affect the lateral transport of IAA, which in turn regulates cell elongation. Although both the flowering plant shoot and the fungus *Phycomyces* are positively phototropic, the optical mechanism for measuring light direction is different for the two cases. *Phycomyces* uses a refraction mechanism to establish the gradient in light intensity and measure light direction, while the gradient in light intensity is established in flowering plants by a screening mechanism.

11.4. ECOLOGICAL SIGNIFICANCE

How light-mediated behavior benefits a species is often obvious by the nature of the response. There are some situations, however, where the ecology of a species may be influenced in very subtle ways by the organism's behavior. For example, phototaxis for the flagellates such as *Chlamydomonas* is an ecologically important response because it places the photosynthetic cell in an optimal light environment. On the other hand, one can argue that it is to the advantage of the *Stentor* cell to exhibit a step-up photophobic response, thereby avoiding intense light that could be photodynamically detrimental, although it is not clear what the advantage is to the cell to contain a powerful photosensitizing dye.[108]

Phototropism permits the plant to orient its shoots and leaves with respect to light. The leaves contain the photosynthetic machinery on which the plant depends, so it seems clear that plants with the ability to array that machinery and thereby optimize light capture in photosynthesis would have a survival advantage.

The vertical migrations of several marine dinoflagellates correspond with the daylight cycle; the organisms rise near the surface during the daylight hours and descend at night.[109] If these cells were not phototactically active and remained at some fixed depth as determined by the physical topography, would they be functionally capable of reproducing themselves? Would they be able to carry on their photosynthetic processes efficiently enough to produce the energy "surplus" necessary for fission? This can be tested and requires further investigation before we can understand the role of phototaxis in the dynamics of population growth. One can ask other questions of this system as well. For example, what environmental factors can alter the "normal" photoresponse? How does temperature affect the photoresponse at the behavioral, physiological, or biochemical level? These questions have not been studied systematically. The interactions between the two branches of photobiology, photoecology and photomovement, will play an important role in answering these questions. It is in the examination of ecological problems that the universal importance of the light behavior of organisms will be appreciated.

11.5. REFERENCES

1. R. M. Macnab and S. -I. Aizawa, Bacterial motility and the bacterial flageller motor, *Annu. Rev. Biophys. Bioeng.* **13,** 51–84 (1984).
2. K. L. Poff, in: *Sensory Perception and Transduction in Aneural Organisms* (G. Colombetti, F. Lenci, and P. -S. Song, eds.), pp. 299–307, Plenum Press, New York (1985).
3. W. Haupt, in: *Sensory Perception and Transduction in Aneural Organisms* (G. Colombetti, F. Lenci, and P. -S. Song, eds.), pp. 1–18, Plenum Press, New York (1985).
4. H. Machemer, in: *Sensory Perception and Transduction in Aneural Organisms* (G. Colombetti, F. Lenci, and P. -S. Song, eds.), pp. 179–209, Plenum Press, New York (1985).
5. R. P. Blakemore and R. B. Frankel, Magnetic navigation in bacteria, *Sci. Am.* **245,** 58–65 (1981).
6. W. Nultsch, Phototaxis and photokinesis, in: *Primitive Sensory and Communication Systems* (M. J. Carlile, ed.), pp. 29–90, Academic Press, New York (1975).
7. K. V. Thimann and G. M. Curry, Phototropism and phototaxis, in: *Light and Life.* (W. McElroy and B. Glass, eds.), pp. 647–670, The Johns Hopkins Press, Baltimore (1961).
8. W. R. Briggs, The phototropic responses of higher plants, *Annu. Rev. Plant Physiol.* **14,** 311–352 (1963).
9. D. S. Dennison, Phototropism, in: *Physiology of Movements. Encyclopedia of Plant Physiology,* New Ser., Vol. 7 (W. Haupt and M. E. Feinleib, eds.), pp. 506–566, Springer-Verlag, Berlin (1979).
10. R. D. Firn and J. Digby, The establishment of tropic curvatures in plants, *Ann. Rev. Plant Physiol.* **31,** 131–148 (1980).
11. H. Senger and W. R. Briggs, The blue light receptor(s): Primary reactions and subsequent metabolic changes, *Photochem. Photobiol. Rev.* **6,** 1–38 (1981).
12. K. L. Poff and C. B. Hong, Photomovement and photosensory transduction (yearly review), *Photochem. Photobiol.* **36,** 749–752 (1982).
13. W. Haupt, Light-mediated movement of chloroplasts, *Annu. Rev. Plant Physiol.* **33,** 205–233 (1982).
14. P. Galland and E. D. Lipson, Photophysiology of *Phycomyces blakesleeanus* (yearly review), *Photochem. Photobiol.* **40,** 795–800 (1984).
15. U. Pohl and V. E. A. Russo, Phototropism, in: *Membranes and Sensory Transduction* (G. Colombetti and F. Lenci, eds.), pp. 231–329, Plenum Press, New York (1984).
16. H. Senger (ed.), *Blue Light Effects in Biological Systems.* Springer-Verlag, Berlin (1984).
17. B. Diehn, M. E. Feinleib, W. Haupt, E. Hildebrand, F. Lenci, and W. Nultsch, Terminology of behavioral responses of motile microorganisms, *Photochem. Photobiol.* **26,** 559–560 (1977).
18. D. -P. Häder, in: *Physiology of Movements. Encyclopedia of Plant Physiology,* New Ser., Vol. 7 (W. Haupt and M. E. Feinleib, eds.), pp. 268–309, Springer-Verlag, Berlin (1979).
19. C. Darwin, *The Power of Movements in Plants,* John Murray, London (1880).
20. D. S. Dennison, in: *Advanced Plant Physiology* (M. B. Wilkins, ed.), pp. 149–162, Pitman, London (1984).
21. W. Haupt and M. E. Feinleib (eds.), in: *Physiology of Movements. Encyclopedia of Plant Physiology,* New Ser., Vol. 7, Springer-Verlag, Berlin (1979).
22. M. E. Feinleib, in: *Sensory Perception and Transduction in Aneural Organisms* (G. Colombetti, F. Lenci, and P. -S. Song, eds.), pp. 119–146, Plenum Press, New York (1985).

23. N. M. L. Morel-Laurens and M. E. Feinleib, Photomovement in an "eyeless" mutant of *Chlamydomonas, Photochem. Photobiol.* **37,** 189–194 (1983).
24. E. Batschelet, *Statistical Methods for Analysis of Problems in Animal Orientation and Navigation.* AIBS, Washington, D.C. (1965).
25. E. Batschelet, Statistical methods in the analysis of problems in animal orientation and certain biological rhythm, in: *NASA Symposium on Animal Orientation and Navigation* (S. R. Galles, K. Schmidt-Koenig, G. J. Jacob, and R. F. Belleville, eds.), pp. 61–91, NASA, Washington, D. C. (1972).
26. P. -S. Song, D. -P. Häder, and K. L. Poff, Step-up photophobic responses in the ciliate, *Stentor coeruleus, Arch. Microbiol.* **126,** 181–186 (1980).
27. P. -S. Song, D. -P. Häder, and K. L. Poff, Phototactic orientation by the ciliate, *Stentor coeruleus, Photochem.* **32,** 781–786 (1980).
28. P. Halldal, Factors affecting light response in phototactic algae, *Physiol. Plant.* **12,** 742–752 (1959).
29. Th. W. Engelmann, Bacterium photometricum. Ein Betrage zur vergleichenden Physiologie des Licht- und Farbensinnes, *Pflugers Arch. Ges. Physiol.* **30,** 95–124 (1883).
30. B. Diehn, Phototactic responses of *Euglena* to single and repetitive pulses of actinic light: Orientation time and mechanism, *Exp. Cell Res.* **56,** 375–381 (1969).
31. T. -H. Chang, E. B. Walker, and P. -S. Song, Effects of ionophores and ruthenium red on the phototaxis of *Stentor coeruleus* as measured by simple devices, *Photochem. Photobiol.* **33,** 933–936 (1981).
32. E. B. Walker, M. Yoon, and P. -S. Song, The pH dependence of photosensory responses in *Stentor Coeruleus* and model system, *Biochim. Biophys. Acta* **634,** 298–308 (1981).
33. K. Bergman, P. B. Burke, E. Cerda-Olmedo, C. N. David, M. Delbruck, K. W. Foster, E. W. Goodell, M. Heisenberg, G. Meissner, M. Zalokar, D. S. Dennison, and W. Shropshire, Jr., *Phycomyces, Bacteriol. Rev.* **33,** 99–157 (1969).
34. W. Hand, in: *The Science of Photobiology* (K. C. Smith, ed.), 1st ed., pp. 313–328, Plenum Press, New York (1977).
35. W. G. Hand and D. Davenport, in: *Photobiology of Microorganisms* (P. Halldal, ed.), pp. 253–282, Wiley-Interscience, New York (1970).
36. P. -S. Song, The electronic spectroscopy of photoreceptors (other than rhodopsin), in: *Photochem. Photobiol. Rev.*, Vol. 7 (K. C. Smith, ed.), pp. 77–140, Plenum Press, New York (1983).
37. P. -S. Song, in: *Sensory Perception and Transduction in Aneural Organisms* (G. Colombetti, F. Lenci, and P. -S. Song, eds.), pp. 47–59, Plenum Press, New York (1985).
38. W. Nultsch and D. -P. Häder, Photomovement of motile microorganisms, *Photochem. Photobiol.* **29,** 423–438 (1979).
39. W. Nultsch, Der Einfluss des Lichtes auf die Bewegung der Cyanophyceen. I. Mitt. Phototopotaxis von *Phormidium autumnale, Planta* **56,** 632–647 (1961).
40. W. Nultsch, Der Einfluss des Lichtes auf die Bewegung der Cyanophyceen. II. Photokinesis bei *Phormidium autumnale, Planta* **57,** 613–632 (1962).
41. W. Nultsch, Der Einfluss des Lichtes auf die Bewegung der Cyanophyceen. III. Photophobotaxis von *Phormidium uncinatum, Planta* **58,** 647–663 (1962).
42. W. Nultsch, H. Schuchart, and M. Hohl, Investigation on the phototactic orientation of *Anabaena variabilis, Arch. Microbiol.* **122,** 85–91 (1979).
43. W. Nultsch, Effects of sodium azide on phototaxis of the blue-green alga *Anabaena variabilis* and consequences to the two-photoreceptor systems-hypothesis, *Arch. Microbiol.* **134,** 33–37 (1983).
44. W. Nultsch, in: *Topics in Photobiology* (H. O. Kim and P. -S. Song, eds.), pp. 95–107, Jeju National University, Jeju (1983).
45. F. Lenci, D. -P. Häder, and G. Colombetti, in: *Membranes and Sensory Transduction* (G. Colombetti and F. Lenci, eds.), pp. 199–229, Plenum Press, New York (1984).
46. W. Nultsch, in: *Sensory Perception and Transduction in Aneural Organisms* (G. Colombetti, F. Lenci, and P. -S. Song, eds.), pp. 147–164, Plenum Press, New York (1985).
47. D. -P. Häder, Electrical and proton gradients in the sensory transduction of photophobic responses in blue-green alga, *Phormidium uncinatum, Arch. Microbiol.* **130,** 83–86 (1981).
48. D. -P. Häder, Extracellular and intracellular determination of light-induced potential changes during photophobic reactions in blue-green algae, *Arch. Microbiol.* **119,** 75–79 (1978).
49. D. -P. Häder, Effects of inhibitors and uncouplers on light-induced potential changes triggering photophobic responses, *Arch. Microbiol.* **120,** 57–60 (1979).

50. D. -P. Häder and K. L. Poff, Dependence of the photophobic response of the blue-green alga, *Phormidium uncinatum*, on cations, *Arch. Microbiol.* **132,** 345–348 (1982).
51. J. Spudich, in: *Sensory Perception and Transduction in Aneural Organisms* (G. Colombetti, F. Lenci, and P. -S. Song, eds.), pp. 113–118, Plenum Press, New York (1985).
52. E. Hildebrand and A. Schimz, in: *Sensory Perception and Transduction in Aneural Organisms* (G. Colombetti, F. Lenci, and P. -S. Song, eds.), pp. 93–111, Plenum Press, New York (1985).
53. E. Hildebrand and N. Dencher, Two photosystems controlling behavioral responses of *Halobacterium halobium, Nature* **257,** 46–48 (1975).
54. R. A. Bogomolni and J. L. Spudich, Identification of third rhodopsin-like pigment in phototactic *Halobacterium halobium, Proc. Natl. Acad. Sci. USA* **79,** 6250–6254 (1982).
55. H. C. Berg, Chemotaxis in bacteria, *Annu. Rev. Biophys. Bioeng.* **4,** 119–136 (1975).
56. A. Schimz and E. Hildebrand, Response regulation and sensory control in *Halobacterium halobium* based on an oscillator, *Nature* **317,** 641–642 (1985).
57. P. -S. Song, Protozoan and related photoreceptors: Molecular aspects, *Annu. Rev. Biophys. Bioeng.* **12,** 35–68 (1983).
58. V. Massey, in: *Photoreception and Sensory Transduction in Aneural Organisms* (F. Lenci andG. Colombetti, eds.), pp. 253–270, Plenum Press, New York (1980).
59. M. F. Land, The physics and biology of animal reflectors, *Progr. Biophys.* **24,** 77–105 (1972).
60. S. O. Mast, *Light and the Behavior of Organisms,* Wiley, New York (1911).
61. M. Melkonian and H. Robenek, Eye spot membranes of *Chlamydomonas reinhardii*: A freeze-fracture study, *J. Ultrastr. Res.* **72,** 90–102 (1980).
62. W. H. Miller, Ocular optical filtering, in: *Invertebrate Photoreceptors Handbook of Sensory Physiology,* Vol. VII/6A (H. Autrum, ed.), pp. 69–143, Springer-Verlag, Berlin (1979).
63. A. Lui, Photoresponse of *Paramecium* provoked by photosensitizing dyes, *Biol. Glasn. Jugosl.* **8,** 11–32 (1956).
64. L. K. Lozina-Lozinsky, Adaptive behavior of *Paramecium caudatum* to action of photodynamic dyes, *Acta Protozool.* **18,** 609–628 (1979).
65. C. J. Brokaw and P. Verdugo (eds.), *Progress in Clinical and Biological Research, Vol. 80, Mechanism and Control of Ciliary movement,* A. L. Liss, New York (1982).
66. I. -H. Kim, R. K. Prusti, P. -S. Song, D. -P. Häder, and M. Häder, Phototaxis and photophobic responses in *Stentor coeruleus*: Action spectrum and role of Ca^{2+} fluxes, *Biochim. Biophys. Acta* **799,** 298–304 (1984).
67. P. -S. Song, D. -P. Häder, and K. L. Poff, Phototactic orientation by the ciliate, *Stentor coeruleus, Photochem. Photobiol.* **32,** 781–786 (1980).
68. P. -S. Song, K. J. Tapley, Jr., and J. D. Berlin, in: *The Biology of Photoreception* (D. Cosens and D. Vince-Prue, eds.), pp. 503–520, Cambridge University Press, Cambridge (1983).
69. P. -S. Song, E. B. Walker, R. A. Auerbach, and G. W. Robinson, Proton release from *Stentor* photoreceptors in the excited states, *Biophys. J.* **35,** 551–555 (1981).
70. K. Iwatsuki and P. -S. Song, Deuterium oxide (D_2O) enhances the photosensitivity of *Stentor coeruleus, Biophys. J.* **48,** 1045–1048 (1985).
71. R. K. Prusti, P. -S. Song, D. -P. Häder, and M. Häder, Caffeine-enhanced photomovement in the ciliate, *Stentor coeruleus, Photochem. Photobiol.* **40,** 369–375 (1984).
72. S. J. Britz, Chloroplast and nuclear migration, in: *Physiology of Movements. Encyclopedia of Plant Physiology,* New Ser., Vol. 7 (W. Haupt and M. E. Feinleib, eds.), pp. 170–205, Springer-Verlag, Berlin (1979).
73. G. Fischer-Arnold, Untersuchungen über die Chloroplastenbewegung bei *Vaucheria sessilis, Protoplasma* **56,** 495–520 (1963).
74. E. Schonbohm, Untersuchungen über die Starklichtbewegung des *Mougeotia*-Chloroplasten, *Z. Bot.* **51,** 233–276 (1963).
75. W. Haupt and R. Gartner. Die Chloroplasten-Orientierung von *Mesotaenium* in starkem Lict, *Naturwiss.* **53,** 411 (1966).
76. F. Mayer, Lichtorientierte Chloroplasten-Verlagerung bei *Selaginella martensii, Z. Bot.* **52,** 346–381 (1964).
77. J. Zurzycki, The action spectrum for the light dependent movements of the chloroplasts in *Lemna trisulca L., Acta Soc. Bot. Polon.* **31,** 489–538 (1962).

78. K. Seitz, Zur Frage der Jodid-Wirkung auf die Starklichtbewegungen der Chloroplaten von *Vallisneria spiralis* ssp. *torta, Z. Pflanzenphysiol.* **62,** 63–69 (1970).
79. W. Haupt, Die Chloroplastendrehung bei *Mougeotia.* I. Mitteilung: über den quantitativen und qualitativen Lichtbedarf der Schwachlicht bewegung, *Planta* **53,** 484–501 (1959).
80. W. Haupt and G. Wagner, in: *Membranes and Sensory Transduction* (G. Colombetti and F. Lenci, eds.), pp. 331–375, Plenum Press, New York (1984).
81. E. Schonbohm, in: *Blue Light Syndrome* (H. Senger, ed.), pp. 69–96, Springer-Verlag, Berlin (1980).
82. W. Haupt, in: *Handbook of Phycological Methods: Development and Cytological Methods* (E. Gantt, ed.), pp. 196–204, Cambridge University Press, Cambridge (1980).
83. P. -S. Song, Q. Chae, and J. G. Gardner, Spectroscopic properties and chromophore conformations of the photomorphogenic receptor: Phytochrome, *Biochim. Biophys. Acta* **476,** 479–495 (1979).
84. W. Rudiger, F. Thummler, E. Cmiel, and S. Schneider, Chromophore structure of the physiologically active form (Pfr) of phytochrome, *Proc. Natl. Acad. Sci. USA* **80,** 6244–6248 (1983).
85. N. G. A. Ekelund, C. Sundqvist, P. H. Quail, and R. D. Vierstra, Chromophore rotation in 124-kilodalton *Avena* phytochrome as measured by light-induced changes in linear dichroism, *Photochem. Photobiol.* **41,** 221–223 (1985).
86. T. R. Hahn, P. -S. Song, P. J. Quail, and R. D. Vierstra, Tetranitromethane oxidation of phytochrome chromophore as a function of spectral form and molecular weight, *Plant Physiol.* **74,** 755–758 (1984).
87. I. -S. Kim, E. S. Kim, and P. -S. Song, Interactions of phytochrome with detergents and liposomes, *Biochim. Biophys. Acta* **747,** 55–64 (1983).
88. M. Furuya, J. H. Freer, A. Ellis, and K. T. Yamomoto, Electrostatic binding of proteins and phytochrome to differently charged liposomes, *Plant Cell Physiol.* **22,** 135–144 (1981).
89. I. -S. Kim and P. -S. Song, Binding of phytochrome to liposomes and oat protoplasts, *Biochemistry* **20,** 5482–5489 (1981).
90. G. Hermann, A. Hattenbach, and E. Müller, Further characterization of the interaction between phytochrome and liposomes, *Biochem. Physiol. Pflanzen* **181,** 61–67 (1986).
91. G. Wagner and K. Klein, Differential effect of calcium on chloroplast movement in *Mougeotia, Photochem. Photobiol.* **27,** 137–140 (1978).
92. E. M. Dreyer and M. H. Weisenseel, Phytochrome-mediated uptake of calcium in *Mougeotia* cells, *Planta* **146,** 31–39 (1979).
93. G. M. Curry and H. E. Gruen, Action spectra for the positive and negative phototropism of *Phycomyces* sporangiophore, *Proc. Natl. Acad. Sci. USA* **45,** 797–804 (1959).
94. G. V. Thimann and G. M. Curry, Phototropism and phototaxis. in: *Comparative Biochemistry: A Comprehensive Treatise,* Vol. 1 (M. Florkin and H. S. Mason, eds.), pp. 243–309, Academic Press, New York (1960).
95. M. Delbruck and W. Shropshire, Jr., Action and transmission spectra of *Phycomyces, Plant Physiol.* **35,** 194–204 (1960).
96. G. M. Curry, K. V. Thimann, and P. M. Ray, The base curvature of *Avena* seedlings to the ultraviolet. *Physiol. Plant.* **9,** 429–440 (1956).
97. W. Shropshire, Jr., and R. B. Withrow, Action spectrum of phototropic tip-curvature of *Avena. Physiol. Plant.* **33,** 360–365 (1958).
98. B. H. Davies, Carotenoids, in: *Chemistry and Biochemistry of Plant Pigments,* Vol. 2 (T. W. Goodwin, ed.), pp. 38–165, Academic Press, New York (1976).
99. P. -S. Song and T. A. Moore, On the photoreceptor pigment for phototropism and phototaxis: Is a carotenoid the most likely candidate?, *Photochem. Photobiol.* **19,** 435–441 (1974).
100. H. Beinert, Flavin coenzymes, in: *The Enzymes,* Vol. 2 (P. D. Boyer, H. Lardy, and K. Myrback, eds.), pp. 339–416, Academic Press, New York (1960).
101. R. Matthews and V. Massey, Forms of old yellow enzyme, in: *Flavins and Flavoproteins* (H. Kamin, ed.), pp. 329–348, University Park Press, Baltimore, and Butterworths, London (1971).
102. M. Sun, T. A. Moore, and P. -S. Song, Molecular luminescence studies of flavins. I. The excited states of flavins, *J. Am. Chem. Soc.* **94,** 1730–1740 (1972).
103. R. D. Vierstra and K. L. Poff, Mechanism of specific inhibition of phototropism by phenylacetic acid in corn seedling, *Plant Physiol.* **67,** 1011–1015 (1981).
104. Y. N. Jan, Properties and cellular localization of chitin synthetase in *Phycomyces blakesleeanus. J. Biol. Chem.* **249,** 1973–1979 (1974).

105. M. Iino and W. R. Briggs, Growth distribution during first positive phototropic curvature of maize coleoptiles, *Plant Cell Environ.* **7,** 97–104 (1984).
106. W. Shropshire, Jr., The lens effect and phototropism of *Phycomyces. J. Gen. Physiol.* **45,** 949–958 (1962).
107. R. D. Vierstra and K. L. Poff, Role of carotenoids in the phototropic response of corn seedlings. *Plant Physiol.* **68,** 798–801 (1981).
108. K. C. Yang, R. K. Prusti, P. S. Song, M. Watanabe, and M. Furuya, Photodynamic action in *Stentor coeruleus* sensitized by endogenous pigment stentorin, *Photochem. Photobiol.* **43,** 305–310 (1986).
109. R. B. Forward, Jr., Light and diurnal vertical migration: Photobehavior and photophysiology of plankton, in: *Photochem. Photobiol. Rev.* Vol. 1 (K. C. Smith, ed.), pp. 157–209, Plenum Press, New York (1976).

12

Photosynthesis

David C. Fork • Department of Plant Biology, Carnegie Institution of Washington, Stanford, California 94305.

12.1. INTRODUCTION

Photosynthesis, the conversion of light energy into stabilized chemical energy, involves the absorption of light by a pigment, energy transfer, energy trapping or stabilization by reaction centers, and the initiation of chemical reactions from donor to acceptor molecules. The process continues with a sequence of oxidation–reduction reactions that compose the electron transport that leads to the formation of reduced nicotinamide adenine dinucleotide phosphate (NADPH) in plants or reduced NAD (NADH) in bacteria and adenosine triphosphate (ATP). Reactions leading to the fixation of carbon are powered by the energy available in the molecules of NADPH (or NADH) and ATP.

The study of photosynthesis is receiving ever-expanding attention. The Sixth International Congress held in Brussels in 1983 was attended by more than 1100 scientists. The proceedings of this congress alone resulted in the publication of the proceedings in four volumes containing over 3400 pages and authored by more than 1200 contributors.[1] The Third Congress held three years earlier in Greece produced six volumes[2] and earlier congresses also yielded a wealth of new results.[3–6]

Photosynthesis is a multifaceted subject that can be studied from many points of view, including those of the physicist, physical chemist, biochemist, biophysicist, biologist, ecologist, and agronomist. This chapter is an introduction to a complex process that is both challenging and important to study and understand. A number of reviews and books have appeared which, in addition to the above-mentioned works, can be consulted for more detailed information than can be presented here.[7–25]

12.1.1. How Did Photosynthesis Evolve?[26,27]

Conditions on the primitive earth were drastically different from those that exist today. The Russian biochemist Oparin[28] suggested that no O_2 existed in the atmosphere, but rather hydrogen and methane were the dominant components with lesser amounts of ammonia, nitrogen, hydrogen sulfide, and water being present. When a mixture of these substances is exposed to UV radiation, radioactivity, heat, or electric discharges, a wide variety of substances can be produced, such as HCN, sugars, amino acids, peptides, nucleosides, and nucleotides, as well as polyphosphates and porphyrins.[29,30]

Since no living organisms would have been around to metabolize these substances on primeval earth, it is conceivable that substantial amounts could have accumulated over long periods of time. When amino acids are heated they form proteinlike molecules that can aggregate into small spheres about the size of bacteria. Nobody knows just how living forms may have arisen from such aggregations, but they may have appeared as early as 4 billion years ago.[31,32]

These primitive organisms could obtain all the energy they required simply by utilizing substances that were found abundantly in the environment. With an increase in the number of such "living organisms," the supply of these compounds would eventually run out. Compounds from which it is easy to obtain energy would be used (oxidized) first. Survival depended on the primitive bacterialike cells being able to oxidize the remaining compounds that were more difficult to oxidize, such as H_2S and H_2O. In order to do this, these "cells" would be greatly assisted if a method could be developed by which the energy in light could be used. Thus, a primitive photosynthetic system probably evolved that could utilize light energy to oxidize an electron donor and thereby synthesize an energy-rich substance for later metabolic use. Since water is one of the most difficult compounds to oxidize, the ability of an organism to accomplish the photosynthetic oxidation of water granted a great evolutionary advantage over other life forms, since it would no longer have to depend on the availability of preformed energy rich compounds.

The Oparin idea of the origin of life in the primitive ocean has been challenged by Woese,[31,32] who proposed that life arose in the atmosphere of early earth because the extremely hot surface of the earth had almost no liquid water. Violent storms could cause dust to be carried into an atmosphere rich in CO_2 and thought to contain appreciable quantities of water vapor and some methane, hydrogen sulfide, ammonia, as well as N_2. This atmosphere could act like a chemical reflux column where droplets containing many absorbed and dissolved substances would provide the necessary surface area for a "membrane" type of chemistry to occur. Macromolecules would partition at interfaces where some types could absorb and convert solar energy into chemical energy. The synthesis of these molecules would tend to be autocatalytic. Woese[31,32] proposed that organisms capable of metabolizing methane (methanogens, one of the most primative known organisms) brought about the formation of water from the large amounts of CO_2 and H_2 in the atmosphere.

Photosynthesis using H_2O as the donor of electrons to reduce carbon to carbohydrate and releasing O_2 as a waste product appears to have evolved very early in the earth's history. The fossil record shows that the O_2-evolving cyanobacteria (blue-green algae) appeared about 3 billion years ago. The ability of the blue-green algae to use H_2O as their electron donor gave them great flexibility in the habitats that they could occupy and provided them with a great evolutionary advantage over photosynthetic bacteria that were confined to areas where they could obtain simple organic compounds or H_2S. Oxygen-evolving photosynthesis became the dominant type of energy-converting system of plants, giving rise to the accumulation of O_2 in the atmosphere and, eventually, to the protective ozone (O_3) layer in the upper atmosphere via the photochemical alteration of O_2, thereby forever changing the course of evolution on earth.

12.1.2. An Overview of Photosynthesis

The time scale for photosynthesis is large. The absorption of a photon produces the exicted state in photosynthetic pigments in about a femtosecond ($1\ \text{fs} = 10^{-15}\ \text{s}$) and charge separation occurs in about 1 picosecond (1 ps = 10^{-12} s). The enzymatic reactions of CO_2 fixation and cellular synthesis take place within seconds or minutes. The very first step in photosynthesis may be considered to be the light-harvesting act. The absorption of a photon raises chlorophyll (Chl) to an excited state, whereupon it may lose its energy by radiationless transition to the ground state, by emitting fluorescence (having a longer wavelength than the photon absorbed), by transferring its energy to another pigment, or by inducing photochemistry. It is the latter process, in a specialized Chl protein termed a reaction center, that brings about the conversion of light to chemical energy. Upon receiving excitation energy, the reaction center Chl expels an electron and becomes oxidized. The electrons expelled by reaction centers are used to reduce compounds termed primary acceptors. The reaction centers containing a specialized Chl or bacteriochlorophyll (BChl) (or dimers of these pigments) are composed of specific proteins that span the width of the photosynthetic membrane. Since reaction centers are relatively few in number, they are associated with light-harvesting complexes (LHC) that either are located within the photosynthetic membranes, form layers, or are assembled in more complex structures such as the phycobilisomes of the red and blue-green algae.[33–35] A photon absorbed by any of the pigments in the antenna is transferred to other pigments having longer wavelength absorption maxima, and eventually reaches the reaction center where a charge separation occurs across the membranes. An electron is transferred from the primary donor to a transient primary acceptor, and from there usually to a stable quinone-type acceptor. The reduced quinones reduce other electron carriers and eventually transfer an electron back to the oxidized primary donor of the reaction center, in the case of photosynthetic bacteria (Section 12.4). Plants and cyanobacteria have two photosystems (I and II). In photosystem II (PSII), reduced quinones transfer electrons to a different reaction center, photosystem I (PSI) (Section 12.5). In PSI an absorption of another photon induces a charge separation across the membrane, the reduction of a different acceptor (an iron-sulfur protein), and, finally, the reduction of $NADP^+$ to NADPH. The energy-rich end products, NADPH and ATP, are used in the dark reactions of CO_2 fixation to yield carbohydrates (CH_2O). Thus, photosynthesis in green plants and cyanobacteria removes electrons (and hydrogen) from water and adds them to CO_2, leaving O_2 as a waste product. The electron flow, either cyclic or noncyclic, induces the transport of H^+ across the photosynthetic membrane. The movement of the H^+ out of the membrane is coupled to a system that results in the formation of ATP.

Bacterial photosynthesis uses only one light conversion step that is connected to an electron transport sequence involving several components whereby an electron is transferred back to the oxidized primary donor of the reaction center, as noted above. This cyclic flow also induces H^+ transport across the photosynthetic membrane, which is coupled to ATP synthesis. In addition to the cyanobacteria, certain species of bacteria, such as the green and purple bacteria, can reduce CO_2 to CH_2O. Green and purple photosynthetic bacteria are unable to oxidize water, but instead must use compounds of low redox potential such as H_2S, H_2, and $Na_2S_2O_3$ as their electron (and proton) supply, and thus do not evolve O_2.

An equation was provided by Van Niel[36] as a general formulation of photosynthesis in plants and bacteria, by which oxidizing and reducing power produced by the photochemical steps leads to the subsequent oxidation of a hydrogen donor and the reduction of CO_2:

$$H_2A + CO_2 \xrightarrow[\text{Chl or BChl}]{\text{light}} CH_2O + A$$

For green plants and cyanobacteria H_2A represents water, and for photosynthetic bacteria H_2A either represents an inorganic sulfur compound such as H_2S, organic acids, or other organic hydrogen donors. Measurements[37,38] with oxygen isotopes are in agreement with Van Niel's concept that the O_2 released in photosynthesis originates from water and not CO_2.[32,33]

12.2. THE PHOTOSYNTHETICALLY ACTIVE PIGMENTS[39–41]

The light reaching the earth's surface has a wavelength distribution extending from about 290 to 1100 nm. Biologically injurious UV light below about 290 nm is screened out of sunlight by the ozone layer, while wavelengths above about 1100 nm are absorbed by water in the atmosphere. As outlined below, there are pigments in the different types of photosynthetic organisms that absorb over this entire wavelength range.

12.2.1. The State of Chlorophyll *in Vivo*[42]

Chemically, Chl consists of four pyrrole rings joined to form a flat ring that is complexed with magnesium in its center (Fig. 12-1A). Removal of the Mg and substitution of H produces a substance termed pheophytin (Ph). The tetrapyrrole ring of Chl has attached to it a long and almost completely saturated hydrocarbon phytol tail. The flat head part of the Chl molecule is more soluble in water (hydrophilic), whereas the phytol tail is more fat-soluble (hydrophobic).

Chl a has the CH_3 side chain in ring II in the position of X (Fig. 12-1A), while Chl b has the CHO group in this position. In bacteriochlorophyll a (BChl a) the CH_3 group occurs in position X, and it substitutes for CH_2CH_3 in ring II. A CH_3CO group substitutes for $H_2C{=}CH$ in ring I, and rings II and IV are reduced. As in Chl a and b, BChl a is esterified with phytol, but in some species geranylgeraniol is found instead. Some bacteria contain BChl b, which has its absorption in the near-infrared region *in vivo* near 1020 nm one of the longest wavelength absorption bands known for a chlorophyll.

Chl extracted by organic solvents yields absorption spectra that are characteristic of the species of Chl present and the particular solvent used.[43] If one measures the absorption spectra of intact material, the absorption spectrum of Chl is complex, being composed of several bands, each absorbing at slightly different wavelengths. These so-called forms of Chl apparently result from the association of Chl with specific protein molecules.

The ability of plants to elaborate Chl–protein complexes absorbing at many different wavelengths is a very important feature of the photochemical apparatus, for a photon will be absorbed and transferred by a Chl having a short wavelength maximum to a Chl having a longer wavelength maximum (Section 12.3.2). In addition, the so-called accessory pigments such as Chl b and Chl c, the red and blue phycobilin pigments of red and blue-green algae, as well as certain carotenoids can absorb short wavelengths of light and

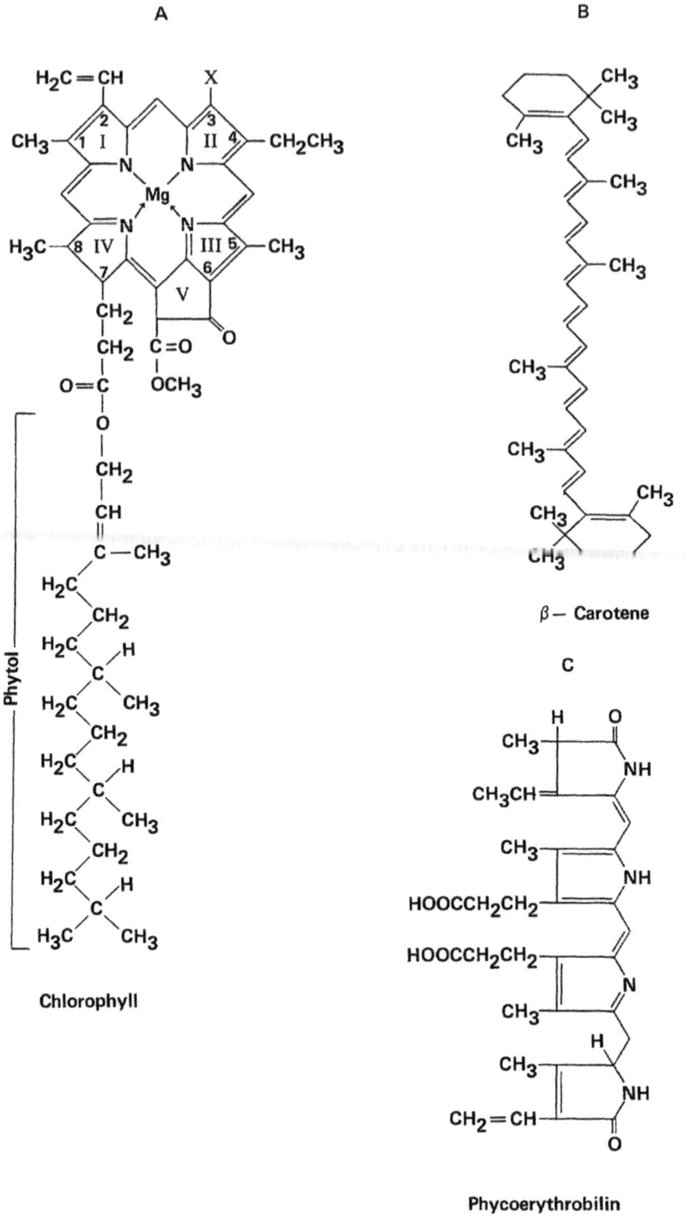

Fig. 12-1. Molecular structures of chlorophyll (A), β-carotene (B), and phycoerythrobilin (C).

transfer them to Chl a, which absorbs at longer wavelengths. The different accessory pigments as well as the different associations between Chl and protein thus function as an effective antenna to collect quanta from a large assemblage of pigments absorbing at shorter wavelengths, and to transfer them to a reaction center Chl protein that absorbs at the longest wavelength and initiates the photochemical reaction.

12.2.2. Distribution of Pigments in Photosynthetic Organisms[39,44,45]

12.2.2.1. Purple and Green Photosynthetic Bacteria

BChl serves as the photosynthetically active pigment in bacteria. Five distinct molecular species of BChl are known: BChl a, b, c, d, and e. These pigments derive their characteristic absorption spectra from the various side groups that are attached to several of the carbon atoms in the structure of Chl a. Most purple bacteria contain BChl a, which has its red absorption bands *in vivo* between 800 and 900 nm and its blue band between 350 and 400 nm (Fig. 12-2A). Some species of purple bacteria contain BChl b, which has its main absorption band *in vivo* at 800–1020 nm. Photosynthetically active carotenoids, such as spheroidene, are commonly found in photosynthetic bacteria. The contribution to absorption of carotenoids can be clearly seen as the triple peaks in the blue region between 440 and 540 nm (Fig. 12-2A).

BChls c, d, or e, whose characteristics are more closely related to higher plant Chl

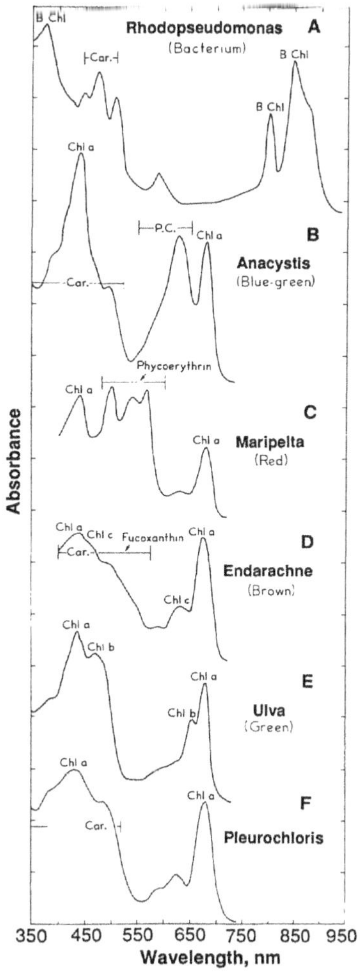

Fig. 12-2. Absorption spectra made *in vivo* for the major kinds of photosynthetic organisms. Spectrum A was measured using chromatophores of the nonsulfur purple bacterium *Rhodopseudomonas spheroides*. Spectrum B was obtained using cells of the cyanobacterium (blue-green alga) *Synechococcus leopoliensis (Anacystis nidulans)*. Spectrum C was measured using the thallus of the deep-water red alga *Maripelta rotata* which grows in submarine canyons off California at depths up to 30 m. Spectrum D was obtained from the thallus of the intertidal brown alga *Endarachne binghamiae* and spectrum E from the green alga *Ulva lobata*. Spectrum F was obtained from cells of the yellow-green alga *Pleurochloris* sp. The approximate absorption maxima and, in some cases, the range of absorption of photosynthetic pigments are shown. Car., carotenoid, PC, phycocyanin. (From Fork, unpublished.)

rather than BChl, are present in the green bacteria. Thus, the main red absorption band of the green bacterium *Chlorobium* is found between 700 and 800, and the blue maximum occurs between 400 and 500 nm.

12.2.2.2. Cyanobacteria (Blue-Green Algae)

These primitive algae probably evolved from a progenitor of the photosynthetic bacteria and, like the bacteria, do not have their photosynthetic pigments localized in chloroplasts as do other kinds of algae and higher plants. Blue-green algae have flattened sacklike structures termed lamellae or thylakoids extending throughout the cytoplasm of the cell.

The blue-green algae contain Chl a as their main photosynthetic pigment. The maximum absorption bands of Chl a occur *in vivo* near 680 nm in the red and around 440 nm in the blue part of the spectrum. Figure 12-1B shows the absorption curve for cells of the blue-green alga *Anacystis*. In addition to β-carotene (Fig. 12-1C), which has been found in all photosynthetic plants, the blue-green algae contain myxoxanthophyll, a carotenoid not found in other organisms. The blue-green algae also contain phycocyanin (PC), absorbing maximally near 626 nm, and allophycocyanin (APC), absorbing at 650 nm, which serve as "accessory pigments." PC as well as the pigment phycoerythrin (PE) belong to a class of water-soluble proteins called the phycobilins (algal bilins). PC, APC, and PE constitute the major accessory pigments of blue-green and red algae, and are present in other less well-known algal groups as well. The chromophores of phycobilins are related to Chl, but here the four pyrrole rings are arranged linearly (Fig. 12-1C). This arrangement is the fundamental structure for the bilin pigments, so called because they were first isolated from animal bile. These pigments serve by absorbing green to orange light, wavelengths not effectively absorbed by Chl.

12.2.2.3. Red Algae

Red algae live largely in marine habitats, and frequently occur in deep waters where red and blue wavelengths are largely filtered out by the upper water layers. Red algae containing the phycobilin pigment PE can absorb the remaining green light in deep water and use it effectively for photosynthesis. Red algae contain, in addition, varying amounts of the pigments PC and APC. Red algae living in deep water have more PE than PC. The combination of PE with Chl a permits red algae to absorb almost all the wavelengths in the visible part of the spectrum (Fig. 12-2C). The absorption bands of Chl a *in vivo* at 440 and 678 nm can be seen, as can the peaks at 500, 540, and 566 nm produced by PE absorption. In addition to β-carotene, the red algae contain α-carotene that has been found in no other plants examined so far, except for certain kinds of green algae (*Siphonales*). Lutein is the principal xanthophyll of red algae.

12.2.2.4. The Brown Algae, Diatoms, and Dinoflagellates[46]

Diatoms and dinoflagellates living in the upper ocean layers account for about a third of the world's photosynthesis. In addition to Chl a, these algae contain Chl c as well as the characteristic xanthophylls fucoxanthin (brown algae and diatoms) and peridinin (dinoflagellates) that serve as antenna pigments for photosynthesis (for a review of carotenoids functioning in photosynthesis, see Ref. 47). The absorption maxima of Chl c occur *in vivo* near 460 and 640 nm, while fucoxanthin and peridinin absorption maxima are near 490 nm (Fig. 12-2D). When in its native state, probably attached to a protein, the absorption maxima of fucoxanthin and peridinin are shifted by about 40 nm to longer wavelengths, compared to the absorption maxima of these pigments in organic solvents. Together with Chl c, these xanthophylls make available a large fraction of the green light that otherwise would be lost for photosynthetic purposes.

12.2.2.5. Green Algae and Higher Plants

These plants constitute the most commonly seen photosynthetic organisms in everyday life. In these plants, Chl a is found in conjunction with Chl b, the latter having blue and red maxima *in vivo* near 470 and 650 nm, respectively (Fig. 12-2E). Unlike the blue-green algae, diatoms, and dinoflagellates, the higher plants and green algae do not have special pigments to absorb light between about 500 and 600 nm, and so reflect these wavelengths and appear green to the eye. In addition to the universal β-carotene, the higher plants and green algae contain lutein as their major xanthophyll, and lesser amounts of violaxanthin and neoxanthin. Unlike xanthophylls such as fucoxanthin and peridinin, the xanthophylls of higher plants and green algae are not particularly active in absorbing light used for photosynthesis.

There are certain algae that contain only Chl a. An example is given in Fig. 12-2F for *Pleurochloris* (a yellow-green alga in the class Eustigmatophyceae). This spectrum clearly shows the contribution to absorption of only Chl a and carotenoids.

12.3. LIGHT HARVESTING AND PHOTOCHEMICAL CONVERSION[14,48]

12.3.1. Absorption of Light and Fluorescence of Pigments

In order for a light reaction to take place, it is necessary for a pigment to absorb light. When a substance absorbs a photon of light, it acquires additional energy, and is described as being in the electronic excited state. According to the laws of quantum mechanics, the transition from the ground state to the excited state can take place only via certain discrete energy states. To be absorbed, the energy content of the photon must be exactly the same as the difference in energy between two electronic states. For this reason, only certain wavelengths of light can be absorbed by a particular substance (Chapter 1).

Usually the energy levels of the first excited singlet state and the ground states are not overlapped by sublevels, as are the first and higher excited states. Therefore, a transition from the ground to the first excited singlet state requires the absorption of a photon, and likewise, the transition from the first excited singlet state to the ground state results in the emission of a photon of light. This light emission is termed fluorescence. Fluorescence emission comes only from the lowest excited singlet state and not from any higher excited state because internal conversions (transitions producing heat) are rapid and take place within a few ps, while fluorescence emission occurs in about a ns (1 ns = 10^{-9} s). Also, small rapid internal conversions (energy losses) occur between sublevels of the lowest excited singlet state. For these reasons, fluorescence is seen at longer wavelengths than absorption. The photosynthesis sensitized by the absorption of high-energy blue-light quanta is no more of an advantage to the plant than is the absorption of lower energy red-light quanta because the higher energy state produced, e.g., by absorption of blue light, is degraded rapidly to the lower energy state before the slower photochemical reactions have a chance to take place.

12.3.2. Energy Transfer[49–52]

Before a quantum that is absorbed by a Chl molecule can be trapped by a reaction center, it can be transferred from one molecule to another, again and again, much like one billiard ball hitting another until it eventually falls into a pocket. A photon that is absorbed by any of the 400 or so Chl molecules that compose the "antenna" molecules for the photochemical trapping center is rapidly transferred from one Chl molecule to another, and is eventually trapped by a reaction center; the whole process takes place within 10^{-9} s.

Energy transfer can be demonstrated experimentally in a green alga such as *Chlorella* by illuminating it with a wavelength of light (i.e., 650 nm) that is absorbed in large measure by Chl b, and by measuring the fluorescence of Chl a. The overlap between the absorption of red bands of Chl b and Chl a is somewhat like the diagrammatic representation given in Fig. 12-3. The absorption maximum of Chl b occurs *in vivo* at a shorter wavelength (650 nm) than the absorption maximum of Chl a (678 nm). In the above experiment, only the characteristic fluorescence of Chl a is seen; no fluorescence can be measured that corresponds to that of Chl b, even though this substance is strongly fluorescent in solution. In fact, studies done *in vivo* with *Chlorella* demonstrate that the transfer of energy from Chl b to Chl a is nearly 100% efficient. Similarly, the phycobilins can transfer their energy to Chl with efficiencies ranging from 80 to 90%. By contrast, the efficiency of energy transfer from carotenoids to Chl range from 20 to 50%, depending on the particular carotenoid involved.

Förster[53,54] provided one of the theoretical foundations for understanding the mechanics of energy transfer between molecules. He suggested that substantial overlap is needed between the fluorescence of the donor molecule and the absorption of the acceptor molecule (Fig. 12-3), and that the transfer rate is proportional to the reciprocal of the sixth power of the distance between the two molecules, so that the rate of transfer has an extremely great dependence on the separation between molecules. The rate of transfer also

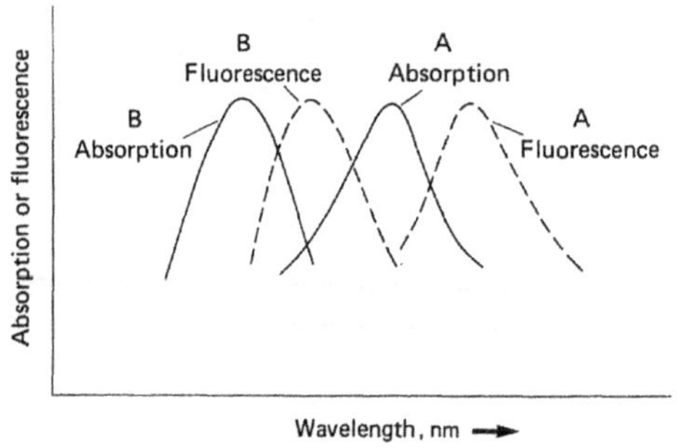

Fig. 12-3. Diagrammatic representation of pigments having absorption and fluorescence bands close together as is the case, e.g., with Chl b and Chl a in green algae and higher plants. Since the fluorescence of B overlaps the absorption of A, there is a good probability that a photon absorbed by B will be transferred to A. This can be demonstrated by exciting B and observing only the fluorescence of A (Section 12.3.2).

depends on the mutual orientation of the molecules (55,56, for reviews see 57,58). For a probability of 50% for energy transfer to occur from one molecule to another, the calculated distance between molecules should be of the order of 5 nm. In chloroplasts the distances are much smaller than this, so that the probability of transfer is very high.

12.3.3. The Reaction Center

The photochemical reactions that convert the energy of an absorbed photon into a stabilized chemical potential take place in specialized Chl proteins in the photosynthetic apparatus called reaction centers. All reaction centers are embedded within the membrane matrix, and receive photons directly by absorption or indirectly from other pigments that act as an antenna to absorb and transfer their absorbed energy to the reaction centers. An electron in Chl, elevated from the ground level to a higher state after absorption of a photon, can be received by a suitable acceptor. The transfer of an electron to an acceptor causes the latter to become reduced and Chl to become oxidized. Chl can be reduced when it, in turn, accepts an electron from a suitable donor in an energetically downhill process. If the acceptor molecule is at a lower electrochemical potential than the excited state of Chl, this transfer will also be with the thermodynamic gradient (downhill), and proceed spontaneously.

There are essentially four kinds of reaction centers: those found in the purple and green bacteria (Section 12.4) and those of PSI and PSII in higher plants, algae, and cyanobacteria (Sections 12.5.1.1, 12.6.2.1). A common feature of all the reaction centers is the occurrence of a Chl (or BChl) that acts as the primary electron donor and is associated with a specific protein. The reaction centers are termed P870 and P840 for purple and green bacteria, respectively, or P680 for PSII and P700 for PSI in plants, algae, and cyanobacteria, on the basis of the wavelength of their maximum absorption in the near-infrared region (e.g., P680 has its maximum absorption at 680 nm). In most purple and green bacteria a dimer of two molecules of BChl a acts as the reaction center for P870 and P840, respectively. By analogy to the bacterial reaction center, the P680 in PSII of plants, algae, and cyanobacteria may be a dimer of Chl a. However, P700 of PSI is apparently a Chl a monomer, at least in its oxidized form.

Evidence suggests that either Chl (BChl) or Ph (or BPh) acts as the first transient electron acceptor for the initial charge-separation event that occurs in the reaction center within about 1 ps. The electron is then transferred within a few hundred picoseconds to a secondary acceptor that acts to stabilize the separated charge and prevent it from recombining again with the primary donor.

Perhaps the best understood reaction center is that found in the purple bacterium *Rhodopseudomonas viridis*. Crystallized reaction centers that are photochemically active have been prepared from this bacterium (Section 12.4).

12.3.4. Delayed Fluorescence[59–62]

If a plant is transferred to the dark after a period of illumination, it can continue to emit a faint red glow. This glow is only detectable with sensitive light-sensing equipment, and was discovered unexpectedly by William Arnold and Bernard Strehler in 1951 in an experiment intended to demonstrate the light-induced formation of ATP during photosynthesis in algae using an extraction of firefly tails that glows when ATP is present. They found that these algae emitted a glow even when no firefly extract was added. It appears that all photosynthetic organisms emit this "afterglow" or luminescence that has been termed "delayed fluorescence" (or delayed light emission) to distinguish it from "prompt fluorescence," which stops within about 10^{-9} s after the exciting light is turned off.

The two types of emission have the same spectral distribution, indicating that they both originate from Chl (or BChl). Delayed fluorescence appears to be produced by the back-reaction of primary (or secondary) electron acceptors with the oxidized Chl that is formed by light at the reaction centers. This back-reduction of Chl apparently returns it to its singlet excited state, from which light can be emitted. Moreover, delayed fluorescence seems to be a phenomenon that is associated predominantly with the reaction center of PSII (as is prompt fluorescence). This is the case because delayed fluorescence is still seen in algal mutants that lack the PSI reaction center P700 (see Section 12.5.1.1 for a discussion of reaction centers and the two photochemical systems). Moreover, delayed fluorescence in dark-adapted plants shows oscillations like those seen for O_2 evolution (described in Section 12.5.2.7) when a sequence of flashes is provided.

12.3.5. The Photosynthetic Unit

Plants contain large amounts of pigments, but only a small fraction of these pigments function in the crucial photochemical steps of photosynthesis. Depending on the particular photosynthetic organism, there can be from 50 to more than 1000 Chl (or BChl) and other light-gathering pigments that serve as an antenna to funnel excitation energy into one reaction center. This concept has been known for over 40 years from the experiments of Emerson and Arnold,[63,64] who measured O_2 evolution from the green alga *Chlorella* exposed to brief but strong and repetitive flashes of light. They used very short flashes of about 10μs (1 μs = 10^{-6} s) of red light that were absorbed by Chl, and were able to excite every photochemical reaction center only once. They measured the amount of O_2 evolved per flash as a function of the dark time between flashes. If the intervals between flashes were sufficiently long, there was enough time for the dark reactions to process the products made during the light reactions. If the flashes were spaced too quickly, however, the dark reactions were unable to keep pace, and photosynthesis, as measured by O_2 evolution, slowed down. The dark time needed for the enzymes to react with the photoproducts was found to be about 0.4 s at 1°C, and 0.02 s at 25°C.

When the intensity of the flashes was sufficiently high so that every Chl molecule was excited at least once, then one O_2 was evolved for every 2400 Chl excited. This result was a surprise, since the authors had expected that under optimum conditions (optimum dark reaction, and all centers turning over) there would have been one O_2 for every Chl present. Thus, the concept of the "photosynthetic unit" was born. The assemblage of 2400 Chl per reaction center enzyme was termed the photosynthetic unit (2400 Chl/O_2). In later experiments it was found that the minimum quantum requirements (number of quanta needed to fix one CO_2 or release one O_2) was around 8. It now appears that 2400/8 or 300 Chl molecules cooperate to "process" one photon. This collection of 300 Chl molecules with its assemblage of dark reactions may now be termed a photosynthetic unit (300 Chl per reaction center).

In green plants one commonly sees photosynthetic unit sizes of 200–400 molecules of antenna Chl for each reaction center, but the antenna for shade plants is often larger. Photosynthetic bacteria have about 50–100 BChl depending on the growing conditions, whereas green photosynthetic bacteria that can grow rapidly in dim light can have about 1000 molecules of BChl associated with each reaction center. In green plants that have two photosystems, PSI and PSII (Section 12.5), it appears that each of these photosystems has its own particular set of antenna pigments in addition to a larger antenna that can serve, or partially serve, both reaction centers.

Calculations show that, even in full sunlight, each Chl molecule in a leaf absorbs only a few photons per second. The experiments of Emerson and Arnold(63,64) show how well plants have adapted to this situation by employing a set of collector pigments to capture and funnel energy to a single specialized Chl that does the photochemical conversion. With this antenna arrangement to concentrate light energy, the dark reactions are thus not limited by the rate at which photochemical conversions take place.

12.4. ANOXYGENIC PHOTOSYNTHESIS OF GREEN AND PURPLE BACTERIA(65)

The purple and green bacteria have only one photosystem, and are unable to use water as an electron donor and produce oxygen (anoxygenic photosynthesis) as do the cyanobacteria, algae, and higher plants (Section 12.5). These bacteria use substances of lower redox potential than water, such as H_2S, H_2, or simple organic compounds. These organisms contain BChl in addition to other photosynthetic pigments. The pigment content of the photosynthetic bacteria can be regulated by the light intensity available during growth. In low light these bacteria elaborate vast amounts of antenna pigments to better harvest the few available quanta. In the dark, all of the photosynthetic bacteria can oxidize stored polysaccharides to obtain energy. The two major groups of photosynthetic bacteria, the purple and green bacteria, are each subdivided into two subgroups—the purple sulfur and purple nonsulfur bacteria, and the green sulfur and green gliding bacteria.

12.4.1. Purple Bacteria

Some of the purple bacteria possess flagellar motility which, combined with phototactic and chemotactic responses, allows them to optimize their growth conditions. The purple bacteria contain BChl a and b in addition to a large variety of carotenoids such as spirilloxanthin, lycopene, and spheroidene, which gives them a wide range of colors such as yellow, brown, red, and purple.(66)

The photosynthetic reactions in the purple bacteria occur within intracytoplasmic membranes that are folded in and are continuous with the cytoplasmic membranes. Disruption of these bacteria causes these folded regions to break up and form vesicles (about 30–100 nm in diameter) termed chromatophores.

The BChl a of purple bacteria has its main absorption band in the near-infrared *in vivo* between 800 and 950 nm (Fig. 12-2A). Some purple bacteria contain BChl b, which absorbs at somewhat longer wavelengths (800–1020 nm). All of the photosynthetic pigments of bacteria (as well as those of algae and higher plants) are bound to specific proteins that are embedded within the lipid bilayers of the photosynthetic membrane.

Most of the BChl is associated with proteins that act as the light-harvesting antenna to absorb and pass excitation to the small number of pigment molecules that serve as the reaction center where primary photochemical reactions take place. In purple bacteria, the cyclic flow of electrons is linked to proton movement that is coupled to ATP formation.

The enzymes of the Calvin–Benson–Bassham carbon reduction cycle or reductive pentose phosphate cycle (Section 12.8.1) have been found in nearly all of the photosynthetic bacteria. These bacteria are capable of fixing CO_2 when they contain ATP and NADH that has been obtained directly or indirectly from photosynthetic electron transport.

12.4.1.1. Purple Sulfur Bacteria (Chromatiaceae)

These bacteria, as the name implies, oxidize sulfide or elemental sulfur to sulfate and fix CO_2 under anaerobic, obligately phototrophic (light-requiring) conditions. Molecular hydrogen can also serve as an electron donor for photosynthetic CO_2 fixation. Some of these bacteria can grow in the absence of sulfide under anaerobic reducing conditions if simple organic compounds are available. Some examples of purple sulfur bacteria are *Chromatium, Thiocapsa, Thiocystis,* and species containing gas vacuoles such as *Lamprocystis* and *Thiodictyon.*

12.4.1.2. Purple Nonsulfur Bacteria (Rhodospirillaceae)

These bacteria are unable to oxidize elemental sulfur to sulfate for phototropic CO_2 assimilation, but instead can use sulfide or thiosulfate, which can be oxidized to sulfate without the intermediate production of sulfur. The purple nonsulfur bacteria are capable of photoassimilating a large number of different organic compounds such as succinate, malate, acetate, pyruvate, and dicarboxylic acids. Many of these organisms can also grow phototrophically using H_2 as their electron donor. Examples of purple nonsulfur bacteria include *Rhodospirillum rubrum, Rs. molischianum, Rhodopseudomonas spheroides, Rp. capsulata,* and *Rp. viridis.*

The great advantage of using photosynthetic bacteria for studies of photosynthesis is that these organisms have only one photosystem, and purified reaction centers devoid of secondary electron carriers and antenna pigments can be prepared that can perform the initial photochemical charge separation reactions of photosynthesis with high efficiency.

With the development of spectroscopic techniques having subpicosecond resolution, it has become possible to elucidate the initial photochemical events occurring within the reaction center of photosynthetic bacteria. Upon absorption of a photon of light in the reaction center of a purple bacterium such as *Rp. sphaeroides,* there is a decrease of BChl a absorbance (bleaching) that is maximum near 870 nm. This bleaching results from the initial charge transfer produced in the bacterial reaction center by the initial primary photochemical event, which produces $P870^+$ BPh^- within a few picoseconds, where $P870^+$ represents the cationic free radical and BPh^- is the anionic free radical of Bph. The production of BPh^- can also be detected spectrophotometrically. Whether BChl reduction precedes BPh reduction is an open question. P870 apparently exists as a dimer $(BChl)_2$ of BChl (or "special pair"). The electron moves next, within several hundred picoseconds, to a quinone molecule (menaquinone) Q_A. All of these reactions are almost

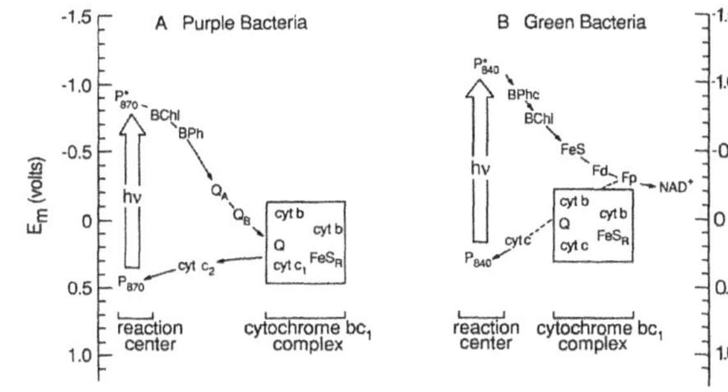

Fig. 12-4. Electron transport pathways in the purple (e.g., *Rhodopseudomonas spheroides*, A) and green (e.g., *Chlorobium limicola*, B) photosynthetic bacteria arranged to show the redox potentials of the reacting components. The upward arrows represent the raising of an electron against the thermodynamic gradient to a higher energy level upon absorption of a photon of light by the reaction center. P870, P840 reaction center bacteriochlorophylls; BChl, bacteriochlorophyll; BPh, bacteriopheophytin; Q, quinone; cyt, cytochrome; FeS_R, Rieske iron-sulfur protein; FeS, bound iron-sulfur centers; Fd, ferredoxin; Fp, flavoprotein or ferredoxin-NADP reductase. (Courtesy of R. E. Blankenship; see Ref. 78.)

temperature-independent and proceed with high efficiency, indicating that all the components are held within close proximity to each other within the confines of the reaction center.

After the initial reactions, the electron is transferred at a much slower rate (~100 μs) to a second quinone acceptor (ubiquinone) Q_B, which becomes reduced to its semiquinone form (Q_B^-). The absorption of a second photon and electron transfer sequence reduces Q_B^- to the fully reduced Q_B^{2-} form. In bacteria, protons derived from inside the cell then convert Q_B^{2-} to dihydroquinone (Q_BH_2). Electrons are subsequently removed from Q_BH_2 and used to reduce $P870^+$ again by a sequence of reactions that are complex and involve b- and c-type cytochromes, a ubiquinone, and an Fe-S protein (Fig. 12-4A). Only the electrons from Q_BH_2 are used to reduce $P870^+$; the H^+ remaining is released at the exterior of the cell. Thus, the cyclic flow of electrons around P870 results in an increased alkalinity inside the cell (as well as increased negative charge). Proton movement back into the cell occurs via a coupled ATPase system that generates ATP. An analogous situation occurs in plants, except that protons are pumped *inside* the thylakoid as a result of the light reactions and pass to the exterior through the ATP-generating coupling factor (Section 12.6.2). In the purple bacteria, the reduced NADH needed for CO_2 reduction is produced by various pathways from the energy available in ATP or from the electrochemical gradient.

Some species of purple bacteria, e.g., *Rp. viridis*, use BChl b molecules as their primary donor (termed P960). A great step forward was recently made by H. Michel and coworkers who prepared crystallized reaction centers from *Rp. viridis*[67] and obtained X-ray analysis[68] of single crystals at a resolution of 3 Å (1 Å = 10^{-8} cm). These studies revealed the locations of four BChl b, two BPh b, one nonheme iron, one menaquinone, and four heme groups (cytochromes); a second quinone and ubiquinone being partially lost during isolation.[69] Although the reaction center BChl of *Rp. viridis* is BChl b instead of the usual BChl a, it is likely that the structural data of Michel and coworkers has applicability to other bacteria. It may even have applicability to PSII of plants.[70] The sequence of events thought to occur[68,69] in preparations of crystalline reaction centers[67] of *Rp. viridis* is diagrammed in Fig. 12-5. Here the primary electron donor P960, a BChl b dimer (BChl b_2), is suggested to pass an electron to BChl b to form BChl b^- in 1

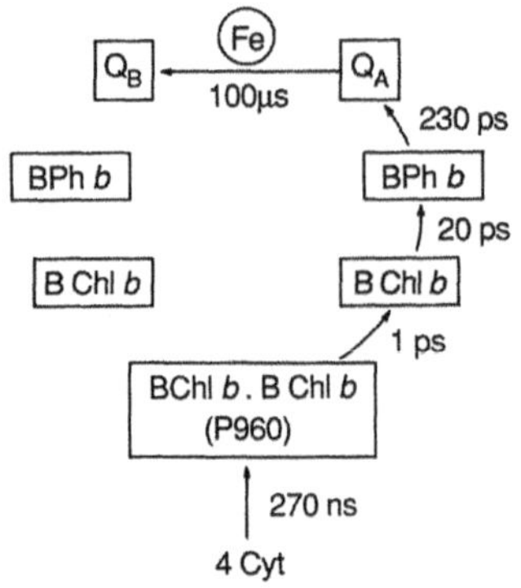

Fig. 12-5. Electron transfer pathway and transfer times for the components of crystallized reaction centers from the purple nonsulfur bacterium *Rhodopseudomonas viridis*. The location of the components is thought to reflect their relative positions in the reaction center. The reaction center containing the dimer of BChl b is indicated as P960. BChl b, bacteriochlorophyll b; BPh b, bacteriopheophytin b; Fe, circled, represents a nonheme iron center; Q_A, menaquinone; Q_B, ubiquinone. Whether BChl b is an intermediate between P960 and BPh b remains to be established.[71] (Modified after Ref. 69 by permission.)

ps, and to BPh to form BPh^- within 20 ps. The involvement of BChl b in this reaction has been questioned, since some experiments show no evidence for the formation of BChl b^- prior to the formation of BPh^-.[(71)] Q_A reduction follows within 230 ps. Spectroscopic evidence suggests that only the pathway proceeding to the right, as shown by arrows in Fig. 12-5, occurs in the reaction center. A strongly bound quinone (Q_A) is located only within the "right" branch of the reaction center structure. A more loosely bound quinone, Q_B, seems to be located on the "left" side of the structure. X-ray analysis showed that the closest distance between Q_A and Q_B is 14 Å. The inorganic iron is located midway between Q_A and Q_B. The function of the BChl and BPh in the left side of the reaction center is not clear. The oxidized special pair of P960 are reduced by another electron-transporting chain containing four heme groups of c-type cytochromes. The closest heme is 21 Å away from the reaction center, and may account for the "slow" transfer time from Cyt to P960 (270 ns) compared to the transfer time of a few ps from the special pair P960 to BPh.

The reaction center of *Rp. viridis* is composed of four different protein components, one being a c-type cytochrome mentioned previously (molecular mass = 38 kDa) and the other protein subunits designated as L (light), M (medium), and H (heavy), whose molecular masses using SDS-PAGE (sodium dodecyl sulfate polyacrylamide gel electrophoresis) are 24, 28, and 38 kDa, respectively. X-ray analysis has revealed relationships between the components of the reaction centers and these protein subunits.[(69,70)] L and M are intertwined within the chromophore region, whereas H is on the opposite, cytoplasmic side to that of the cytochromes.

12.4.2. Green Bacteria

Except for the gliding green bacteria (Section 12.4.2.2), the green bacteria are nonmotile. In green bacteria the photosynthetic pigments, as well as most of the carotenoids, are located in extrinsic pigment-protein structures termed chlorosomes[(72–74)] that are attached to the surface of the cytoplasmic membrane (Fig. 12-6). Green bacteria contain BChl c, d, and e (sometimes called *Chlorobium* chlorophylls) in addition to BChl a. As with the purple bacteria, the green bacteria regulate the amount of BChl and the numbers of chlorosomes on the basis of the light intensity. The reaction centers located in the cytoplasmic membranes contain specialized molecules of BChl, which act as electron

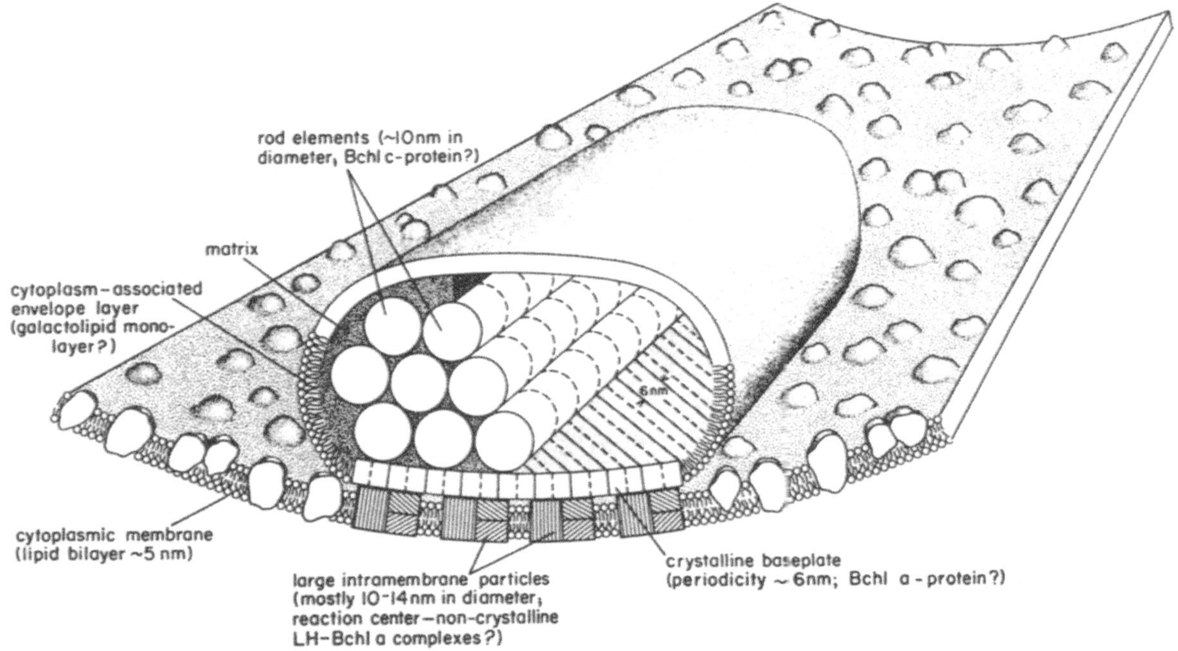

Fig. 12-6. A model of the chlorosome and the associated cytoplasmic membrane of a green photosynthetic bacterium (*Chlorobium limicola*). (From Ref. 73, reprinted by permission.)

donors when they receive energy transferred to them by the antenna carotenoids and by BChl c, d, and e.

12.4.2.1. Green Sulfur Bacteria (Chlorobiaceae)

These nonmotile bacteria are strictly anaerobic and need both H_2S and light for their growth. Most green sulfur bacteria can use molecular hydrogen while others can use thiosulfate as an electron donor for CO_2 reduction. Some members of this group can also use organic compounds such as acetate, butyrate, or pyruvate as electron donors. Unlike the purple bacteria, the green bacteria seem to use a reductive carboxylic acid cycle instead of the Calvin–Benson–Bassham cycle (Section 12.8.1) as the main method of assimilating CO_2.(75) In *Chlorobium limicola* and *Prosthecochloris aestuarii*, commonly studied representatives of this group, the antenna size is around 1500 BChl molecules per reaction center(76,77) instead of 50–100 as found in purple bacteria. This large antenna may have hampered the isolation of pure reaction center preparations from green sulfur bacteria.

The primary electron donor, P840, of the green sulfur bacteria is BChl a which, as in purple bacteria, may also be a dimer. Photooxidation of this pigment produces a maximum decrease of absorbance at 840 nm. There is some evidence from ps-spectroscopy that the primary electron acceptor is BChl a. BPh c may serve as the next acceptor and it, in turn, appears to be oxidized by an Fe-S protein having a midpoint potential of about 550 mV(78) (Fig. 12-4B). The green bacteria have been found to contain another Fe-S protein functioning in electron transport that has a redox potential near 420 mV. There is a very large variety of b- and c-type cytochromes, both membrane-bound and soluble, in the green sulfur bacteria, and their participation in cyclic electron flow is not completely understood. The green sulfur bacteria can reduce NAD^+ directly, analagous to the reduction of $NADP^+$ by PSI (Section 12.5.1.3). A comparison of Fig. 12-4 with Fig. 12-7 will show that there are more similarities between the electron transport in green bacteria and PSI than between these bacteria and the purple bacteria.

So far it has not been possible to observe light-induced ATP formation in the green bacteria. However, a membrane-bound ATPase has been found(79) in *Chlorobium*, which suggests that these bacteria may be able to use cyclic electron flow to generate ATP as in the purple bacteria.

12.4.2.2. Green Gliding Bacteria (Chloroflexaceae)

The filamentous green bacterium *Chloroflexus aurantiacus* was discovered only in 1977,(80) it inhabits hot springs around the world where it lives at its optimum temperature ranging between 50 and 60°C. Unlike the green sulfur bacteria, the green bacterium *Chloroflexus* exhibits gliding motility much like that seen in some species of blue-green algae (*Oscillatoria*) and in the flexibacteria.

Under anaerobic conditions, this bacterium grows best photosynthetically when organic substrates such as acetate, lactate, pyruvate, glycerol, or glucose are available to serve as electron donors. It may also be able to use H_2S. Under aerobic conditions, the synthesis of BChl stops and the organism reverts to the respiratory metabolism of simple organic substrates. It is unable to grow in the dark under anaerobic conditions. *Chloroflexus* seems to have a photosynthetic electron transport much like that found in purple

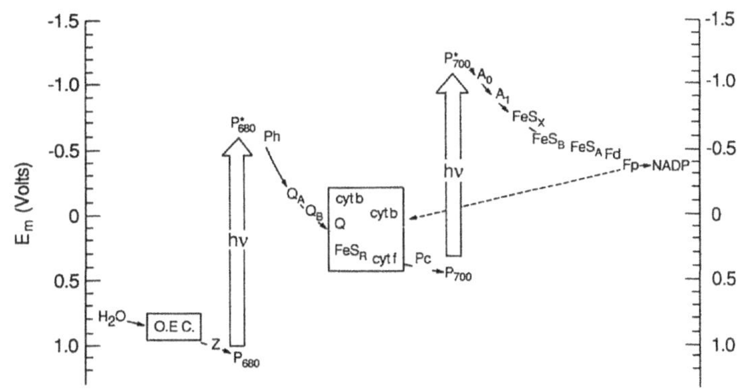

Fig. 12-7. The light reactions and electron transport pathways for oxygenic photosynthesis (the so-called Z scheme) showing excited state redox potentials. The upward arrows represent the light reactions taking place in the reaction centers. OEC, oxygen-evolving complex; Z, donor to PSII; P_{680}, reaction center chlorophyll of PSII; Ph, pheophytin acceptor of PSII; Q_A, Q_B, quinone compounds; cyt, cytochrome; FeS_R, Rieske iron–sulfur proteins; PC, plastocyanin; P_{700}, reaction center Chl of PSI; A_O (Chl a?), A_1 (quinonelike compound?), early acceptors of PSI (quinone-chl compounds?); FeS_X, FeS_B, FeS_A, bound iron sulfur protein acceptors of PSI where FeS_B and FeS_A may operate in parallel; Fd, ferredoxin, Fp, flavoprotein (ferredoxin-NADP reductase); cyclic electron flow around PSI is indicated by the dashed line. A description of the pathways of electron flow in the cyt b_6/f complex (box) are given in Ref. 170. (Adapted from Ref. 82 by permission.)

bacteria. It contains BChl c in addition to BChl a, and has its antenna pigments within chlorosomes. Its primary electron donor, a form of BChl, designated as P870, apparently exists as a dimer as in other photosynthetic bacteria. Excitation of P870 reduces BPh a. Two distinct species of menadione are reduced in sequence. Cytochromes of the c-type link electron flow back to P870 to complete the cyclic electron flow.[78,81]

12.5. OXYGENIC PHOTOSYNTHESIS OF CYANOBACTERIA, ALGAE, AND HIGHER PLANTS

It is possible that the joining together of an electron transport chain of the type found in the green sulfur photosynthetic bacteria (which resembles photosystem I; Section 12.5.1) with the type found in the purple photosynthetic bacteria (which has similarities to PSII; Section 12.5.2) by means of an iron–sulfur protein complex that contained quinones and cytochromes (Section 12.5.1.4) produced organisms that could ultimately use water as their oxidant to reduce CO_2 to the level of carbohydrate (Fig. 12-7). After further evolution of the redox chemistry of the reaction center and of the water-oxidizing system, such a union would provide the newly evolved organism with two light reactions that could operate in a concerted way to elevate an electron from +0.8 V (the potential of the H_2O/O_2 couple) to −0.4 V (the potential of (CO_2/glucose) for a total span of 1.2 V.[82]

12.5.1. Photosystem I[69]

As mentioned previously, this photosystem has homologies to the green sulfur photosynthetic bacteria. Photosystem I (PSI) is able to elevate an electron to a sufficiently low

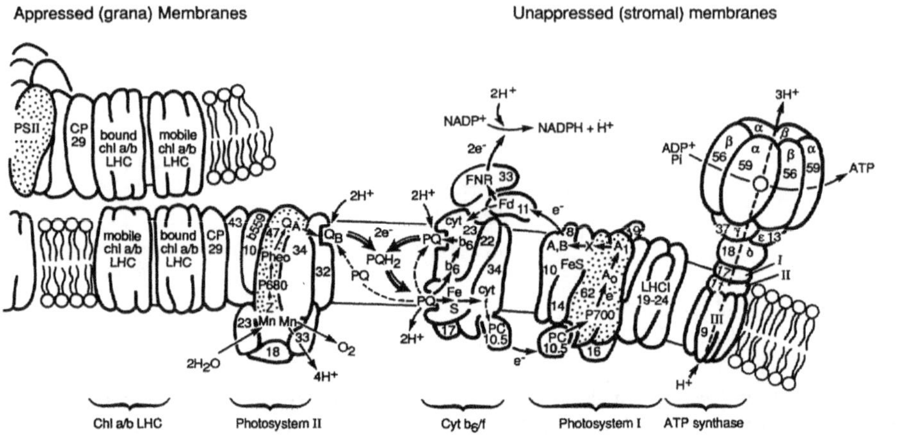

Fig. 12-8. The physical organization of PSI and PSII and their associated antenna Chl, electron transport components, and the ATP-synthase (CF_1-CF_0) as they may be held in the thylakoid membrane of a green plant. FNR, ferredoxin-$NADP^+$ reductase; LHC, light-harvesting complex; PQ, plastoquinone. The numbers refer to the masses (kDa) of the proteins indicated. Other abbreviations used are given in Fig. 12-7. (Adapted by permission from T. G. Dunahay and L. A. Staehelin.)

potential (~1.0 V) so that the reduced primary electron acceptors reduce iron-sulfur proteins and ferredoxin which, in turn, reduce $NADP^+$ to NADPH.

The reaction center "core" of PSI is a hydrophobic, membrane-bound pigment protein complex composed of several polypeptides. Two of these polypeptides, having molecular masses from 62 to 70 kDa (Fig. 12-8), are thought to contain P700 and the acceptor compounds (for a review, see Refs. 83 and 84). A light-harvesting complex, termed LHC-I,[85–87] contains about 40% of the total Chl of PSI. The LHC-I complex contains some Chl b but much less than LHC-II (Section 12.7.2), which has a Chl a/b ratio of about 3.7 vs. 1.2 for LHC-I. The LHC-II complex contains subunits having molecular masses of 21 and 23.5 kDa, respectively. Carotenoids are present in both the core complex of PSI as well as in LHC-I.

12.5.1.1. P700[83]

The primary electron donor of PSI, P700, is a Chl molecule situated within a specialized protein complex that comprises the reaction center. It was designated P700 because upon its oxidation by light it exhibits a maximum bleaching (absorbance decrease) in the light around 700 nm.[88] Detailed flash-induced difference spectra from 250 to 850 nm have been measured[89] that for spinach show negative peaks at 430 and 682 nm in addition to the 700-nm peak. The differential absorption coefficient of P700 at 700 nm is 64 mM^{-1}. P700 has a reasonably low redox potential (+0.48 V), and is therefore rather easy to oxidize chemically (e.g., with ferricyanide) as well as by light. A correlation has been seen between oxidized P700 signals measured optically with those measured by light-induced ESR (electron spin resonance) spectroscopy. The ESR signal corresponds to a free radical centered at G = 2.0025, has a peak-to-peak width of about 7.5 gauss,[90] and can be seen at low temperature (e.g., 77°K).

Upon absorbing a quantum of light, P700 reduces an acceptor having a redox potential of about −1.0 V (Fig. 12-7). It is uncertain as to whether P700 exists in the reaction center as a monomer or a dimer, but most evidence suggests that in its oxidized state it is in the monomeric form.

12.5.1.2. Primary Acceptor of Photosystem I

A flash-induced absorbance decrease at 430 nm, termed P430, was attributed to the reduction of the primary electron acceptor of PSI.[91] This substance may be a semiquinone molecule. It appears that an even earlier acceptor preceeds P430 and, judging from its spectrum, it appears to be a Chl anion. The reduction of the Chl anion takes place at cryogenic temperatures in less than 10 ps,[92–95] and P430 becomes reduced within about 200 ps. Subsequent to the reduction of these primary acceptors, certain iron-sulfur proteins act as tertiary acceptors and become reduced within about 200 ns.

The primary reactions of PSI share homologies with primary reactions of PSII as well as with the green and purple photosynthetic bacteria. A common feature of the reaction center in all these photosystems is the occurrence of Chl like pigments in monomeric or dimeric form acting as the primary donor as well as the primary acceptor. Apparently only Chl a, BChl a, and in a few cases BChl b, function in this capacity as primary reaction center pigments.

12.5.1.3. Ferredoxin and Ferredoxin-NADP$^+$ Reductase

Ferredoxin is a type of nonheme iron-sulfur protein that has been shown to occur in all organisms investigated so far, from obligate anaerobic bacteria to plants and animals. These proteins serve as electron transfer agents for a number of different types of reactions.[96]

Plant ferredoxin has a low molecular mass of 12 kDa and consists of a single polypeptide chain containing a high percentage of acidic amino acids. There are many similarities in the amino acid sequences of plant and bacterial ferredoxins, suggesting that higher plants and photosynthetic bacteria had a common ancestor in the anaerobic bacteria, such as the present day *Clostridium*.

Ferredoxin is reduced during electron transfer in PSI and it in turn can reduce cytochrome in the Cyt b_6/f complex (see below) or participate in the reduction of NADP$^+$ to NADPH (Fig. 12-7). This reaction is the last step of photosynthetic electron transport, and it requires reduced ferredoxin and a flavoprotein enzyme termed ferredoxin-NADP$^+$ reductase (termed FNR or Fp). This flavin enzyme has been isolated from spinach and crystallized.

12.5.1.4. The Cytochrome b/c Complex[97,98]

A protein complex containing both b- and c-type cytochromes and an FeS compound is present in cyanobacteria and plants as well as in purple bacteria and, perhaps, in green sulfur bacteria as well. In oxygenic (O_2-evolving) photosynthesis, this complex mediates electron transport between PSII and PSI, and in purple bacteria the complex functions in cyclic electron transport. This complex is similar to the well-known complex III, an ubiquinol-cyt c oxidoreductase that functions in mitochondrial electron transport.[99]

The cyt b/c complex functioning in photosynthesis contains two molecules of cyt b_6 and one cyt c (which, in the case of green plants, is replaced by cyt f) and one FeS protein (called the Rieske FeS protein after its discoverer). Considerable work has been done on the protein composition as well as the source for the encoding of the subunits of this complex.[101,102] The cyt b_6/f complex also functions as a link in cyclic electron flow around PSI. The electrons produced by the photoreaction of PSI, if not used to reduce $NADP^+$, can be passed to this complex (Fig. 12-7). This cyclic electron flow is coupled to proton translocation and can be used as a means of generating ATP when the reduction of $NADP^+$ has ceased.

The cyt b_6 in this membrane-spanning protein seems to be located at the site nearer to the "outside" of the membrane, while the FeS center and cyt f are located more toward the inner side of the thylakoid membrane. The function of this protein is to oxidize the plastoquinol (PQH_2) reduced by PSII and reduced plastocyanin (Fig. 12-8). The mechanism of the plastoquinol oxidation is currently under investigation, but a so-called Q-cycle mechanism[103] seems to be a likely candidate.[104–106] A Q cycle brings about a sequential two-step oxidation of plastoquinol by the cyt b_6/f–FeS complex, giving rise to the translocation of two protons across the membrane for every electron transferred to P700. So far Q cycles have been more clearly shown to occur in photosynthetic bacteria and mitochondria than in plants.

12.5.2. Photosystem II

The function of photosystem II (PSII) is to generate sufficient oxidizing power so that water can be oxidized and plastoquinone reduced. This oxidation of water and the concomitant evolution of O_2 takes place in distinct protein domains that are tightly coupled to other separate protein domains where the primary photochemistry of PSII occurs (Fig. 12-8).

Considerable progress has been made recently on the protein composition of the PSII complex.[107,108] The central core of PSII contains a 47-kDa pigment–protein complex that spans the membrane and apparently contains the photochemical machinery of PSII. This protein holds an array of light-harvesting Chl a molecules and is suggested to contain P680, the primary electron donor of PSII. This protein is also suggested to contain the primary electron acceptors of PSII. In addition, the 47-kDa protein may also hold the secondary electron plastoquinone acceptor Q_A (which is apparently complexed with iron) and the secondary electron donor termed Z (which appears to be a species of plastohydroquinol) (Fig. 12-7). The 47-kDa protein has a fluorescence band at 695 nm at 77°K, and shows a low temperature thermoluminescence band (the latter originating by a reversal of charge separation).

The membrane-spanning 47-kDa protein has three hydrophilic proteins tightly associated with it on the inner side of the thylakoid lumen, whose molecular masses are 33, 23, and 18 kDa, respectively.[107] There is a strong possibility that four Mn ions, required for O_2 evolution, may be held at the interface of the 47- and 34-kDa proteins, and that the water-oxidizing reaction is mediated by the 34-kDa protein, perhaps at this interface. In contrast, there are suggestions that the Mn ions are on the interface of the extrinsic 33-kDa protein and an intrinsic 34-kDa protein (also called the D_2 protein), and the two are essential for O_2 evolution. The 23- and 18-kDa proteins may function in some way to shield the O_2-evolving complex from reductants other than water. The 23-kDa protein may hold Ca^{2+} as well as Cl^- ions, and the 18-kDa protein may bind Cl^- ions. These ions apparently play an essential role in PSII, but more work is needed before a better understanding of their functioning will be obtained.

Several other proteins are associated with the PSII complex: a 43-kDa protein having about 20 Chl a molecules (the so-called core antenna of PSII, which has a flourescence band at 685 nm at 77°K), and a 29-kDa Chl protein having mostly Chl a and, in addition, a small amount of Chl b. The light-harvesting Chl a/b (LHC-II) protein contains the bulk of the pigments that are used for light absorption by PSII. Figure 12-8 portrays diagrammatically the structure of the O_2-evolving photosynthetic apparatus as it might be arranged in the chloroplast thylakoid membrane. There is at present no general consensus on the composition and identification of many of the polypeptide components. Many recent studies are aimed at understanding the detailed topography of the proteins in thylakoid membranes.[(109)]

Two intrinsic proteins having molecular masses of 32 kDa (lysine-poor) and 34 kDa (lysine-rich), respectively, are thought to be closely associated with the 47-kDa reaction center protein. Except for the hypothesis mentioned above, the function of the 34-kDa protein (perhaps equivalent to the so-called D_2 protein) is not well understood. It may, however, play a role on the oxidizing side of PSII.[(110–112)] The 32-kDa protein (equivalent to the D_1 protein) contains the secondary plastoquinone acceptor of PSII (Q_B) and is known as the herbicide-binding protein (Q_B),[(113)] since it is the site where potent inhibitors of photosynthesis such as DCMU [3-(3,4-dichlorophenyl)-1,1-dimethylurea] and Atrazine bind to this protein and prevent normal electron flow from Q_A to Q_B (Section 12.5.2.3). It appears that this is also the protein where anions, particularly bicarbonate, modulate electron flow.[(114)] PSII also contains two proteins having molecular masses of about 10 kDa, one of which appears to bind cyt b_{559} (Section 12.5.2.6).

12.5.2.1. P680

P680 has a high redox potential that has been estimated to be near 1.1 V[(115)] (Fig. 12-7). For this reason it is difficult to oxidize P680 chemically, as can be done with P700. Detailed light-induced difference spectra have been obtained[(116)] for P680 that show major negative bands at 430 and 674 nm. In the near-infrared, a small, broad, positive peak occurs near 825 nm. An ESR signal attributed to oxidized P680 was seen[(117)] that had a g value of 2.0025 and ΔH of 8G. Although it has not been measured, it is likely that, like the oxidation of P700, the oxidation of P680 takes place within a few picoseconds. It is not certain whether P680 exists as a monomer or a dimer.

12.5.2.2. Pheophytin

Pheophytin a (Ph a) has been considered to be the primary acceptor for the electron provided by the photooxidation of P680.[(118)] Ph a appears to exist as a monomer on the basis of its ESR (g = 2.0333, ΔH = 13G)[(119)] and ENDOR (electron-nuclear double resonance) characteristics.[(120)] Ph a has been estimated to have a redox potential of −0.61 V. The function of Ph a was detected in experiments where Q_A (Section 12.5.2.3) was reduced chemically in the dark by adding dithionite ($Na_2S_2O_4$). Under this condition, the reversible photochemical activity (of Ph a) could still be observed if an electron donor were also provided for P680.

The existence of an acceptor prior to Ph a has been proposed.[121] Wraight[122] pointed out that a species of Chl a may function as the primary acceptor in PSII. If so, there would be a conservation of function in all photosystems from bacteria to higher plants in that only Chl-type molecules (Chl or BChl) serve as the primary photochemical reactants in photosynthesis. This idea awaits proof that BChl serves as the primary acceptor in bacteria.

12.5.2.3. Q_A and Q_B

Q_A is the major electron acceptor between Ph and Q_B, the second quinone electron acceptor of PSII. The redox potential of Q_A is around zero volts. There is evidence for another form of Q_A having a lower redox potential (−0.25 V).[123] Q_A is apparently a protein-bound Fe/quinone complex,[124] since its ESR signal is like that seen in bacterial reaction centers. If Q_A is reduced, fluorescence is high and if it is oxidized, fluorescence is low,[125] i.e., fluorescence has a variable yield depending on the redox state of Q_A. A more recent unproven concept for the variable fluorescence is that it is a rapid delayed fluorescence generated upon charge recombination between reduced Ph and P680.[126]

A light-induced absorbance decrease at 550 nm has been used as a measure of primary reactions in PSII (reduction of Q_A) and is apparently caused by a bandshift of Ph.[127,128] Q_A reduction can also be followed as absorbance increases in the ultraviolet region near 320 nm.[127] Q_A is reduced by Ph (or Chl) within 200–400 ps (for a review, see Ref. 129).

The electron acceptor designated Q_B is a plastoquinone molecule attached to a 32-kDa protein that is folded back and forth across the thylakoid membrane 5 or 7 times. This Q_B protein accepts two electrons from Q_A within 600 μs before it is reduced to Q_BH_2. Further electron flow involves an exchange of the Q_BH_2 molecule from the plastoquinone (PQ) pool (Section 12.5.2.3). Several anions (such as formate, acetate, and nitrite) inhibit this reaction, but bicarbonate ion reverses this inhibition and restores efficient electron flow from Q_A to Q_B and from Q_B to PQ.[114,129] The Q_B protein is known as the herbicide-binding protein since many herbicides such as DCMU and Atrazine bind to this protein and prevent normal electron flow from Q_A to Q_B.

Q_B requires two electrons for its reduction, whereupon it is apparently released, in its protonated form, within 1 ms to the mobile PQ pool (see below).[130] It thus functions as a two-electron gate[131–133] to transfer electrons in pairs to the large pool of plastoquinone molecules. The reduction of Q_A by the reaction center is a single electron transfer process. Q_B is a form of plastoquinone, since it shows a maximum light-induced absorbance increase in the ultraviolet region at 320 nm, characteristic of reduced semiquinone, after the first and third of a series of flashes.[134] Other electron carriers may also function between Ph and quinones.[135]

12.5.2.4. Plastoquinone

Early experiments by Amesz[136] and by Rumberg *et al.*[137] showed that in intact algae there was a large pool of PQ that was reduced by PSII and oxidized by PSI. The size of this PQ pool is about 5–10 molecules per PSII reaction center, each capable of storing two electrons (for a review, see Ref. 138). As noted earlier, doubly reduced Q_B after protonation exchanges with a PQ pool (Fig. 12-8). The proton uptake from outside the thylakoid is too slow (~60 ms) to account for this protonation. Thus, it is believed that the protonation is derived from the H^+ obtained from the neighboring protein.

12.5.2.5. Donors to P680[139]

The nature of the primary electron donor(s) to P680 has not yet been elucidated completely. A donor, termed Z, is apparently located on the 47-kDa reaction center polypeptide. Other possibilities cannot yet be disproven. There is also the possibility that more than one species of Z donates electrons to PSII (for a review, see Ref. 129). Interestingly, Z has been suggested to be a species of plastohydroquinol, whereas Q, the primary acceptor (Section 12.5.2.3) of PSII, is the oxidized form of the same PQ. It has been suggested that a cation quinone radical such as Z^+ would have a redox potential near +0.95 V at physiological conditions, while an anion quinone such as Q would be near 0.0 V.[140] An ESR signal (termed signal II) that has the characteristics of an oxidized semiquinone (perhaps complexed with two protons or a metal ligand to give it a high redox potential) seems to be produced by a donor to P680. The ESR signal II has been designated as signal II_{vf} (very fast), as that seen in intact samples, and signal II_f (fast) as that seen in samples treated by washing with Tris buffer.[141] Z can transfer its electron to photooxidized P680 in about 40 ns.[142]

12.5.2.6. Cytochrome b_{559}[139,143]

Cytochrome b_{559} undergoes low-temperature oxidation mediated by PSII. Cyt b_{559} cannot be the primary donor to PSII; prior to freezing, this cyt can be chemically oxidized, e.g., by the addition of potassium ferricyanide; then at low temperature, the C550 change (Ph reduction; Section 12.5.2.2) can still be seen even though cyt b_{559} is oxidized. Cyt b_{559} may function by mediating cyclic electron flow around PSII, which may act as a protection mechanism for the photochemical system of PSII by preventing the accumulation of the strong oxidizing equivalents that could induce photoinhibition.

12.5.2.7. Oxygen Evolution[139]

There has been progress recently in opening the black box that contains the mystery of how plants can cleave water to release electrons, hydrogen, and O_2 all under physiological conditions. In order to release one molecule of O_2, electrons must be removed from two molecules of water. Since the absorption of one quantum produces one free electron by the reaction center, it is necessary to repeat this process four times before O_2 can be released. It has long been known that a very short and intense flash of light given to plants kept in the dark for a sufficiently long time yields no O_2 evolution. Joliot *et al.*[144] and Kok *et al.*[145] provided the key to understanding the mechanism of O_2 evolution. In their experiments, sensitive polarographic techniques were used to study the O_2 evolved from dark-adapted algae illuminated with successive brief flashes of light strong enough to excite every reaction center. In algae kept in the dark for a long period, the first flash

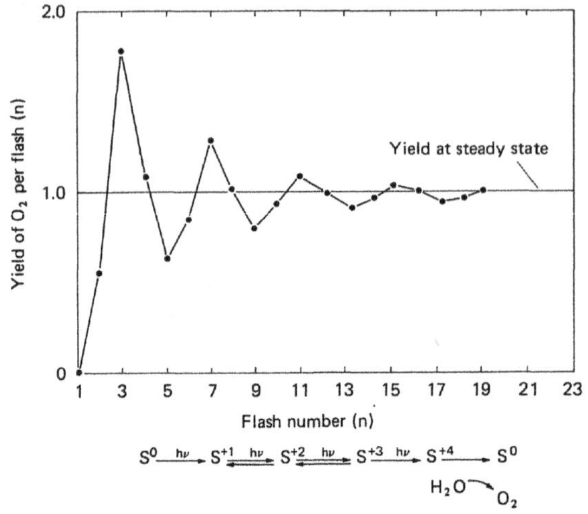

Fig. 12-9. Relative yield of O_2 per flash in dark adapted spinach chloroplasts in relation to the steady-state value. (Adapted from Ref. 144 with permission of P. Joliot.)

yielded no O_2 evolution, and the second flash only a negligible amount. The third flash, however, yielded the highest amount, and the fourth flash somewhat less (Fig. 12-9). This sequence continued with the next two flashes showing low yields of O_2 evolution followed by two flashes having high yields. Interestingly, the peaks of highest O_2 yield occurred after every fourth flash, except for the first peak of O_2 evolution that was seen after the third flash. Kok *et al.*[145] proposed a model by which the oxygen-evolving complex (OEC) can exist in several redox states, the so-called S states; the complex accumulates 4+ charges from the reaction center of PSII. This complex in its fully oxidized state (S^{4+}) oxidizes water releasing a molecule of O_2 and returns the system to the fully discharged state (S^0). The intermediate oxidation states, S^{2+} and S^{3+}, are unstable and decay in the dark to the stable S^{1+} state. Thus, in dark-adapted algae given three flashes, O_2 evolution is maximum, and continued flashes give maximum O_2 evolution with a periodicity of four. This periodicity continues as a damped oscillation with a period of four.

Kok's model[145] provides a useful framework to explain the mechanism of O_2 evolution, and some new results have clarified the nature of the OEC. Part of the difficulty of studying this system is that it is very labile: it is destroyed, for example, by mild heating (a few minutes at 50°C), by washing with Tris buffer, or by UV irradiation.

Mn plays an important role in O_2 evolution.[146–148] Removal of about four Mn atoms/400 Chl from the native site leads to the complete inhibition of O_2 evolution. Mn is not involved in the light reactions of PSII since Mn-deficient samples that can no longer evolve O_2 can still function photochemically if suitable chemical donors are provided.[149,150] This result suggests that Mn plays a role in the enzyme involved in water oxidation.[147]

The protein responsible for some of the binding of Mn in PSII has only recently been suggested[151,152] to be the extrinsic protein having a molecular mass of 33 kDa; it is still possible that a contaminant protein was the nonbinding protein in these experiments. Mn may also be associated with the 34-kDa protein (Fig. 12-8). This Mn protein may represent the enzyme that accumulates the four charges necessary for O_2 evolution. However, the possible role of the intrinsic 34-kDa protein in the O_2 evolution mechanism needs to be elucidated by further studies (see, e.g., Ref. 110). It has also been seen that Ca^{2+} and Cl^- ions must be bound to some sites on the OEC in order to have maximum PSII activity.[107]

12.6. PHOTOSYNTHETIC PHOSPHORYLATION[153,154]

12.6.1. Noncyclic and Cyclic Photophosphorylation

The job of the two light reactions with their coupled electron transport in oxygenic photosynthesis is completed with the formation of NADPH and the production of ATP. These two substances are required in order to drive CO_2 reduction reactions to form carbohydrates. We have seen how electron transport leads, as its end result, to the formation of NADPH, but this flow yields ATP as well. The coupling of ATP formation with the one-way transport of electrons to NADP has been termed "noncyclic" photophosphorylation. It has been shown that under certain experimental conditions ATP is the only product formed in the light when adenosine diphosphate (ADP) and inorganic phosphate (Pi) are added to the reaction mixture containing chloroplasts[155,156] or chromatophores.[157] This process is known as "cyclic photophosphorylation."

The light and dark phases of phosphorylation can be separated.[158] After first illuminating chloroplasts and then transferring them to a solution kept in the dark that contains ADP, Pi, and Mg^{2+}, it is still possible to form some ATP. It is clear from this type of experiment that something is formed in the light, with a lifetime of several seconds, that can lead to the formation of ATP in the dark.

12.6.2. The Chemiosmotic Hypothesis[159–161]

Mitchell[160,161] was honored with a Nobel Prize for his "chemiosmotic" hypothesis whereby the electrochemical potential generated across the photosynthetic membrane as a result of the light reactions and electron transport gives rise to movements of H^+ from the outside to the inside of the membrane. The sum of the pH gradient (ΔpH), resulting from the translocated H^+, and the membrane potential ($\Delta\psi$), resulting from the gradient of electric potential across the membrane, is called the proton motive force (pmf).

The sequence of events according to the chemiosmotic hypothesis begins with the O_2-evolving system located on the inside of the thylakoid, which releases one proton for each electron extracted from water by PSII. The reduction of Q_A by PSII requires the transfer of one e^-. Q_B, which acts as a two electron "gate," transfers $2e^-$ to PQ, and requires the uptake of $2H^+$ from outside the thylakoid. Oxidation of plastoquinone is achieved by removal only of its electrons by the reactions of electron transport. The protons released from plastoquinol preferentially diffuse into the inside space of the thylakoid. ATP formation results when the internal H^+ ions pass out of the membrane through the ATPase enzyme or "coupling factor" (CF_0-CF_1 ATP-synthase; see Fig. 12-8) resulting in the dehydration of ADP and Pi to form ATP.[162–164] Only the CF_0 part of the ATPase enzyme is embedded within the membrane. The major portion (termed CF_1) is extrinsic to the membrane and contains five different subunits.[165]

Boyer[166] and Slater[167] provided a mechanism to explain how the coupling factor may work. Boyer's model, often termed the "energy-linked binding change mechanism," suggests that the coupling enzyme contains three reactive sites whose binding constants for substrates and products change as a result of energy-dependent conformational changes induced by the energy of protons that pass out through the coupling factor.[168] There is evidence for localized domains of H^+ in chloroplasts. Protons involved in coupling are not always in equilibrium with the protons in the surrounding bulk water phase enclosed by the thylakoid.[169–171]

There are many results that favor the Mitchell hypothesis. For example, it is known that in order to have ATP formation it is necessary to have a membrane that maintains a separation between an inner and an outer space so that differences in ion concentration can be achieved. Water movement as a result of such ion transport gives rise to volume changes in chloroplasts that can be detected as large light-scattering changes. Interestingly, illuminated chloroplasts cause the pH of their bathing medium to rise, presumably as a result of H^+ being taken up by the illuminated chloroplasts. This H^+ disappearance is reversible. The pH of the external medium decreases again (H^+ goes out) after the chloroplasts are put in the dark. The separation of the light and dark phases of phosphorylation can be explained by the generation of a high internal H^+ concentration produced by the light reactions. In the dark, these H^+ pass out through the coupling factor giving rise to ATP formation.

Even more interesting, and perhaps providing more support for Mitchell's hypothesis, are experiments[172] showing that the action of light could be substituted for by a dark soaking of chloroplasts in a medium at pH 4, so that the concentration of H^+ inside the chloroplasts was increased. Raising the pH to a value of about 8 in the presence of ADP and Pi also gave ATP formation.

The Mitchell hypothesis explains the experimental observation that electron transport is coupled to phosphorylation in that the rate of electron transport is controlled by the rate at which ATP formation can take place. If phosphorylation is "uncoupled," then electron transport can proceed at a much faster rate. Many different types of substances act as uncouplers. Uncouplers act according to the chemiosmotic hypothesis by permitting the passive movements of H^+ and other ions so that the pH and/or the electrochemical gradients are dissipated. The uncoupling action of ammonia or of an amine such as methylamine, apparently, results from the ability of these compounds to diffuse freely through the membrane and combine with the H^+ on the inside space of the thylakoid to form NH_4^+ or the positive amine ion.

There are a number of antibiotic-like compounds, termed ionophores, that act as ion-translocating substances.[173,174] The antibiotic nigericin allows the exchange of H^+ for K^+ and other alkali metal cations, or vice versa, thus affecting phosphorylation but not diminishing the electric charge across the membrane that results from the unequal ion concentration on the inside and outside spaces. Addition of nigericin and K^+ to chloroplasts allows the H^+ inside the thylakoid to be replaced by K^+. Another antibiotic, gramicidin, and the compound carbonylcyanide *m*-chlorophenylhydrazone (CCCP) apparently make "holes" in the membrane, allowing free diffusion of H^+ in and out, thus abolishing the charge across the membrane and uncoupling phosphorylation. The antibiotic valinomycin causes membranes to be permeable to ions such as K^+ or NH_4^+, but not to H^+. Thus, a membrane potential will either be formed or dissipated depending on the direction of electron flow.

12.7. STRUCTURAL ORGANIZATION OF PHOTOSYNTHESIS[175,176]

12.7.1. The Chloroplast

Except for photosynthetic bacteria and blue-green algae, the higher plants and other algae have their photosynthetic apparatus localized in definite structures termed chloroplasts. Chloroplasts vary widely in shape, depending on the plant species, from the cup-shaped structures in *Chlorella,* to spiral bodies in *Spirogyra,* to star-shaped structures in other algae. Spinach chloroplasts are flattened spheres with a diameter of about 5–7 μm.

A chloroplast is surrounded by a continuous double membrane. The outer membrane is connected to the endoplasmic reticulum. Apparently the inner part of the double membrane is not connected to other membranes of the cell but is involved with the formation of the extensive lamellar system found inside the chloroplasts. This envelope acts as a selective barrier to the passage of metabolites in and out of the chloroplast.

12.7.2. Lamellae, Grana, and the Stroma[177,178]

A characteristic internal feature of higher plant chloroplasts is the occurrence of a complex structural arrangement of lamellae (or thylakoids). Frequently, the lamellae are closely layered one upon the other to form structures termed grana, which look somewhat like stacks of pancakes. Portions of the lamellae may extend out from the grana stacks into the surrounding space, called the stroma, which is composed of a proteinaceous matrix. The light reactions and coupled electron transport are localized in the lamellae and grana structures. The products of the light reactions, ATP and NADPH, diffuse out of these structures into the stroma where the enzymes of CO_2 fixation are located.

Chloroplasts can be prepared with and without their surrounding double membranes. The so-called class I chloroplasts, with intact surrounding membranes, retain their soluble stroma components and can fix CO_2 at high rates. However, the class II chloroplasts, which lack surrounding membranes, lose their soluble components from the stroma and require the readdition of these lost components in order to fix CO_2.

The stroma also contains ribosomes and DNA, which are involved in chloroplast regulation and replication. Starch grains are also commonly seen in the stroma region, as are globules containing the lipophilic chloroplast quinones, such as plastoquinone, vitamin K_1, and both α-tocopherol and α-tocopherylquinones. These globules do not play a direct role in photosynthesis but appear to serve as a storage pool for these substances.

Since grana stacks are found in almost all higher plant chloroplasts, it would seem that they play some significant role in these plants. All the factors that lead to grana stack formation are not understood, and chloroplasts lacking grana show all the photochemical activities thus far measured in chloroplasts containing grana. For example, the blue-green algae carry on complete photosynthesis, including O_2 evolution, yet have no grana stacks, but only have unstacked lamellae extending throughout the chloroplasts. Studies comparing photochemical activities have shown that PSI appears early in the development of chloroplast lamellae. The appearance of PSII activity seems to be correlated with the later formation of grana.

The grana and stroma lamellae of chloroplasts of higher plants, such as spinach, can be separated from each other by mechanically disrupting the chloroplasts. Differential centrifugation yields a dense fraction containing almost pure grana and a less dense fraction consisting of intergrana lamellae (or stroma lamellae). The intergrana lamellae have a high Chl a/b ratio (about 5–6), very low Chl a fluorescence, and a high concentration of P700. This fraction does not evolve O_2 or show other reactions that are associated with PSII.

In contrast to the less dense intergrana lamellae that contain mostly PSI, the denser grana fraction contains both PSI and PSII. This fraction has a low Chl a/b ratio (2.5), has a high Chl a fluorescence, evolves O_2, and has cyt b_{559} and P680.

There are at least five supramolecular protein complexes held within the thylakoid membranes: the PSI and PSII pigment–protein complex, the cyt b/c complex, the coupling factor, and the light-harvesting Chl–protein complex. Most of the latter complex, termed the Chl a/b light-harvesting complex (LHC-II), belongs to PSII and contains nearly all of the Chl b. There is evidence that PSI has specific Chl a/b proteins also (LHC-I; Section 12.5.1).

The supramolecular complexes are embedded in the lipid bilayers that compose the thylakoid membranes and are composed of about 50% lipid. About 80% of the lipids of higher plant thylakoids is composed of neutral galactolipids monogalactosyldiacylglycerol (MGDG) and digalactosyldiacylglycerol (DGDG) whose acyl chains are composed of linolenic acid, a fatty acid having a high degree of unsaturation which is very fluid at physiological temperatures.[179]

As stated earlier, most of the PSII complexes and the PSII light-harvesting complexes are located in the appressed grana membranes with only about 10–20% in the stroma-exposed membranes.[177,178] It is considered likely that the PSII units in the stroma lamellae are different from those in the grana region. Most of the PSI complex as well as the coupling factor complex is found on the stroma thylakoids. The cyt b/c complex, by contrast, seems to be rather evenly distributed between stroma and grana thylakoids.

This uneven distribution of the photosystems between grana and stroma suggests that there must be some mobile substance capable of transferring electrons from PSII to PSI. Plastoquinone has been suggested to be one of the two mobile electron carriers that can shuttle electrons from PSII to PSI by migrating within the fluid matrix of the thylakoid membrane. Plastocyanin is the other carrier that plays the role of an electron shuttle. Since most of the light-harvesting chlorophylls of PSII are separated from PSI, it would seem that extra light energy absorbed by PSII cannot be made available to PSI. The redistribution (or "spillover") of energy from PSII to PSI has been postulated[180] to be brought about by a mechanism involving a reversible phosphorylation of the LHC-II, which increases their negative charge causing them to move out of the grana toward the PSI-containing stromal regions, thereby allowing more of the light absorbed by PSII to reach PSI. Many differentially pigmented plants are able to redistribute their absorbed quanta between PSII and PSI. However, not all plants use a mechanism involving phosphorylation of LHC-II. In the red and blue-green algae, a different mechanism involving cyclic electron flow in PSI is apparently involved (for a review, see Ref. 181).

12.7.3. Granal and Agranal Chloroplasts

Some plants that possess the so-called C_4-dicarboxylic acid pathway for CO_2 fixation (Section 12.8.2) have two different types of chloroplasts, depending on their location in the plant. In leaves of a C_4 plant, e.g., maize (*Zea mays*), the cells that surround the vascular bundles contain almost no grana but only stroma lamellae, while chloroplasts contained in the mesophyll regions have extensive grana as well as stroma lamellae. Initially it was suggested that the chloroplasts obtained from vascular bundles that have no grana would have no PSII activity, by analogy to intergrana lamellae of spinach, and that the mesophyll chloroplasts of maize would contain both PSI and PSII. A good deal of evidence now suggests that the agranal bundle sheath chloroplasts do have high PSI activity and are somewhat deficient in PSII, having perhaps only about 40% of the activity seen in mesophyll chloroplasts. These results suggest that a direct correlation cannot be made between the extent of grana stacking and the level of PSII activity.

12.8. CARBON DIOXIDE FIXATION[182,183]

12.8.1. The Calvin–Benson–Bassham or C_3 Carbon Cycle

CO_2 fixation requires the NADPH and ATP that is formed by the light reactions of photosynthesis, and proceeds via a number of dark reactions. In general, there are two major types of CO_2 fixation: the 3-carbon (C_3) and the 4-carbon (C_4) pathways.

The C_3 pathway, or reductive pentose phosphate (RPP) cycle, is also termed the Calvin–Benson–Bassham cycle after its discoverers. Melvin Calvin was awarded the Nobel Prize for chemistry in 1961 for this work that elucidated the pathway of carbon in photosynthesis. The crucial reaction of the C_3 pathway is the addition of CO_2 to the 5-carbon sugar ribulose bisphosphate (RuBP) to form two molecules of phosphoglyceric acid (PGA). The reducing power of NADPH along with ATP is used to change the acid group in PGA into the aldehyde group of the 3-carbon sugar phosphoglyceraldehyde (triose phosphate). Energy storage of photosynthesis is achieved with the formation of this 3-carbon sugar. The primary carboxylase in all photoautotrophs is ribulose-1,5-bisphosphate carboxylase (RuBPCase). Since this enzyme also acts as an oxygenase, it is often abbreviated RuBISCO, where RuBIS stands for ribulose bisphosphate, C for carboxylase, and O for oxygenase. This ubiquitous protein accounts for about a quarter of the protein content of plants, and is thus the most abundant protein on earth.

The other steps of the Calvin–Benson–Bassham cycle serve to regenerate the initial CO_2 acceptor, RuBP, via a complex series of reactions involving 3-, 4-, 5-, 6-, and 7-carbon sugar phosphates. Although sugars and carbohydrates are formed as end products of photosynthesis, other compounds can be formed as well. Substances such as fatty acids, lipids, amino acids, and organic acids can also be synthesized in photosynthetic CO_2 fixation. The formation of these substances seems to be partially under the control of environmental factors such as light intensity and the concentrations of CO_2 and O_2. A better understanding of how these factors interact may lead to the ability to control growth conditions so as to produce plants that have the desired amount of sugars, lipids, or proteins.

12.8.2. The C_4 Cycle[183–186]

There are a number of plants widely scattered throughout the plant kingdom that seem to thrive in hot and arid environments. The vascular bundles in these plants are

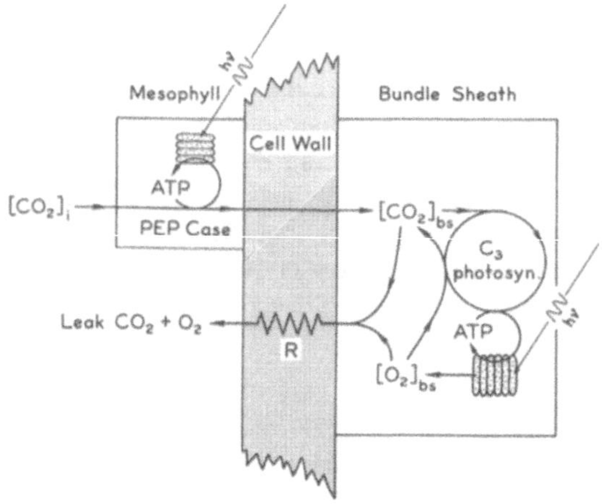

Fig. 12-10. The mechanism used by C_4 plants to concentrate CO_2 in the bundle sheath cells $(CO_2)_{bs}$ above the concentration found in the atmosphere surrounding the mesophyll cells $(CO_2)_i$. Loss of CO_2 is retarded by the resistance to diffusion (R) presented by the cell wall. PEPCase, phosphoenol pyruvate carboxylase. (From Ref. 186, reprinted by permission.)

tightly surrounded by the so-called bundle sheath cells containing chloroplasts that can be distinguished from mesophyll cells located outside and around the bundle sheath cells (Fig. 12-10). In these plants the initial fixation of CO_2 does not occur via condensation with RuBP, as in the Calvin–Benson–Bassham cycle, but via an intermediate cycle termed the C_4 cycle by which CO_2 first combines with the 3-carbon acid phosphoenolpyruvate (PEP) to form the 4-carbon acids, malic and aspartic in mesophyll chloroplasts (Fig. 12-10). These reactions proceed via the intermediate 4-carbon oxaloacetic acid and are catalyzed by the enzyme PEP carboxylase. The entire pathway is often called the Hatch–Slack–Kortschak pathway, after its discoverers. The 4-carbon acids, malic and aspartic, cannot be converted into sugars, however, until they are decarboxylated (lose the CO_2 fixed previously) to form the 3-carbon pyruvic acid. By reaction with ATP formed photosynthetically, the pyruvic acid is again converted into PEP that can again accept a CO_2 molecule. Essentially what this cycle accomplishes is to pump CO_2 from the outside and release it into the inside of the plant. Once inside the plant, the released CO_2 reacts in the usual manner with RuBP via the Calvin–Benson–Bassham cycle in the bundle sheath chloroplasts. This may seem like a useless endeavor for the plant especially when it is considered that an extra ATP must be used to operate this "CO_2 pump." Plants having the C_3 cycle require 3 ATP and 2 NADPH molecules, but C_4 plants require 4 molecules of ATP and 2 of NADPH. As mentioned in Section 12.7.3, C_4 plants have a characteristic anatomy and two quite distinct types of chloroplasts.

It appears that the CO_2 pump of C_4 plants is located in the mesophyll cells and functions efficiently by trapping any CO_2 that diffuses through the stomatal pores on the epidermis of the leaf. This trapped CO_2 is then released into the interior bundle sheath cells where it is fixed via the Calvin–Benson–Bassham cycle. The CO_2 pump operates efficiently to keep the concentration of free CO_2 in the mesophyll cell so low that CO_2 tends to diffuse from the atmosphere, where it is present in concentrations of about 300 ppm (0.03%), into the inside space of the leaf even though the stomates may be almost closed. The ability to obtain CO_2 even with stomates partially closed means that less water will be lost, a vital requirement for plant survival in climates that are both hot and dry. It thus appears that it is worthwhile for the plant to spend the extra ATP needed to operate the CO_2 pump, since it enables it to collect sufficient CO_2 but not to lose substantial quantities of water in the process. In any case, this extra ATP may be rather easily obtained by cyclic photophosphorylation, since plants living in arid regions usually receive an ample amount of light.

The ability of C_4 plants to photosynthesize and grow in arid regions has important practical consequences for agricultural production in these regions. Until now, most of the agricultural effort has been directed to developing crop plants that are suited to temperate regions. In the future, more effort will undoubtedly have to be devoted to developing crop plants that are well suited to grow in the earth's vast arid regions. Plants with the C_4 pathway may play an important role in this undertaking, since they have higher rates of photosynthesis at high temperatures and high light intensities compared to C_3 plants.

12.8.3. Crassulacean Acid Metabolism (CAM)[187]

Plants having crassulacean acid metabolism (CAM) fix CO_2 in a way somewhat like that of C_4 plants, except that C_4 plants use separate compartments for the initial CO_2 fixation and the subsequent Calvin–Benson–Bassham cycle, whereas the CAM plants fix CO_2 into malic acid and store it in a vacuole during the night. The Calvin–Benson–Bassham cycle operates during the day using the CO_2 released upon enzymatic decarboxylation of malic acid. Thus, the C_4 plants separate CO_2 fixation and the Calvin–Benson–Bassham cycle spatially, whereas CAM plants separate them temporally. Typically CAM plants, such as those in the family Crassulaceae from which the name is derived, are adapted to arid environments where water conservation is essential.[188] Typically CAM plants are from regions (such as South Africa) where the environment has alternating wet periods and hot, dry periods. The CAM plants are protected by a thick cuticle and a reduced number of stomates. CAM plants do not avoid photorespiration (see next section), but the CO_2 that is produced by photorespiration is not lost, since it is prevented from escaping from the plant during the day because stomates are mainly closed during this time.

12.8.4. Photorespiration[189,190]

In addition to CO_2 as the substrate for RuBPCase, O_2 may act as an alternate substrate in which case there is the production of phosphoglycolate rather than phosphoglyceric acid. Thus, the enzyme is also called RuBISCO where O stands for oxygenase, as noted earlier. Photorespiration is an energy-consuming process, and is seen mainly in C_3 plants and competes with the fixation of CO_2. Although photorespiration appears to be a wasteful process that C_3 plants must endure because of the high O_2 concentration in the atmosphere, there is evidence that photorespiration may protect plants from photooxidative damage.[191] If ways were found to eliminate photorespiration, then it might be possible to produce a considerable increase in crop productivity. C_4 plants such as maize, sorghum, and other important crop plants as well as CAM plants do not suffer from photorespiration, since the CO_2 is sequestered within the plant where it can compete effectively with O_2 for RuBP. As was noted earlier, the C_4 plants are superior to C_3 plants under full-sun, high-temperature, and dry conditions, but C_3 plants are better under conditions of low light and cool temperatures.

12.9. PHOTOSYNTHESIS AS A SOURCE OF ENERGY[17,192–194]

There is increasing interest in the possibility of using photosynthetically converted light energy as a means of expanding our energy supply. As has been pointed out by Hall,[193] the magnitude of energy stored by photosynthesis (about 10 times the world's annual use of energy) is not generally appreciated. Only about 10% of agricultural production is used for food. "Waste" products from agriculture and forestry represent an enormous energy potential. Stalks and chaff from grain crops, sawdust, wood chips, and paper contain energy trapped by the photosynthetic process. Wood chips can be converted by pyrolysis into charcoal and numerous valuable hydrocarbon products. Plant material can be fermented into methane, converted into wood alcohol, or burned efficiently to release energy.

It has been suggested[195] that the sugar (as well as the cellulose) produced photosynthetically by cane or sugar beet plants could be economically converted into ethyl alcohol with almost no energy loss. Alcohol can be added to gasoline to reduce the demand for petroleum. It may be economically feasible to harvest hydrocarbons directly from plants rather than obtaining them indirectly, such as by the conversion of carbohydrates into ethanol. The rubber tree (*Hevea*) or guayule (*Parthenium*), for example, might serve as a direct source of photosynthetically produced hydrocarbons that could substitute for petroleum hydrocarbons currently being used for chemicals and for starting materials for many synthetic reactions. Some algae, such as *Dunaliella,* inhabit waters of extremely high salinity, such as the Dead Sea, and produce large quantities of the valuable compounds glycerol and β-carotene.[196]

In principle, it is also possible to use solar energy trapped by bacteria or hydrogen-adapted algae to make hydrogen, a nonpolluting and easily transportable fuel. The anaerobic purple photosynthetic bacteria do not evolve O_2, and since they cannot oxidize water they must depend on a large variety of organic or inorganic substances for their metabolic needs. The photosynthetic production of H_2 requires that the appropriate bacteria be supplied with an organic (or inorganic) electron donor, the right amounts of amino acids (or limiting NH_4^+), and that they be in a condition where there is a high activity of energy conversion in relation to the cell's biosynthetic needs. Under these conditions, the bacteria regulate themselves to produce H_2 by energy-dependent means. Of course, any large-scale system based on bacterial H_2 production will also require large amounts of inexpensive and easily available electron donors. This requirement may impose a severe limitation on the economic feasibility of this type of system.

The most readily available electron donor is, of course, water, but only higher plants and algae can extract electrons from water with the operation of their two-pigment systems. We have seen that PSI can reduce compounds that have redox potentials near -1.0 V (Section 12.5.1). Since at the physiological pH of 7 the E_0' of the hydrogen electrode is about -0.42 V, there is sufficient potential to produce molecular hydrogen. But algae do not normally produce H_2 with the reductant generated by PSI unless they have been hydrogen-adapted. Hydrogen adaptation is achieved by keeping algae such as *Chlorella* in the dark for 10–12 h under nitrogen or hydrogen. Hydrogen production can apparently be seen after appropriate adaptation conditions in a large number of variously pigmented algae, including the marine red and brown algae, in addition to the green algae.

The adaptation procedure apparently activates a hydrogenase enzyme, but this enzyme is inactivated if O_2 is present. The photoproduction of hydrogen is mediated only by PSI, and the O_2-evolving system of PSII is not operative under anaerobic conditions. The substrate for H_2 production may be a product of anaerobic metabolism that accumulates during the adaptation period at a very slow rate in the algal cell. This slow formation of the substrate for H_2 production may be the cause of the low rate of H_2 evolution that usually amounts to only about 10% of the maximum rate of O_2 produced photosynthetically.

To utilize the ability of algae to oxidize water, release O_2, and at the same time produce hydrogen requires considerable reengineering of the photosynthetic machinery. A separation of incompatible reactions has already been achieved by certain blue-green algae that fix nitrogen in their resting cells (heterocysts). These special cells provide physical protection of the nitrogen-reducing enzymes against the destructive effects of O_2 by removing it quickly by means of a high respiratory activity in the heterocysts.

It may be possible to isolate from the living cell the various subsystems of photosynthesis such as the photochemical reaction centers, the O_2-evolving system, and the H_2-evolving system, and use them *in vitro*.(197,198) For example, it might be possible to absorb photosynthetic components on appropriate supports so that the products of one system would be brought close enough to be used by the next enzyme system, and so on until the desired products could be obtained. In order to avoid inactivation of a hydrogenase enzyme by O_2, it may be possible to separate across opposite sides of a synthetic membrane the reactions yielding O_2 from those producing H_2. This may be accomplished, for example, by a system employing microencapsulation, but at present our knowledge about the separation of components of the photosynthetic system is rudimentary at best.

A laboratory scale production of H_2 has been demonstrated by coupling the reducing power produced by chloroplasts with an algal or bacterial hydrogenase. A problem with this system is that, again, O_2 production inactivates the hydrogenase system. This problem may be alleviated by removing the O_2 from the reaction mixture, e.g., with the addition of glucose and glucose oxidase. Unfortunately, the yield of H_2 from these composite systems is disappointingly low (only a few percent of the average rate of normal photosynthesis), and these systems are very labile and expensive.

It is clear that the energy stored and the organic matter made available from photosynthesis are enormous. Many economic obstacles currently hinder the utilization of this vast potential, and hopefully these problems can be overcome. Continued research directed toward obtaining a better fundamental understanding of the process will aid in our efforts to tap this vast energy source.

12.10. LOOKING AHEAD

It may seem at first that photosynthesis has been so intensively investigated that little remains to be done. However, a closer look into almost any of the topics covered here will show that many rewarding results await future investigators.

A great step forward has recently been made in our understanding of the structure of the bacterial reaction center with the crystallization of isolated reaction centers from *Rhodopseudomonas viridis*. This achievement will allow the development of a better understanding of how the primary energy conversion of photosynthesis takes place.

Moreover, since these reaction centers have features similar to those of green plants and algae, which can cleave water, it is likely that these purified centers can serve as models for further study of plant reaction centers.

The recent availability of PSII membrane preparations retaining high rates of O_2 evolution have allowed the isolation and the possible identification of the function of certain membrane proteins that are part of the O_2-evolving complex. The function of several of the other proteins known to be associated with PSII remains to be elucidated, as does the role of calcium and chloride ions in the reaction centers of PSII. The manganese-containing enzyme that functions eventually to oxidize water remains to be isolated and characterized.

PSII in chloroplasts of higher plants and algae is apparently not homogeneous, but seems to be heterogeneous on both the oxidizing and reducing sides of the PSII reaction center. This heterogeneity of PSII comes mainly from measurements of the yield and induction kinetics of Chl a fluorescence from PSII, and is currently not completely understood (for a review, see Ref. 199).

The role of cyt b_{559} remains an enigma. Some evidence suggests that this cytochrome functions to prevent excess quanta from damaging PSII by mediating a cyclic flow of electrons around PSII. This idea has yet to be proved conclusive, however.

Electron transport in photosynthesis requires the transfer of electrons and hydrogen ions across the thylakoid membranes. This flow of electrons is mediated by three integral membrane-spanning protein complexes, PSI, PSII, and the Cyt b_6/f complex. Future research is needed to reveal the detailed topography of these proteins, which can be obtained from determinations of their amino acid sequences, as well as from computer analysis to predict the folding patterns of the proteins from hydrophobicity measurements (hydropathy plots). The unraveling of the details of light-driven proton gradient formation[200] as well as the gaining of a better understanding of the molecular mechanism of photophosphorylation[201] provide many challenges for future investigators.

The way in which proteins are contained within the lipid matrix of the thylakoid membrane is not yet clearly understood. The polar lipid monogalactosyldiacylglycerol (MGDG) constitutes about half of the dry weight of the thylakoid membrane, and yet this lipid does not form the bilayer configuration that is thought to be important in all biological membranes. This major component of the thylakoid membrane may provide the necessary fluid environment to support the dynamic movement of intermediates involved in electron transport, and may as well serve to anchor the large intrinsic proteins within the thylakoid bilayer. More work is needed to elucidate why MGDG is such a major component of the thylakoid membrane, and how it interacts with proteins and other lipids.

These are just a few of the problems that remain to be answered. As is usually the case, the answering of one question leads to many more. We have a long way to go before we understand nature's elaborate process of converting solar energy into stable compounds that are vitally important to all life.

12.11. REFERENCES

1. C. Sybesma (ed.), *Advances in Photosynthesis Research,* Vols. 1–4, Martinus Nijhoff/Dr. W. Junk, The Hague (1984).

2. G. Akoyunoglou (ed.), *Photosynthesis,* Vols. 1–6, Balaban International Science Services, Philadelphia (1981).
3. D. O. Hall, J. Coombs, and T. W. Goodwin (eds.), *Proceedings of the Fourth International Congress on Photosynthesis,* The Biochemical Society, London (1978).
4. M. Avron (ed.), *Proceedings of the Third International Congress on Photosynthesis,* Vols. 1–3, Elsevier, Amsterdam (1975).
5. G. Forti, M. Avron, and A. Melandri (eds.), *Proceedings of the Second International Congress on Photosynthesis Research,* Dr. W. Junk, The Hague (1972).
6. H. Metzner (ed.), *Progress in Photosynthesis Research,* Vols. 1–3, International University of Biological Sciences, Tübingen (1969).
7. R. K. Clayton, *Molecular Physics in Photosynthesis,* Blaisdell, New York (1965).
8. L. P. Vernon and G. R. Seely (eds.), *The Chlorophylls,* Academic Press, New York (1966).
9. E. Rabinowitch and Govindjee, *Photosynthesis,* Wiley, New York (1969).
10. O. V. S. Heath, *The Physiological Aspects of Photosynthesis,* Stanford University Press, Stanford (1969).
11. R. K. Clayton, *Light and Living Matter, Vol. 1, The Physical Part,* McGraw-Hill, New York (1970).
12. R. P. F. Gregory, *Biochemistry of Photosynthesis,* Wiley-Interscience, London (1971).
13. R. K. Clayton, *Photosynthesis: How Light Is Converted to Chemical Energy,* Addison-Wesley, New York (1974).
14. Govindjee and R. Govindjee, The absorption of light in photosynthesis, *Sci. Am.* **231,** 68–82 (1974).
15. Govindjee (ed.), *Bioenergetics of Photosynthesis,* Academic Press, New York (1975).
16. J. Barber (ed.), *Primary Processes of Photosynthesis,* Elsevier, Amsterdam (1977).
17. A. Mitsui, S. Miyachi, A. San Pietro, and S. Tamura (eds.), *Biological Solar Energy Conversion,* Academic Press, New York (1977).
18. R. K. Clayton, *Photosynthesis: Physical Mechanisms and Chemical Patterns,* Cambridge University Press, Cambridge (1980).
19. D. O. Hall and K. K. Rao, *Photosynthesis,* 3rd ed., Edward Arnold, New York (1981).
20. M. D. Hatch and N. K. Boardman (eds.), *The Biochemistry of Plants, Vol. 8, Photosynthesis,* Academic Press, New York (1981).
21. Govindjee (ed.), *Photosynthesis: Energy Conversion by Plants and Bacteria,* Vol. 1, Academic Press, New York (1982).
22. Govindjee (ed.), *Photosynthesis: Development, Carbon Metabolism, and Plant Productivity,* Vol. 2, Academic Press, New York (1982).
23. S. M. Danks, E. H. Evans, and P. A. Whittaker, *Photosynthetic Systems: Structure, Function and Assembly,* Wiley, Chichester (1983).
24. C. H. Foyer, *Photosynthesis,* Wiley, New York (1984).
25. Govindjee, J. Amesz, and D. C. Fork (eds.), *Light Emission from Plants and Bacteria,* Academic Press, Orlando (1986).
26. E. Broda, *The Evolution of the Energetic Process,* Pergamon Press, Oxford, (1975).
27. E. Broda and G. A. Peschek, Did respiration or photosynthesis come first? *J. Theor. Biol.* **81,** 201–212 (1979).
28. A. I. Oparin, The origin of life and origin of enzymes, *Adv. Enzymol.* **27,** 347–380 (1965).
29. J. P. Ferris and D. A. Usher, Origins of life, in *Biochemistry* (G. Zubay, ed.), pp. 1190–1241, Addison-Wesley, Reading, Mass. (1983).
30. R. E. Dickerson, Chemical evolution and the origin of life, *Sci. Am.* **239,** 70–85 (1978).
31. C. R. Woese, A proposal concerning the origin of life on the planet earth, *J. Mol. Biol.* **13,** 95–101 (1979).
32. C. R. Woese, *The Origin of Life* (J. S. Head, ed.), Carolina Biology Readers, Carolina Biological Supply Co., Burlington (1984).
33. E. Gantt, Phycobilisomes, Light-harvesting pigment complexes, *Bioscience* **25,** 781–788 (1975).
34. E. Gantt, Phycobilisomes, *Annu. Rev. Plant Physiol.* **32,** 327–347 (1981).
35. A. N. Glazer, Phycobilisome, A macromolecular complex optimized for light energy transfer, *Biochim. Biophys. Acta* **768,** 29–51 (1984).
36. C. B. van Niel, The present status of the comparative study of photosynthesis, *Annu. Rev. Plant Physiol.* **13,** 1–26 (1962).

37. R. Radmer and O. Ollinger, Isotopic composition of photosynthetic O_2 flash yields in the presence of $H_2^{18}O$ and $H_2^{18}O_3^-$, *FEBS Lett.* **10,** 57–61 (1980).
38. A. Stemler and R. Radmer, Source of photosynthetic oxygen in bicarbonate-stimulated Hill reaction, *Science* **190,** 457–458 (1975).
39. A. W. D. Larkum and J. Barrett, Light harvesting processes in algae, in: *Advances in Botanical Research,* Vol. 10 (H. W. Woolhouse, ed.), pp. 1–219, Academic Press, New York (1983).
40. T. W. Goodwin (ed.), *Chemistry and Biochemistry of Plant Pigments,* Academic Press, London (1976).
41. T. W. Goodwin (ed.), *Chemistry and Biochemistry of Plant Pigments,* Vols. 1 and 2, Academic Press, London (1976).
42. J. J. Katz, L. L. Shipman, and J. R. Norris, Structure and function of photoreaction-centre chlorophyll, in: *Chlorophyll Organization and Energy Transfer in Photosynthesis,* Ciba Found. Symp. 61 (new series), pp. 1–40, Excerpta Medica, Amsterdam (1979).
43. H. H. Strain and W. A. Svec, Extraction, separation, estimation, and isolation of the chlorophylls, in: *The Chlorophylls* (L. P. Vernon and G. R. Seely, eds.), pp. 21–66, Academic Press, New York (1966).
44. J. H. C. Smith and C. S. French, The major and accessory pigments in photosynthesis, *Annu. Rev. Plant Physiol.* **14,** 181–224 (1963).
45. Govindjee and B. Z. Braun, Light absorption, emission, and photosynthesis, in: *Algal Physiology and Biochemistry* (W. D. P. Stewart, ed.), pp. 346–390, University of California Press, Berkeley (1974).
46. J. M. Anderson and J. Barrett, Chlorophyll-protein complexes of brown algae: P700 reaction centre and light-harvesting complexes, in *Chlorophyll Organization and Energy Transfer in Photosynthesis,* Ciba Found. Symp. (new series), pp. 81–104, Excerpta Medica, Amsterdam (1979).
47. D. Siefermann-Harms, Carotenoids in photosystem I. Location in photosynthetic membranes and light harvesting function, *Biochim. Biophys. Acta* **811,** 325–355 (1985).
48. K. Sauer, Primary events and the trapping of energy, in: *Bioenergetics of Photosynthesis* (Govindjee, ed.), pp. 115–181, Academic Press, New York (1975).
49. L. N. M. Duysens, Transfer of excitation energy in photosynthesis, Ph.D. Thesis, Univ. of Utrecht (1952).
50. L. N. M. Duysens, Photosynthesis, in: *Progress in Biophysics and Molecular Biology,* Vol. 14 (J. A. V. Butler and H. E. Huxley, eds.), pp. 1–104, Pergamon Press, Oxford (1964).
51. G. Hoch and R. S. Knox, Primary processes in photosynthesis, *Photophysiology* **3,** 225–251 (1968).
52. R. S. Knox, Excitation energy transfer and migration: Theoretical considerations, in: *Bioenergetics of Photosynthesis* (Govindjee, ed.), pp. 183–221, Academic Press, New York (1975).
53. T. Förster, Zwischenmolekulare Energiewanderung und Fluoreszenz, *Ann. Physik.* **2,** 55–75 (1948).
54. T. Förster, Delocalized excitation and excitation transfer, in: *Modern Quantum Chemistry, Part III, Action of Light and Organic Molecules* (O. Sinanoglu, ed.), pp. 93–137, Academic Press, New York (1965).
55. R. van Grondelle and J. Amesz, Excitation energy transfer in photosynthetic systems, in: *Light Emission from Plants and Bacteria* (Govindjee, J. Amesz, and D. C. Fork, eds.), Academic Press, Orlando (1986).
56. G. R. Seely and J. S. Connolly, Fluorescence of photosynthetic pigments in vitro, in: *Light Emission from Plants and Bacteria* (Govindjee, J. Amesz, and D. C. Fork, eds.), Academic Press, Orlando (1986).
57. G. Hoch and R. S. Knox, Primary processes in photosynthesis, in: *Photophysiology,* Vol. 3 (A. C. Giese, ed.), pp. 225–251, Academic Press, New York (1968).
58. R. S. Knox, Theory and modeling of excitation delocalization and trapping, in: *Encyclopedia of Plant Physiology,* Vol. 19, new series, (L. A. Staehelin and C. J. Arntzen, eds.), pp. 286–298, Springer-Verlag, Berlin (1986).
59. J. Lavorel, Luminescence, in: *Bioenergetics of Photosynthesis* (Govindjee, ed.), pp. 223–317, Academic Press, New York (1975).
60. S. Malkin, Delayed luminescence, in: *Primary Process of Photosynthesis* (J. Barber, ed.), pp. 349–431, Elsevier, Amsterdam (1977).
61. Govindjee and P. A. Jursinic, Photosynthesis and fast changes in light emission by green plants, in: *Photochem. Photobiol. Rev.,* Vol. 4 (K. C. Smith, ed.), pp. 125–205, Plenum Press, New York (1979).
62. J. Lavorel, J. Lavergne, and A. L. Etienne, A reflection on several problems of luminescence in photosynthetic systems, *Photochem. Photobiophys.* **3,** 287–314 (1982).
63. R. Emerson and W. Arnold, A separation of the reactions in photosynthesis by means of intermittent light, *J. Gen. Physiol.* **15,** 391–420 (1932).

64. R. Emerson and W. Arnold, The photochemical reaction in photosynthesis, *J. Gen. Physiol.* **16,** 191–205 (1932).
65. R. K. Clayton and W. R. Sistrom (eds.), *The Photosynthetic Bacteria,* Plenum Press, New York (1978).
66. N. Pfennig, General physiology and ecology of photosynthetic bacteria, in: *The Photosynthetic Bacteria* (R. K. Clayton and W. R. Sistrom, eds.), pp. 3–18, Plenum Press, New York (1978).
67. H. Michel, Three-dimensional crystals of a membrane protein complex. The photosynthetic reaction centre from *Rhodopseudomonas viridis, J. Mol. Biol.* **158,** 567–572 (1982).
68. J. Deisenhofer, D. Epp, E. Miki, R. Huber, and H. Michel, X-ray structure analysis of a membrane protein complex. Electron density map at 3Å resolution and a model of the chromatophores of the photosynthetic reaction center from *Rhodopseudomonas viridis, J. Mol. Biol.* **180,** 385–398 (1984).
69. J. Deisenhofer, H. Michel, and R. Huber, The structural basis of photosynthetic light reactions in bacteria, *Trends in Biochem. Sci.* **10,** 243–248 (1985).
70. J. Deisenhofer, O. Epp, K. Miki, R. Huber, and H. Michel, Structure of the protein subunits in the photosynthetic reaction centre of *Rhodopseudomonas viridis* at 3Å resolution, *Nature* **318,** 618–624 (1985).
71. N. W. Woodbury, M. Becker, D. Middendorf, and W. W. Parson, Picosecond kinetics of the initial photochemical electron-transfer reaction in bacterial photosynthetic reaction centers, *Biochemistry* **24,** 7516–7521 (1985).
72. J. M. Olson, Chlorophyll organization in green photosynthetic bacteria, *Biochim. Biophys. Acta* **594,** 33–51 (1981).
73. L. A. Staehlin, J. R. Golecki, and G. Drews, Supramolecular organization of chlorosomes (*Chlorobium* vesicles) and of their membrane attachment sites in *Chlorobium limicola, Biochim. Biophys. Acta* **589,** 30–45 (1980).
74. T. Wechsler, F. Suter, R. C. Fuller, and H. Zuber, The complete amino acid sequence of the bacteriochlorophyll c binding polypeptide from chlorosomes of the green photosynthetic bacterium *Chloroflexus aurantiacus, FEBS Lett.* **181,** 173–178 (1985).
75. B. B. Buchanan, P. Schürmann, and K. T. Shanmugam, Role of the reductive carboxylic acid cycle in a photosynthetic bacterium lacking ribulose 1,5-diphosphate carboxylase, *Biochim. Biophys. Acta* **283,** 136–145 (1972).
76. C. F. Fowler, N. A. Nugent, and R. C. Fuller, The isolation and characterization of a photochemically active complex from *Chloropseudomonas ethylica, Proc. Natl. Acad. Sci. USA* **68,** 2278–2282 (1971).
77. B. K. Pierson and R. W. Castenholz, Photosynthetic apparatus and cell membrane of the green bacteria, in: *The Photosynthetic Bacteria* (R. K. Clayton and W. R. Sistrom, eds.), pp. 179–197, New York (1978).
78. R. E. Blankenship, Electron transport in green photosynthetic bacteria, *Photosyn. Res.* **6,** 317–333 (1985).
79. D. D. Burns and M. Midgley, Localization and possible role of an adenosine triphosphatase in *Chlorobium thiosulfatophilum, Eur. J. Biochem.* **67,** 323–333 (1976).
80. B. K. Pierson and R. W. Castenholz, A phototrophic gliding filamentous bacterium of hot springs, *Chloroflexus aurantiacus,* gen. and sp. nov., *Arch. Microbiol.* **100,** 5–24 (1974).
81. J. Amesz, Electron transport in green bacteria, in: *Advances in Photosynthesis Research,* Vol. 1 (C. Sybesma, ed.), pp. 621–628, Martinus Nijhoff/Dr. W. Junk, The Hague (1984).
82. R. E. Blankenship and R. C. Prince, Excited-state redox potentials and the Z scheme of photosynthesis, *Trends Biochem. Sci.* **10,** 382–383 (1985).
83. A. W. Rutherford and P. Heathcote, Primary photochemistry in photosystem I, *Photosyn. Res.* **6,** 295–316 (1985).
84. J. Amesz, Photosynthesis: Structure of the membrane and membrane proteins, *Prog. Bot.* **47,** 87–104 (1985).
85. J. E. Mullet, J. J. Burke, and C. J. Arntzen, Chlorophyll proteins of photosystem I, *Plant Physiol.* **65,** 814–822 (1980).
86. P. Haworth, J. L. Watson, and C. Arntzen, The detection, isolation and characterization of a light-harvesting complex which is specifically associated with photosystem I. *Biochim. Biophys. Acta* **724,** 151–158 (1983).
87. J. M. Anderson, A chlorophyll a/b-protein complex of photosystem I, *Photobiochem. Photobiophys.* **8,** 221–228 (1984).

88. B. Kok, On the reversible absorption change at 705 mμ in photosynthetic organisms, *Biochim. Biophys. Acta* **22,** 399–401 (1956).
89. T. Hiyama and B. Ke, Difference spectra and extinction coefficients of P700, *Biochim. Biophys. Acta* **267,** 160–172 (1972).
90. H. Beinert, B. Kok, and G. Hoch, The light induced electron paramagnetic resonance signal of photocatalyst P700, *Biochem. Biophys. Res. Commun.* **7,** 209–212 (1962).
91. T. Hiyama and B. Ke, A new photosynthetic pigment "P430": Its possible role as the primary electron acceptor of photosystem I, *Proc. Natl. Acad. Sci. USA* **68,** 1010–1013 (1971).
92. J. M. Fenton, M. J. Pellin, Govindjee, and K. J. Kaufmann, Primary photochemistry of the reaction center of photosystem II, *FEBS Lett.* **100,** 1–4 (1979).
93. K. Kanogawa, A. Namiki, N. Nakashima, K. Yoshihara, and I. Ikegami, Picosecond transient behavior of reaction-center particles of photosystem I isolated from spinach chloroplasts. Energy and electron transfer upon single and multiple photon excitation, *Photochem. Photobiol.* **34,** 511–516 (1981).
94. V. A. Shuvalov, B. Ke, and E. Dolan, Kinetic and spectral properties of the intermediary electron acceptor A_1 in photosystem I, *FEBS Lett.* **100,** 5–8 (1979).
95. V. A. Shuvalov, A. V. Klevanik, A. V. Sharkov, P. G. Kryukov, and B. Ke, Picosecond spectroscopy of photosystem I reaction centers, *FEBS Lett.* **107,** 313–316 (1979).
96. D. O. Hall, R. Cammack, and K. K. Rao, Non-haem iron proteins, in: *Iron in Biochemistry and Medicine* (A. Jocob and M. Worwood, eds.), Academic Press, New York (1974).
97. G. Hauska, E. Hurt, N. Gabellini, and W. Lockau, Comparative aspects of quinol-cytochrome c/plastocyanin oxidoreductases, *Biochim. Biophys. Acta* **726,** 97–133 (1983).
98. J. Bennett, The simple art of making photosynthetic complexes, *Trends in Biochem. Sci.* **6,** VI–VII (1981).
99. J. S. Rieske, Composition, structure, and function of complex III of the respiratory chain, *Biochim. Biophys. Acta* **456,** 195–247 (1976).
100. J. Alt, P. Westhoff, B. B. Sears, N. Nelson, E. Hurt, G. Hauska, and R. G. Herrmann, Genes and transcripts for the polypeptides of the cytochrome b_6/f complex from spinach thylakoids, *EMBO J.* **2,** 979–986 (1983).
101. W. Heinemeyer, J. Alt, and R. G. Herrmann, Nucleotide sequence of the clustered genes for apocytochrome b_6 and subunit 4 of the cytochrome b/f complex in spinach plastid chromosome, *Curr. Gen.* **8,** 543–549 (1984).
102. J. Alt and R. G. Herrmann, Nucleotide sequence of the gene for preapocytochrome f in the spinach plastid chromosome, *Curr. Gen.* **8,** 551–557 (1984).
103. P. Mitchell, Possible molecular mechanisms of the protonmotive function of cytochrome systems, *J. Theor. Biol.* **62,** 327–367 (1976).
104. M. A. Selak and J. Whitmarsh, Kinetics of the electrogenic step and cytochrome b_6 and f redox changes in chloroplasts, *FEBS Lett.* **150,** 286–292 (1982).
105. P. Joliot and A. Joliot, Proton pumping and electron transfer in the cytochrome b/f complex of algae, *Biochim. Biophys. Acta* **849,** 211–222 (1986).
106. P. R. Rich, Mechanisms of quinol oxidation in photosynthesis, *Photosyn. Res.* **6,** 335–348 (1985).
107. N. Murata and M. Miyao, Extrinsic membrane proteins in the photosynthetic oxygen-evolving complex, *Trends Biochem. Sci.* **10,** 122–124 (1985).
108. D. M. Ghanotakis and C. F. Yocum, Polypeptides of photosystem II and their role in oxygen evolution, *Photosyn. Res.* **7,** 97–114 (1985).
109. W. A. Cramer, W. R. Widger, R. G. Herrmann, and A. Trebst, Topography and function of thylakoid membrane proteins, *Trends Biochem. Sci.* **10,** 125–129 (1985).
110. T. Kambara and Govindjee, Molecular mechanism of water oxidation in photosynthesis based on the functioning of manganese in two different environments, *Proc. Natl. Acad. Sci. USA* **82,** 6119–6123 (1985).
111. J. G. Metz and D. Miles, Use of a nuclear mutant of maize to identify components of photosystem II, *Biochim. Biophys. Acta* **681,** 95–102 (1982).
112. J. G. Metz and M. Siebert, Presence in photosystem II core complexes of a 34-kilodalton polypeptide required for water photolysis, *Plant Physiol.* **76,** 829–832 (1984).
113. D. J. Kyle, The 32000 dalton Q_B protein of photosystem II, *Photochem. Photobiol.* **41,** 107–116 (1985).
114. J. J. Eaton-Rye, D. J. Blubaugh, and Govindjee, Action of bicarbonate on photosynthetic electron

transport in the presence or absence of inhibitory anions, in: *Ion Interactions in Energy Transfer Biomembranes* (G. C. Papageorgiou, J. Barber, and S. Papa, eds.), pp. 263–278, Plenum Press, New York (1986).

115. P. Jursinic and Govindjee, Temperature dependence of delayed light emission in the 6 to 340 microsecond range after a single flash in chloroplasts, *Photochem. Photobiol.* **26,** 617–628 (1977).
116. H. J. van Gorkom, M. P. J. Pulles, and J. S. C. Wessels, Light-induced changes of absorbance and electron spin resonance in small photosystem II particles. *Biochim. Biophys. Acta* **408,** 331–339 (1975).
117. H. J. van Gorkom, J. J. Tamminga, and J. Haveman, Primary reactions, plastoquinone and fluorescence yield in subchloroplast fragments prepared with deoxycholate, *Biochim. Biophys. Acta* **347,** 417–438 (1974).
118. V. V. Klimov and A. A. Krasnovsky, Pheophytin as the primary electron acceptor in photosystem 2 reaction centers, *Photosynthetica* **15,** 592–609 (1981).
119. V. V. Klimov, E. Dolan, and B. Ke, EPR properties of an intermediary electron acceptor (pheophytin) in photosystem II reaction centers at cyogenic temperatures, *FEBS Lett.* **112,** 97–100 (1980).
120. J. Fajer, M. S. Davis, A. Forman, V. V. Klimov, E. Dolan, and B. Ke, Primary electron acceptors in plant photosynthesis, *J. Am. Chem. Soc.* **102,** 7143–7145 (1980).
121. A. W. Rutherford, EPR evidence for an acceptor functioning in photosystem II when the pheophytin acceptor is reduced, *Biochem. Biophys. Res. Commun.* **102,** 1065–1070 (1981).
122. C. A. Wraight, Current attitudes in photosynthesis research, in: *Photosynthesis: Energy Conversion by Plants and Bacteria*, Vol. 1 (Govindjee, ed.), pp. 17–61, Academic Press, New York (1982).
123. W. F. J. Vermaas and Govindjee, The acceptor side of photosystem II in photosynthesis, *Photochem. Photobiol.* **34,** 775–793 (1981).
124. J. H. A. Nugent, B. A. Diner, and M. C. W. Evans, Direct detection of the electron acceptor of photosynthesis II, *FEBS Lett.* **124,** 241–244 (1981).
125. L. N. M. Duysens and H. E. Sweers, Mechanism of two photochemical reactions in algae as studied by means of fluorescence, in: *Studies on Microalgae and Photosynthetic Bacteria,* pp. 353–372, University of Tokyo Press (1963).
126. V. V. Klimov, S. I. Allakhverdiev, and V. Z. Pashchenko, Measurement of the activation energy and lifetime of fluorescence of photosystem 2 chlorophyll, *Dokl. Akad. Nauk. SSSR* **242,** 1204–1207 (1978).
127. H. J. van Gorkom, Identification of the reduced primary electron acceptor of photosystem II as a bound semiquinone anion, *Biochim. Biophys. Acta* **347,** 439–442 (1974).
128. V. V. Klimov, A. V. Klevanik, V. A. Shuvalov, and A. A. Krasnovsky, Reduction of pheophytin in the primary light reaction of photosystem II, *FEBS Lett.* **82,** 183–186 (1977).
129. Govindjee, Photosystem II: The oxygen evolving system of photosynthesis, in: *Advances in Photosynthesis Research,* Vol. 1 (C. Sybesma, ed.), pp. 227–238, Martinus Nijhoff/Dr. W. Junk, The Hague (1984).
130. W. A. Cramer and A. R. Crofts, Electron and proton transport, in: *Photosynthesis: Energy Conversion by Plants and Bacteria,* Vol. 1, (Govindjee, ed.), pp. 387–467, Academic Press, New York (1982).
131. B. R. Velthuys, Electron-dependent competition between plastoquinone and inhibitors for binding to photosystem II, *FEBS Lett.* **126,** 277–281 (1981).
132. B. R. Velthuys and J. Amesz, Charge accumulation at the reducing side of system 2 of photosynthesis, *Biochim. Biophys. Acta* **333,** 85–94 (1974).
133. B. Bouges-Bocquet, Electron transfer between the two photosystems in spinach chloroplasts, *Biochim. Biophys. Acta* **314,** 250–256 (1973).
134. M. P. J. Pulles, H. J. van Gorkom, and J. G. Willemsen, Absorbance changes due to the charge-accumulating species in system 2 of photosynthesis, *Biochim. Biophys. Acta* **449,** 536–540 (1976).
135. M. C. W. Evans, Y. E. Atkinson, and R. C. Ford, Redox characterization of the photosystem II electron acceptors. Evidence for two electron carriers between pheophytin and Q, *Biochim. Biophys. Acta* **806,** 247–254 (1985).
136. J. Amesz, Spectrophotometric evidence for the participation of a quinone in photosynthesis of intact blue-green algae, *Biochim. Biophys. Acta* **79,** 257–265 (1964).
137. B. Rumberg, P. Schmidt-Mende, and H. T. Witt, Different demonstrations of the coupling of two light reactions in photosynthesis, *Nature* **201,** 466–468 (1964).
138. J. Amesz and L. N. M. Duysens, Primary and associated reactions of system II, in: *Primary Processes of Photosynthesis,* Vol. 2 (J. Barber, ed.), pp. 149–185, Elsevier, Amsterdam (1977).

139. Govindjee, T. Kambara, and W. Coleman, The electron donor side of photosystem II: The oxygen evolving complex, *Photochem. Photobiol.* **42,** 187–210 (1985).
140. P. M. Wood and D. S. Bendall, The reduction of plastocyanin by plastoquinol-1 in the presence of chloroplasts, *Eur. J. Biochem.* **61,** 337–344 (1976).
141. G. T. Babcock, R. E. Blankenship, and K. Sauer, Reaction kinetics for positive charge accumulation on the water side of chloroplast photosystem II, *FEBS Lett.* **61,** 286–289 (1976).
142. J. A. van Best and P. Mathis, Kinetics of reduction of the oxidized primary electron donor of photosystem II in spinach chloroplasts and in *Chlorella* cells in the microsecond and nanosecond time ranges following flash excitation, *Biochim. Biophys. Acta* **503,** 178–188 (1978).
143. J. M. Briantais, C. Vernotte, M. Miyao, N. Murata, and M. Picaud, Relationship between O_2 evolution and cytochrome b-559 high-potential form in photosystem II particles, *Biochim. Biophys. Acta* **808,** 348–351 (1985).
144. P. Joliot, G. Barbieri, and R. Chabaud, Un nouveau modèle des centres photochimiques du système II, *Photochem. Photobiol.* **10,** 309–329 (1969).
145. B. Kok, B. Forbush, and M. McGloin, Cooperation of charges in photosynthetic O_2 evolution. I. A linear four step mechanism, *Photochem. Photobiol.* **11,** 457–475 (1969).
146. G. M. Cheniae, Photosystem II and O_2 evolution, *Annu. Rev. Plant Physiol.* **21,** 467–498 (1970).
147. J. Amesz, The role of manganese in photosynthetic oxygen evolution, *Biochim. Biophys. Acta* **726,** 1–12 (1983).
148. G. C. Dismukes, The metal centers of the photosynthetic oxygen evolving complex, *Photochem. Photobiol.* **43,** 99–115 (1986).
149. G. M. Cheniae and I. F. Martin, Sites of function of manganese within photosystem II. Roles in O_2 evolution and system II, *Biochim. Biophys. Acta* **197,** 219–239 (1970).
150. P. Homann, Effects of manganese on the fluorescence of chloroplasts, *Biochem. Biophys. Res. Commun.* **33,** 229–234 (1968).
151. J. Barber, Has the mangano-protein of the water splitting reaction of photosynthesis been isolated? *Trends Biochem. Sci.* **9,** 79–80 (1984).
152. D. A. Abramowicz, T. K. Raab, and G. C. Dismukes, Manganese proteins from spinach thylakoids and O_2-evolving photosystem-II enriched membranes (OEPII): Their role in oxygen evolution, in *Advances in Photosynthesis Research,* Vol. 1 (C. Sybesma, ed.), pp. 349–353, Martinus Nijhoff/Dr. W. Junk, The Hague (1984).
153. R. E. McCarty, Photosynthetic phosphorylation by chloroplasts of higher plants, in: *Photochemical and Photobiological Reviews,* Vol. 5 (K. C. Smith, ed.), pp. 1–48, Academic Press, New York (1980).
154. D. R. Ort and B. A. Melandri, Mechanism of ATP synthesis, in: *Photosynthesis, Energy Conversion by Plants and Bacteria,* Vol. 1 (Govindjee, ed.), pp. 537–587, Academic Press, New York (1982).
155. D. I. Arnon, M. B. Allen, and F. R. Whatley, Photosynthesis by isolated chloroplasts, *Nature* **174,** 394–396 (1954).
156. D. I. Arnon, The discovery of photosynthetic phosphorylation, *Trends Biochem. Sci.* **9,** 258–262 (1984).
157. A. Frenkel, Light induced phosphorylation by cell-free preparations of photosynthetic bacteria, *J. Am. Chem. Soc.* **76,** 5568–5569 (1954).
158. G. Hind and A. T. Jagendorf, Separation of light and dark stages in phosphorylation, *Proc. Natl. Acad. Sci. USA* **49,** 715–722 (1963).
159. J. Barber (ed.), *Electron Transport and Photophosphorylation,* Elsevier, Amsterdam (1982).
160. P. Mitchell, Coupling of phosphorylation to electron and hydrogen transfer by a chemi-osmotic type of mechanism, *Nature* **191,** 144–148 (1961).
161. P. Mitchell, Chemiosmotic coupling in oxidative and photosynthetic phosphorylation, *Biol. Rev.* **41,** 445–502 (1966).
162. G. G. Hammes, Mechanisms of ATP synthesis and coupled proton transport: Studies with purified chloroplast coupling factor, *Trends Biochem. Sci.* **8,** 131–134 (1983).
163. R. E. McCarty and C. M. Nalin, Structure, mechanism, and regulation of the chloroplast H^+-ATPase (CF_1-CF_0), in: *Encyclopedia of Plant Physiology,* new series, Vol. 19 (L. A. Staehelin and C. J. Arntzen, eds.), pp. 576–583, Springer-Verlag, Berlin (1986).
164. H. Strotmann, Evaluation of results on nucleotide-binding sites of the ATPase complex, in: *Encyclopedia of Plant Physiology,* new series, Vol. 19 (L. A. Staehelin and C. J. Arntzen, eds.), pp. 584–594, Springer-Verlag, Berlin (1986).

165. H. Strottmann and S. Bickel-Sandkötter, Structure, function, and regulation of chloroplast ATPase, *Annu. Rev. Plant Physiol.* **35,** 97–120 (1984).
166. P. D. Boyer, Coupling mechanisms in capture, transmission and use of energy, *Annu. Rev. Biochem.* **46,** 957–966 (1977).
167. E. C. Slater, Mechanism of oxidative phosphorylation, *Annu. Rev. Biochem.* **46,** 1015–1026 (1977).
168. P. D. Boyer and W. E. Kohlbrenner, The present status of the binding-change mechanism and its relation to ATP formation by chloroplasts, in: *Energy Coupling in Photosynthesis* (B. R. Selman and S. Selman-Reimer, eds.), pp. 231–240, Elsevier, Amsterdam (1981).
169. D. R. Ort and B. A. Melandri, Mechanism of ATP synthesis, in: *Photosynthesis: Energy Conversion by Plants and Bacteria,* Vol. 1 (Govindjee, ed.), pp. 537–587, Academic Press, London (1982).
170. B. A. Melandri and G. Venturoli, Local and delocalized interactions in energy coupling, in: *Encyclopedia of Plant Physiology,* new series, Vol. 19 (L. A. Staehlin and C. J. Arntzen, eds.), pp. 560–569, Springer-Verlag, Berlin (1986).
171. R. A. Dilley, Proton-membrane interactions in chloroplast thylakoids, in: *Encyclopedia of Plant Physiology,* new series, Vol. 19 (L. A. Staehelin and C. J. Arntzen, eds.), pp. 570–575, Springer-Verlag, Berlin (1986).
172. A. T. Jagendorf and E. Uribe, ATP formation caused by acid-base transition of spinach chloroplasts, *Proc. Natl. Acad. Sci. USA* **55,** 170–177 (1966).
173. R. A. Dilley, Coupling of ion and electron flow in chloroplasts, *Curr. Top. Bioenerg.* **4,** 237–271 (1971).
174. H. Baltscheffsky and M. Baltscheffsky, Energy conversion reactions in bacterial photosynthesis, *Curr. Top. Bioenergy.* **4,** 273–325 (1971).
175. C. J. Arntzen and J. M. Briantais, Chloroplast structure and function, in: *Bioenergetics of Photosynthesis* (Govindjee, ed.), pp. 51–113, Academic Press, New York (1975).
176. S. Kaplan and C. J. Arntzen, Photosynthetic membrane structure and function, in: *Photosynthesis: Energy, Conversion by Plants and Bacteria,* Vol. 1 (Govindjee, ed.), pp. 65–151, Academic Press, New York (1982).
177. J. M. Andersson, Consequences of spatial separation of photosystem 1 and 2 in thylakoid membranes of higher plant chloroplasts, *FEBS Lett.* **124,** 1–10 (1981).
178. J. M. Andersson and B. Andersson, The architecture of photosynthetic membranes: lateral and transverse organization, *Trends Biochem. Sci.* **7,** 288–292 (1982).
179. C. Hitchcock and B. W. Nichols, *Plant Lipid Biochemistry,* Academic Press, London, New York (1971).
180. J. Bennett, K. E. Steinback, and C. J. Arntzen, Chloroplast phosphoproteins: Regulation of excitation energy transfer by phosphorylation of thylakoid membrane proteins. *Proc. Natl. Acad. Sci. USA* **77,** 5253–5257 (1980).
181. D. C. Fork and K. Satoh, The control by state transitions of the distribution of excitation energy in photosynthesis, *Annu. Rev. Plant Physiol.* **37,** 335–361 (1986).
182. J. A. Bassham, Kinetic studies of photosynthetic carbon reduction cycle, *Annu. Rev. Plant Physiol.* **15,** 101–120 (1964).
183. J. A. Bassham and B. B. Buchanan, Carbon dioxide fixation pathways in plants and bacteria, in: *Photosynthesis: Development, Carbon Metabolism, and Plant Productivity,* Vol. 2 (Govindjee, ed.), pp. 141–189, Academic Press, New York (1982).
184. M. D. Hatch and C. R. Slack, Photosynthetic CO_2 fixation pathways, *Annu. Rev. Plant Physiol.* **21,** 141–162 (1970).
185. O. Björkman and J. Berry, High efficiency photosynthesis, *Sci. Am.* **229,** 80–93 (1973).
186. J. A. Berry and G. Farquhar, The CO_2 concentrating function of C_4 photosynthesis. A biochemical model, in: *Proc. Fourth Int. Congr. on Photosynthesis* (D. O. Hall, J. Coombs, and T. W. Goodwin, eds.), pp. 119–131, The Biochemical Society, London (1978).
187. M. Kluge, Crassulacean acid metabolism, in: *Photosynthesis: Development, Carbon Metabolism, and Plant Productivity,* Vol. 2 (Govindjee, ed.), pp. 141–189, Academic Press, New York (1982).
188. C. B. Osmond, Crassulacean acid metabolism: A curiosity in context, *Annu. Rev. Plant Physiol.* **29,** 379–414 (1978).
189. W. L. Ogren and R. Chollet, Photorespiration, in: *Photosynthesis: Development, Carbon Metabolism, and Plant Productivity,* Vol. 2 (Govindjee, ed.), pp. 191–230, Academic Press, New York (1982).
190. I. Zelitsch, *Photosynthesis, Photorespiration, and Plant Productivity,* Academic Press, New York, London (1971).

191. C. B. Osmond and O. Björkman, Simultaneous measurements of oxygen effects on net photosynthesis and glycolate-metabolism in C_3 and C_4 species of *Atriplex, Carnegie Inst. Wash. Year Book* **71,** 141–148 (1972).
192. J. S. Connolly (ed.), *Photochemical Conversion and Storage of Solar Energy,* Academic Press, New York (1981).
193. D. O. Hall, Photosynthesis for Energy, in: *Advances in Photosynthesis Research,* Vol. 2 (C. Sybesma, ed.), pp. 727–740, Martinus Nijhoff/Dr. W. Junk, The Hague (1984).
194. L. N. M. Duysens, Specially developed photosynthetic organisms as future possibility for large-scale low-cost conversion of solar radiation into useful high-grade energy, in: *Advances in Photosynthesis,* Vol. 2 (C. Sybesma, ed.), pp. 741–744, Martinus Nijhoff/Dr. W. Junk, The Hague (1984).
195. M. Calvin, Solar energy by photosynthesis, *Science* **184,** 375–381 (1974).
196. M. Avron, The outdoor cultivation of the halotolerant alga *Dunaliella*: A model for biosolar energy utilization for useful chemical products, in: *Advances in Photosynthesis Research,* Vol. 2 (C. Sybesma, ed.), pp. 745–753, Martinus Nijhoff/Dr. W. Junk, The Hague (1984).
197. S. G. Boxer, Model reactions in photosynthesis, *Biochim. Biophys. Acta* **725,** 265–292 (1983).
198. J. Barber (ed.), *Photosynthesis in Relation to Model Systems,* Topics in Photosynthesis, Vol. 3, Elsevier, Amsterdam (1979).
199. M. T. Black, T. H. Brearley, and P. Horton, Heterogeneity in chloroplast photosystem II, *Photosyn. Res.* **8,** 193–207 (1986).
200. R. C. Prince, Redox-driven proton gradients, *Bioscience* **35,** 22–26 (1985).
201. P. D. Boyer, B. Chance, L. Ernster, P. Mitchell, E. Racker, and E. C. Slayter, Oxidative phosphorylation and photophosphorylation. *Annu. Rev. Biochem.* **46,** 955–1026 (1977).

13

Bioluminescence

13.1. INTRODUCTION

For most people, bioluminescence is represented by the flash of the firefly or the "phosphorescence" that frequently occurs on agitating the surface of ocean water. Because of the ease of obtaining material, the firefly's bioluminescence reaction was one of the first to receive detailed biochemical study, with the result that this system is the archetype of a

John Lee • Bioluminescence Laboratory, School of Chemical Sciences, Department of Biochemistry, University of Georgia, Athens, Georgia 30602.

variety of enzymatic processes that produce light in the many bioluminescent organisms, ranging from the marine bacteria to the large luminous beetles from South America. What is usually understood by the term *bioluminescence* is a cold-light emission of high efficiency that has a biological function for the organism concerned, although what this function is is still conjectural in many cases. In contrast to bioluminescence, a great number of biological processes have been observed to be accompanied by the emission of a very low level of light, so-called biological chemiluminescence. This observation and the fact that bioluminescence is so widespread among many phyla (although rarely are many members of a phylum bioluminescent) has led to the suggestion that low-level light emission arose very early in evolution, and that the efficient bioluminescence was a secondary adaptation that enabled some species to compete more effectively in their biological niches.

13.1.1. Occurrence

More than half of all phyla in the animal kingdom contain members that are bioluminescent.[1] Many bioluminescent organisms have a worldwide distribution and some are found only in certain areas of the world. Table 13-1 lists the organisms that have been well studied.

In the first half of this century, E. N. Harvey of Princeton University made a lifelong

TABLE 13-1. Properties of Bioluminescent Organisms

Type and examples	Bioluminescence maxima range *in vivo* λ_B(nm)	Occurrence
Bacteria, *Photobacterium, Vibrio*	472–545	Marine
Fungi	525–550	Terrestrial
Protozoans, dinoflagellates	470–480	Marine
Coelenterates	455–508	Marine
Hydrozoan jellyfish (*Aequorea*)		
Ctenophores (*Mnemiopsis*)	480–505	Marine
Anthozoans (*Renilla*)	505–510	Marine
Annelids		
Polychaete worms	465–520	Marine
Earthworms	500–570	USA, Australasia
Molluscs		
Limpet (*Latia*)	535	Freshwater, New Zealand
Bivalve (*Pholas*)	490	Mediterranean
Squids	416–540	Marine
Crustaceans (*Cypridina*)	465	Marine, Japan
Euphausiids and decapods	455–475	Marine
Millipedes	496	USA
Insects		
Firefly	552–582	Terrestrial
Glow-worm (*Arachnocampa*)	488	Australasia
Railroad worm (*Phrixothrix*)	560,630	Brazil
Fish	455–507	Marine

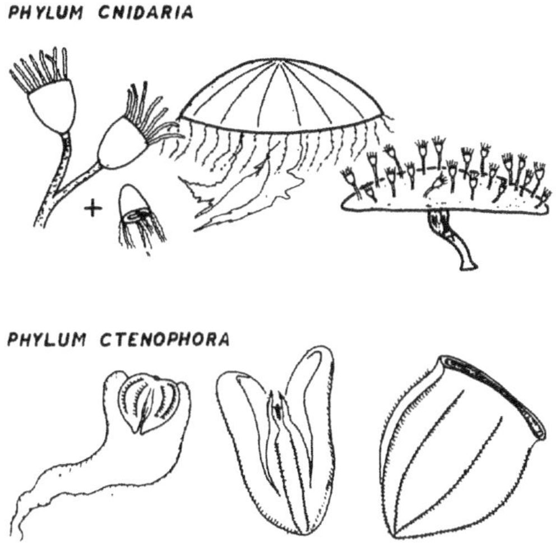

Fig. 13-1. Some bioluminescent coelenterates. Left to right, Cnidaria: Hydrozoa (two types), Scyphozoa, and Anthozoa (*Renilla*); Ctenophora: Tentaculata (*Pleurobrachia* and *Mnemiopsis*) and Nuda. [From W. W. Ward, Energy Transfer Processes in Bioluminescence, *Photochem. Photobiol. Rev.* **4,** 1–57, (1979); see this reference for a full description.]

study of bioluminescence, especially of its biological aspects. Over 40 years ago, he wrote:

> Apparently, there is no rhyme or reason in the distribution of luminescence throughout the plant or animal kingdom. It is as if the various groups had been written on a blackboard and a handful of damp sand cast over the names. Where each grain of sand strikes, a luminous species appears. . . . Several large groups contain no luminous forms whatever. It is an extraordinary fact that one species in a genus may be luminous and another closely allied species contain no trace of luminosity.[2]

Bioluminescence occurs in many terrestrial forms but is most common in the sea, particularly in the deep ocean where the majority of species are luminescent.[1] On land, in addition to the firefly, there exist luminous fungi (''foxfire''), glowworms, freshwater snails (*Latia*) found only in New Zealand, earthworms, and many insects including the ''railroad worm'' found in South America. In the ocean, protozoa are mainly responsible for the so-called phosphorescence, but the coelenterates contain the most bioluminescent species, i.e., the soft corals (Anthozoa), jellyfish (Hydrozoa), and comb-jellies (Ctenophora). Some representatives from the coelenterates are illustrated in Fig. 13-1.

There are many types of luminous fish some of which derive their luminescence from a culture of symbiotic bacteria. There are marine worms (*Chaetopterus, Balanoglossis, Odontosyllis*), a clam (*Pholas*), crustacea (*Cypridina* or *Vargula*), squid (*Watasenia*), shrimp (*Holophorus*), and echinoderms (sea stars and sea urchins). Bioluminescence is not found in any organisms higher than fish.

13.1.2. Light Organs and Organization

In most cases, the bioluminescence is manufactured within specialized cells called photocytes. Many higher vertebrates have a highly organized collection of these cells making up a light organ or photophore. Harvey observed that ''there seems to have been no development of luminosity along direct evolutionary lines, although a more or less definite series of gradations with increasing structural complexity may be traced among the forms with highly developed luminous organs.''[2]

In those fish that utilize the bioluminescence from symbiotic luminous bacteria, the photophore is like a culture vessel. Many other fish use a different bioluminescence system to produce light involving either intra- or intercellular processes. There also exist a number of highly specialized accessory structures to the light organ, in some cases reaching a degree of complexity comparable to, and in fact resembling, the vertebrate eye.

13.1.3. Origin and Function of Bioluminescence[3–5]

The biological purpose of bioluminescence is to be seen, so it could not have been until the development of primitive visual processes that evolutionary pressure favored the diversion of valuable metabolic energy into light production. Energy conservation is perhaps the reason that bioluminescence systems are found to be very efficient. One proposal for the origin of bioluminescence is that the first bioluminescence system developed out of a biological chemiluminescence reaction involving an oxygenase enzyme acting on a hydrocarbon. This mechanism is a free-radical one. The accumulation of a suitably fluorescent metabolite, e.g., the highly fluorescent proteins that are found in the bioluminescent bacteria and coelenterates, would then provide a means to sensitize or amplify this light production at a minimal energy cost. We will see shortly how such a mechanism operates in the bacterial and coelenterate bioluminescences. Higher on the evolutionary scale, selection was made of reactions that would produce highly fluorescent product molecules directly in their excited singlet state, and with everything done on only one enzyme. This is the mechanism of the firefly, *Cypridina*, and other higher forms.

It is a reasonable (though controversial) conjecture that the bacteria were the first organisms to "discover" bioluminescence and a niche in which it could be of advantage. This is based on the facts that bacteria lie lowest on the evolutionary scale, they can mutate and evolve rapidly if benefit is to be derived from it, and they already possess the substrates for bioluminescence. Although luciferase, the enzyme needed to generate the bioluminescence, is unique to the bioluminescent bacteria, it has been postulated that luciferase might also have a terminal oxidase function in a respiratory pathway operative at a very low oxygen level, i.e., one approaching anaerobiosis.

The alternative view held by most scientists in the field is that bioluminescence arose independently in many organisms. This is because the property is spread over such a diversity of organisms that it seems a relatively recent event in evolution, and the luciferins often have very different chemical structures, even among closely related organisms. Additionally, there is no structural property of the luciferases that would suggest a common lineage.

The biological aspect of bioluminescence function that has received the most study and is best understood is that of the firefly system.[6] The flash is a sexual signaling and is made up of a series of short flashes (Fig. 13-2), a "Morse code" that is characteristic of the species. The flying male flashes and may be answered after a definite time interval, about 2 s, by the female. In one exceptional though interesting case, the female of one species is insectivorous, feeding on the males of other species and luring its prey to it by imitating the female response appropriate to the prey species.

There is also a visually discernible difference in the color of bioluminescence between certain species of firefly. This appears to be an adaptation since the yellow bioluminescence (Fig. 13-3c) is from fireflies active at dusk, the yellow light contrasting

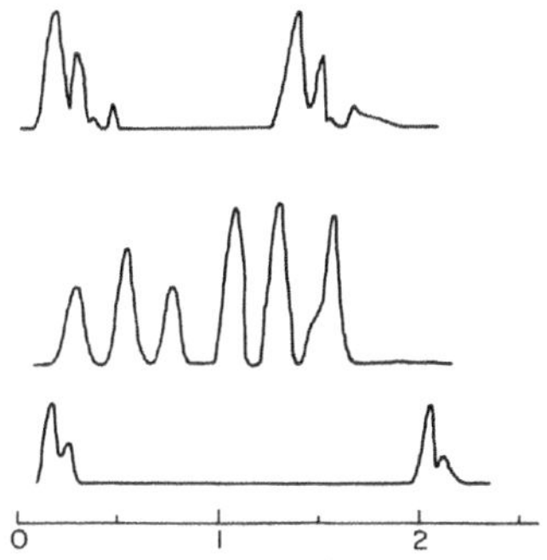

Fig. 13-2. Flash sequence from live fireflies; the vertical direction represents relative light intensity. (Top) *Photinus gracilobus*, (middle) *Photinus melanuris*, (lower) *Photinus nothus*. [Redrawn from H. H. Seliger and W. D. McElroy, *Light: Physical and Biological Action*, Academic Press, New York (1965).]

against the green background of foliage. The night-active fireflies emit at shorter wavelengths (Fig. 13-3a), the green light more efficiently matching the visual sensitivity of their species.(7)

It has recently been shown that starting with a mixed population of luminous and nonluminous dinoflagellates and their predators under controlled laboratory conditions, the luminous types show an increase in survival over nonluminous types. The luminous flash probably deters the predator, conferring survivorship on the species that possess this property. Other escape flashes are evident in shrimp, squid, and the *Cypridina*, but have received little controlled study.

Some luminous fish have a light organ that houses a symbiotic culture of luminous bacteria. Some fish (the "flashlight" fish) have even adapted a lid to move over the light organ to create a flashing of the otherwise continuous emission. Other fish appear to contain a luminous system that is the same as in *Cypridina*, i.e., the luciferins and luciferases cross-react. The luminescence probably serves as a mating signal, or as a lure.

In the Australasian glowworm, which is the larva of a dipteran (fly), in contrast to the firefly, which is a coleopteran (beetle), the luminescence has a lure function. The larva hangs in the middle of a web and emits a blue glow. Small, winged insects are attracted to the glow, ensnared in the web, and devoured.

Bioluminescence, therefore, appears to be a biological chemiluminescence that has been adapted and refined to function in ways advantageous for all species that possess it.

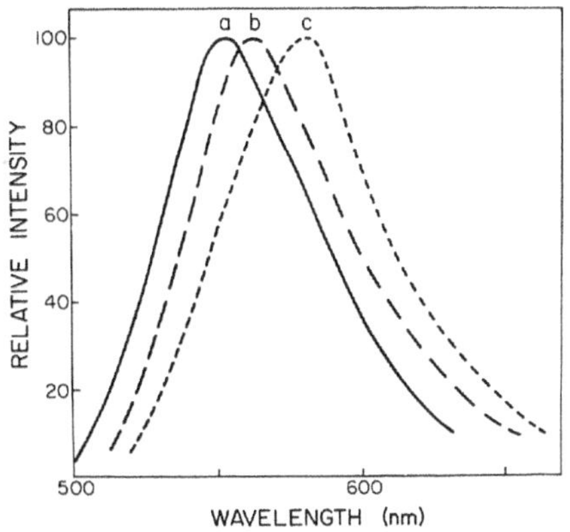

Fig. 13-3. Firefly bioluminescence spectra *in vivo*. (a) *Photurus pennsylvanicus;* (b) *Photinus pyralis;* (c) *Pyrophorus plagiothalamus*. [Redrawn from H. H. Seliger and W. D. McElroy, *Light: Physical and Biological Action*, Academic Press, New York (1965).]

Seliger summarized the biological functions of bioluminescence by the "four P's": predation, protection, prenuptial (firefly communication), and perfidy (the insectivorous firefly).

13.1.4. Mechanism

The color of the emitted light ranges from the red of the railroad worm through the deep blue characteristic of most of the marine creatures. Thus, the first questions that might arise in the mind of an investigator are: What is the chemical reaction and the identity of the molecule responsible for the light emission? Why are the colors different? Are all the bioluminescence systems the same or similar in mechanism?

It is relatively straightforward to measure the spectral emission of the bioluminescence from an organism *in vivo,* and this is usually characterized by the wavelength position of the spectral emission maximum, which we will designate by the symbol λ_B. Table 13-1 shows that many types of organism have a range of λ_B; often the λ_B is found to be species-dependent.

However, it is considerably more difficult to obtain extracts from an organism that in solution will react to produce the same bioluminescence. Extensive efforts are usually needed to obtain the chemically pure components. If the pure bioluminescence reaction has high efficiency, as measured by the quantum yield of bioluminescence (see Section 13.2), then the steps leading to light emission can be characterized in a chemical sense. Since a chemiluminescence reaction is one that proceeds with sufficient release of energy to produce a molecule in the system in an electronically excited state, bioluminescence can be viewed as enzyme-catalyzed chemiluminescence.

To return to the first of the above questions, it was discovered by Boyle in the 17th century that air was required for the bioluminescence of bacteria and fungi, and it is now known that oxygen in some form is involved in all bioluminescence systems. At the end of the last century, Dubois found that two components could be extracted from the firefly light organ, one with hot water and the other with cold water. The cold-water extract, which was heat-labile, he named luciferase, and the hot-water extract, luciferin. He made a similar observation with the clam *Pholas* but found that a cross-reaction between clam luciferin and firefly luciferase, and vice versa, did not occur.[(2)] Sometimes it has been found that the luciferins (designated LH_2 in Table 13-2) and the luciferases of bioluminescent members of a class do cross-react; this is also occasionally observed between classes, and even in more distant relationships. The ability to cross-react or not is a consequence of the chemical structure of the luciferin, which is quite different, say, between the clam and firefly, and therefore the luciferin from the one is not a substrate for the luciferase of the other species.

In Table 13-2 the properties of some of the better studied bioluminescence systems are given. It needs to be emphasized here that the symbol LH_2 is used for the organic molecule substrate in the reaction. It is a generic symbol and we will see later that the chemical structure of this molecule is usually different between the types of bioluminescent organisms. Table 13-2 also shows the remarkable biochemical variability from one type of reaction to the next. Most exhibit the classical enzyme (luciferase)–substrate reaction, but some have a precharged system called a "photoprotein," which will be explained later. Some systems only involve luciferase and luciferin with O_2; others use several proteins. Some luciferases are small proteins, whereas others are large, multi-

TABLE 13-2. Properties of Bioluminescence Systems

Source	Substrates	Proteins involved (M_r; subunits)
Bacteria	$FMNH_2$, RCHO,O_2	Luciferase (76,000; α, β)
		Lumazine protein (21,200)
Dinoflagellates	LH_2, O_2, H^+	Luciferase (130,000)
		LH_2-binding protein
Coelenterates: Hydrozoa, Ctenophores	Ca^{2+}, O_2	Photoprotein (21,000)
		Green-fluorescent protein (30,000)
Anthozoa	LH_2, Ca^{2+}, O_2	Luciferase (35,000)
		Green-fluorescent protein (52,000; α_2)
		LH_2-binding protein (18,500)
Crustaceans: *Cypridina*	LH_2, O_2	Luciferase (68,000; α_6)
Firefly	Lh_2, ATP, O_2	Luciferase (60,000)

subunit enzymes. Also, each luciferase is specific for each type of organism, and no relatedness has been observed between the luciferases.

Thus, the biochemistry of bioluminescence is analogous to the biological potpourri remarked on by Harvey, i.e., each system has found its own pathway to the final product, light emission. Many believe that this argues for bioluminescence being a case of convergent evolution. However, what has emerged in recent years is that, in contrast to the diversity of its biology and biochemistry, the subject is unified by its chemical mechanism. To expose this common mechanistic thread that runs through each system, we will first need to consider some basic physical principles, and then consider in depth several of the well-studied bioluminescence mechanisms.

13.2. CHEMILUMINESCENCE REACTIONS[8]

Many chemical reactions may proceed with the release of sufficient free energy to excite one of the participants into an excited electronic state, but visible light does not result unless certain conditions are met. In the gas phase at low pressure, the rate of energy release on bond formation can be faster than the rate of dissipation of excess energy by collisions with the surrounding molecules. What is called "nonadiabatic crossing" to the excited state manifold of products is a likely occurrence, with the consequent generation of visible or infrared chemiluminescence.

In solution, the time scales for collision with solvent molecules and for atomic motions are the same. In other words, the vibrational interaction with the medium is so predominant that fast cooling by collision with solvent molecules prevents any highly energetic or "hot" species from being formed as the result of reactions involving the rearrangement of atoms. Therefore, in 1939, Weiss proposed that the only process fast enough to produce chemiluminescence excitation in solution is electron transfer, and this idea has received much experimental support.[9] An example is the one-electron oxidation of a donor molecule (D) by an acceptor (A) where there is a large redox energy difference between the pair. On placing two electrodes into a solution of D and A, electron transfer pairs of opposite charge can be generated under very precisely controlled conditions, and their mutual annihilation produces "electrochemiluminescence" if there is sufficient energy to place one of the products in its excited electronic state. There now exists a detailed understanding of electrochemiluminescence reactions, and a good example is the electron transfer between the radical anion of diphenylanthracene, $DPA^{\bullet -}$, and its radical cation, $DPA^{\bullet +}$, where about 10% of the products appear in the excited state DPA*):

$$DPA^{\bullet +} + DPA^{\bullet -} \rightarrow 0.2\ DPA^{*} + 1.8\ DPA \tag{13-1}$$

The asterisk is used here to mean that a compound is in an excited singlet state, the same state from which fluorescence occurs.

Bioluminescence reactions are of course not electrochemical, and so we must look elsewhere for chemical reactions that can be used as bioluminescence models. The appropriate model reactions are the oxygen oxidations of certain organic compounds that give good yields of chemiluminescence, and it is well established that some of these proceed through the formation of an intermediate cyclic peroxide, often a dioxetanone (I), which contains a high chemical energy content by virtue of its highly strained ring. The decomposition of dioxetanones, however, usually generates the product ketone more in its triplet than singlet state:

(I)

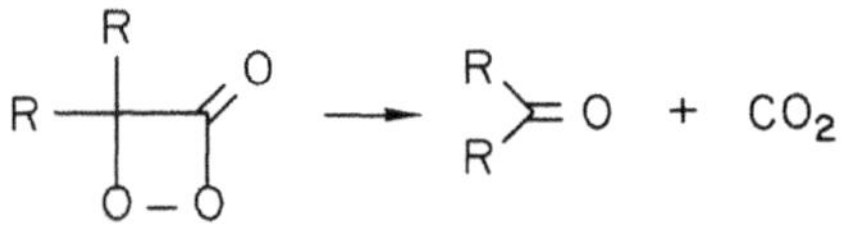

(13-2)

A triplet product does not yield chemiluminescence, but rather its energy is degraded to heat.

Therefore, it was proposed by Schuster and by McCapra (see Ref. 8) that an electron transfer is the necessary excitation step in the chemiluminescent breakdown of dioxetanes or other cyclic peroxides that give excited singlets in substantial yield. The best evidence for this is in the chemiluminescence of diphenoyl peroxide (II), as catalyzed by the presence of an easily oxidizable and fluorescent "activator" such as rubrene. The mechanism has been shown to be recombination of a rearranged radical ion pair within a solvent cage:

II

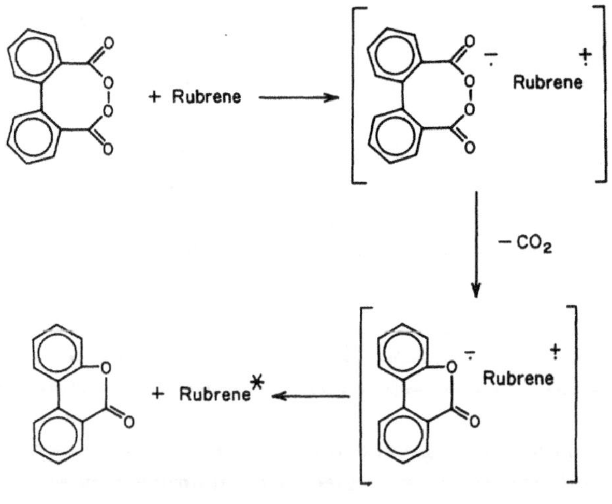

(13-3)

Yields of the excited activator up to 10% have been reported. This mechanism has been dubbed "chemically induced electron exchange luminescence" (CIEEL). It is believed that the decomposition of dioxetanes yields excited states by a similar mechanism, but this is not so firmly established.

Prior to knowledge of the chemical mechanism, it is often useful to categorize chemiluminescence reactions on the basis of spectra. If the emitting molecule is a product of the reaction, the process is called "direct chemiluminescence":

$$A + B \rightarrow C^* \tag{13-4}$$

$$C^* \rightarrow C + h\nu_1 \tag{13-5}$$

The spectral distribution of direct chemiluminescence corresponds to the fluorescence of C [Eq. (13-5)]. The excitation energy, however, may end up on another molecule D, which, although present in the solution, is neither a reactant nor a product of the chemistry. This is called "indirect" or "sensitized" chemiluminescence by analogy to sensitized photochemistry and sensitized fluorescence:

$$A + B + D \rightarrow C + D^* \tag{13-6}$$

$$D^* \rightarrow D + h\nu_2 \tag{13-7}$$

The chemiluminescence emission in this case is the same as the fluorescence of D. D is called the chemiluminescence "sensitizer" to correspond to the photochemical sensitizer in the reverse of Eqs. (13-7) and (13-6). In both cases the sensitizer is the component unchanged by the chemiluminescent or photochemical process.

There are some cases of sensitized chemiluminescence in which the nature of the primary excited species C is established or strongly inferred, and the excitation of D is by one of the energy transfer processes discussed in Chapter 1. Electron transfer interactions between excited molecules are also possibilities. In most cases the identity of C is not established; the most outstanding example is the reaction of oxalyl chloride and hydrogen peroxide in the presence of a dye, which is one of the most efficient chemiluminescence reactions known. This is the reaction used in the commercially available "light-stick."

The quantum yield is an important characteristic of chemiluminescence reactions, just as it is for photochemical reactions (Chapter 2). In order to preserve the Second Law of Photochemistry, i.e., the correspondence between one photon and one molecule, we define the chemiluminescence quantum yield with respect to one of the chemical participants, e.g., molecule A in Eq. (13-4), $\phi_C(A)$, as the number of photons emitted divided by the number of A reacted. Note that this is the inverse of the definition of quantum yield when photons are entering the system. For a bioluminescence quantum yield, we use the symbol $\phi_B(A)$, or $\phi_B(C)$ if the quantum yield is with respect to the number of molecules of C produced, etc.

The ϕ_C for direct chemiluminescence is the product of two efficiencies, one for the fluorescence of the emitter C, $\phi_F(C)$ [Eq. (13-5)], and the other ϕ_E, the efficiency for putting the product into its excited state [Eq. (13-4)]. We can then write:

$$\phi_C(A) = \phi_E \cdot \phi_F(C) = \phi_C(B) = \phi_C(C) \tag{13-8}$$

since, as written, Eq. (13-4) has unit stoichiometry.

We have already seen in Chapter 1 how a number of competing processes reduce ϕ_F below unity. For reasons that are not understood, ϕ_E is found to be much less than unity for most chemiluminescence reactions in solution, but in the gas phase, and also surprisingly for some bioluminescence reactions for which reliable information is at hand, ϕ_E can be very nearly unity. Thus, any mechanism proposed for excitation in bioluminescence has to account for this important observation.

The ϕ_F obviously cannot exceed unity, and this means that $\phi_F(C) > \phi_C(C)$. In other words, any proposed emitter, as a necessary qualification, has to have a fluorescence quantum yield that is greater than the chemiluminescence quantum yield. Additionally, the parameters describing its fluorescence spectral distribution, i.e., the position of the maximum λ_F, the width at half-height, etc., have to be the same as for the chemiluminescence spectrum.

13.3. GENERALIZATIONS ABOUT BIOLUMINESCENCE REACTIONS[10,11]

The term *bioluminescence system* will be used to describe the combined sequence of reactions used by an organism to generate light. The present state of knowledge of the details of six or more bioluminescence systems, the chemical structure of the luciferins (LH_2), and the characteristics of the luciferase enzymes (E) suggest a general sequence of reactions that lead to the emission of light. Usually some preliminary steps occur to "prepare" the luciferin for the chemiluminescence reaction, but all the steps have to be untangled from the total system before the excitation step can be isolated for study. Some bioluminescence systems use all of the general sequence of reactions and some omit certain steps.

13.3.1. Activation

After the initial binding of the substrates to luciferase, the first step in the sequence is the activation of luciferin,

$$E:LH_2 + A \rightarrow E:LH_2B + C \tag{13-9}$$

where A is a cofactor and C is some product, e.g., adenosine triphosphate (ATP) and pyrophosphate, respectively, in the case of firefly bioluminescence. We use the symbol E: to designate the substrate interaction with the enzyme, usually known to be noncovalent.

13.3.2. Oxygenation

The second step in our generalized scheme is oxygenation;

$$E:LH_2B + O_2 \rightarrow E:LBHOOH \tag{13-10}$$

The *Cypridina* bioluminescence system leaves out the activation step; oxygenation of its luciferin is its first reaction.

13.3.3. Excitation

The peroxide breaks down via a mechanism described above to produce some excited molecule:

$$E{:}LBHOOH + M \rightarrow E{:}LBO^{*} + P + M \qquad (13\text{-}11)$$

where M is another cofactor that may be needed to set off this process, like Ca^{2+} in the bioluminescent jellyfish extract (aequorin) reaction (see below), and P is other products, like H_2O, or CO_2 for the firefly. Light emission results from deexcitation of E:LBO*.

13.3.4. Turnover

Finally, the organism might want to use the enzyme again, i.e., the luciferase is released ready to react again:

$$E{:}LBO \rightarrow E + LO + B \qquad (13\text{-}12)$$

Apparently the firefly does not need to do this *in vivo,* since it has all the luciferase it needs for a lifetime of one-time reactions. *In vitro,* firefly luciferase and most of the other types of luciferase can be made to turn over.

It is clear that to understand a bioluminescence reaction in chemical terms, the nature of the oxygenated species produced by Eq. (13-10) must first be established, and then the excitation process [Eq. (13-11)] can be proposed on the basis of model chemiluminescence reactions. In the following sections we shall examine the three bioluminescence systems that have received the most extensive studies, and show what sequence of reactions is involved, and where understanding of the excitation reaction has progressed to up to this time.

13.4. FIREFLY BIOLUMINESCENCE[12]

Firefly luciferin (LH_2) is D(-)-2-(6′-hydroxy-2′-benzothiazolyl)-Δ^2-thiazoline-4-carboxylic acid (III):

III

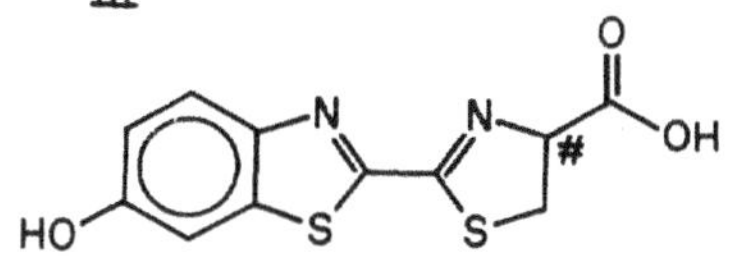

The reference mark (#) indicates the optically active center, and this turns out to be the "business end" of the molecule. The first step in the sequence is the activation, by a luciferase-catalyzed reaction, of the carbonyl group on LH_2 with ATP to form the luciferyl adenosine monophosphate derivative, LH_2AMP, with the release of pyrophosphate, PP_i:

$$E{:}LH_2 + ATP \rightarrow E{:}LH_2AMP + PP_i \qquad (13\text{-}13)$$

13.4.1. Chemical Mechanism

Oxygenation proceeds in several steps, the first possibly being Eq. (13-14):

$$E:LH_2AMP + O_2 \rightarrow E:LHOOHAMP \tag{13-14}$$

and there is ample but indirect evidence that this hydroperoxide cyclizes to the dioxetanone (IV). This has been suggested to decompose by a CIEEL mechanism:

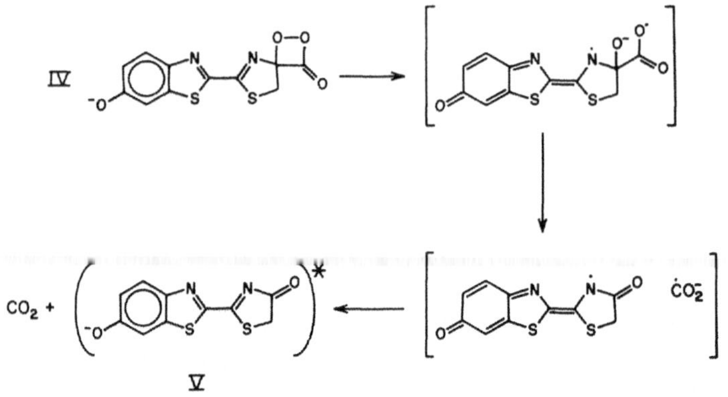

(13-15)

Here the excitation step is an intramolecular electron transfer, the benzothiazolyl portion of the molecule playing the role of the easily oxidizable fluorescent activator required for the CIEEL mechanism.

The intermediacy of the dioxetanone (IV) has been inferred from mass spectrometric or radioactivity analysis of the enzymatic reaction products with heavy-atom or radioactive labels.

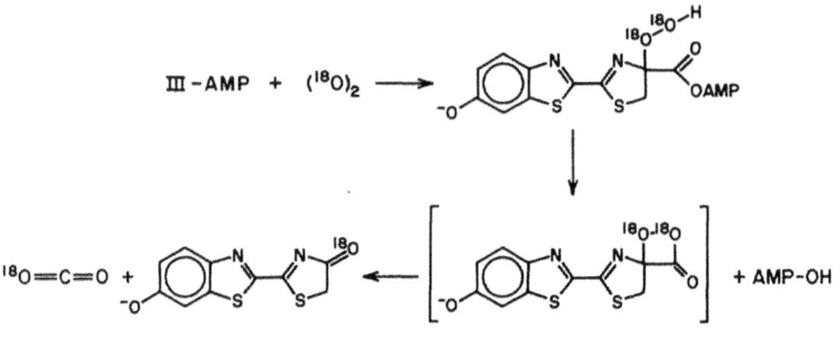

(13-16)

A cyclic peroxide intermediate leads to the product CO_2 containing one atom of ^{18}O. It took many years before these labeling results were agreed upon because the experiments are very difficult to perform.[13]

13.4.2. The Light Emitters

The $\phi_B(LH_2)$ has been measured to be almost unity, the highest found for any bioluminescence system, and the quantum yields for oxygen consumption $\phi_B(O_2)$ and

CO_2 production $\phi_B(CO_2)$ are about the same. We can infer that ϕ_E is unity, but unfortunately the product V [Eq. (13-15)] is too unstable to have its fluorescence measured accurately.

The *in vivo* and *in vitro* bioluminescence emission spectra are the same, but their maxima λ_B are species-dependent. Figure 13-3 shows three bioluminescence spectra representing the range of firefly bioluminescence encountered, $\lambda_B = 552–575$ nm. The *in vitro* spectra show the same species dependence, and since all these reactions use the same synthetic firefly luciferin, the spectral differences must originate in some property of the luciferase. It has been suggested that the binding site on the enzyme must differ slightly from one species of luciferase to the next, e.g., in electrostatic charge or dielectric constant. This produces what is analogous to a "solvent" effect on the fluorescence spectrum of the emitting product V (Chapter 2).

Firefly luciferin can be made to chemiluminesce if it is reacted with oxygen in an aprotic solvent, i.e., one like dimethylsulfoxide that has no available protons for hydrogen bonding. If the solution is kept very dry and made very basic, the chemiluminescence is yellow-green, with practically the same emission spectrum as the bioluminescence. In a weaker base, a red emission of lower quantum yield is found. The yellow-green emission ($\lambda_C = 562$ nm) comes from VI, a dianion that in solution may add a proton at a rate in competition with the radiative rate to form the red-emitting species ($\lambda_C = 615$ nm) V.[14]

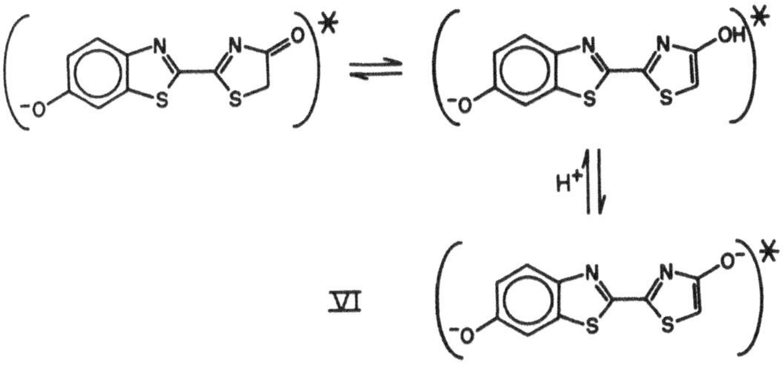

VI

(13-17)

Thus, the yellow bioluminescence is believed to come from VI. In the active site on luciferase, the protonation reaction VI → V is hindered unless the pH is lowered below 6 (Fig. 13-4). Raising the temperature or adding heavy metal ions (Zn^{2+} or Cd^{2+}) also induces the red emission band, and these are all conditions that must relieve the hindrance caused by luciferase on the protonation reaction [Eq. (13-17)].[15]

Before we go on to consider the other bioluminescence systems, let us first summarize what we have learned about the firefly reaction. A single protein, firefly luciferase, possesses all the required catalytic properties to generate yellow light: activation, oxygenation, and excitation. In addition, the protonation of the emitter is slowed to the point that it does not compete with the emission rate (about 10^8 s^{-1}). Firefly luciferin was the first one to have its structure fully determined. The luciferin is activated by reaction with ATP. The bioluminescence reaction products are H_2O, CO_2, and a molecule whose fluorescence spectral properties correspond to the bioluminescence. The fluorescence of product V also has an acid–base relationship between two emitting species, as does the chemiluminescence of luciferin in dimethylsulfoxide. Luciferases purified from different firefly species give slightly different emissions, ranging from yellow through orange, suggesting that the product environment on each enzyme is not the same. We shall encounter these same or modified properties when we deal with the bioluminescence systems from the lower orders.

13.5. *CYPRIDINA* AND COELENTERATE BIOLUMINESCENCE[10,16]

A substantial body of chemical knowledge is now available for many marine bioluminescence systems. Many of them use an imidazopyrazine luciferin (VII):

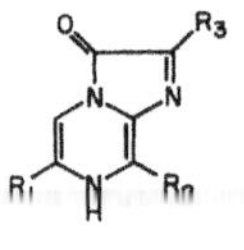

VII

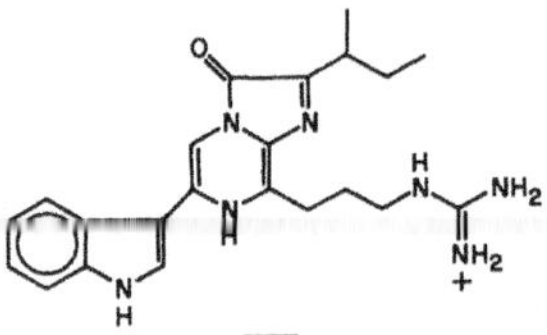

VIII

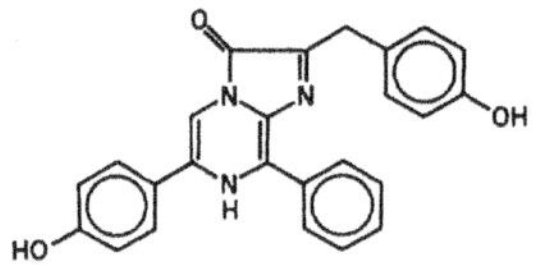

IX

Although the mechanism of oxidation is the same, the substituent groups (R) differ between *Cypridina* (VIII) and "coelenterate-type" (IX) luciferin (also called "coelenterazine"). A remarkable fact about these bioluminescence systems that use imidazopyrazine luciferin is that they have achieved their biological adaptations by certain changes at the molecular level that we can easily study and understand.

We should further contrast these bioluminescence systems with the firefly in that the latter is only found in insects that are terrestial, whereas systems utilizing the imidazopyrazine luciferin are found in seven phyla (out of a recognized 30 total animal phyla) all in the marine environment. Firefly luciferase alone contains all the functions required for the bioluminescence system on one protein, but other systems, the coelenterates in particular, utilize a variable number of enzymes and proteins with an additional variety of packaging.

13.5.1. Chemical Mechanism

Spectral data to be discussed later, and multilabeling experiments similar to those alluded to in the firefly discussion, have now established the oxidation mechanism of Eq. (13-18):

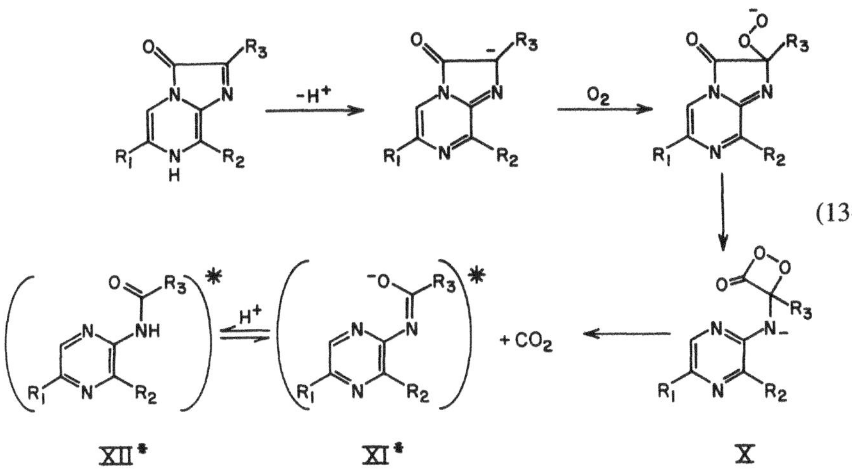

(13-18)

In basic aprotic solvents a chemiluminescence reaction occurs with the natural luciferin or with synthetic analogs, e.g., R_1 = *p*-hydroxyphenyl, phenyl, or H; R_2 = methyl or H; R_3 = methyl. With *Cypridina* luciferase only the natural substituents give bioluminescence, but the coelenterate luciferases appear to be less restrictive. The chemiluminescence spectra are green to yellow, and are distinct from the bioluminescence spectra, which range from blue to green depending on the substituents used.

13.5.2. Enzymology

The enzymology of the *Cypridina* system is the simpler. The system is named from its primary source, a small ostracod, *Cypridina hilgendorfii,* a crustacean that is particularly abundant in Japanese waters. The system is also found in many species of luminous fish. When stimulated, the ostracod secretes the luciferin and luciferase from separate glands into the surrounding seawater, so it can be said that the *in vivo* and *in vitro* reactions are the same. There are three reactions that constitute this system, and the one enzyme luciferase catalyzes them all: oxygenation, excitation, and release for turnover:

$$E + LH_2 \rightarrow E:LO^* + CO_2 \qquad (13\text{-}19)$$

The coelenterates are a diverse group of organisms made up of two main phyla, Cnidaria and Ctenophora. Within Cnidaria are the classes Anthozoa, such as the soft coral, *Renilla reniformis* ("sea pansy"), Hydrozoa, e.g., the jellyfish *Aequorea aequorea,* and Scyphozoa. The bioluminescence systems from *Renilla* and *Aequorea* are the ones that have received the most study. The bioluminescence systems of all of the coelenterates involve at least three of the general reactions of Eqs. (13-19) and (13-12). Although the chemical mechanism is apparently the same for all of them, they differ among the classes of the Cnidaria in the steps of the activation reaction, and among the phyla as to the nature of the excitation step. In fact these systems now require us to introduce more complications into the simple picture represented by Eqs. (13-9) to

(13-12). Their study also revealed, for the first time, the phenomenon of "sensitized bioluminescence" [cf. Eqs. (13-6) and (13-7)].

The first step in anthozoan bioluminescence is activation:

$$E_S : LH_2S + PAP \rightarrow LH_2 + PAPS + E_S \qquad (13\text{-}20)$$

Luciferyl sulfate, LH_2S (LH_2 is IX), and a sulfokinase, E_S, have been obtained from *Renilla*. The activating reaction [Eq. (13-20)] requires the coenzyme 3′,5′-diphosphoadenosine, PAP, and liberates the corresponding sulfate, PAPS. Free luciferin is rapidly autooxidized in solution, but in the cell it is protected by a binding protein, E_B, from which it is released by the addition of Ca^{2+}.

$$E_B : LH_2 + Ca^{2+} \rightarrow E_B : Ca^{2+} + LH_2 \qquad (13\text{-}21)$$

It is thought that this Ca^{2+}-binding step controls the light flash in the whole animal.

The next steps for coelenterate luciferin are oxygenation and excitation, both on the anthozoan luciferase, E.

$$E : LH_2 + O_2 \rightarrow E : LHOOH \qquad (13\text{-}22)$$

$$E : LHOOH \rightarrow E : LO^* + CO_2 \qquad (13\text{-}23)$$

The product of oxygenation is not detectable, as it quickly breaks down to the excited final product, $E : LO^*$.

There is a remarkable difference between the Anthozoa and the Hydrozoa in this last step, so much so that the protein involved in the latter is not called luciferase but "photoprotein," specifically "aequorin," "mnemiopsin," etc., depending on its source. What is extracted from the jellyfish is the oxygenated species $E_A : LHOOH$, the photoprotein (aequorin). The Hydrozoa do not use the activation or binding reactions [Eqs. (13-20) and (13-21)]. Adding Ca^{2+} to photoproteins triggers the bioluminescence:

$$E_A : LHOOH + Ca^{2+} \rightarrow E_A : LO^* + CO_2 \qquad (13\text{-}24)$$

The stable oxygenated complex may be regenerated by adding LH_2 (IX) to apoaequorin.

13.5.3. The Light Emitters

For the reaction of *Cypridina* luciferin in basic aprotic solvents, the chemiluminescence emission is from the monoanion XI, as the R_2 guanidino group is not positively charged under these basic conditions. For luciferin of structure IX, the emitting product is a dianion or trianion by virtue of dissociation of the R_1 or R_3 to the phenolates. These chemiluminescence spectra have maxima at much longer wavelengths ($\lambda_C > 500$ nm) than the bioluminescence spectra.

The *in vitro* reaction of VIII with *Cypridina* luciferase is very efficient, $\phi_B = 0.3$, and the blue bioluminescence is a broad featureless spectrum with λ_B at 460 nm. This spectrum is the same as the fluorescence of XII bound to luciferase. In aqueous solution

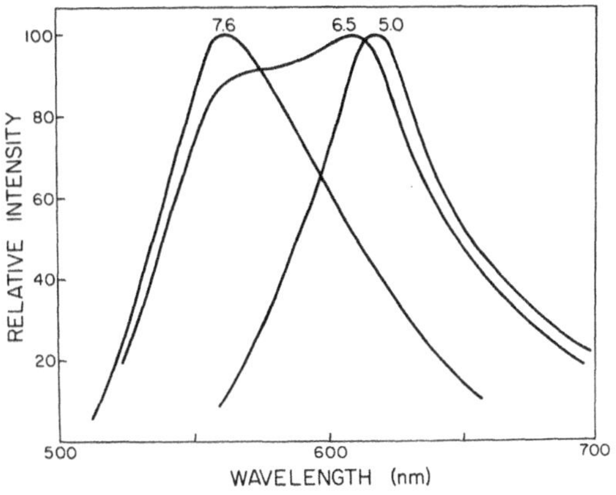

Fig. 13-4. Effect of pH on the firefly bioluminescence spectra *in vitro*. The spectrum with a maximum at 616 nm is normalized to unity, but in absolute terms its contribution is only one-half that of the band having its maximum at 562 nm. The pH values are indicated on each spectrum. [Redrawn from H. H. Seliger and W. D. McElroy, *Light: Physical and Biological Action*, Academic Press, New York (1965).]

free XII is not fluorescent. Thus, the binding site on *Cypridina* luciferase must be one that allows rapid protonation of the initially formed product to produce XII*, and that inhibits the nonradiative loss of its excited state energy.

The *in vitro* bioluminescence of coelenterate luciferin has a relatively low efficiency, $\phi_B = 0.07$. The emission is broad with $\lambda_B = 480$ nm (Fig. 13-4). However, the fluorescence of the neutral species (XII) derived from natural coelenterate luciferin has $\lambda_F = 402$ nm in aqueous solution. From a number of studies with luciferin analogs it has been concluded that, in contrast to the *Cypridina* reaction, the monoanion (IX) is the emitter of coelenterate bioluminescence. In basic aprotic solvent, IX has $\lambda_F = 480$ nm and $\phi_F = 0.07$. The active site of coelenterate luciferase must then hinder the excited protonation of the initially formed XI to XII; a similar property has been described for firefly luciferase. Whether protonation of the excited product is slowed or not is a mechanism for adaptation that would tune the bioluminescence spectrum to satisfy some biological function. For example, a spectral shift might satisfy a requirement for optimal transmission of the bioluminescence signal in the marine optical milieu, or for a better match to the spectral sensitivity of the target organism, etc.

13.5.4. Sensitized Bioluminescence

There is a remarkable difference between the *in vivo* bioluminescence spectra from the Ctenophores and some of the Cnidaria. The latter have a narrow structured emission, $\lambda_B = 509$ nm, identical to the fluorescence of a "green-fluorescent protein" (GFP) that can be purified from extracts of these organisms. The *in vivo* and *in vitro* bioluminescences and this fluorescence are compared in Fig. 13-5. The green-fluorescent protein is very fluorescent, $\phi_F = 0.8$, and has no luciferase activity. If present in the *in vitro* reaction at a concentration of more than 10^{-5} M, the bioluminescence emission corresponds exactly to the green fluorescence and also to the *in vivo* λ_B, and the ϕ_B is enhanced to over 0.2. Thus, it is evident that the green-fluorescent protein sensitizes the bioluminescence.

The Ctenophores, on the other hand, have a broad bioluminescence emission, $\lambda_B =$

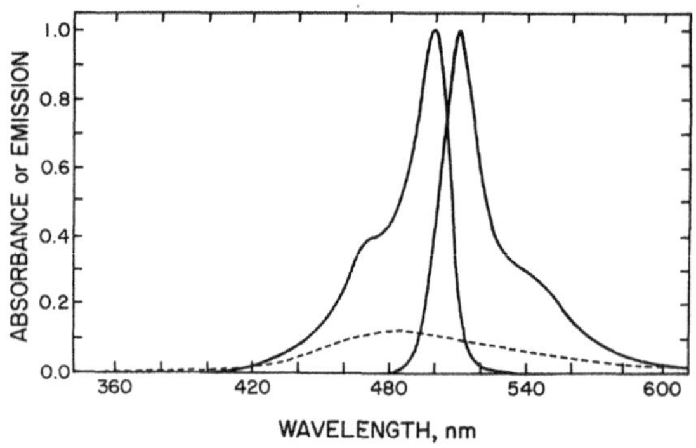

Fig. 13-5. Absorption (left full line) and fluorescence (right full line) spectra of the green-fluorescent protein from *Renilla*. The dashed line is the *in vitro* bioluminescence with *Renilla* luciferase and a synthetic luciferin. On addition of green-fluorescent protein (1 μM), the relative bioluminescence yield and spectrum change to match that of the right full line. [Redrawn from W. W. Ward, Energy transfer processes in bioluminescence, *Photochem. Photobiol. Rev.* **4,** 1–57 (1979).]

490 nm, similar to the *in vitro* spectrum with luciferin (IX). No green-fluorescent protein is found in the Ctenophores or the Scyphozoans.

The mechanism of the sensitization process is unknown. Dipole-dipole energy transfer has been suggested [Eq. (13-25)], but it is also possible that the green-fluorescent protein functions as a fluorescent activator in a CIEEL mechanism [see Eq. (13-3)].

$$E_A:LO^* + GFP \rightarrow E_A:LO + GFP^* \qquad (13\text{-}25)$$

The fluorophore of the green-fluorescent protein is XIII. Its structure suggests that it is related to the luciferin. The biosynthesis of all these molecules (VIII, IX, and XIII) is probably cyclization of natural amino acids whose side chains can be recognized in the R substituents.

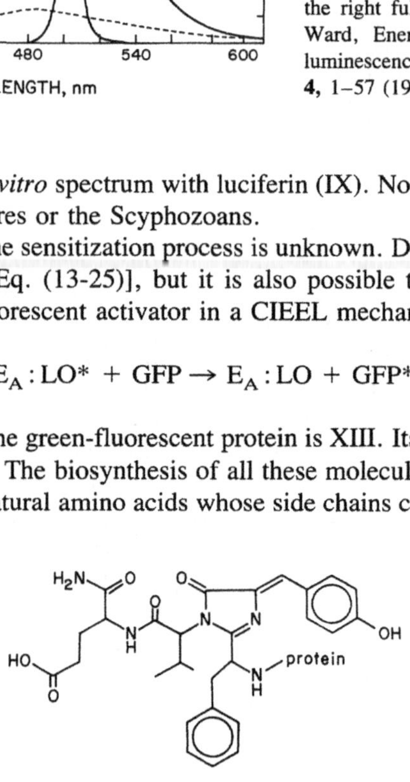

XIII

13.5.5. Control[17]

If Ca^{2+} is added to a mixture of *Renilla* luciferin bound to E_B [Eq. (13-21)] and luciferase, the bioluminescence intensity (Fig. 13-6, soluble) rises in a few tenths of a second and falls very slowly, unlike the *in vivo* flash (Fig. 13-6) obtained on stimulating the animal. This discrepancy is explained by the observation that the light-giving organ, or photocyte, of *Renilla* is made up of much smaller structures called "lumisomes," membrane-enclosed vesicles about 0.2 μm in diameter, containing E_B, LH_2, E, and GFP. On adding Ca^{2+} to a preparation of the lumisomes, a flash much more like the *in vivo* bioluminescence (Fig. 13-6) is observed. It is suggested that the proteins are packaged within the lumisome so that the transient concentration of Ca^{2+} is controlled in such a way as to produce the *in vivo* bioluminescence kinetics.

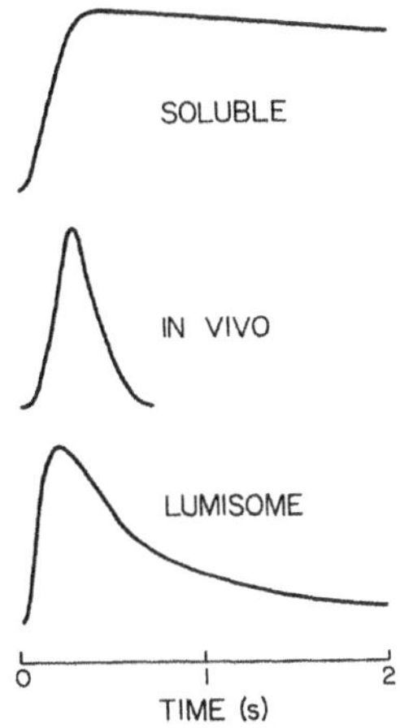

Fig. 13-6. Light flash from the *Renilla* bioluminescence system on the addition of Ca^{2+} at zero time. [Redrawn from J. M. Anderson and M. J. Cormier, Lumisomes, the cellular site of bioluminescence in coelenterates, *J. Biol. Chem.* **248,** 2937–2943 (1973).]

Thus, we see in organisms that lie lower on the evolutionary scale than the firefly, a bioluminescence system involving a multienzyme package. Adaptations of the bioluminescence spectrum are also evident and are understood at the molecular level. These adaptations involve excited state protonation kinetics as well as energy transfer to a sensitizing acceptor molecule. In the firefly system, all the reactions have been "packaged" onto the same macromolecule, i.e., activation, light reaction, and spectral adaptation. We shall now look at bioluminescence in a lower form of life, the bacteria.

13.6. BACTERIAL BIOLUMINESCENCE[18,19]

The luminous bacteria are mostly found in the ocean, both free living and symbiotic with certain fish. Most of the marine types are classified into two genera, *Photobacterium* and *Vibrio,* and these are the ones that have received the most biochemical study. It can be said that bacterial bioluminescence is one of the longest studied mechanisms in biochemistry, dating from the 17th century, as mentioned in the Introduction, when Boyle observed what we now know as the oxygen requirement. In the bioluminescence field, this system is the one that has received the most attention, probably because the mechanism defies elucidation and remains a challenge for future research. An advantage in the study of bacterial bioluminescence is that the cells can be grown in kilogram quantities by routine fermentation procedures, the yield of luciferase is high, and the luciferase is relatively easy to purify.

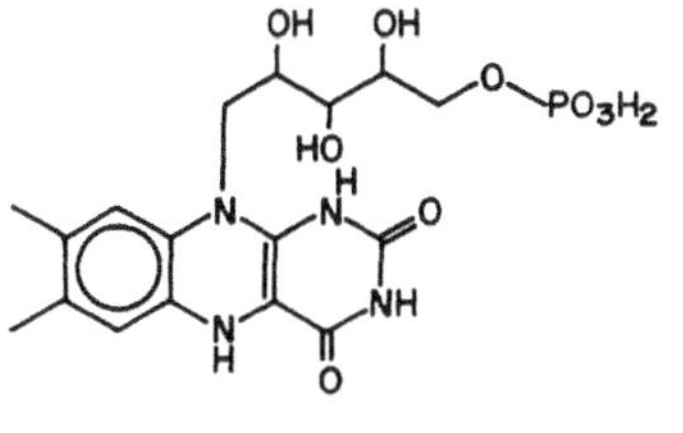

XIV

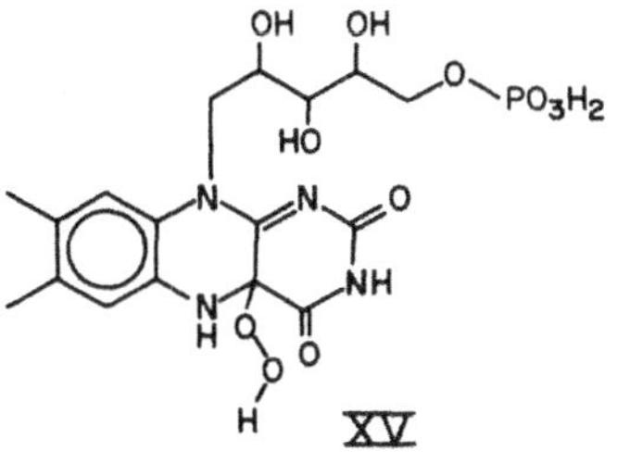

XV

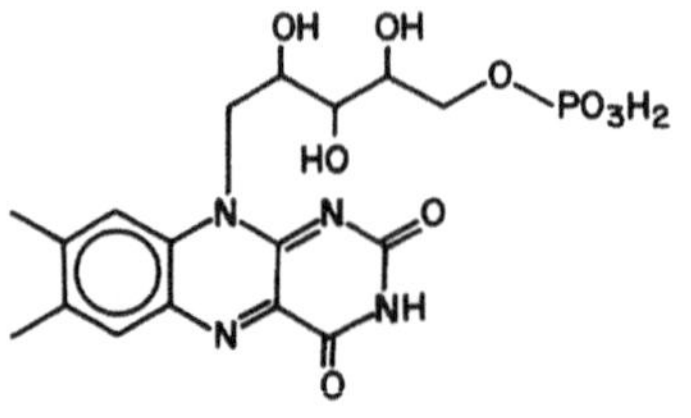

XVI

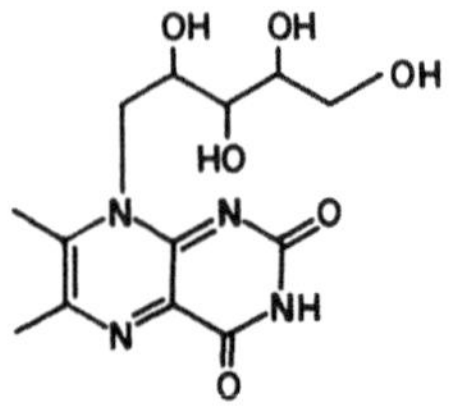

XVII

13.6.1. Chemical Mechanism

Light emission can be obtained *in vitro* if reduced riboflavin monophosphate (i.e., flavin mononucleotide, $FMNH_2$, XIV) is added to a solution of luciferase and a long-chain aliphatic aldehyde. *In vivo,* tetradecanal is probably the aldehyde "luciferin," and the flavin is not free but bound to a protein, probably luciferase itself. Two known products of the *in vitro* reaction are FMN (XVI) and an aliphatic acid of the same chain length as the aldehyde.

No CO_2 can be detected in the products, so a dioxetanone intermediate is not considered to occur. Instead the intermediate species of high energy content [Eq. (13-22)] is believed to be some type of strained peroxide. Indeed, if aldehyde is omitted from the reaction, a flavin–oxygen–luciferase complex is formed that has a distinct absorption spectrum, $\lambda_A = 370$ nm, and it can be stabilized for as long as 24 h. Because many dihydroflavin derivatives have similar absorption spectra, it was proposed that this complex was a dihydroflavin hydroperoxide substituted at the 4a position (XV in Scheme 4). This idea has now been proven by measuring the ^{13}C NMR of the complex using synthetic $FMNH_2$ containing ^{13}C at specific positions around its isoalloxazine ring. The resonance from the 4a C is the only one that shows a distinct chemical shift in the NMR of the complex (Fig. 13-7), and this can be interpreted as due to a highly strained peroxide substitution at this position.[20]

In all the bioluminescence systems this complex is the only oxygenated species to be directly characterized in this way. Furthermore, in other flavoenzyme oxidation reactions this same species has been proposed to be a key hydroxylating species, but the luciferase complex is the only one stable enough for NMR characterization. The luciferase-bound

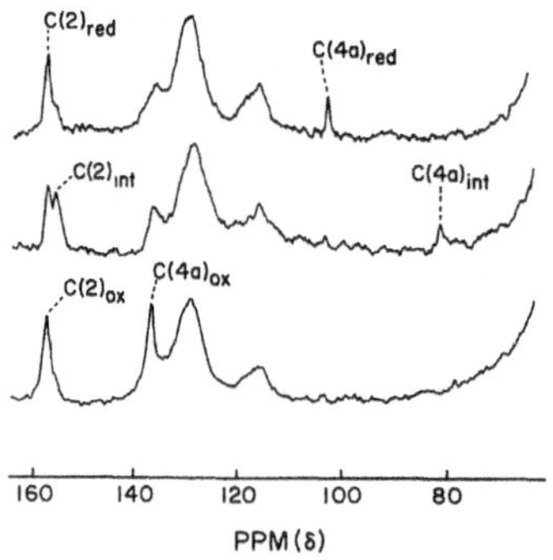

Fig. 13-7. Carbon-13 nuclear magnetic resonance spectra of flavin enriched at the 2 and 4a positions with ^{13}C, in the presence of bacterial luciferase. The top is for $FMNH_2$ and the middle is after adding O_2, and this represents the intermediate II. Only the resonance from the 4a–C is significantly affected. The lowest spectrum is after full oxidation to FMN. The unlabeled bands are the contributions from the natural abundance ^{13}C in the luciferase molecule. (Redrawn from Ref. 20.)

species XV is called "intermediate II." It reacts with aldehydes to produce bioluminescence possibly by the CIEEL mechanism.

13.6.2. Enzymology

Within the cell we can consider activation to be the generation of $FMNH_2$ coupled to the oxidation of NADH by an oxidoreductase, E_R:

$$E_R : NADH + FMN + H^+ \rightarrow E_R : NAD^+ + FMNH_2 \tag{13-26}$$

The luminous bacteria contain very high concentrations of luciferase, and thus the $FMNH_2$ is rapidly shuttled into the oxygenation reaction before it has a chance to be oxidized directly and uselessly by oxygen:

$$E + FMNH_2 \rightarrow E : FMNH_2 \overset{O_2}{\rightarrow} E : FMNH_2O_2 \tag{13-27}$$

Although not as stable, this luciferase-bound peroxide is very analogous to the photoproteins (e.g., aequorin) discussed in the last section.

13.6.3. Lumazine Protein[21]

The *in vitro* reaction of bacterial luciferase yields a broad emission spectrum with λ_B about 495 nm. However, for *Photobacterium* cells *in vivo* the spectrum is distinctly different with λ_B around 475 nm (Fig. 13-8). This is due to the sensitization of the reaction by a second protein identified in these cells and named "lumazine protein" because the ligand is 6,7-dimethyl-8-ribityllumazine (XVII). This lumazine derivative is a natural metabolite, the precursor to riboflavin in procaryotes.

If lumazine protein is included in the *in vitro* reaction, the bioluminescence spectrum

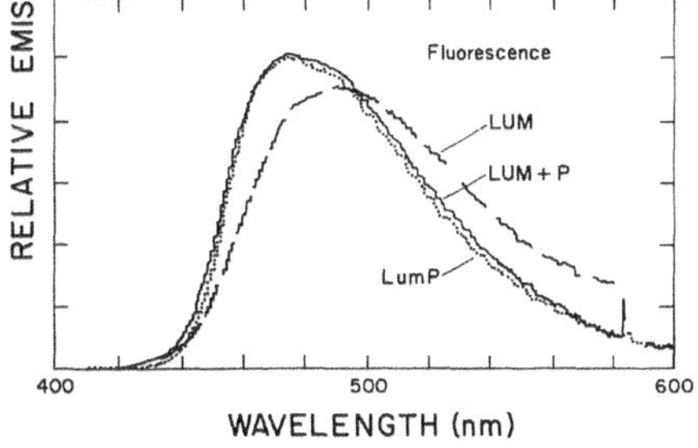

Fig. 13-8. (top) Shift of the *in vitro* bioluminescence (dashed curve) on including lumazine protein in the reaction (reformed from the apoprotein, 25 μM and the lumazine derivative, 8 μM). (lower) Fluorescence of the lumazine derivative (dashed curve), lumazine protein (dots), and the reconstituted hololumazine protein (solid line). [From J. Lee, L. A. Carreira, R. Gast, R. M. Irwin, P. Koka, E. D. Small, and A. J. W. G. Visser, Properties of a lumazine protein from the bioluminescent bacterium *Photobacterium phosphoreum*, in: *Bioluminescence and Chemiluminescence* (M. A. DeLuca and W. D. McElroy, eds.), pp. 103–112, Academic Press, New York (1981).]

is blue-shifted, and if the lumazine protein concentration is at least 10 μM, the spectrum becomes identical to that of the *in vivo* bioluminescence, which is also the same as the fluorescence of lumazine protein itself, $\lambda_F = 475$ nm, thus proving the sensitizer function of lumazine protein. The mechanism of sensitization is not known, but it is obviously not the same energy transfer process as in the coelenterate reaction [Eq. (13-25)], where the *in vitro* bioluminescence emitter is proposed to be the energy donor with GFP the acceptor. The important differences are that the bioluminescence spectral shift caused by lumazine protein is to a higher energy, not a lower one (Fig. 13-8), and there is only a small overlap between the absorption spectrum of lumazine protein and the *in vitro* bioluminescence spectrum. Furthermore, the inclusion of lumazine protein can enhance the ϕ_B up to three times with values as high as $\phi_B = 0.25$ being observed.

The FMN product is fluorescent, but its spectrum with $\lambda_F = 535$ nm disqualifies it from being the origin of the bioluminescence emission. On the other hand, in the course of the bioluminescence reaction another highly fluorescent but transient species has been detected. It has a fluorescence spectrum identical to the bioluminescence spectrum and its concentration bears a kinetic relationship to the bioluminescence intensity (Fig. 13-9). A number of reduced compounds related to $FMNH_2$ also produce bioluminescence with luciferase and aldehyde, but with a shifted bioluminescence spectrum, within the range 475–560 nm, depending on the compound. These reactions also show a fluorescent transient with a spectrum corresponding to the shifted bioluminescence spectrum, proving that this transient is indeed the *in vitro* bioluminescence emitter and that its structure, presently unknown, must be derived from the reduced substrate used.

To account for the complex kinetics of the *in vitro* bioluminescence, it has been proposed that the reaction is sensitized by the fluorescent transient, but this still remains to be thoroughly established. Again, it is an attractive possibility to consider that a CIEEL process occurs with either the fluorescent transient or lumazine protein, if present, as fluorescent activators.

Bioluminescent bacteria produce a continuous light emission at a level controlled only by the metabolic status of the organism, e.g., oxygen concentration, carbon sources, and the like. In the brightest bacteria a large fraction of the total metabolic energy is allocated to the generation of bioluminescence. How bioluminescence has evolved in the procaryotes is a fascinating question.[22] The small molecules involved are common: $FMNH_2$, the lumazine derivative, and the fatty aldehydes. Flavin peroxide intermediates are believed to occur in many flavoenzyme reactions. The lumazine apoprotein could be

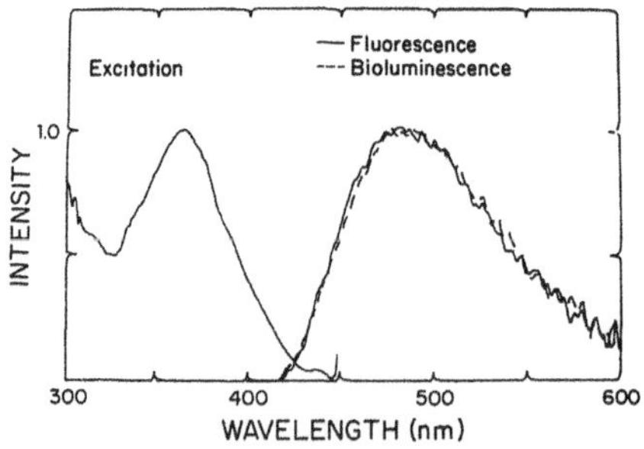

Fig. 13-9. Excitation (left curve) and fluorescence spectra of the fluorescent transient formed in the bacterial bioluminescence reaction. This fluorescence and the bioluminescence spectra (dashed curve) are identical. [From I. B. C. Matheson and J. Lee, Kinetics of bacterial bioluminescence and the fluorescent transient, *Photochem. Photobiol.* **38**, 231–240 (1983).]

the binding protein for the lumazine metabolite. Oxidoreductase is also ubiquitous in the procaryotes. The bioluminescent bacteria seem to distinguish themselves from other procaryotes primarily by the presence of luciferase. By the application of DNA recombinant techniques it may soon become possible to trace the lineage of the bacterial luciferases and answer the question of how and why the bacteria "invented" this enzyme.

No new principles have come out of studies of the many other bioluminescence systems. Some of these systems use H_2O_2 instead of O_2, and a few others have had their luciferins characterized: they are aldehydes in the earthworm and snail systems, and a tetrapyrolle in the dinoflagellates. Sensitization of the bioluminescence is found in the shrimp system and probably the earthworm.

13.7. LOW-LEVEL LUMINESCENCE, BIOLOGICAL CHEMILUMINESCENCE(23,24)

Although we have been discussing the terms light emission, chemiluminescence, and quantum yield, we have not yet considered the important device that has made the accurate measurement of these quantities possible. This is the photomultiplier tube (Chapter 1), a vacuum tube device containing a photosensitive cathode coating on the inside of the front window, which emits an electron on absorption of a photon, the electron then being attracted by a voltage gradient to a series of electroemissive dynodes. These dynodes, made of a Ag/Mg alloy, have a low work function, and when one electron hits the surface of the dynode it will knock off another two or more electrons. These are attracted by a positive voltage to the next dynode where they multiply again, and so on down the line of 10 or more dynodes until out of the anode comes a burst of 10^6 or more electrons, a negative pulse of current, corresponding to that initial impinging photon.

The photomultiplier serves as the "eye" of a television camera, and was developed primarily for this purpose and for scintillation counting in nuclear physics in the early 1940s. Since that time, photomultipliers have been developed to be more and more sensitive to light, and to have a wide spectral sensitivity. At the present time it is possible to make a photomultiplier device that can detect less than 100 photons/s coming from a chemical reaction contained in a volume of about 1 cm^3. This means that light emission would be measurable for a fast reaction with a ϕ_C of below 10^{-17}. It has been observed that a great many exothermic chemical reactions, and a number of biological systems, do indeed emit light at a very low level.

The human eye is also a very good light detector, comparable in sensitivity, but not accuracy, to the photomultiplier. Therefore, weak light emission from a number of chemical reactions has been known for hundreds of years. Many investigations, particularly by groups of Russian workers during the 1930s, have shown that a variety of living systems give weak visible and ultraviolet luminescence. They used Geiger tubes for detection, and this luminescence is sometimes referred to as mitogenetic or Gurvich radiation, after its discoverer. This luminescence is usually associated with rapidly growing or respiring cells, e.g., onion root tips, dividing yeast cells, white blood cells (leukocytes), liver mitochondria and microsomes, contracting skeletal and heart muscle, etc.

In order to distinguish this low-level luminescence from what we have been discussing up to now, we shall use the term "biological chemiluminescence." Unlike most cases of bioluminescence, where the light serves some definite biological function, such as communication in the fireflies, biological chemiluminescence has no apparent advantage to the organism and may just be the result of a very minor energy wastage through an inevitable Maxwell–Boltzmann probability of populating an excited state of a product molecule in a very exergonic reaction. It could be, on the other hand, that it is a very efficient chemiluminescent reaction, but occurring to only a minor extent in the total reaction. Oxygen and radical intermediates appear to be involved in biological chemiluminescence.

Bioluminescence and biological chemiluminescence can be better compared in terms of the light emission from the functional biological unit, i.e., from a single bacterium or the organelle of a higher organism. Luminous bacteria each typically emit 10^3 photons/s, while a lumisome from *Renilla* will produce a flash of peak intensity about 10^2 photons/s, and a single dinoflagellate, which probably contains thousands of organelles, when stimulated will give off more than 10^8 photons. In contrast, the rapidly respiring mitochondria or leukocytes undergoing phagocytosis have an average light level far less than 1 photon/s per cell. Actually, the distinction between bioluminescence and biological chemiluminescence is blurred by the fact that the so-called dark mutants of luminous bacteria are only dark because they give off 10,000-fold less light than the wild-type strain, i.e., at levels that could be regarded as that of biological chemiluminescence.

Not much is known about the mechanism of biological chemiluminescence, but since the oxidation of lipids and other unsaturated hydrocarbons gives rise to a chemiluminescence of low efficiency, it is proposed that such a process is probably the basis for biological chemiluminescence. This luminescence requires molecular oxygen and can often be stimulated by the addition of hydrogen peroxide. A mechanism based on the chain oxidation of a hydrocarbon is

$$R^{\bullet} + O_2 \rightarrow ROO^{\bullet} \qquad (13\text{-}28)$$

$$ROO^{\bullet} + RH \rightarrow ROOH + R^{\bullet} \qquad (13\text{-}29)$$

$$ROO^{\bullet} + ROO^{\bullet} \rightarrow RO^{*} + RO + O_2 \qquad (13\text{-}30)$$

The RO^* is an excited ketone, and the fluorescence yields of these are extremely low, resulting in the very low overall yield of light. However, this can be enhanced up to 10^6 times by adding certain fluorescent substances.

13.8. APPLICATIONS OF BIOLUMINESCENCE[25]

13.8.1. Ultrasensitive Analysis

Some efficient chemiluminescence reactions are widely used as emergency light sources and as novelties. The use of bioluminescence reactions in this way, however, is impractical due to the cost of preparation and the difficulty of storing enzymes over long periods. However, bioluminescence reactions are finding increasing application in the analysis of biological materials. The detection and assay of ATP by the firefly reaction, for instance, is more sensitive than any other method. Often a substance may be assayed indirectly by coupling it via some enzyme to a bioluminescence cofactor like ATP or NADPH. Together with bacterial luciferase the latter has wide applicability since many assays in the clinical laboratory are presently made by NAD(P)H coupling, and the luciferase technique can improve sensitivity by many orders of magnitude.

Bioluminescence-linked immunoassays also offer advantages over alternative techniques in sensitivity, stability, simplicity of measurement, and avoidance of the use of radioisotopes with their accompanying hazards and disposal problems. For a bioluminescent immunoassay an antibody (Ab) has to be produced for the substance to be measured (the hapten, H), and also the hapten has to be covalently attached to a component of the bioluminescence system, e.g., a protein (E). The assay reaction is

$$E\text{–}H : Ab + H \rightarrow H : Ab + E\text{–}H \tag{13-31}$$

Only the free E–H will have activity in the bioluminescence reaction. A potential advantage of these systems is in the development of homogeneous assays where the complete assay is achieved without separation steps. This goal is being sought using sensitized bioluminescence systems, i.e., where the acceptor protein (A–H) also has to be freed for the correct bioluminescence to be generated.

The use of the firefly system to assay ATP is the most widely used bioluminescence method. We will describe this one in detail to show how these procedures are done. For very low amounts of ATP, the firefly assay is practically the only available technique. If a limiting quantity of ATP is added to a solution of firefly luciferin and luciferase, a flash of light is produced, which, if detected by a photomultiplier linked to a current meter and recorder, produces a curve like that shown in Fig. 13-10. A straight line results when total photons emitted in the first 15 s is plotted against the amount of ATP added, and this is the calibration line. For an unknown sample, the ATP content can be calculated from its light yield. This method has been used to find out how much ATP is present in nerve cells, plant seeds, and bacteria, and modification of the procedure by enzyme coupling enables the assay of other important cofactors like AMP, cyclic AMP, and pyrophosphate. Enzymes that use ATP can be determined, e.g., creatine phosphokinase. One of the most important applications has been to the assay of biomass via the ATP content, to which it is

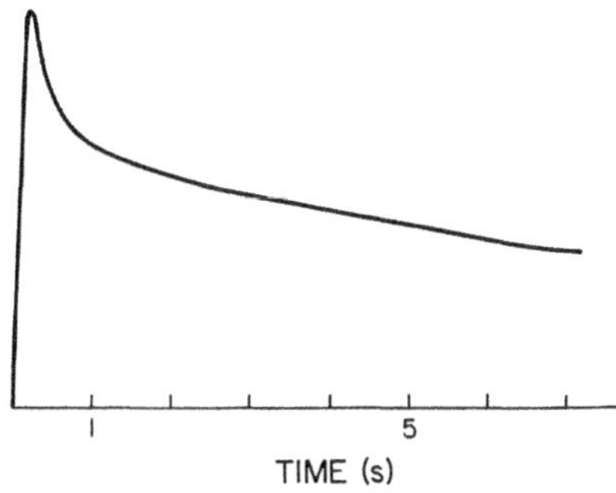

Fig. 13-10. Light flash from the firefly bioluminescence *in vitro* reaction obtained by injecting ATP into a mixture of luciferase and luciferin. The vertical direction represents relative light intensity. For 10^{-9} g of ATP the flash maximum (I_o) has an intensity of about 10^{10} photons/s.

directly related. In a hospital laboratory, for example, a blood or urine specimen can be rapidly measured for infectious levels of bacteria in this way.

Whole cells of bioluminescent bacteria are used in a method for determining toxic materials in foods and stream water, since the light emission is quenched by many poisons. Also, dark mutants of these bacteria can be designed that will back-mutate in response to specific mutagens, and this has been proposed as a replacement for the Ames test for carcinogens. Also under this subject of whole-cell luminescence, some clinical laboratories are using the biological chemiluminescence associated with leukocytes to monitor phagocytic competence.

The aequorin reaction is able to detect Ca^{2+} concentrations down to 10^{-12} M. Many physiological processes involve rapid changes in Ca^{2+} concentration, and the aequorin reaction has therefore found use in studying such things as Ca^{2+} ion transients in muscular contraction. In this context, an application of a different type can be performed that is highly dependent on bioluminescence, i.e., coupling an image intensifier tube to a microscope to observe the localization and time responses of light emission in a bioluminescent organism. For nonluminescent ones microinjection of aequorin allows observation of Ca^{2+} concentration fluxes in single organisms. By this method nerve impulses across single dinoflagellates and waves of Ca^{2+} concentration changes on fertilization of single eggs have been visualized.

13.8.2. Recombinant DNA Techniques

The techniques of modern molecular biology are providing new approaches to solving problems in the bioluminescence field just as they are in most of the biological sciences. The primary sequence of aequorin has been determined by cloning as well as by chemical methods. However, the luciferases from bacteria and firefly were resistant to having their sequence done by chemical means, and so their primary structures have had to be inferred from their DNA base sequences. Recombinant DNA methods are also making available several of these proteins in much larger quantities than have hitherto been extractable from natural sources. Also as with other proteins, specific amino acid alterations can be introduced as a means for highlighting those residues that influence the bioluminescence activity.

While there are no unique applications of recombinant DNA methods to bioluminescence problems at this time, it is in the reverse direction that some novel developments are occurring. Bacterial luciferase has been used as a marker for detecting which colonies of treated *E. coli* have been transformed to contain a certain foreign gene. The two genes for the luciferase subunits, called *LuxA* and *LuxB*, are inserted into the transforming DNA, between the promoter of the foreign gene and the gene itself. Cells that are transformed will manufacture both luciferase and the foreign gene product when the promoter is turned on, and will be luminescent. The method is rapid and sensitive.[26]

In another application of bacterial luciferase, the expression of nitrogenase genes in *Rhizobium* during stages of its infection of plant cells, has been followed by fusing the *Lux* genes after the nitrogenase promoter, and detecting luminescence. Recently, both the *Lux* and the firefly genes have been inserted into the genomes of plant cells themselves. Since the production of luciferase can measured by its bioluminescence yield, then in these transformed plants the bioluminescence might provide the ability to rapidly and relatively precisely measure the rates of gene expression.[27]

13.9. REFERENCES

1. E. N. Harvey, *Bioluminescence,* Academic Press, New York (1952).
2. E. N. Harvey, *Living Light,* Princeton University Press, Princeton, N.J. (1940).
3. H. H. Seliger, Origin of bioluminescence, *Photochem. Photobiol.* **21,** 335–361 (1975).
4. J. Buck, Functions and Evolutions of Bioluminescence, in: *Bioluminescence in Action* (P. J. Herring, ed.), pp. 419–460, Academic Press, New York (1978).
5. J. W. Hastings, Biological diversity, chemical mechanisms, and the evolutionary origins of bioluminescent systems, *J. Mol. Evol.* **19,** 309–321 (1983).
6. J. E. Lloyd, Bioluminescence and Communication, in: *How Insects Communicate* (T. A. Sebeok, ed.), pp. 164–183, Indiana University Press, Bloomington (1977).
7. H. H. Seliger, A. B. Lall, J. E. Lloyd, and W. H. Biggley, The colors of firefly bioluminescence. I. Optimization model. II. Experimental evidence for the optimization model, *Photochem. Photobiol.* **36,** 673–688 (1982).
8. K.-D. Gundermann and F. McCapra, *Chemiluminescence in Organic Chemistry,* Springer-Verlag, Berlin (1987).
9. L. R. Faulkner and R. S. Glass, Electrochemiluminescence, in: *Chemical and Biological Generation of Excited States* (W. Adam and G. Cilento, eds.), pp. 191–227, Academic Press, New York (1982).
10. W. W. Ward, General Aspects of Bioluminescence, in: *Chemi- and Bioluminescence* (J. G. Burr, ed.), pp. 321–358, Marcel Dekker, New York (1985).
11. P. J. Herring (ed.), *Bioluminescence in Action,* Academic Press, New York (1978).
12. M. A. DeLuca and W. D. McElroy, Purification and properties of firefly luciferase, *Methods Enzymol.* **57,** 3–15 (1978).
13. O. Shimomura, Mechanism of bioluminescence, in: *Chemical and Biological Generation of Excited States* (W. Adam and G. Cilento, eds.), pp. 249–276, Academic Press, New York (1982).
14. E. H. White, J. D. Miano, C. J. Watkins, and E. J. Breaux, Chemically produced excited states, *Angew. Chem. (Engl.)* **13,** 229–243 (1974).
15. L. J. Bowie, R. Irwin, M. Loken, M. DeLuca, and L. Brand, Excited-state proton transfer and the mechanism of action of firefly luciferase, *Biochemistry* **12,** 1852–1857 (1973).
16. F. H. Johnson and O. Shimomura, Introduction to the bioluminescence of Medusae, with special reference to the photoprotein aequorin, *Methods Enzymol.* **57,** 271–291 (1978).
17. M. J. Cormier, K. Hori, and J. M. Anderson, Bioluminescence in Coelenterates, *Biochim. Biophys. Acta* **346,** 137–164 (1974).
18. M. M. Ziegler and T. O. Baldwin, Biochemistry of Bacterial Bioluminescence, in: *Current Topics in Bioenergetics,* Vol. 12 (R. D. Sanadi, ed.), pp. 66–113, Academic Press, New York (1980).
19. J. Lee, The mechanism of bacterial bioluminescence, in: *Chemi- and Bioluminescence* (J. G. Burr, ed.), pp. 401–437, Marcel Dekker, New York (1985).
20. J. Vervoort, F. Muller, J. Lee, W. A. M. Van der Berg, and C. T. W. Moonen, Identification of the true carbon-13 nuclear magnetic resonance spectrum of the stable intermediate II in bacterial luciferase, *Biochemistry* **25,** 8062–8067 (1987).
21. A. J. W. G. Visser and J. Lee, Excited state probes in bacterial bioluminescence, in: *Excited State Probes in Biochemistry and Biology* (A. G. Szabo and L. Masotti, eds.), Plenum Press, New York (1988).
22. H. H. Seliger, The evolution of bioluminescence in bacteria, *Photochem. Photobiol.* **45,** 291–297 (1987).
23. R. C. Allen, Biochemiexcitation: Chemiluminescence and the study of biological oxygenation reactions, in: *Chemical and Biological Generation of Excited States* (W. Adam and G. Cilento, eds.), pp. 309–344, Academic Press, New York (1982).
24. D. Slawinska and J. Slawinski, Low level luminescence from biological objects; and, Applications of bioluminescence and low-level luminescence from biological objects, in: *Chemi- and Bioluminescence* (J. G. Burr, ed.), pp. 495–531, 533–601, Marcel Dekker, New York (1985).
25. L. J. Kricka, P. E. Stanley, G. H. G. Thorpe, and T. P. Whitehead (eds.), *Analytical Applications of Bioluminescence and Chemiluminescence,* Academic Press, Orlando (1984).
26. J. Englebrecht, M. Simon, and M. Silverman, Measuring gene expression with light, *Science* **227,** 1345–1347 (1985).
27. D. J. O'Kane, W. L. Lingle, J. E. Wampler, M. Legocki, R. P. Legocki, and A. A. Szalay, Visualization of bioluminescence as a marker of gene expression in rhizobium-infected soybean root nodules, *Plant Mol. Biol.* **10,** 387–399 (1988).

Index

GPSR Compliance

The European Union's (EU) General Product Safety Regulation (GPSR) is a set of rules that requires consumer products to be safe and our obligations to ensure this.

If you have any concerns about our products, you can contact us on ProductSafety@springernature.com

In case Publisher is established outside the EU, the EU authorized representative is:

Springer Nature Customer Service Center GmbH
Europaplatz 3
69115 Heidelberg, Germany

Batch number: 10762209

Printed by Printforce, the Netherlands